Springer Series in

SOLID-STATE SCIENCES 172

Springer Series in
SOLID-STATE SCIENCES

The Springer Series in Solid-State Sciences consists of fundamental scientific books prepared by leading researchers in the field. They strive to communicate, in a systematic and comprehensive way, the basic principles as well as new developments in theoretical and experimental solid-state physics.

Please view available titles in *Springer Series in Solid-State Sciences* on series homepage http://www.springer.com/series/682

Daniele Sanvitto
Vladislav Timofeev
Editors

Exciton Polaritons in Microcavities

New Frontiers

With 178 Figures

Springer

Editors

Daniele Sanvitto
NNL, Istituto Nanoscienze - CNR
Via Arnesano, 73100 Lecce, Italy

and

Istituto Italiano di Technologia
IIT-Lecce, Via Barsanti, 73010 Lecce, Italy
daniele.sanvitto@nano.cnr.it

Vladislav Timofeev
Institute of Solid State Physics
Institutskaya 2, 142432 Chernogolovka, Russia
timofeev@issp.ac.ru

Springer Series in Solid-State Sciences ISSN 0171-1873
ISBN 978-3-642-44320-6 ISBN 978-3-642-24186-4 (eBook)
DOI 10.1007/978-3-642-24186-4
Springer Heidelberg Dordrecht London New York

Printed on acid-free paper

Springer is part of Springer Science+Business Media (www.springer.com)

Preface

The importance of the exciton–polariton phenomenon in the optical properties of high purity bulk semiconductors was recognized in the late 1950s, with the pioneering works of Pekar and Hopfield. However, bulk polaritons had three strong limitations: they were not observable inside the luminescence light cone, the light field as well as the exciton density in the bulk material could not be changed at will, and finally 3D polaritons did not show energy minimum for the ground state. Neverthless in the early 1990s a new stage in polariton physics appeared when confinement of electronic (quantum wells, wires and quantum dots) and photonic states was starting to be strongly studied in the scene of optical phenomena in semiconductors. This was the time when the first vertical cavity surface emitting lasers (VCSELs) were showing to work under continuous wave regime.

In this context, in 1992, Weisbuch and coworkers realized that in a high finesse semiconductor VCSEL, the electromagnetic field of increased density inside the cavity could interact strongly with the excitons of a quantum well, so to lose their independent character as excitons or photons, to give rise to new eigenstates: the two-dimensional exciton–polariton modes.

This observation paved the way to the study of a new panorama of polaritonic effects via optical excitation and detection which was previously impossible in bulk semiconductors. Indeed, these new half-light/half-matter states had all the advantages that were lacking in their 3D predecessors. On the one hand the strength of the coupling was now easy to control by the cavity length, number of QWs or light confinement, while on the other, the presence of such a peculiar polariton *dispersion* could offer a plethora of new interesting physical observations.

State-of-the-art semiconductor microcavities using ultrahigh reflecting distributed Bragg mirrors (DBRs) can reach Q-factors approaching values as high as 10^5, yet polaritons have a relatively short lifetime compared to their excitonic counterparts, and all their dynamics is studied within a few tenths of picoseconds in the best samples. However, one of the very important consequences of their very short lifetimes is that the photon part leaks out of the cavity at a very high rate, carrying out all the important information on the exciton–polariton states. Therefore, a direct and unique one-to-one correspondence between the quantum state of a polariton inside the microcavity and the emitted photon emerges. As a result, exciton–polariton states, their fast dynamics, interactions, unusual collective phenomena under nonequilibrium conditions, and their related quantum effects can be studied directly in optical experiments such as photoluminescence, reflectivity, transmission, and inelastic resonant light scattering. This situation contrasts strongly with that of polaritons in bulk semiconductors or simply excitons in quantum wells, and it is the principal reason why the new exciton–polariton physics has acquired such a strong interest.

Since the first observation of polaritons in microcavities, it took almost 10 years to unravel an extremely important consequence of this light-matter mixture: the exceptionally high nonlinearity inherited by the electronic Coulomb interaction. In the year 2000, it was found that polaritons can undergo a parametric process converting efficiently two particles resonantly excited at a finite $\boldsymbol{k}$ vector into a signal at the bottom of the lower polariton branch (LPB) and an idler at higher $\boldsymbol{k}$ (OPO). This was the first turning point demonstration of the great new possibilities offered by exciton–polaritons due to their peculiar dispersion and their strong χ^3 nonlinearities.

In the following years, the quest for pure quantum effects originating from parametric oscillations has been the topic of many papers, and several geometries—using coupled cavities or 0D systems—have been studied. However, until now, no clear proof of true quantum phenomena, related to single pair scattering, has been experimentally given, and the semi-classical picture could still hold to describe polariton phenomenology.

However, a true breakthrough which has extended the study of polaritons beyond its own domain was set by the clear-cut demonstration of a coherent polariton state, in all aspects similar to a Bose Einstein condensate achieved by stimulation of bosonic polariton particles accumulating in the ground state. It is interesting to note that the very same effect, at that time called "exciton boser," was proposed in 1996. Yet, it took some time to get the present understanding that exciton–polaritons behave in a wide density range as good composite bosons and another 10 years to convince the international community of the strong stunning analogy between the two phenomena. Nevertheless, for their intrinsic dissipative nature, and fast dynamics of relaxation and decay, polaritons are also similar to a standard photonic laser, while the similarity with a BEC is mainly in their interparticle interactions. For this reason there is still now some debate, which has been quite strong in the past, on whether condensation is an appropriate term for exciton–polariton quasiparticles. However, beyond the adequate terminology to use, the dissipative and interacting nature of polaritons, their nonequilibrium character, their peculiar dispersion and strong nonlinearities render these bosonic quasiparticles definitely unique and extremely peculiar, with a lot of similarities and differences to both semiconductor lasers and standard atomic BECs.

It is undoubtful, however, that the exciton–polariton Bose gas is a very interesting system for studying collective many-body phenomena and for possible application in the field of quantum photonics. First of all, exciton–polariton systems start to exhibit the quantum behavior at relatively high temperatures, around tens and even hundreds of Kelvin, due to a very small effective mass of the polariton quasiparticle. Besides, polaritons constitute an ideally two-dimensional or even one- and zero-dimensional system because motion in the other directions could be practically frozen due to the advanced semiconductor growth and processing techniques. In the recent years, we have witnessed the startling observation of new phenomena associated with this quantum phase transition, some of which have shown to be unique of this system.

We believe that a book that was collecting almost all the work done on this subject, which puts together all the advances and recent discoveries on polariton phenomena made by the major groups who are active in this field, would definitely be of great interest to both physicists approaching this subject for the first time, as well as a wide audience of experts in other disciplines who want to be updated on this fast moving field at the frontier between nonlinear optics and quantum collective phenomena of condensed particles.

The content of this book reflects the most interesting and important up-to-date achievements in the field, and it is organized as follows:

In the first chapter, an introduction to the physics of exciton–polaritons in microcavity with historical sequences of the most important turning points and outcomes is made by *F.P. Laussy* which dedicates this chapter to the formation of coherence in a Bose gas with a full quantum derivation which is extended from 2D to 0D polaritons. Interestingly, in the second part of this chapter, it is shown that even in the simplest picture of a linear regime, exciton–polaritons can still show fascinating effects impossible to observe in other systems (such as atomic BEC) like propagation without scattering and diffusion-less motion.

On the contrary, Chap. 2 by *V.D. Kulakovskii*, *S.S. Gavrilov*, *S.G. Tikhodeev*, and *N.A. Gippius*, concerns with the strong nonlinearities, which are fundamental to determine the parametric scattering regime (OPO). A full theoretical model based on the Ginzburg–Landau–Gross–Pitaevskii (GLGP), describing the formation and decay dynamics of the process under different polarization and excitation conditions, is used to burn out the experimental results showing the complex and fascinating physics of parametrical pair processes in microcavity polaritons, such as polarization instabilities and hysteresis effects.

In Chap. 3, B. *Deveaud-Plédran* and *K.G. Lagoudakis* review the fundamental evidences which finally led to the compelling assertion that exciton–polaritons, as composite bosons undergo a phase transition in every aspect similar to what happens in a Bose Einstein condensate. Indeed, given the dissipative nature of this quantum gas, spontaneous formation of pinned vortices, due to pumping and decay, emerges. In this chapter, an insight into the origin and nature of these vortices, such as the independent vorticity of different polariton spin population, is provided. In the last part of the chapter, the authors describe the effect of random potential fluctuations on the dynamics of spontaneously appearing vortices.

A detailed microscopic picture of the two-dimensional steady-state exciton–polariton bose-condensate is described by *G. Roumpos* and *Y. Yamamoto* in Chap. 4. The authors present their recent observations of a power-law decay of the spatial coherence together with the appearance of a bound pair of vortex and antivortex, under certain excitation conditions. These outcomes are associated with the characteristic features of the Berezinskii–Kosterlitz–Thouless (BKT) phase transition. The results suggest that vortex–antivortex pairs can appear in a condensed exciton–polariton gas without rotation or stirring, just as an intrinsic feature of the 2D nature of the polariton condensate.

Temporal coherence and the second order correlation function (g^2) of a condensate of exciton–polaritons are studied in Chap. 5 by *D.N. Krizhanovskii*, *D.M. Whittaker*, *M.S. Skolnick*, and *M. Wouters*. Interestingly, the coherence time and the bunching on the g^2 are independent of the material used to achieve strong exciton–photon coupling (CdTe or GaAs semiconductors) or the way condensation is obtained (via nonresonant or OPO configuration). What is truly affecting these characteristics is the intrinsic particle fluctuations introduced by the laser used to create the polariton population. Indeed, the effects of particle fluctuations on polariton–polariton interactions and the non-equilibrium character of the polariton system determine the coherent properties of the condensate and the behavior of the first and second order correlation functions.

One of the hallmarks of quantum bosonic fluid is the irrotational character which leads to the formation of metastable states of quantized vortices with the appearance of a phase singularity at their center. The theory and the experiments on the stability of externally induced vortices in a polariton condensate, their interactions and propagation in a quantum fluid are presented in Chap. 6 by *F.M. Marchetti* and *M.H. Szymańska*. After a complete introduction on the OPO steady state, its instabilities, and superfluid currents, the concept of triggered optical parametric oscillator (TOPO) is introduced with the purpose to inject vortices in a steady-state condensate of polaritons. In this case, a phase patterned pulsed laser pump is used under the TOPO regime, and polariton dynamics, following the excitation, is observed and studied.

It has been anticipated that exciton–polariton condensates in semiconductor microcavities should behave as an unusual quantum fluid with unique properties due to its nonequilibrium nature. Experiments on superfluidity effects in resonantly created exciton–polariton states are discussed by *A. Amo* and *A. Bramati* in Chap. 7. Using natural defects present on the sample, the transition from a regime of superfluidity with no apparent scattering of the condensate to a perturbed state, evidenced by the appearance of shock waves around the obstacle, is demonstrated. The experimental observations are in qualitative agreement with the generalized Gross–Pitaevskii theory and show how exciton–polaritons in microcavities are an extraordinary interesting system for exploring the rich physics of non-equilibrium quantum fluids. In the wide field of quantum gases, exciton–polariton condensates in semiconductor microcavities have provided novel experimental and theoretical insights into the physics of superfluidity and hydrodynamics, in condensed matter, the nature of which, sometimes, is still under debate.

Polaritons are also interesting bosonic quasiparticles because they possess a defined state of spin. In Chap. 8, *A. Kavokin* addresses some of the effects derived from this spinor nature of polariton condensates and shows the peculiar phenomenology of the specific spin-dependent features which make the collective polariton state a unique object to study spin effects in a gas of interacting Bosons. Moreover, a brief introduction to possible spinoptronic devices is described. Several spin-dependent phenomena are discussed and analyzed, such as the spin-Meissner effect, polarization multistability, spin switching, and spin rings. The future perspectives of exciton–polariton spin superfluidity in microcavities are also considered.

Random disorder is unavoidable and naturally exists in real semiconductor microcavity structures. The impact of fluctuating disorder potential on the dynamical properties of exciton–polariton condensates is the focus of the presentation by G. *Malpuech* and *D. Solnyshkov* of Chap. 9. In the first part, the authors consider a stable condensate propagating in a disorder potential and find that the condensate forms either a glassy insulating phase, at low polariton density (regime of strong localization), or a superfluid phase above the percolation threshold. Then, the case of a propagating noninteracting condensate which is always localized due to Andersen localization is analyzed. Finally, in the last section, the chapter closes with the case of a propagating interacting condensate where the three regimes of strong localization, Anderson localization, and superfluid behavior are all accessible. The calculated localization length is found to be strongly dependent on the system parameters: infinite in the superfluid regime, but drastically reduced in the regime where the gas flows at supersonic speeds.

Exciton–polariton kinetics almost always occurs under thermal non-equilibrium, in the presence of a driving source and subject to dissipation and relaxation processes. To account for the most recent experiments on polariton dynamics, a theoretical formalism capable of modeling the phase and amplitude dynamics of polariton quantum fluids fed by an incoherent reservoir is described in Chap. 10. The theoretical formalism introduced by *M. Wouters* and *V. Savona* is based on a truncated Wigner approximation (TWA) where the polariton quantum fluid is described by a classical field subject to stochastic forces and coupled to the reservoir that produces gain, losses, and energy relaxation mechanisms. Calculations of polariton distribution functions as well as the first order off-diagonal correlation functions, below and above condensation threshold, are shown to be rather effective for describing some recent experimental results on polariton dynamics.

In Chap. 11, the coupling of exciton–polaritons with acoustic phonon is presented by *E. Cerda-Méndez*, *D.N. Krizhanovskii*, *M. Wouters*, *K. Biermann*, *R. Hey*, *M.S. Skolnick*, and *P.V. Santos*. The acoustic phonons are generated via surface acoustic waves (SAW) from the top of the microcavity samples. The coupling with polaritons allows the formation of mini-gaps which serve as 1D potentials for the localization of the condensate generated via OPO. This novel approach leads to the formation of arrays of quasi one-dimensional interacting polariton condensates which are similar to the optical lattices of atomic BEC. Using first and second order correlations, it is demonstrated that the moving deformation, modulated by surface acoustic waves, affects both the energetic configuration and the spatial coherence length of polariton condensates.

One of the most controversial paradigms which is still under debate in the community is to which extent a polariton condensate is different from a standard photonic laser. Exciton–polariton condensation and lasing in a standard VCSEL exhibit many common properties: both emit coherent light, both have density onset for optical gain, and exhibit rather high temporal and in-plane coherence. So, the reasonable question appears: what is the principal difference between them? The detailed analysis of similarities and differences between exciton–polariton condensation in microcavities and lasing is discussed by *D. Snoke* in Chap. 12. Experiments in which each transition can be observed independently and in the same physical system help to determine differences in the behavior of each regime. These examples are reviewed also in view of the recent experimental observations of photon condensation for a dye molecular system in an optical cavity, where the system is always in the weak coupling regime.

The prospects of technological applications of polariton condensates for a new generation of low threshold coherent light emitters are considered by *J.J. Baumberg* and *G. Christmann* in Chap. 13. The authors overview the nearest prospects of room temperature polariton condensates and discuss the importance of the material choice, as well as sample design, for the realization of low threshold lasers. These advances are supported by their successful realization of polariton lasing at room temperature using a III-nitride semiconductor microcavity with GaN as active medium.

Another approach for the realization of room temperature polariton lasers is the use of organic semiconductors which have a very strong coupling with light and big exciton binding energy. Organic semiconductors have both electrical and structural properties that are inherently different from their inorganic counterparts. The exciton of organic molecular crystals is of Frenkel' type, for instance, so strongly localized in the molecular aggregates but can be stable up to room temperature and beyond. Strong exciton–photon coupling and polariton lasing at room temperature in organic single crystal microcavities and hybrid organic–inorganic microcavities are discussed by *S. Kena-Cohen* and *S.R. Forrest* in Chap. 14. A robust exciton–polariton state in organic materials is shown to be easily observed at room temperature with a neat increase of its coherence for high pumping values. These experimental results lead the way to the realization of inexpensive polariton lasing devices.

Finally, the recent progress in the rapidly developing field of electrically driven polariton systems is highlighted by *S.I. Tsintzos*, *N.T. Pelekanos*, and *P.G. Savvidis* in Chap. 15. The threshold of polariton lasers is few orders of magnitude lower than that of a conventional semiconductor photon lasers operating in the weak coupling regime. This makes polariton lasers extremely promising, if not for other fundamental properties, as low-power sources of coherent light. For this reason, the possibility to observe polariton emission out of electrical injection is one of the main tasks to follow. This challenge is the subject of this last chapter in which the authors describe their recent results on microcavity emission, via electrical rather than optical excitation, up to room temperature and keeping the system still under strong coupling regime.

We would like to thank all the authors of this work for their dedication and for spending much of their time in contributing to the successful realization of this book.

Lecce, Chernogolovka

Daniele Sanvitto
Vladislav Timofeev

Contents

Contributors

Alberto Amo CNRS-Laboratoire de Photonique et Nanostructures, Marcoussis, France

Jeremy J. Baumberg Nanophotonics Centre, Cavendish laboratory, University of Cambridge, Cambridge, UK

Klaus Biermann Paul-Drude-Institut für Festkörperelektronik, Berlin, Germany

Alberto Bramati Laboratoire Kastler Brossel, Université Pierre et Marie Curie-Paris 6, École Normale Supérieure et CNRS, Paris, France

Edgar Cerda-Méndez Paul-Drude-Institut für Festkörperelektronik, Berlin, Germany

Gabriel Christmann Nanophotonics Centre, Cavendish Laboratory, University of Cambridge, Cambridge, UK

Benoit Deveaud-Pledran Ecole Polytechnique Fédérale de Lausanne (EPFL), Lausanne, Switzerland

Stephen R. Forrest Departments of Physics and Electrical Engineering and Computer Science, University of Michigan, Ann Arbor, MI, USA

Sergey Gavrilov Institute of Solid State Physics, RAS, Chernogolovka, Russia

Nikolay A. Gippius A.M. Prokhorov General Physics Institute, RAS, Moscow, Russia

LASMEA, UMR 6602 CNRS, Université Blaise Pascal, Aubière, France

Rudolf Hey Paul-Drude-Institut f¨ur Festk¨orperelektronik, Berlin, Germany

Alexey Kavokin Laboratoire Charles Coulomb, CC074, Universite de Montpellier II, Place Eugene de Bataillon, Montpellier, cedex, France

Physics and Astronomy School, University of Southampton, Highfield, Southampton, UK

Stéphane Kéna-Cohen Department of Physics, Imperial College, London, UK

Dmitry N. Krizhanovskii Department of Physics and Astronomy, University of Sheffield, Sheffield, UK

Vladimir. D. Kulakovskii Institute of Solid State Physics, RAS, Chernogolovka, Russia

Konstantinos G. Lagoudakis Ecole Polytechnique Federale de Lausanne (EPFL), Lausanne, Switzerland

Fabrice P. Laussy Walter Schottky Institut, Technische Universität München, Garching, Germany

Gerome Malpuech LASMEA, Nanostructures and Nanophotonics group Clermont Université, Université Blaise Pascal, CNRS, France

Francesca M. Marchetti Departamento de Física Teórica de la Materia Condensada, Universidad Autónoma de Madrid, Madrid, Spain

Nikolaos T. Pelekanos Department of Materials Science and Technology, University of Crete, Crete, Greece

FORTH, Heraklion, Greece

Georgios Roumpos E. L. Ginzton Laboratory, Stanford University, Stanford, CA, USA

JILA, National Institute of Standards and Technology, and the University of Colorado, Boulder, CO, USA

Paulo Santos Paul-Drude-Institut für Festkörperelektronik, Berlin, Germany

Vincenzo Savona Institute of Theoretical Physics, Ecole Polytechnique Fédérale de Lausanne EPFL, Lausanne, Switzerland

Pavlos G. Savvidis Department of Materials Science and Technology, University of Crete, Crete, Greece

FORTH, Heraklion, Greece

Maurice S. Skolnick Department of Physics and Astronomy, University of Sheffield, Sheffield, UK

David Snoke Department of Physics and Astronomy, University of Pittsburgh, Pittsburgh, PA, USA

D. Solnyshkov LASMEA, Nanostructures and Nanophotonics group Clermont Université, Université Blaise Pascal, CNRS, France

Marzena H. Szymańska Department of Physics, University of Warwick, Coventry, UK

Sergey G. Tikhodeev A.M. Prokhorov General Physics Institute, RAS, Moscow, Russia

Simeon I. Tsintzos Department of Materials Science and Technology, University of Crete, Crete, Greece

FORTH, Heraklion, Greece

David M. Whittaker Department of Physics and Astronomy, University of Sheffield, Sheffield, UK

Michiel Wouters Ecole Polytechnique Federale de Lausanne (EPFL), Lausanne, Switzerland

Department of Physics, University of Antwerpen, Antwerp, Belgium

Yoshihisa Yamamoto E. L. Ginzton Laboratory, Stanford University, Stanford, CA, USA

National Institute of Informatics, Chiyoda-ku, Tokyo, Japan

Chapter 1
Quantum Dynamics of Polariton Condensates

Fabrice P. Laussy

Abstract We illustrate the rich and fundamental physics that is accessible with the semiconductor implementation of the quantum superposition of light and matter: *exciton–polaritons*. The short lifetime of polaritons makes them an out-of-equilibrium system. Their dynamic is an important ingredient in their behaviour and properties. Their peculiar dispersion also allows a rich engineering of various processes, tuning the system from light- to matter-like. Finally, the exciton–exciton interaction turns them into a non-linear system. The interplay of all these factors makes polaritons one of today's most versatile and fruitful research arena, both theoretically and experimentally. In this chapter we give a rather general picture of these specificities that we isolate in various dimensionalities (0, 1, and 2D). One of the most intensively researched area in the semiconductor implementation of the polariton physics is related to Bose–Einstein condensation. We solve exactly a configuration of relaxation from the Rayleigh circle into the ground state in the framework of quantum Boltzmann master equations and show how coherence builds up spontaneously in the system, by copying in a single quantum state statistical features characteristic of the macroscopic system. In this way, we extend to higher order correlations the historical reasoning of Einstein, who predicted the phenomenon by arguments on the mean populations. We show how lifetime and pumping allow a simpler treatment by reducing the required number of states, for which we present a full quantum treatment. We contrast this condensate build-up with the 0D case where the reduced complexity allows an exact numerical treatment. The coherence build-up in this cavity QED limit manifests as lasing with a sharp line in the cavity mode that produces a variation of the Mollow triplet in the exciton emission, as the cavity effectively replaces the laser in the conventional resonance fluorescence scenario. We show how lasing also arises in this case as a condensation

F.P. Laussy (✉)
Walter Schottky Institut, Technische Universität München, Am Coulombwall 3,
85748 Garching, Germany
e-mail: fabrice.laussy@gmail.com

D. Sanvitto and V. Timofeev (eds.), *Exciton Polaritons in Microcavities*,
Springer Series in Solid-State Sciences 172, DOI 10.1007/978-3-642-24186-4_1,

of polaritons, and can be substituted in the case of vanishing intensities by a coherent field formed when strong coupling is optimum. This zero-dimensional limit also provides an exact picture of the transition from the quantum to the classical regime, a universal process of unsuspected complexity. Finally, we illustrate the recent development of polariton quantum hydrodynamics with propagation of coherent wave packets. The short lifetime allows a continuous observation of this dynamics in real space, a picture completed with the observation of their emission spectra in energy–momentum space. The peculiar polariton dispersion is the source of interesting behaviours even when described by the most fundamental and simplest equation of quantum physics: the Schrödinger equation.

1.1 Introduction

Quantum condensates—the quantum states formed by a coherent gathering of a large number of particles—are one of the chief scientific developments of modern science. They hold a privileged status that accounts for the excitement they stir in a wide and multidisciplinary community, in that they wed the quantum (or microscopic) and the classical (or macroscopic) in a unified paradigm, the condensate wave function ψ. At the quantum level only, the wave function is an amplitude of probability that collapses upon measurement and with such counter intuitive properties as entanglement or non-locality. At the classical level, the wave function is replaced by a continuous and macroscopic field, that embeds and provides all the information on the object. Quantum condensates are macroscopic fields that magnify the effect of a single particle by superimposing a large number of them in the same and identical state, with effect to retain much of their quantum character. This leads to striking phenomena such as superfluidity and superconductivity. As obvious from the examples we just took, we shall adopt a rather general definition of "condensation," where superconductivity is seen as a condensation or Cooper pairs and lasing as a condensation of photons. We will also consider the Bose–Einstein condensate in the most traditional sense of the term.

The main topic of this volume, the polariton, is another peculiar object that arises from two protagonists lying at opposite extremes—light and matter—wed by the quantum superposition of their wave functions:

$$|\psi_\pm\rangle = \chi_{\mathrm{C}}^{\pm} |\psi_{\mathrm{C}}\rangle + \chi_{\mathrm{X}}^{\pm} |\psi_{\mathrm{X}}\rangle , \tag{1.1}$$

where ψ_{C} is for the photon—"C" stands for "cavity"—and ψ_{X} for the material excitation—"X" stands for "exciton," the correlated electron-hole pair excitation of the semiconductor crystal [1]. $\chi_{\mathrm{C,X}}^{\pm}$ are the so-called Hopfield coefficients which weigh the polaritons with their photon and exciton character [2]. The "mélange" is maximum when $\chi_{\mathrm{C}}^{\pm} = \pm\ \chi_{\mathrm{X}}^{\pm} = 1/\sqrt{2}$. By changing the ratio of these two components, one can interpolate smoothly between light and matter. This can be achieved by detuning the two modes in energy, which is usually experimentally straightforward, either by tuning the cavity (by designing it with a wedge or by injection of a gas) or the exciton (by temperature or with an electrical field). Attributes from light that can be transferred to the polaritons are those such as high degree of coherence, very small effective mass (the cavity photon acquires a small mass due to its confinement), or optical interfacing with the external world, enabling the experimentalist a great ease in both manipulation and detection. Those attributes inherited from the "matter" part are strong interactions between particles and large effective masses. Interactions make the polariton physics non-linear and thus both highly non-trivial, with behaviours of strongly correlated systems, and extremely rich in terms of applications and engineering. Depending on the particular application, various combinations of these characteristics play the major role in a polariton device. Both the light and the matter parts can be optimally implemented in a semiconductor environment. This makes the system easily configurable, quickly enhancing its figures of merits as technology progresses and with great promises for large-scale integration and massive technological deployment. Maybe the most obvious illustration of this flexibility are the several dimensionalities in which polaritons can be realised. Figure 1.1 displays various star realisations.

Three-dimensional systems were the first to elicit interest for condensates in the solid state, with excitons as the candidate bosons [3], but they have been hampered by various problems such as a too short lifetime or, on the other hand, difficulty of detection for dark states with a long lifetime. The technological breakthrough of heterostructures [4] allowed to reduce the dimensionality and solve most of these problems, for instance, the short exciton lifetime, has been addressed by indirect excitons in coupled quantum wells [5]. The polariton case also met with difficulties in 3D that were also circumvented in lower dimensions, such as the absence of an energy minimum for the bulk polariton since there is no $k = 0$ bulk photon.

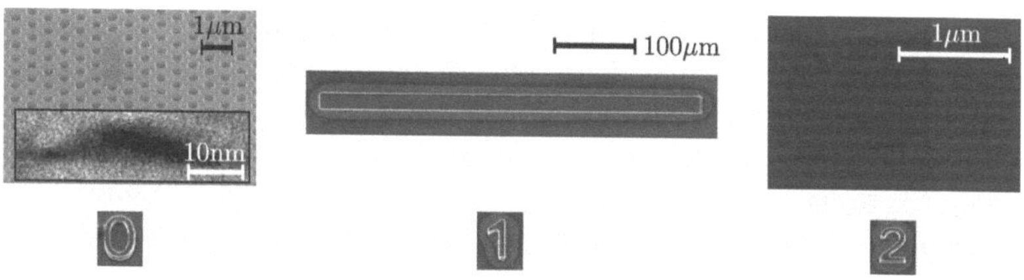

Fig. 1.1 Stages for polariton physics in three dimensionalities as seen by an experimentalist: examples are given in 0, 1 and 2D. Other realisations are possible. Those specifically highlighted are as follows: (*0D*) [*left panel*] a quantum dot (*QD*) in an L3 photonic crystal (courtesy of *A. Laucht*). Active QDs inside the structure are not visible, but an AFM image is provided for one of them before overgrowth. (*1D*) [central panel] a stripe, or etched 2D cavity (courtesy of *P. Savvidis*) as seen from the top. Propagation is constrained to 1D inside the stripe, in the plane of the page. (*2D*) [*right panel*] section of a distributed Bragg mirror (courtesy of *D. Sanvitto*), as seen from the side. Propagation is constrained to 2D in the plane perpendicular to the page. The labels 0, 1 and 2 (also from *P. Savvidis*), used here to label the dimension of each case, are grown also to label characteristics of the samples, such as the dimension of a structure. This illustrates how technologically advanced is the growth process

The microcavity polariton field was born in 1992 when Weisbuch et al. reported strong coupling in planar cavities [6] (much like that in Fig. 1.1c). Although they had the physics of VCSELs in mind, they identified their finding as the realisation in the solid state of a much deeper achievement: the cavity QED paradigm of *strong coupling* [7]. Anecdotically, they misquote the first author of [7] (M. G. Raizen), attributing the paper to a co-author (R. J. Thompson), maybe because the latter published a few months earlier in the same journal a significant improvement of the 1989 atomic experiment, reducing the number of atoms to the quantum limit [8].[1] The quoted paper is however the correct one since Weisbuch et al. had not reached the quantum limit—in a sense that will be shortly discussed—and there is still no evidence to this date that this has ever been achieved with 2D polaritons or even 1D polaritons. We will discuss at length what peculiar insights can be learned from the 0D case which arrived there first [9, 10].

The polariton device of Weisbuch et al. [6] quickly demonstrated its great ease of control by angle-resolved excitation and detection [11]. The next breakthrough came with the experiments of Savvidis et al. [12] and Baumberg et al. [13] who demonstrated spectacularly the Bose character of 2D polaritons in a pump and probe experiment, with stimulated scattering triggered by a pulse. This pointed towards the feasibility of a "polariton laser," a mechanism proposed earlier by Ĭmamoḡlu et al. [14] where coherent light is emitted from a polariton condensate. There has been ample debates ever since on the adequacy of the terminology of "Bose–Einstein condensates" to refer to optical systems where coherence is provided foremost by the photonic component and is degraded by the excitonic one or in cases where excitation is resonant, and thus directly transferred in the cavity by the laser [15, 16]. Crowning a series of accumulating evidence from various groups [17, 18], Kasprzak et al. [19] reported Bose–Einstein condensation of polaritons in late 2006. The next step from there was superfluidity, a concept discussed at an early stage in the background of polariton lasers [20] and attacked theoretically by Carusotto and Ciuti [21]. Experimentally, superfluidity of polariton condensates has been approached from various directions, all pointing towards a positive outcome [22–28]. In these works, dimensionality is of a more or less fundamental character, from the mere requirement to provide a space–time background for propagation of the fluid [23] to its link with universality in the theory of phase transitions [24] or for topological reasons [22, 27]. Depending on the nature of the propagation of the condensate, dimensionality might play or not a critical role. For instance, the propagation of a wave packet in the linear Schrödinger equation is essentially unrelated to dimensionality and can be, with little loss of generality, projected onto one dimension. The propagation of a soliton, on the other hand, which stability is assured by non-linearity, depends crucially on local topology, being stable in 1D and unstable in 2D. In both cases, however, a spatial degree of freedom exists into which the quantum dynamics can unravel as a function of time.

[1] Or maybe because of a simple oversight since Raizen had a different affiliation.

There is equally rich dynamics of condensation in other degrees of freedom than real space as a function of time, such as the field amplitude or its phase as a function of pumping power. The problem of nucleation, or formation, of Bose–Einstein condensates, for instance, is an illustrious example of a much-investigated dynamics of condensates [29–33]. In microcavity polaritons, it has practical rather than fundamental implications since the short lifetime of the particles and the existence of a bottleneck in their relaxation are two strong obstacles for the formation of a condensate. The problem has first been attacked with semi-classical rate equations, of the Boltzmann type, by Tassone et al. [34], Porras et al. [35], Malpuech et al. [36] and others. Coherence build-up from a master equation approach was investigated by Rubo et al. [37], Laussy et al. [38], Cao et al. [39] and Doan et al. [40]. As quantum correlations extend into many degrees of freedom, the problem is extremely complicated. Techniques that bring together the quantum nature of a ground state with spatial and temporal extensions with classical reservoirs have recently been proposed by Wouters and Savona [41] and Savenko et al. [42]. It also has enjoyed much insights from quantum field theories, for instance by Keeling et al. [43, 44], Szymanska et al. [45], and Marchetti et al. [46]. Reducing the dimensionality of the system is a way to simplify considerably the system, allowing to solve it essentially exactly (in the sense of making no approximation, even if it solved numerically). This is where the ultimate limit of 0D polaritons enters in the condensation problem.

The first reference in Weisbuch et al. seminal paper [6] is to the pillar of cavity QED, the Jaynes and Cummings model [47], that describes the full-field quantisation interaction between light, described as an harmonic mode, and matter, described as a two-level system. As already alluded to before, the 2D case does not, however, correspond to this scenario. The strong coupling of quantum well excitons does not rely explicitly on their quantisation, and in the linear regime, it can in fact also be interpreted as a classical effect—more appropriately denominated "normal mode coupling"—in the sense that it is equivalently described by a classical theory of the electromagnetic field, in this case, Maxwell equations coupled to the excitonic susceptibility [48, 49]. The Jaynes–Cummings physics, on the other hand, hinges intrinsically upon the concept of quanta: one excitation more or less makes a difference in the system, by shifting by a discrete amount its energy, with detectable consequences in the power spectra or the counting statistics of the emitted photons. As a result, the two-polariton state does not consist of the direct product (properly symmetrised) of two states from (1.1) as is the case in higher dimension according to the conventional Bose condensation scenario. Instead, a new polariton state is formed, which general form reads:

$$|n\pm\rangle = \chi_{\mathrm{C}}^{\pm(n)} |n, \mathrm{g}\rangle + \chi_{\mathrm{X}}^{\pm(n)} |n-1, \mathrm{e}\rangle \,, \tag{1.2}$$

where n photons interact with the ground $|\mathrm{g}\rangle$ and excited $|\mathrm{e}\rangle$ state of the 0D exciton. Following Cohen-Tannoudji [50], we will at times speak of "dressed states" rather than polaritons. The case $n = 1$ recovers the previous one of (1.1). In (1.2), the condensed state appears to be through the photon fraction alone. By contrast, the linear regime of normal mode coupling in higher dimensions distributes the excitation throughout various emitters, which behave collectively as another harmonic oscillator [51], and condensation can be regarded as the accumulation of one-polariton states, as if the latter were fundamental particles.

The proper solid-state counterpart to cavity QED is provided by quantum dots (QDs) in microcavities, which achieved the strong-coupling regime in late 2004, early 2005 [52–54] and shortly thereafter confirming its quantum nature [9, 10]. The quantum regime, ruled by the few quanta of excitations, is interesting in itself [55], but in the wake of the previous discussion, we shall focus on the opposite limiting case of the large number of excitations, where the quantisation of the field breaks down and a continuous, coherent field overtakes. This transition exhibits a spectacular crossover from the quantum to the classical regime, that puts lasing as a condensation of the 0D polaritons [56, 57].

In summary of this rapid overview, polaritons—quantum superposition of light and matter—are privileged objects to study various aspects of quantum condensates, where a large and coherent accumulation of them in a single state gives rise to a macroscopic quantum state. Dimensionality is one of the factors that allows to focus more particularly on various aspects of the quantum dynamics. In the following, after a succinct introduction in the formalism and notations in Sect. 1.2, we address first, in Sect. 1.3, the dynamics of the condensate formation in a gas of bosons. Dimensionality will play its role here in justifying that polaritons are good bosons: in 2D, their excitonic fraction is also bosonic. Although we shall assume a particular configuration, the density of states will also be that of a 2D system. Because of the complexity of the problem, much approximations will be made, for instance, the weakly interacting gas will be assumed ideal, an approximation also best justified in 2D. In Sect. 1.4, we address the same problem but in 0D, which allows us to keep track of all quantum correlations. Even in this much simplified picture, we shall see that a full quantum description of the problem is extremely complicated. In Sect. 1.4, we will turn to another aspect of the dynamics of polariton condensates, made possible by experiments such as those of Amo et al. [23] and Sanvitto et al. [58], namely, the space–time dynamics of wave packets, a configuration now shared by the 2D and 1D systems. For brevity, we will consider the 1D case only, with little loss of generality in the approximations we shall make. Experiments have also been conducted in this configuration [59] which makes it valuable for its own sake. A full 2D description with proper attention to its specifics has been given by Marchetti and Szymanska [58] and also in their contribution to this volume. In Sect. 1.6, we summarise the main statements and give an outlook of our approach where, among other things, we try to provide a unified picture to the two concepts of condensation and strong light-matter coupling, that transcends or, on the opposite, hangs upon dimensionality.

1.2 Modeling of the Polariton Dynamics

The basic Hamiltonian for strong light–matter coupling in a field theory reads, in real space, $H = \int \psi^\dagger(\mathbf{r}) D \psi(\mathbf{r})\, d\mathbf{r}$ where ψ is a two-component field [1]:

$$\psi = \begin{pmatrix} \psi_C \\ \psi_X \end{pmatrix}, \tag{1.3}$$

with ψ_C and ψ_X two fields annihilation operators at point $\mathbf{r}$, where $\mathbf{r}$ has the dimension of the problem (0, 1 or 2D). The free energy operator D is most simply expressed in the reciprocal ($\mathbf{k}$) space, as the linear coupling with strength g of two free particles of mass m_C and m_X:

$$D = \begin{pmatrix} \omega_C + \mathbf{k}^2/(2m_C) & g \\ g & \omega_X + \mathbf{k}^2/(2m_X) \end{pmatrix}. \tag{1.4}$$

Strictly speaking, the cavity photon is parabolic only at small momenta and recovers the free photon linear dispersion at large momenta, but we will keep the description in terms of an effective photon mass for simplicity and generality. This provides the basic description of polaritons, where they appear as the eigenstates of H. The polariton energies, that is, the corresponding eigenvalues, are shown in Fig. 1.2. There is an imbalance between a small m_C and a large m_X that causes the asymmetry in the dispersion of the lower polariton. The zero-dimensional case freezes all the remaining degrees of freedom: $\mathbf{r}$ is fixed at the point of interaction $\mathbf{r}_0$ where the photon energy becomes constant too. The Hamiltonian reduces in 0D to:

$$H_{0D} = \omega_C a^\dagger a + \omega_X \sigma^\dagger \sigma + g(a^\dagger \sigma + a \sigma^\dagger), \tag{1.5}$$

where we introduced for convenience the notation $a = \psi_C(\mathbf{r}_0)$ and $\sigma = \psi_X(\mathbf{r}_0)$. The excitonic field ψ_X loses its bosonic character in 0D because the confinement of the electron-hole pair in the QD results in their separate quantisation, overtaking the weaker Coulomb interaction which, in higher dimensions, correlates them into a bound state that behaves approximately like a boson, the better the smaller the exciton density [60]. This prevents the presence of two electrons or two holes by Pauli exclusion, and in the ideal 0D limit, the light–matter Hamiltonian therefore couples a Bose and a Fermi 0D field, thereby realizing the Jaynes–Cummings Hamiltonian: light is modelled by an annihilation operator a that follows Bose algebra, $[a, a^\dagger] = aa^\dagger - a^\dagger a = 1$, and matter by an annihilation operator σ that follows Fermi algebra, $\{\sigma, \sigma^\dagger\} = \sigma\sigma^\dagger + \sigma^\dagger\sigma = 1$. Dissipation is conveniently and accurately described in this formalism in the Lindblad form [61]:

$$\mathcal{L}_O \rho \equiv 2O\rho O^\dagger - O^\dagger O \rho - \rho O^\dagger O\,, \tag{1.6}$$

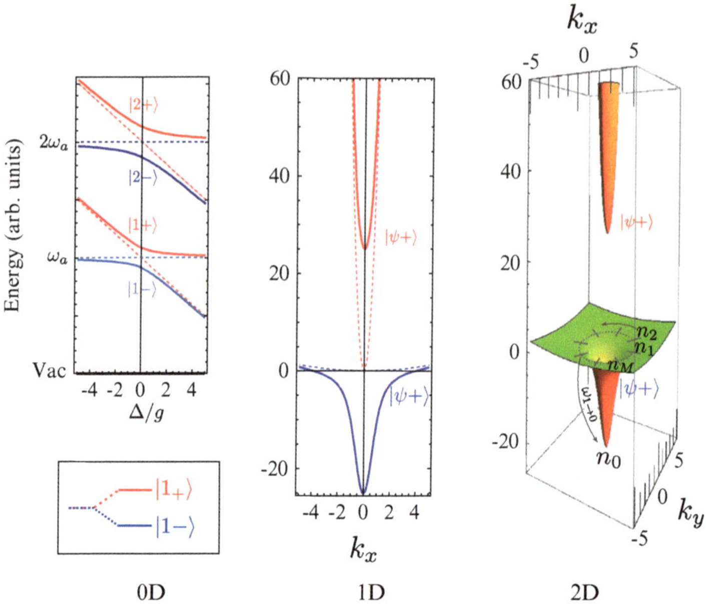

Fig. 1.2 Polaritons in three dimensionalities as seen by a theorist. The 0D case realises in its most fundamental form the quantum superposition of (1.1). The exciton and photon fractions can be changed, detuning by Δ the energy between the bare modes. In this zero-dimensional case, each quantum should be accounted for through higher excited states leading to new polariton states, (1.2), shown here up to two quanta. The 1D case has an additional degree of freedom. Because energy changes with momentum, detuning of the bare mode is intrinsically built in the polariton dispersion, giving rise to a most peculiar shape in particular for the lower polariton. The 2D case adds still another degree of freedom resulting in the 2D dispersion that is generated by rotation of the 1D dispersion. The bottom of the lower polariton dispersion in k-space is the final state where to accommodate the condensate. In 0 and 1D, the bare modes are represented as well in dashed lines. In 2D, the particular configuration of states on the Rayleigh circle selected to describe Bose condensation analytically in Sect. 1.2 are represented along with one particular relaxation towards the ground state at rate $\omega_{1\to 0}$

for any operator O. Decay and pumping of mode c at rates γ and P, respectively, follow from this as $(\gamma/2)\mathcal{L}_c$ and $(P/2)\mathcal{L}_{c^\dagger}$ added to Liouville–von Neumman equation $\partial_t\rho = i[H,\rho]$. In a full Hamiltonian picture, lifetime is conveniently included as the imaginary part of the energy, by promoting the Hamiltonian H to $H - i\gamma/2$, which is justified in part by the Lindblad form above (the procedure is correct within limits and is therefore more a trick than a rigorous step; for instance, it is well known that commutation relation are violated by this procedure). If $\psi(t)$ is the solution to the Schrödinger equation with potential H, then $\psi(t)\mathrm{e}^{-\gamma t/2}$ is the solution to the version with the finite lifetime introduced in this way, since $\psi_{\gamma=0}(t) = \exp(-\mathrm{i}Ht)\psi(t=0)$ becomes $\psi(t) = \exp(-\mathrm{i}Ht)\exp(-\gamma t/2)\psi(0) = \exp(-\gamma t/2)\psi_{\gamma=0}(t)$.

In higher dimensions, the second quantization notation in reciprocal space is the most convenient; it allows to track all processes in a physically transparent way, such as scattering $\psi^{\dagger}_{\mathbf{k}+\mathbf{q}}\psi_{\mathbf{k}}\phi^{\dagger}_{\mathbf{q}}$ involving a phonon field ϕ or polariton–polariton interaction $\psi^{\dagger}_{\mathbf{k}_1+\mathbf{q}}\psi^{\dagger}_{\mathbf{k}_2-\mathbf{q}}\psi_{\mathbf{k}_1}\psi_{\mathbf{k}_2}$ [62]. We will rely only implicitly on this form, which can be found in full details elsewhere [1]. We will introduce the rest of the notation as needed.

1.3 Formation of Coherence in a Gas of Boson (2D Polaritons)

The dynamics of a quantum gas in the context of kinetic theory finds its simplest description by including quantum statistics in the transition rates of Boltzmann equation [63], an approach pioneered by Uehling and Uhlenbeck [64] which amounts to upgrading the rate $\omega_{i\to j}$ to $(1 \pm n_j)\omega_{i\to j}$ with $+$ for bosons (describing Bose stimulation) and $-$ for fermions (describing Pauli blocking). In both cases, the term "1" corresponds to spontaneous emission. With the number of particles in the initial state available for scattering, the resulting quantum Boltzmann equation acquires a beautiful symmetry in momentum space:

$$\dot{n}_{\mathbf{k}} = \sum_{\mathbf{k}'} \Big(\omega_{\mathbf{k}'\to\mathbf{k}}(n_{\mathbf{k}}+1)n_{\mathbf{k}'} - \omega_{\mathbf{k}\to\mathbf{k}'}n_{\mathbf{k}}(n_{\mathbf{k}'}+1)\Big), \tag{1.7}$$

with $n_{\mathbf{k}}$ the polariton population in the state $\mathbf{k}$, and $\omega_{\mathbf{k}\to\mathbf{k}'}$ the transition rate between states $\mathbf{k}$ and $\mathbf{k}'$, that depends on the underlying interaction. The microscopic derivation of these coefficients for some of the dominant processes has been performed by Ciuti et al. [65] and Malpuech et al. [66] and is now compiled in textbooks [1]. It consists typically of a matrix element of the interaction calculated with the Fermi golden rule (with a δ energy-conserving term) and weighted by the density of particles mediating the interaction, such as phonons or electrons at their lattice temperature. By lowering temperature, rates are increased for relaxation towards the ground state and condensation may follow, triggered by Bose stimulation. We do not consider here polariton–polariton interaction although it is known to help and in some cases to be necessary for condensation.

Because it is such a superficial doctoring up of the classical equation—although to monumental consequences—(1.7) is usually called the "semi-classical Boltzmann equation." Its algorithmic complexity is the same as the classical version and thus it can be simulated for complex and realistic systems with many degrees of freedom and a large number of particles. In the case of polaritons, extensive numerical simulations have been performed by Tassone et al. [34], Porras et al. [35], Malpuech et al. [36], Cao et al. [39], and others, and a thorough understanding of the kinetics of condensation in various limits has now been achieved [67]. Recently, Hartwell and Snoke could even reproduce quantitatively the experimental data and fit it with the semi-classical Boltzmann equations [68], so the dynamics is essentially understood and is becoming a problem for engineering to achieve better condensation within the polariton lifetime.

What (1.7) is missing is the quantum aspect of the dynamics, in particular, to describe and to take into account the effect of coherence. That the term coherence is so loosely defined is partly a misfortune, partly an indication of a deep connection between various fundamental concepts, that the term "coherence" is so loosely defined. There is quantum coherence, on the one hand, in the sense of purity of state (related to the existence of wave function or of a density matrix whose square traces close to unity), and there is, on the other hand, optical coherence that can be of temporal, spatial or spectral character (among others) and is related to the ability of a field to produce interferences. Glauber was awarded the Nobel prize (in 2005) for identifying what characterises quantum coherence of the optical field [69], namely, the nth-order ($n \in \mathbf{N}$) properties of quantum correlators of the type:

$$G^{(n)}(t_1, \ldots, t_n) = \langle a^\dagger(t_1) \cdots a^\dagger(t_n) a(t_n) \cdots a(t_1) \rangle, \tag{1.8}$$

where a is the annihilation operator of the bosonic field. This includes as particular cases of interest the first-order correlator:

$$G^{(1)}(t, \tau) = \langle a^\dagger(t) a(t+\tau) \rangle , \tag{1.9}$$

whose τ-Fourier transform is related to the power spectrum (that can be probed in the photoluminescence emission, for instance) and the second-order correlator:

$$G^{(2)}(t, \tau) = \langle a^\dagger(t) a^\dagger(t+\tau) a(t+\tau) a(t) \rangle , \tag{1.10}$$

that is related to photon counting statistics. In the following, we will also use their normalised version $g^{(1)}(t, \tau) = G^{(1)}(t, \tau)/n_a(t)$ and $g^{(2)}(t, \tau) = G^{(2)}(t, \tau)/\big(n_a(t) n_a(t+\tau)\big)$. The major characteristics of quantum coherence are provided by the case of zero time delay, $\tau = 0$, being the field intensity to first order and two-photon coincidences to second order. Glauber discussed how and why a proper definition of quantum field coherence is a factorisation to all orders n of the correlators:

$$\langle a^{\dagger n} a^n \rangle = \langle a^\dagger a \rangle^n \tag{1.11}$$

and how such a condition is satisfied for the so-called *coherent state*:

$$|\alpha\rangle = \exp\left(-\frac{|\alpha|^2}{2}\right) \sum_{n=0}^{\infty} \frac{\alpha^n}{\sqrt{n!}} |n\rangle , \tag{1.12}$$

with $\alpha \in \mathbf{C}$. The coherent state minimises quantum uncertainty, such as position and momentum. As the phase of this state is exactly determined (by the phase of α in the complex plane), the particle number—which is its quantum conjugate—has a fluctuation, which one can see from (1.12) is Poissonian, and that indeed satisfies $g^{(2)} = 1$. The photons from a coherent field are uncorrelated in their detection time, in stark contrast with a Fock state $|n\rangle$ or a thermal state:

$$\rho = (1 - \theta)\theta^{a^\dagger a} . \tag{1.13}$$

The discovery, by Hanbury Brown and Twiss [70], that thermal light (from a star) exhibits the statistical properties of (1.13) has been one of the major achievements of quantum optics.

From another point of view, in the theory of phase transitions, it is known since Landau [71] that an "order parameter" characterises the various phases of second-order transitions and that the order parameter of a quantum condensate (a Bose–Einstein condensate in the thermodynamic limit, a superfluid or a superconducting phase) is the quantum average of the field annihilation operator: $\langle\psi(\mathbf{r})\rangle$ from (1.3), or in the spatially homogeneous case, simply $\langle a\rangle$. A theory of phase transitions for the polariton condensate building upon the semi-classical Boltzmann equations above, could therefore look forward to upgrading the ground state to the quantum level while keeping the complex relaxation dynamics at the semi-classical level. This approach has been taken by Rubo et al. [37], who, isolating the dynamics of the quantum mode from the others, arrive to the master equation:

$$\partial_t \rho = -\frac{1}{2}[W_{\mathrm{out}}(t)(a^\dagger a\rho + \rho a^\dagger a - 2a\rho a^\dagger) + W_{\mathrm{in}}(t)(aa^\dagger\rho + \rho aa^\dagger - 2a^\dagger \rho a)], \tag{1.14}$$

where ρ is the density matrix for the ground state alone and $W_{\mathrm{in/out}}(t)$ are time-dependent transition rates provided by the semi-classical Boltzmann equations. Rubo et al. [37] obtained the exact time-dependent solution of (1.14) which is, starting from the vacuum, a thermal state in the form of (1.13) with a time-dependent θ which is a complicated function of $W_{\mathrm{in,out}}(t)$, for which there exists a closed-form expression [37] that we need not reproduce here. The density matrix being thermal has important (exponential) fluctuations in the number of particles when not close to the vacuum, and its order parameter is zero, regardless of the intensity. At such, the quantum promotion of the ground state only is insufficient: isolating the ground state from the rest of the system does not bring more information than the semi-classical picture, in particular, the statistics of the state is the same; it remains thermal even when the population of the ground state grows much above 1, thanks to an incoming rate overcoming (with the help of Bose stimulation) decay rates. The ground-state population builds up, which is the semi-classical feature, but its coherence does not. The condensate nucleation has to be put "by hand," by introducing a "seed" $|\alpha_0\rangle$ in the form of (1.12) as the initial condition, that is, a coherent state with small but non-zero amplitude α_0, in which case the dynamics become $D(\alpha_0)\rho(t)D^\dagger(\alpha_0)$ where $D(\alpha) = \exp\left(\alpha a^\dagger - \alpha^* a\right)$ is the coherent-state displacement operator. This shows that the system behaves as a *polariton amplifier*, able to magnify coherence, but unable to create it.

Laussy et al. [72] and Rubo [73] considered non-linear corrections to (1.14) able to circumvent this shortcoming. In laser theory, coherence builds up, thanks to the positive feedback of the cavity that injects back the photons from the growing field into the inverted population, which is also, by itself, merely an amplifier. Sarchi and Savona [74] also studied nucleation of the polariton condensate, thanks to exciton–exciton non-linearities. The role of non-linearities is certainly important, but a mechanism of condensate formation can be found in the spirit of Rubo et al. [37]—upgrading the Boltzmann equation to the quantum level—even for the weakly interacting or the non-interacting gas. This mechanism was proposed by Laussy et al. [38] and coupled to the semi-classical Boltzmann equations of a realistic polariton system by Laussy et al. [75], showing their spontaneous condensation and spontaneous coherence growth. We now discuss it in more detail.

The type of quantum Boltzmann equations that can give rise to the sought behaviour of coherence build-up has been derived and investigated in much details by Gardiner et al. [32, 76–81] to describe the dynamics of condensation in atomic traps. The main insight is that the quantum upgrade must not be limited to the ground state, even if only this one will ultimately acquire a quantum character. One must retain at least some of the quantum correlations $\langle a_0^\dagger a_0 a_{\mathbf{k}}^\dagger a_{\mathbf{k}}\rangle$, that, in the semi-classical description, are factored out into products $\langle a_0^\dagger a_0\rangle\langle a_{\mathbf{k}}^\dagger a_{\mathbf{k}}\rangle$. The dynamics of the polariton gas in this context can be described in the framework of a master equation, that is, with rates between quantum states in a configuration space rather than between populations as in (1.7). The equation of motion in this case, that we will call with Gardiner et al., a "Quantum Boltzmann Master Equation" (QBME), is therefore a function of $p(n_0, n_1, n_2, \ldots, n_N) = \langle n_0, n_1, n_2, \ldots, n_N | \rho | n_0, n_1, n_2, \ldots, n_N\rangle$, the joint probability to have n_i particle in the state $|i\rangle$. In a closed system (infinite lifetime and no pumping), it is of the type:

$$\begin{aligned}\dot{p}(n_0, n_1, n_2, \ldots, n_M) = \sum_{i<j}^{M} &(n_i+1)n_j\{\omega_{i\to j}\,p(n_i+1, n_j-1) - \omega_{j\to i}\,p(n_i, n_j)\}\\ &+ n_i(n_j+1)\{\omega_{j\to i}\,p(n_i-1, n_j+1) - \omega_{i\to j}\,p(n_i, n_j)\},\end{aligned} \tag{1.15}$$

where we have used in the sum (and will be using from now on) the abbreviation $p(n_i, n_j)$ for $p(n_0, \ldots, n_i, \ldots, n_j, \ldots, n_M)$. The sum is taken over M states with $M \to \infty$ in a continuous system. Numerical solutions of this equation become rapidly much more demanding than for their semi-classical counterparts but remain tractable for reasonably large systems. Laussy et al. [38] solved analytically the steady state in the case $M = 1$, which contains the gist of the mechanism for the condensate nucleation. Based partly on these results, we will present here the more general solution of the steady state for arbitrary M, in the case where the excited states are degenerate in energy. This will confirm and extend the previous discussions [38, 75].

With little loss of generality into the mechanism at work, we therefore solve the QBME under the assumption that:

$$\omega_{i\to j} = \delta_{j,0}\,\omega_{1\to 0} + \delta_{i,0}\,\omega_{0\to 1}\,, \tag{1.16}$$

with $i \neq j$, that is, we consider relaxation between the ground state (labelled 0) on the one hand and any of the excited states (labelled $i = 1, \ldots, M$) on the other hand with the same rate $\omega_{0\to 1}$ and also the opposite process of relaxation from any one of the excited state to the ground state at a higher (by definition of the ground state) rate $\omega_{1\to 0}$. We introduce the parameter:

$$\xi = \omega_{1\to 0}/\omega_{0\to 1}\,, \tag{1.17}$$

which, as was just said, is larger than one. This situation represents the polariton system sketched in Fig. 1.2c where the ground state lies at the bottom of the lower polariton branch and the excited states are degenerated in a Rayleigh scattering circle. The excited states are not directly linked to each other, but this will not affect the steady state, so we ignore such a term to keep the solution as simple and transparent as possible. What is important is the existence of a large $M \gg 1$ number of states available to feed the condensate. The model will show that coherence grows when a single quantum state—the ground state—acquires characteristic of the system as a whole. Such characteristics are, for instance, a peaked distribution of the total number of particles, which is what determines second-order coherence $g^{(2)} = 1$ when this is realised in a single state. The peaked distribution is otherwise always realised in a macroscopic system as a whole because of the central limit theorem, which states that the sum $\sum_{i=0}^{M} N_i$ of random variables N_i (with a mean) is a normal distribution. In the boson gas system, each N_i—the number of particles in the state i—has thermal fluctuations and a small mean. But the total number of particles has Poissonian fluctuations, as is well known from statistical mechanics in the canonical ensemble. Under conditions suitable for condensation, when relaxation rates are high enough towards the ground state, the accumulation of particles results in a transfer of the macroscopic properties into one single quantum state, the epitome of a quantum condensate. This is in stark contrast with (1.14) where correlations are not retained, and as a result, even when the population is macroscopic, the state has the features of a microscopic degree of freedom, not a macroscopic one. The results below are no less than the historical Einstein result for Bose condensation based on statistical arguments only [82], brought to the quantum coherence picture, thanks to the QBME.

The mathematical details are cumbersome, and we will focus the discussion on the physical picture. We start with the observation that if we assume the factorisation $p(n_0, \ldots, n_M) = p_0(n_0) \ldots p_M(n_M)$ in (1.15), which is the counterpart of breaking out correlators in products of populations, the resulting equation can also be solved, and by the method of detailed balance, one finds:

$$p_i(n) = (1 - \theta_i)\theta_i^n\,, \tag{1.18}$$

with $\theta_i = \langle n_i\rangle/(1 + \langle n_i\rangle)$ and $\langle n_i\rangle = \sum_n n p_i(n)$; the average population for $i = 0, \ldots, M$, showing that, in the absence of correlations, fluctuations are those of a thermal state, (1.13), regardless of the population. We will now show that the probabilities computed from tracing the other degrees of freedom,

$p_i(n) = \sum_{n_j, j\neq i} p(n_0, \ldots, n_j, \ldots n_M)$, can be very different in character when retaining correlations between states. Equation (1.18) will be taken as our starting point, since we have, at the outset, no reason to suppose special correlations between states. That is—under our assumption of a degenerate circle of excited states—we take as the initial condition the thermal equilibrium configuration with $\langle n_0\rangle = \theta_0/(1-\theta_0)$ particles in the ground state and $\langle n\rangle = \theta/(1-\theta)$ in any of the excited state (the same number by reason of symmetry and equilibrium). They are both small numbers, but the total mean in the entire system $N = \langle n_0\rangle + M\langle n\rangle$, is large thanks to a large number of states $M \gg 1$. The initial state is therefore:

$$p(n_0, n_1, \ldots, n_M) = (1-\theta_0)\theta_0^{n_0}(1-\theta)^M\theta^\Sigma\,, \tag{1.19}$$

where $\Sigma = \sum_{i=1}^M n_i$ is the number of non-condensed particles. Equation (1.19) is a solution of the QBME (1.15)–(1.16) provided that $\xi = \theta_0/\theta$. This condition for thermal equilibrium, where states are uncorrelated, is in the non-condensed phase an excellent approximation because of a fast dephasing between states. The mean total N therefore fluctuates according to the probability to have N particles, $P(N) = \sum_{n_j, \Sigma+n_0=N} p(n_0, n_1, \ldots, n_M)$, that one can compute as:

$$P(N) = (1-\theta_0)(1-\theta)^M\theta^N\binom{M+N-1}{M-1}{}_2F_1\left(1, -N, -M-N+1, \frac{\theta_0}{\theta}\right)\,, \tag{1.20}$$

where ${}_2F_1$ is the hypergeometric function [83]. This function is bell-shaped when N is large enough. This result is the exact mathematical statement that a large thermal reservoir has normal fluctuations peaked around a macroscopic average. The solution of (1.15) can be found in a generalisation of the method of Laussy et al. [38] by normalisation of detailed balanced type of equations within manifolds of excitations. We give the result directly:

$$p(n_0, \Sigma) = \frac{P(n_0+\Sigma)}{\left(\frac{\xi}{\xi-1}\right)^M - \frac{1}{\xi^{n_0+1}}\binom{M+\Sigma+n_0}{M-1}{}_2F_1(1, M+\Sigma+n_0+1, \Sigma+n_0+2, \frac{1}{\xi})}\,, \tag{1.21}$$

where the numerator is given by (1.20) and the denominator follows from $\sum_{k=0}^{n_0+\Sigma}\binom{M+k-1}{M-1}\xi^{\Sigma-k}$. As the repartition of particles among the excited states does not affect the result, we introduced the notation $p(n_0, \Sigma)$ for any of the $p(n_0, n_1, \ldots, n_M)$ for which $n_1 + \cdots + n_M$ sums to Σ. The reduced distribution for the ground state $p_0(n_0) = \sum_{n_1,\ldots,n_M}^\infty p(n_0, n_1, \ldots, n_M)$ follows from this notation and (1.21) as $p_0(n_0) = \sum_{\Sigma=0}^\infty\binom{M+\Sigma-1}{M-1}p(n_0, \Sigma)$. Note how combinatorics affect the result by introducing the binomial, which brings the macroscopic into the ground state when M is large enough. It is evident, from this exact result, that the solution is considerably more complex than the thermal expression, (1.19), which it includes as a particular case. The solution is also more complicated than the "co-thermal" ansatz [84], which is a mixture of a thermal and a coherent state (the displaced thermal state discussed above [37] is another example of such a state). Although the co-thermal state provides excellent and compact approximations, the exact result hinges on combinatorics and therefore roots Bose condensation into the peculiar statistical properties of bosons.

We now illustrate with an example how the ideal boson gas condenses based on solution (1.21), which will be the support for all the figures that accompany this discussion. We assume $\langle n_0 \rangle = 0.3$ particle in the ground state (on average) and $\langle n_i \rangle = 0.25$ in any one of the excited state. With these parameters, the total number of particles $N = \langle n_0 \rangle + M \langle n \rangle$ needs $M = 7$ levels in the exciton reservoir so that the system may dispose on average of more than two particles, the minimum required for Bose condensation [38]. With this initial condition, the parameter (1.17) that maintains thermal equilibrium in the form of (1.19) is $\xi \approx 1.15$. This parameter that we introduced as the ratio of transition rates $\omega_{1\to 0}$ from the reservoir to the condensate and the other way around, $\omega_{0\to 1}$ can, from statistical axioms, be related to the temperature as $\xi = \mathrm{e}^{(E_1 - E_0)/kT}$, so increasing values of ξ correspond to diminishing temperatures. In Fig. 1.3, we plot $p(n_0, \Sigma)$ as given by (1.21). As

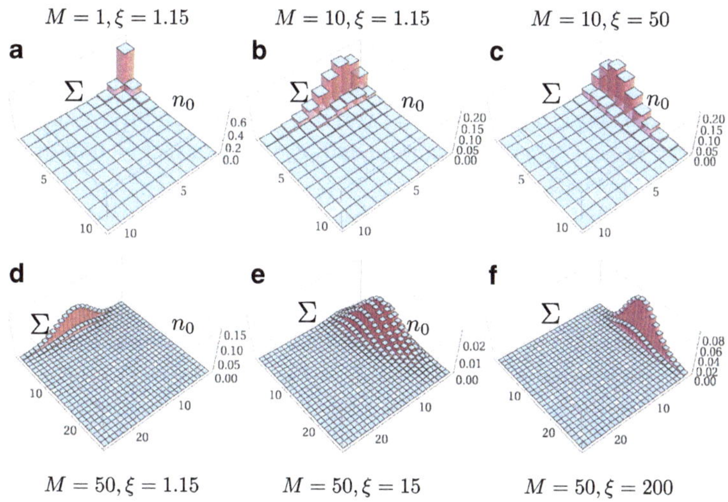

Fig. 1.3 Revisiting the Bose–Einstein condensation of the ideal gas, with spontaneous coherence build-up out of thermal equilibrium when temperature is lowered. The joint probability $p(n_0, \Sigma)$, (1.21), to have n_0 particles in the ground state and Σ distributed in the excited states is plotted as a function of the number of excited states M (in the configuration sketched in Fig. 1.2) and the inverse temperature ξ. The system admits the thermal equilibrium classical solution (1.19) for $\xi \approx 1.15$, with 0.3 particle in the ground state and 0.25 particle in each of the M excited state, all with thermal statistics. As we lower the temperature (increasing ξ), a transfer is observed of the statistics from the macroscopic system—which from the central limit theorem has the form of a normal distribution—to the ground state. This describes Bose condensation. Larger M results in higher coherence of the condensate. Figure (e) shows the condensate nucleation where many, but not most, of the particles are condensed, resulting in a high degree of correlations between the ground state and the rest of the system as a whole, while individual correlations with any given particular state are small

there is no essential information in the way the system distributes its excitations throughout the various states of its reservoir, and only the overall distribution over $\Sigma = \sum_{i=1}^{M} n_i$ needs be recorded, we can conveniently display the solution in a 2D plot with, as it appears in Fig. 1.3, the macroscopic variable Σ on the left-hand side and the microscopic one n_0 on the right-hand side. In the case $M = 1$, with the ground state and another single state for the excitations, the only parameter breaking the symmetry between the states is $\xi > 1$. Even in this simplest of cases, the system can grow some coherence by condensing the bit provided by this one excited state, but, expectedly, the coherence is small. In (a), the system sustains its thermal equilibrium (since $\xi \approx 1.15$) for $M = 1$ and is an uncorrelated product of thermal states. Increasing M to 10 excited states in (b) and to 50 excited states in (d) but retaining thermal equilibrium by keeping ξ constant, one sees that as the system becomes macroscopic, it acquires a peaked distribution of its total number of particles (with an average of $\approx 0.25 \times 10 = 2.5$ and $0.25 \times 50 = 12.5$ particles in these two cases) although each slice in isolation remains thermal as also does, as can be seen on the figure, the ground state. From the macroscopic systems (b) and (d), if we now reduce the temperature, that is, increase ξ, we see a transfer of the statistics from the system as a whole, Σ, to the ground state, n_0. The coherence acquired becomes perfectly Poissonian in the limit $M \to \infty$, as is clear comparing (c) and (f), the latter having much higher coherence, thanks to the larger reservoir. Figure. 1.3e shows the condensate formation when the ground state is in the process of acquiring the coherence in a single state from the macroscopic set of levels as a whole. At this intermediate stage, there are strong correlations between the ground state and the rest of the system. This is a beautiful and essentially exact picture of quantum condensation of the ideal gas.

In Fig. 1.4, we show other quantities, also derived from (1.21), that characterises condensation more succinctly than the full distribution function. In (a), we show the condensate distribution, that is, the reduced probability $p_0(n_0)$ with all the degrees of freedom from the rest of the system traced out. Starting from the thermal case at the point marked (1) by the arrow, the system develops a characteristic peaked fluctuation that characterises coherence in the sense of Glauber. Slices from the density plot are shown in (b), with the thermal (1) and fully grown coherent (5) states filled and also displayed in log scale in (c). The thermal statistics is exponentially decreasing, and thus, although it has a few excitation on average, it fluctuates wildly. In (d), we display still another average, this time the mean population in the ground state, that follows tightly the statistics profile, showing that coherence builds along with the population. We also show $g^{(2)}$ that goes from 2 at equilibrium towards 1 as temperature vanishes. We also plot in dotted line the number $N - n_0$ of uncondensed particles, to compare with the statistics, showing that coherence growth is even steeper than population build-up.

An important feature of polaritons is their dissipative nature. Lifetime and, to compensate, pumping are important factors that are simply included in (1.7) by adding the terms $P_{\mathbf{k}} - \gamma_{\mathbf{k}} n_{\mathbf{k}}$. In a realistic modelisation of the problem, $P_{\mathbf{k}}$ is peaked at high momenta in the exciton reservoir. It is straightforward to upgrade in this

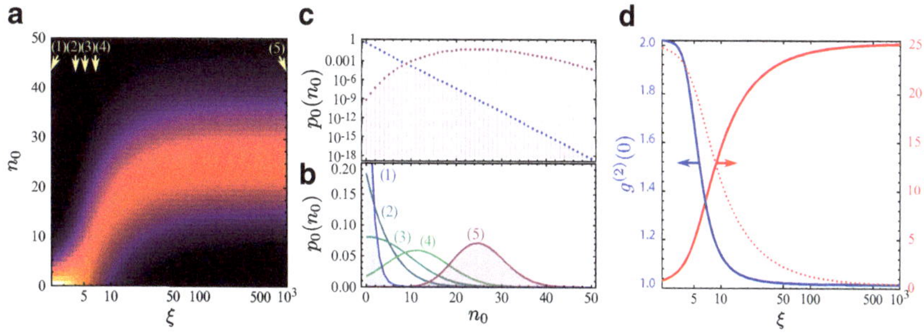

Fig. 1.4 Dynamics of condensation in 2D (in the configuration of Fig. 1.2c) as seen from the particle number statistics. In (**a**), the density plot $p_0(n_0)$ shows the evolution, as temperature is decreased, from a thermal distribution on the left (with $\langle n_0 \rangle = 0.3$ particles in the ground state with the monotonically decreasing distribution shown in (**c**)) to a coherent distribution on the right (with $\langle n_0 \rangle$ close to 25 particles with peaked fluctuations also shown in (**c**)). Slices at the points indicated by the arrows are shown in (**b**). In (**d**), $\langle n_0 \rangle$ (and $N - \langle n_0 \rangle$ dotted) are shown together with $g^{(2)}$ as temperature is decreased. Particles accumulate in the ground state as coherence grows

way the QBME (1.15), by adding to $\dot{p}(n_i)$ terms of the type $\gamma_i[(n_i+1)p(n_i+1) - n_i p(n_i)]$ for decay and $P_i[p(n_i-1) - p(n_i)]$ for pumping of the ith level. A convenient quality of pumping is that it can produce coherence build-up like the macroscopic model where $M \gg 1$, with the two-level system only. We have shown previously how Bose–Einstein condensation can be seen as copying the macroscopic state into a single quantum state, and that a large number of excited states is needed to describe macroscopic properties such as a peaked distribution of the particle number. Pumping allows to "copy" in the ground-state macroscopic properties with only one state to model the rest of the system, since the ground state can keep sucking particles and build its coherence without depleting the excited state that is continuously refilled by pumping. This is seen straightforwardly in the solution of the $M = 1$ QBME with pump (in the excited state) and decay (in the ground state) [38]:

$$p_0(n_0+1) = \frac{\omega_{1\to 0}\langle n_1 \rangle_{n_0}}{\omega_{0\to 1}(\langle n_1 \rangle_{n_0+1} + 1) + \gamma} p_0(n_0) , \tag{1.22}$$

obtained by tracing over the excited state $\sum_{n_1} n_1 p(n_0, n_1) = \langle n_1 \rangle_{n_0} p_0(n_0)$. This has introduced the conditional average $\langle n_1 \rangle_{n_0}$, which is the quantity that retains the correlation needed for the transfer of coherence into the ground state. The approximate closed-form relation $\langle n_1 \rangle_{n_0} = N - n_0$ where N is the total average number of particles in the steady state, which is correct in the mean when $n_0 \ll N$, provides the solution:

$$p_0(n_0) = \left(\frac{\omega_{1\to 0}}{\omega_{0\to 1}}\right)^{n_0} \frac{(1-N)_{n_0}}{{}_2F_1(1, 1-N, 1-N-\frac{\gamma}{\omega_{0\to 1}}, \frac{\omega_{1\to 0}}{\omega_{0\to 1}})(1-N-\frac{\gamma}{\omega_{0\to 1}})_{n_0}} , \tag{1.23}$$

where $(x)_n$ is the Pochhammer Symbol [83]. The pump P has disappeared from the solution, but is implicitly present through the existence of a non-zero N in the steady state for particles with a lifetime. Although (1.21) is exact for arbitrary M, (1.23) is approximate in presence of pump and decay even for $M = 1$, hinting at the difficulty brought by a dissipative system. Equation (1.23) is however a good approximation provided that not most of the particles are in the condensate. This restriction was not encountered in the closed system where condensation in the zero temperature limit was properly described. This limitation is however easily satisfied with large enough lifetime. In more complicated systems, linearization of the conditional averages are used instead and also perform well [75].

The full quantum theory of this restricted two oscillators model is possible. In the following, we note a_1 the state that models the reservoir as a whole. It also obeys Bose statistics, and the physics is thus that of two coupled harmonic oscillators with relaxation described by Lindblad terms $\mathcal{L}$ in the master equation:

$$\partial_t \rho = \Big[\frac{\gamma_1}{2}\mathcal{L}_{a_1} + \frac{\gamma_0}{2}\mathcal{L}_{a_0} + \frac{P_1}{2}\mathcal{L}_{a_1^\dagger} + \frac{\omega_{1\to 0}}{2}\mathcal{L}_{(a_1 a_0^\dagger)} + \frac{\omega_{0\to 1}}{2}\mathcal{L}_{(a_0 a_1^\dagger)}\Big]\rho, \tag{1.24}$$

where $\mathcal{L}$ is given by (1.6). We can solve (numerically) the quantum Boltzmann master equation that follows and characterise each mode ($x = a_0, a_1$) with the average population $n_x = x^\dagger x$ and the second-order correlation function, $g_x^{(2)} = \langle x^\dagger x^\dagger x x\rangle / \langle n_x\rangle^2$ as well as for the cross-correlation function between the modes a_0 and a_1, $g_{01}^{(2)} = \langle n_0 n_1\rangle / (\langle n_0\rangle\langle n_1\rangle)$. These quantities are plotted in Fig. 1.5 as a function of pumping. To link the present full quantum master equation approach with the kinetic model, we derive from (1.24) the steady-state solutions for the average populations that are of the type of a Boltzmann source and sink problem, that is, $\langle n_x\rangle = P_x^{\text{eff}} / \Gamma_x^{\text{eff}}$. This provides us with the effective pumping and intensity decay rates for the three modes:

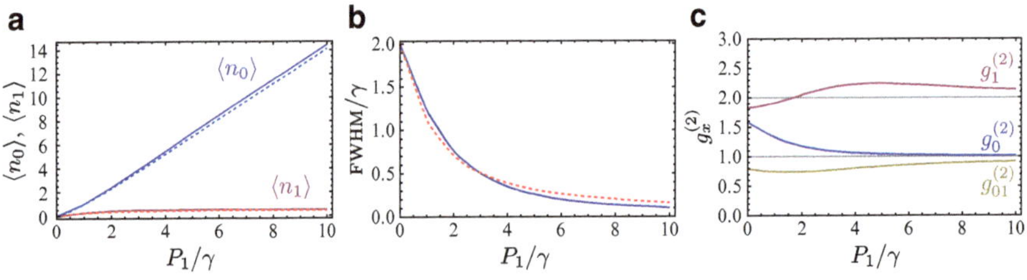

Fig. 1.5 Dynamics of condensation in an open quantum system with two modes, (1.24), with pumping P_1 of the excited state and decay γ of both states. The excited state decays to the ground state with a rate $\omega_{1\to 0} = 5\gamma$ while the opposite process occurs at the smaller rate $\omega_{0\to 1} = \gamma$, allowing condensation in the ground state. The typical phenomenology of condensation is displayed as follows: at large excitation, (**a**) the ground state population grows linearly, (**b**) the spectral linewidth narrows and (**c**) $g^{(2)}$ tends to 1. Approximated semi-classical formulas are plotted with dashed lines

$$P_1^{\text{eff}} = P_1 + \omega_{0\to 1}\langle n_0\rangle\,,\qquad \Gamma_1^{\text{eff}} = \gamma_1 - P_1 + \omega_{1\to 0}(1 + g_{01}^{(2)}\langle n_0\rangle) - \omega_{0\to 1}g_{01}^{(2)}\langle n_0\rangle\,,$$
$$P_0^{\text{eff}} = \omega_{1\to 0}\langle n_1\rangle\,,\qquad \Gamma_0^{\text{eff}} = \gamma_0 + \omega_{0\to 1}(1 + g_{01}^{(2)}\langle n_1\rangle) - \omega_{1\to 0}g_{01}^{(2)}\langle n_1\rangle\,. \tag{1.25}$$

The semi-classical rate equations are recovered by setting the cross-correlation function to unity. Solving exactly such rate equations, one obtains approximate formulas for the populations that, in turn, provide analytical expressions for all the effective parameters above. The expressions are too lengthy to be given here, but we plot them in Fig. 1.5a and b in dashed lines. Their asymptotic behaviour at large pump is constant for $\langle n_1\rangle$ and linear for $\langle n_0\rangle$:

$$\langle n_1\rangle \to \frac{\gamma_0 + \omega_{0\to 1}}{\omega_{1\to 0} - \omega_{0\to 1}}\,, \tag{1.26a}$$

$$\langle n_0\rangle \to \frac{P_1(\omega_{1\to 0} + \gamma_0)^2 - (\omega_{0\to 1} + \gamma_0)[\gamma_1\gamma_0 + \omega_{1\to 0}(\gamma_1 + \gamma_0)]}{(\omega_{1\to 0} - \omega_{0\to 1})\gamma_0(\omega_{1\to 0} + \gamma_0)}\,. \tag{1.26b}$$

In this full quantum derivation, the cross-correlation function is computed numerically and self-consistently and can be, in general, larger (resp. smaller) than unity, showing bunching (resp. anti-bunching) of joint emission in the modes a_0 and a_1. This is shown in Fig. 1.5c (lower line) where cross emission is slightly antibunched, as the destruction of the first one implies the creation of the second one. At very high pumping, the emissions become statistically independent ($g_{01}^{(2)} \to 1$).

A typical manifestation of coherence build-up is line narrowing of the luminescence emission. The FWHM of the condensate mode is given approximately by Γ_0^{eff} in (1.25). Γ_0^{eff} decreases appreciably when $\omega_{1\to 0} \gg \omega_{0\to 1}$. To check that this process is accompanied by the expected line narrowing of the emission in the regime of Fig. 1.5, we compute the photo-luminescence spectrum $S(\omega) \propto \Re\int_0^\infty \langle a_0^\dagger(0)a_0(\tau)\rangle e^{i\omega\tau}d\tau$. The spectrum is single peaked. The associated linewidth is extracted and plotted in Fig. 1.5b. The line narrowing follows remarkably well that predicted by the effective intensity decay rate, (1.25). The value at $P_1 = 0$ corresponds to $\Gamma_0^{\text{eff}}(0) = \gamma_0 + \omega_{0\to 1}$ (2γ in our example). Such a formalism was used by del Valle et al. [85] with a pulsed excitation to describe the experimental observation of time-resolved polariton condensation.

1.4 Formation of Coherence in a Fully Quantized System (0D Polaritons)

In the 0D case, the considerable simplicity achieved by the reduced dimensionality allows an exact treatment of the quantum problem. The simplification even extends not only in the dimensionality of the Hilbert space but in the dynamical equations as well, as one can show [86] that the single-time dynamics, for instance, is entirely contained in the diagonal correlators $\langle a^{\dagger n} a^n \rangle$. They obey the following set of coupled equations:

$$\left[1 + \frac{\Gamma_\sigma + (2n-1)\gamma_a}{\kappa_\sigma} + \frac{n\gamma_a}{\Gamma_\sigma + (n-1)\gamma_a} - \frac{2P_\sigma}{\Gamma_\sigma + n\gamma_a}\right] \langle a^{\dagger n} a^n \rangle$$
$$= \frac{nP_\sigma}{\Gamma_\sigma + (n-1)\gamma_a} \langle a^{\dagger n-1} a^{n-1} \rangle - \frac{2\gamma_a}{\Gamma_\sigma + n\gamma_a} \langle a^{\dagger n+1} a^{n+1} \rangle , \qquad (1.27)$$

for $n \in \mathbf{N}$. We have introduced shortcuts in notations: $\Gamma_\sigma = \gamma_\sigma + P_\sigma$ and $\kappa_\sigma = 4g^2/\gamma_a$, which physical meaning (of pump induced broadening and effective Purcell rate [55, 56, 87]) will not be discussed here. Other non-zero correlators are related explicitly to $\langle a^{\dagger n} a^n \rangle$, such as $\langle a^{\dagger n} a^n \sigma^\dagger \sigma \rangle = \left[P_\sigma \langle a^{\dagger n} a^n \rangle - \gamma_a \langle a^{\dagger n+1} a^{n+1} \rangle\right] / (\Gamma_\sigma + \gamma_a(n-1))$ (that provides the QD probability of excitation n_σ as the particular case $n = 0$) and similarly for $\langle a^{\dagger n} a^{n-1} \sigma \rangle$, which is too lengthy to be written here but is given elsewhere [57]. Although a closed-form solution, if it exists, might not be straightforward (del Valle and Laussy [57] give various limiting closed-form approximations), the problem complexity has been reduced considerably and is readily solved numerically. Solutions for $n_a = \langle a^\dagger a \rangle$ and $g^{(2)}$ from Laussy et al. [88] are shown in Fig. 1.6a and b. There are various regions of interest. At

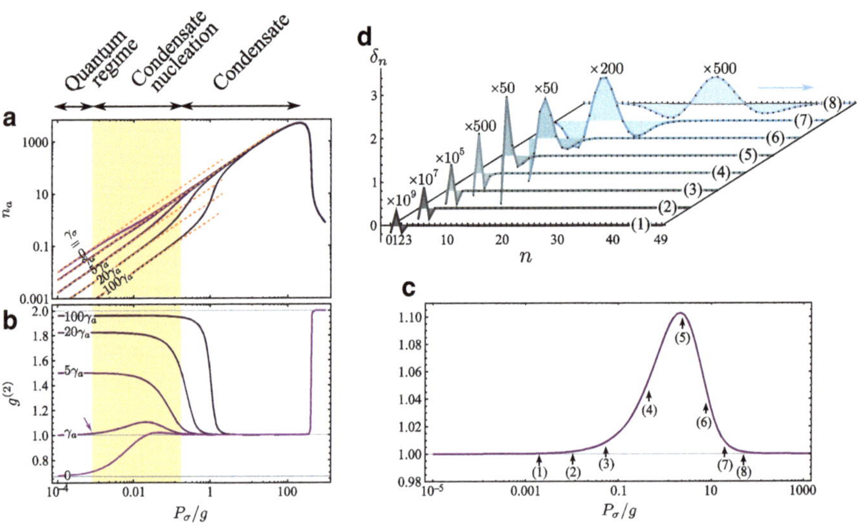

Fig. 1.6 (**a**) Cavity population n_a and (**b**) second-order coherence $g^{(2)}$ as a function of pumping power for several configurations of γ_σ/γ_a, adapted from Laussy et al. [88]. Various regimes are clearly defined, most importantly, a "condensate formation" that separates the quantum from the classical regime. (**c**) Universal curve for the second-order coherence $g^{(2)}$ when going from a coherent field established by strong coupling to one established by stimulated emission lasing and (**d**), deviation of the statistics realised from a Poissonian distribution, (1.33), for the points marked by arrows in (**c**). The maximum value ≈ 1.10282 is the same for any system realising lasing in strong coupling

large pumping,[2] regardless of the system parameters (in this case the ratio γ_σ/γ_a), the solutions $(\gamma_a n_a, g^{(2)})$ converge to a unique one, with full statistical coherence $g^{(2)} = 1$ maintained over a clearly defined plateau. This regime is known in the literature as *one-atom lasing* [89]. It has developed a coherent state in a way that has both similarities and distinctions from the condensate formation discussed in the previous section. They both share the concept of out-of-equilibrium condensation in a single mode and put forward the question of its formation. In the 0D case, however, there is no blueprint of coherence to be found in the macroscopic system. We now describe how the system builds coherence in this fully quantum and microscopic system.

Before the condensate is formed, at vanishing pumping, the system is in what we shall refer to as the "quantum regime," since a few excitations are present and they manifest with the full quantum character of (1.2). In this regime, in stark contrast with the lasing where it is pinned to unity, the photon counting statistics covers a large gamut of values, from anti-bunching (below one) to bunching (above one). It has been shown by Gartner [90] and del Valle and Laussy [57] that the possible values are:

$$g^{(2)}_{P_\sigma \to 0} = 2\frac{\kappa_\sigma(\gamma_a + \gamma_\sigma) + \gamma_\sigma(\gamma_a + \gamma_\sigma)}{\kappa_\sigma(3\gamma_a + \gamma_\sigma) + (\gamma_a + \gamma_\sigma)(3\gamma_a + \gamma_\sigma)}, \tag{1.28}$$

with extrema 0 and 2. This is also very different from the un-condensed phase of the previous section which, being a macroscopic system, was merely a thermal state with $g^{(2)} = 2$ for any given state and, in particular, the ground state. The 0D system is linear in this region; as the population is directly linked to the fraction of time, the system spends in its first excited state rather than in the vacuum, "vac." Therefore, the population increases linearly with pumping, namely:

$$n_a = \frac{\kappa_\sigma}{\kappa_\sigma + \gamma_\sigma}\frac{1}{\gamma_a + \gamma_\sigma}P_\sigma\,. \tag{1.29}$$

The system can be made to operate as a quantum source on the one hand, emitting photons one at a time, a situation realised for an ideal QD which has no spontaneous emission other than in the cavity mode ($\gamma_\sigma = 0$ ideally and up to $\gamma_\sigma < \gamma_a$). On the other hand, by making the QD decay dominant in the modes other than the cavity, $\gamma_\sigma > \gamma_a$, the other situation of bunched emission is realised. In practice, this is less appealing since it is the default situation corresponding to the emission of thermal, or chaotic, sources, or from the un-condensed phase of the 2D system above.

When the condensate is formed in the lasing region at higher pumping, the specifics of the QD are lost in the cavity field. In this case, the increase of population is linked to the cavity photon lifetime only:

[2] But not too large such that the system is quenched, a region which we shall not discuss here, although the breakdown of the condensate is also an interesting transition, which one can see in Fig. 1.6 is furthermore abrupt.

$$n_a = \frac{1}{2\gamma_a} P_\sigma \,. \tag{1.30}$$

By comparing (1.29) and (1.30), one can see that both regimes can be united under the same linear relationship throughout, a situation realised when:

$$\gamma_a = \gamma_\sigma \,. \tag{1.31}$$

This relation is the one that maximises the criterion for strong coupling [87]:

$$|\gamma_a - \gamma_\sigma| \leq 4g \,. \tag{1.32}$$

In this case, of maximum strong coupling for the given decay rates (possibly large), not only is the efficiency rate of growth for the cavity field the same in the quantum and in the lasing regime, it is also the condition that, from (1.28), makes $g^{(2)}$ equal to unity in these two regions. In this case, even if the number of photon is much below unity, the photon counting statistics (accumulated over long enough time to acquire a signal) shows that photons are uncorrelated in their arrival time. Not only $g^{(2)}$ but also all higher order coherence $g^{(n)}$, cf. (1.11), tend exactly to unity in this regime [88], proving that, despite its vanishing population, the field is perfectly coherent in the sense of Glauber [69].

Remarkably, although one can bring the system to perfect Poissonian statistics both in the quantum and in the classical regime, this situation cannot be sustained when crossing from one case to the other. In the former case, coherence is established by a balance of the quantum coupling that maximises strong coupling, (1.31). In the latter, regardless of this criterion, coherence is established according to the single-emitter lasing, with stimulated emission into the cavity mode. In between them, the system passes through a step where the condensate is formed. This intermediate regime manifests itself in both n_a and $g^{(2)}$ as a deviation from these ideal trends, namely, $n_a(P_\sigma)$ exhibits a concavity below the line $1/(2\gamma_a)$ and $g^{(2)}$ exhibits a corresponding jump in the statistics. These are seen in Fig. 1.6a and b, respectively. The case $\gamma_\sigma = 0$ is the most efficient in the quantum regime to grow the cavity field, and there is a drop from its efficiency to that of lasing as the system shifts from strong coupling to stimulated emission. On the opposite, cases where $\gamma_\sigma > \gamma_a$ have worst efficiencies that get bettered by the onset of stimulated emission. This is more apparent in cases where $\gamma_\sigma \gg \gamma_a$, provided that the strong-coupling criterion (1.32) is still maintained. In the optimum case (1.31), there is a loss of efficiency in the rate of growth of the field as the paradigm is changed. This transition region is made obvious in the counting statistics, with a characteristic curve that is magnified in Fig. 1.6c. In Fig. 1.6d, we show explicitly how this difference in the photon statistics occurs in the distribution $p(n) = \langle n|\rho|n\rangle$, by comparing it with the ideal Poissonian statistics with the same mean value n_a:

$$\delta_n = p(n) - \mathrm{e}^{-n_a} n_a^n / n! \,. \tag{1.33}$$

In the vanishing pumping region, the field is exactly coherent. As pumping is increased, a small deviation of statistics develops, that affects the bottom of the Jaynes–Cummings ladder, as is expected since this is where the system is populated. One finds that in the transition, there is lower probability to have two photons than in an ideal laser of the corresponding intensity n_a. This imbalance grows linearly, and in the transition region (4–7), it spreads over many rungs, with excess of photons nearby the maximum of the distribution while neighbouring rungs are depleted to compensate. In the stimulated emission lasing region (8), this perturbation in statistics becomes distributed along the ladder at the same time as it becomes weaker in amplitude, recovering exact Poissonian fluctuations at high intensities (beyond point (7)). A remarkable property of this transition from the quantum to the classical is that it is universal, in the sense that the shape is invariant for the dimensionless parameters P_σ/γ_a and γ_σ/γ_a, for any value of g. Interestingly, such a local maximum of statistics when crossing the thresholds to stimulated emission has been observed in experimental realisations of a few-emitters laser with a shape that resembles the one in Fig. 1.6 [91–94]. However, in the majority of cases, it was linked to an experimental limitation of finite time resolution, whereas it is in our case a manifestation of an intrinsic and universal transition in the system, namely, the formation of the condensate in the 0D system, a picture that we will now substantiate by considering the behaviour of the polaritons.

To achieve lasing with a single QD, one needs very good strong coupling, that is, $\gamma_a \ll g$ on top of satisfying (1.32). In this case, the polaritons in the various rungs of the Jaynes–Cummings ladder, (1.2), are well defined. In particular, they are well separated in energy the ones from the others. The system can then be seen as a gas of un-condensed polaritons. Its power spectra appear as a sequence of sharp peaks located at the transition energies between the rungs of the Jaynes–Cummings ladder, which show that excited states are well defined in isolation. This is shown in Fig. 1.7a–c, with only the Rabi doublet at very low excitation (a), that gets flanked by the peaks from higher excited states as excitation is increased (b and c). We have chosen the QD power spectrum, since it maps better the energy distribution than its cavity counterpart [95]. As more particles are fed into the system, coherence starts to grow in the photon field, with a striking transformation from the quantised picture of sharp peaks [96] to a smooth continuum, for which a compact analytical expression can be found [56]. This transition is well seen in the density plot of Fig. 1.7 where one can follow visually the breakdown of quantisation and the emergence of a smooth triplet. The triplet itself is isolated in Fig. 1.7e. It is a counterpart of the resonance fluorescence spectrum under high excitation, known as the Mollow triplet [97]. Whereas the latter arises when exciting a two-level system with a laser beam, the one developed here [56, 98] arises with no pre-established coherence or imposed from outside, but is self-grown by the condensed cavity photon field. This has interesting consequences for the dynamics of the Rayleigh scattering peak that exists as in the conventional scenario, in the form of the very sharp peak centered at the cavity, as seen in Fig. 1.7f. In this case, the condensate is fully formed and behaves essentially like an external laser field. Its photons scatter off the QD and account for this sharp resonant emission. Its intensity increases with

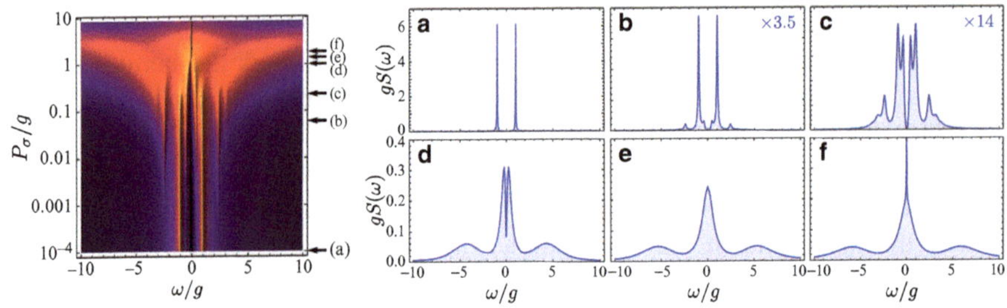

Fig. 1.7 Dynamics of condensation in 0D, as seen through the QD emission spectrum as a function of pumping power. At vanishing pumping, the system is in the linear regime and exhibits the Rabi doublet of strong coupling, (1.1). As pumping is increased, higher excited states probe the quantised structure of the coupled fields and a sequence of peaks appear (**b**) that reproduces the structure of the Jaynes–Cummings ladder. In this non-linear quantum regime, polaritons are well-defined objects in isolations. A new structure emerges in the condensate formation, with more polaritons added to the system and more peaks contributing to the spectrum (**c**) that melts into another structure of much-reduced complexity, a triplet (**d**–**f**) reminiscent of the Mollow triplet [97]. In this phase, on top of the cavity QED counterpart of the Mollow triplet [56], seen most clearly in (**e**), sits a sharp peak that is a Rayleigh scattering peak. Interestingly, it can be positive, as is usual (**f**), describing coherent scattering of the photons off the QD, but also negative, as seen in (**d**), in a stage where the condensate is still forming and excitations are coherently scattered from the QD into the condensate

pumping power. In Fig. 1.7e, we choose the special case where this Rayleigh peak just starts to appear, when the condensate has achieved its formation and is now increasing its magnitude. Below that point, as in the case of Fig. 1.7d, the scattering peak is negative and results in a sharp absorption line in the QD power spectrum. At this stage of the condensate formation, the cavity field is building and does so by sucking up coherently the QD energy, which is the source of excitation: the pumping excites the QD which scatters at the cavity photon energy its excitation into the condensate.

The cavity field is the one that undergoes condensation (the QD being bathed in the condensate) and emits a single and very narrow line that is essentially the Rayleigh peak in the QD spectrum once it is formed. There is an even richer dynamics in the polariton basis. This one is not directly accessible experimentally, but one can compute it theoretically as easily as for the cavity or QD spectra. In Fig. 1.8, we plot such a transition, by showing the position of the resonances in the system, that is, the ω_p where the system can make a transition. More accurately, the power spectrum $S(\omega)$ can be decomposed as a sum of peaks at the said frequencies ω_p (with a broadening γ_p that we shall however not discuss here) [61]:

$$S(\omega) = \frac{1}{\pi} \sum_{p \in \mathbf{N}} \frac{L_p \frac{\gamma_p}{2} - K_p(\omega - \omega_p)}{(\frac{\gamma_p}{2})^2 + (\omega - \omega_p)^2} . \tag{1.34}$$

The emission is Lorentzian weighted by coefficients L_p which determine the strength of the corresponding transitions. In addition to the emission at these

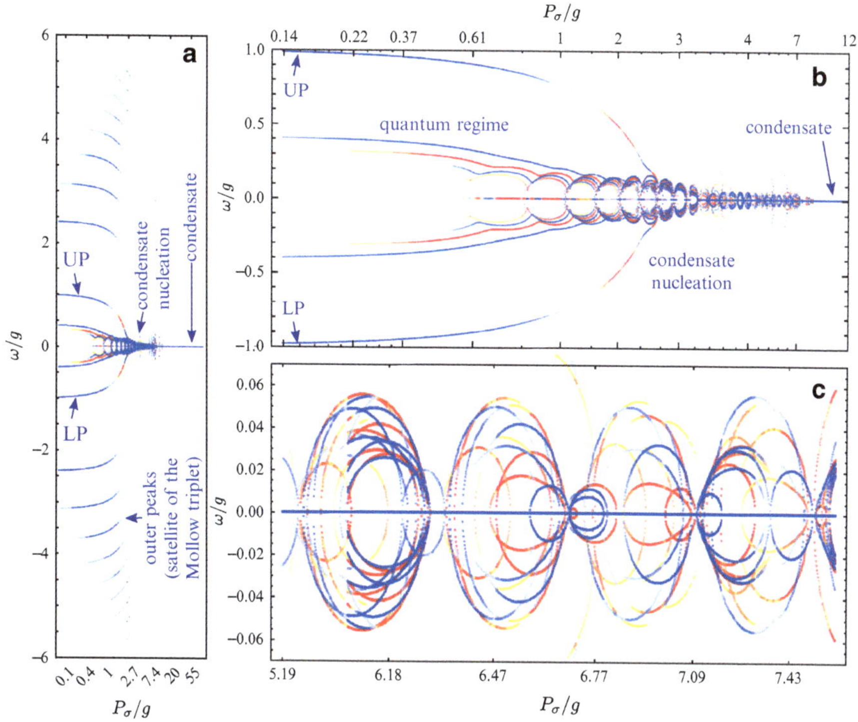

Fig. 1.8 Lasing as a condensation of dressed states (adapted from del Valle and Laussy [57]). The resonances ω_p of the system weighted by the strength L_p of the associated transition (such that a resonance disappears with the vanishing intensity of its emission) are plotted as a function of pumping. In (**a**), an overall view is given with, in particular, the resonances that give rise to the Mollow triplet satellites. In (**b**), a restriction to within the upper (*UP*) and lower (*LP*) polaritons is shown and in (**c**) the plot is further zoomed. With increasing pumping, polaritons collapse all together as the condensate is formed. Unexpectedly, there is an extremely complex structure in the condensate formation

resonances—in the form of spontaneous emission with a characteristic Lorentzian lineshape—there are interference terms between them, weighted by coefficients K_p. These terms are negligible when transitions are well separated the ones from the others, in which case essentially exact results can be obtained with a kinetic theory of dressed states only, computing the density of occupation with rate equations [99]. But when transitions overlap significantly, they affect each other and the second terms in the sum of (1.34) become important. These terms are of the type of the dispersive part in a driven oscillator. The overlap between transitions can be due to the system not being in very strong coupling, in which case it is natural that the dressed state picture by itself is inaccurate, but more to the point, it can also be due to pumping bringing dressed states closer to each others, a situation realised as the system grows coherence and polaritons condense to acquire a common energy. Figure 1.8 displays this process by showing ω_p weighted by K_p. The colour code is such that zero goes to white, so that lines disappear when the transitions vanish. The underlying dynamics to the observable transition, that in Fig. 1.7, is considerably more complicated than a mere collapse of the lines. The system is well defined with a quantised skeleton in the quantum regime, where it exhibits a few lines well separated from the others. In Fig. 1.8a, one can see neatly in particular the outer peaks that come from transitions of the type $|n\pm\rangle \rightarrow |n-1,\mp\rangle$ and which, thanks to their broadening γ_p, will eventually form the side peaks of the Mollow triplet. We have also marked UP and LP the upper and lower polaritons, that arise from the decay of $|1+\rangle$ and $|1-\rangle$. The part beyond the lower and upper polariton behaves as expected. In Fig. 1.8b, we zoom on the region in-between them. As pumping is increased and the system grows coherence, the set of ω_p exhibits an unexpected pattern in the region associated to the formation of the condensate, from the criteria discussed previously. There is a rich and complex "bubbling" of the dressed states, one part of which is magnified in Fig. 1.8c, that gets more and more squeezed towards the condensate energy until, eventually, coherence is fully established and all resonances have collapsed onto a single line. As pumping is further increased, the condensate is destroyed by self-quenching, but we do not further discuss this regime, only do we need to mention that the single line remaining in this case corresponds to weak coupling. The complexity of such a phenomenon has been addressed by del Valle [100] in a simplified model of two coupled qubits, with, therefore, a finite (and small) Hilbert-space, whereas the Jaynes–Cummings system is unbounded. This behaviour is due to channels of coherence flow in the system that are opened by the interplay of excitation and decay. We will not discuss it in more details, but wish to emphasize the extremely rich and unexpected quantum dynamics of condensation in 0D, where one can keep track exactly of all quantum correlations. The quantization of the field breaks down, and a smooth, continuous classical field develops as a result, that can be described with a few degrees of freedom, corresponding to its macroscopic character, as opposed to the quantum regime where a large number of correlators are required to keep track of the possible quantised transitions, manifesting themselves as sharp peaks isolated from their neighbours. In view of how this transition from the quantum to the classical realm is achieved in a full microscopic picture, much understanding remains to be achieved of this fascinating physics.

1.5 Propagation of Polariton Wavepackets (1D Polaritons)

In the two previous sections, we have addressed the dynamics of formation of the condensate. In this section, we will consider dynamics from a different perspective, in a way that has been made possible, thanks to the design of a new type of experiments pioneered in Madrid by Amo et al. [23] and Sanvitto et al. [58], which take advantage of the versatility of polaritons to set a condensate in motion and probe continuously its coherent propagation in space and time through the steady emission of photons. The experiments are fairly recent and more remains to be understood than has been explained so far. The propagation of a condensate brings to mind the problem of superfluidity, and the Madrid experiments indeed display the phenomenology expected from a superfluid such as propagation unaltered by potential fluctuations and defects. The real-space propagation of a localised chunk of condensate is however not the simplest framework to demonstrate and characterise superfluidity. This class of experiment came last with atomic condensates, for instance, which started by evidencing the more conventional and clear-cut phenomenology of helium flow below the λ point, such as two-fluid hydrodynamics or quantisation of the angular momentum absorbed by a condensate as an ordered lattice of an integer number of vortices. The Landau criterion [101]—according to which superfluidity is the result of stability of the condensate to external perturbation below a critical velocity because of the linearisation of the spectrum of excitations—made a lasting impact in the question of defining superfluidity, a sort of counterpart to the Meissner effect in defining superconductivity. The subsequent microscopic derivation of the Landau spectrum by Bogoliubov for the weakly interacting gas [102] (and later by Feynman for helium II [103]) brought compelling support for this criterion. In polariton physics, the Landau criterion has been advocated and studied by Carusotto and Ciuti [21], who proposed an experiment (later carried out by Amo et al. [25]) demonstrating a collapse of the Rayleigh scattering circle. The Landau criterion is however not as clear cut as the Meissner effect for superconductivity which is robust to all possible complications such as various dimensionalities, finite size systems, non-linear regime, etc., and it has been recently criticised by Cancellieri et al. [104] for polaritons and by others for other systems [105, 106]. Back to the experiment which motivates our interest in the space-time dynamics of a condensate [23], another of its remarkable features, beside flowing unaffected past the obstacles naturally present in the structure, is its shape-preserving propagation, with a diffusion much smaller than what would be expected for a wave packet of massive particles with the effective polariton mass. This shape-preserving propagation, on the other hand, evokes a soliton [107], a 1D class of which is known to exist in a constant density Bose–Einstein condensate, either as a depletion in the uniform background (dark soliton) [108] or on the opposite a standing out aggregation (bright soliton) [109], depending on the sign of the interaction (attractive or repulsive, respectively). There is a vast literature documenting solitons in non-linear optics, and theoretical predictions from this community have been made [110] simultaneously or shortly after the Madrid experiment. As an interacting condensate, the polariton fluid should be able to sustain soliton solutions and claims have been recently made in this direction [111]. Keeling and Berloff discuss some of these issues in a wider context with comparison to other systems [112].

We will touch here only the surface of the problem and focus on the new viewpoint into the dynamics of quantum wave packets that polariton condensates offer. The generic problem can be cast as a generalised Schrödinger equation:

$$i\hbar\partial_t |\psi\rangle = (D + V + \mathcal{I} + \Gamma)|\psi\rangle + F\,, \tag{1.35}$$

where D is the dispersion of the free particles, V is an external potential, $\mathcal{I}$ is the self-interaction term (it is italicised because it is a functional that depends on ψ), Γ is the lifetime and F, the only non-homogeneous term, is an external drive. As compared to the two previous sections, here, the problem is fully coherent, even the pumping scheme, and therefore, the system is in a pure state. A typical explicit pumping in 1D reads $F(x,t) = \exp\left(-(x - x_{\mathrm{L}})^2/(2\sigma_{\mathrm{L}})^2\right) e^{i\omega_{\mathrm{L}} t}$, localised in a (Gaussian) spot of extension σ_{L} and injecting particles at the energy ω_{L} of the laser. Experimentally, this is realised by resonant excitation with a coherent laser, which imprints its coherence on the polariton system. On the other hand, in the question of polariton condensation that was previously discussed, one has to be careful that the coherence, allegedly spontaneously built up as a result of Bose condensation, is not in fact the result of a transfer from that of the external laser. In particular, a well-defined non-zero $\langle a \rangle$ is automatic with a state imprinted by an external laser, which has a well-defined phase put "by hand" in the equation. Here, although we will still speak of a "condensate" in the sense outlined in the definition, we fully accept that coherence is inherited from outside. Technically, the problem thus reduces to the dynamics of an initial state,

$$\psi(x, t = 0) = \frac{1}{\sqrt{2\pi}\sigma_x} \exp\left(-\frac{(x - x_0)^2}{2\sigma_x^2}\right) \exp(ik_0 x)\,, \tag{1.36}$$

under the equation of motion (1.35).

The wave-packet propagation has a long history in the quantum theory, starting with Schrödinger himself [113] who was thinking of the packet as a microscopic mechanical object. He had provided an equation of motion for it and was looking solutions that satisfy the correspondence principle. In this way, he derived the coherent state, (1.12), that allows, thanks to confinement, to keep the particle located in space and time. The propagation of the free Gaussian wave packet—the solution of which was probably known to Schrödinger but was first published by Darwin [114]—diffuses and thus eventually results in complete indeterminacy of the position of the particle which is spread over the entire space, a situation not very satisfying when trying to offer an interpretation of the theory. Note that the free Gaussian packet does not compulsorily only expand. It can also contract, for arbitrary amount of time, until it reaches a focal point and then expand again. In any case, however, it is ultimately completely delocalised, and all packets have the same fate under the dynamics of the free Schrödinger equation [115]. It nevertheless quickly appeared that quantum reality is conceptually far more intricate and time-independent properties like eigenvalues are more relevant to describe experiments than the time dynamics of a wave packet. The interest in this dynamical aspect of Schrödinger equation back-dropped, with occasional revivals for its own sake such as in quantum chaos [116], but they are rarely the main focus of some phenomenon.

The situation is drastically opposite with shape-preserving wave packets, at the top of which lies the soliton, first reported by Russell [117] and explained by Korteweg and de Vries [118] as a balance between dispersion and non-linearity. The non-diffusion of solitons (and other properties such as their stability under collisions) makes them identifiable as particles, maybe in the way Schrödinger was expecting. Solitons are not the only class of shape-preserving packets. Instead of being self-sustained, the non-linear balance can be enforced in a much more straightforward way by an external potential, which was the route taken by Schrödinger. Yan showed that the displacement $D(\alpha)$ of any excited state of the harmonic oscillator is shape preserving in a harmonic potential, the coherent state being the particular case of displacing the vacuum [119]. Their energy spectra are fascinating and offer a new meaning to the wave function [103]. The main drawback of such a cohesion of the packet that is enforced externally is that, being static in nature, it works for a stiff class of solutions only. For instance, if the Gaussian (coherent state or displaced vacuum) in the harmonic potential has not the correct width, it pulsates (expanding and contracting) as it propagates harmonically in the potential, giving rise to the so-called *squeezed states* [120]. The constrained trajectory within the potential boundaries, which corrects for diffusion by pushing back the packet, is another limitation of this sort of shape-preserving propagation. Interestingly, Berry and Balazs found an astonishing shape-preserving solution of the free Schrödinger equation, where this time acceleration of the packet accounts for counter acting its diffusion [121]. That a free packet should accelerate is even more surprising, but this was immediately understood by Berry and Balazs as the result of the solution not being normalised and therefore representing not one particle in isolation but a whole family, the Galilean propagation of which causes an emerging curvature (the caustic) in the space–time trajectory.

All the previous discussion applies for the case of a massive particle, with dispersion given by:

$$D = -\nabla^2/(2m). \tag{1.37}$$

for smoothing out the wave function, in much the same way as it does with temperature in the heat equation. If we consider a linear dispersion instead, $E_k = \langle k| D |k'\rangle = sk\delta(k - k')$, with s a real number (the "speed of sound"), one can readily solve the Schrödinger (1.35) that follows, $\mathrm{i}\partial_t\psi(x) = (-\mathrm{i}s\partial_x + V(x) - \mathrm{i}\gamma/2)\psi(x)$, for the Gaussian wavepacket as the initial condition:

$$\psi(x,t) = \frac{1}{\sigma_x\sqrt{2\pi}}\exp\left(-\frac{(x-x_0-st)^2}{2\sigma_x^2}\right)\exp(\mathrm{i}k_0(x-st))\exp(-\gamma t/2)$$
$$\exp\left(-\frac{\mathrm{i}V_0}{2s}\left\{\mathrm{Erf}\left(\frac{x-x_V}{\sqrt{2}\sigma_V}\right)-\mathrm{Erf}\left(\frac{x-x_V-st}{\sqrt{2}\sigma_V}\right)\right\}\right), \tag{1.38}$$

where we included a Gaussian obstacle on the way (at x_V with strength V_0 and width σ_V), but neglected non-linearities (lifetime simply comes as a product $\exp(-\gamma t/2)$ as already discussed). Ignoring for a while the potential on the second line, $V_0 = 0$, one can see that the packet is just translated. With the linear dispersion, the Schrödinger equation becomes a type of "transport equation," as is clear from the solution: it indeed literally transports the wave packet, real and imaginary parts moving rigidly with the wave packet profile $|\psi(x,t)|^2$. Maybe less expected is that

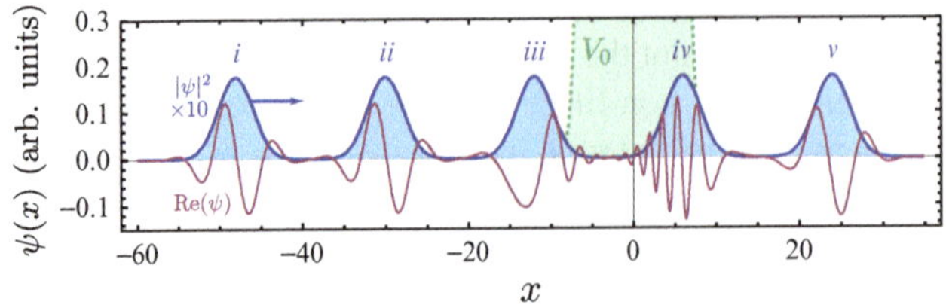

Fig. 1.9 Propagation of a wave packet whose dispersion is exactly linear, at five regular interval of times. In free space, the packet is "transported," both the absolute square $|\psi|^2$ (*thick solid*) and its real (*thin*) and imaginary parts, so *ii* is a translated copy of *i*. As the packet encounters a potential V_0 (here located at the origin with its foot shown as the shaded area with the *dotted boundary*), it passes through unaffected, with no distortion of its shape, regardless of the strength and extension of the potential. The real and imaginary parts are what is reacting to the potential. This is observable in the spectral emission (*not shown*) with a spreading of the emission towards lower k. As there is no final state to scatter, though, the packet continues its course; as the dispersion is linear, it does so unaffected

the presence of the Gaussian potential does not affect this picture as far as the packet profile $|\psi|^2$ is concerned: the term on the second line being a complex exponential, it cancels exactly when taking the absolute square of the wave function. This results in its shape-preserving motion, totally unaffected by the potential regardless of the strength of the latter. This is shown in Fig. 1.9. Here, the spectrum of emission is useful to understand this unexpected behaviour. The second line shows a modulation of the real and imaginary parts when crossing the potential. The Fourier transform of this results in smearing out the energy to lower k, the extent of which can be arbitrarily large, depending on the potential width and strength, but as there is no final state, the propagation in real space is unaffected. If the dispersion would be, for instance, conic, $E_k = s|k|$, a reflected and also shape-preserving packet would appear in presence of the potential, or it would be totally reflected for a large enough potential [122].

A particular case of the linear dispersion is the flat dispersion $E = E_0$, which needs to be solved separately:

$$\psi(x,t) = \frac{\mathrm{e}^{-\frac{(x-x_0)^2}{2\sigma_x^2} + \mathrm{i}k_0 x - \frac{\mathrm{i}t}{2}\left(2E_0 + \exp\left(-\frac{(x-x_V)^2}{2\sigma_V^2}\sqrt{2\pi}V_0\right)/\sigma_V\right)}}{\sqrt{2\pi}\sigma_x}. \tag{1.39}$$

This case completes naturally the linear case since there is still preservation of the shape and an underlying coherent dynamic of the real and imaginary parts, only there is no propagation—the packet is frozen—since the speed of propagation, determined by the slope (being the same for all k), has vanished. Here, also the amplitude has, nevertheless, a dynamic and the presence of a potential results in broadening the energy, without affecting the packet profile $|\psi|^2$. In (1.35), the presence of the driving term F causes the apparition of such a horizontal segment in the energy spectrum $S(k,\omega)$ space, located at the laser. The dispersion of the driven system will be populated too, but the laser is effectively forcing in the system a state very much like (1.39), effectively pinning a condensate. Note also that at large negative detunings (when the photon dispersion "penetrates" deeply into the exciton dispersion), the upper polariton at small k is locally flat, inheriting this property from the heavy exciton. One can thus conveniently create polariton states of the type of (1.39) with a laser shining above the upper polariton at normal incidence.

We now turn to the more general polariton case. The state $|\psi\rangle$ can be either a single-valued field (no spin), projecting to $\psi(x,t)$ in space time (where x is, in our case, a real variable since we address the 1D problem) or a spinor [123] or the exciton–photon coupled system, (1.3), or a combination. Without pumping and decay, and for the scalar field with parabolic dispersion (1.37), (1.35) is a form of Gross–Pitaevskii equation. Its driven version has been studied by Raju et al. [124] and Vyas et al. [125]. In the polariton context, where D is given by (1.4), the system received in-depth attention by Carusotto and Ciuti [21], Egorov et al. [110], Ciuti and Carusotto [126], Szymanska et al. [127] and others. The non-linear term $\mathcal{I}$ is the source of much physics under resonant excitation, in particular with parametric oscillations and scattering [128–130], which is one of the key aspect of the Madrid experiment [23], where a so-called TOPO—for triggered optical parametric oscillator—was used to create the condensate with a pulse (the trigger) and set it in motion by endowing it with a momentum k_0, (1.36), by exciting at an angle. We will not address here this more complicated term but assume directly the end product which is a generation of coherent wave packet with a momentum. Another crucial and fundamental role of non-linearity is how the solution affects and self-adjusts the dispersion, resulting in striking phenomenology. Linearisation of the dispersion was also clearly observed in the Madrid experiment [23] although in this case one was too far from this region of the polariton dispersion for such an effect to be observed without non-linearities. Instead of analysing in details how non-linear terms of the TOPO result in such a phenomenology (Szymanska et al. [127] and Marchetti and Szymanska in this volume focus on this configuration), we will observe it here in the linear regime, by propagating polaritons at various points of the dispersion. The linear case is already interesting, thanks to the peculiar shape of the dispersion, shown in Figs. 1.1 and 1.10. In Fig. 1.10b, the lower branch is reproduced for a small range of k_x, and (c) shows its first and second derivatives. At the point $k_/$ where $\partial_k^2 E = 0$ and where, therefore, $\partial_k E$ is locally constant, the polariton dispersion is linear for a non-vanishing neighbourhood of $k_/$, as seen in (b) where the tangent overlaps neatly with the dispersion. At other points, the contact is more punctual. Adjusting parameters, one can engineer a variety of cases providing a linear dispersion around a given mean momentum and with more or less extension.

We solve the dynamics numerically. There are many ways to integrate (1.35). As we are considering a propagation problem where the solution travels for long times over large distances, the Crank–Nicolson method is not accurate enough [131]. We adapt for the polariton problem an algorithm from Shao and Wang [132] with several orders of magnitudes gain in accuracy over conventional finite difference methods, allowing us to obtain the numerical results in a few seconds on a laptop on grids of $2^{10} \times 2^8$ in space×time. Figure 1.10a shows the energy spectra $S(\omega, k)$

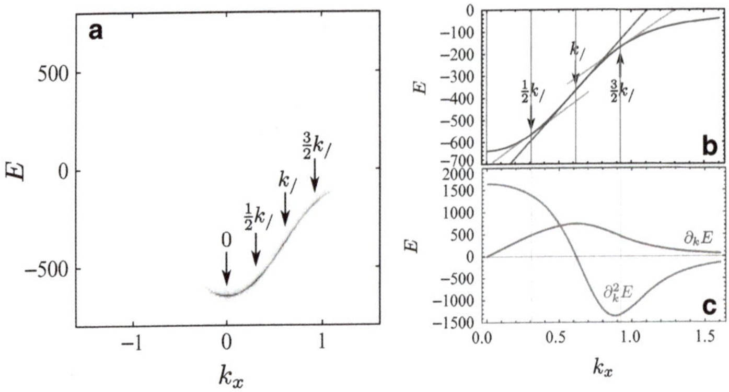

Fig. 1.10 (**a**) The polariton dispersion reconstructed in the energy (E) momentum (k) space by propagating coherent wave packets. We have selected four illustrative cases: $k = 0$ where the condensate only diffuses but does not propagate, $k_/$ where the polariton dispersion is linear over a large neighbourhood of k and $\pm 50\%$ this value, where the dispersion changes concavity, (**b**) reproduces the lower branch with its tangents at $k_/$ and $\pm 50\% k_/$, showing how the linear dispersion can be excellently approximated, thanks to the engineering of the dispersion. (**c**) This engineering is done by analysing the k derivatives of E. When the second derivative vanishes, the first one is locally constant, resulting in the linear increase of energy with momentum

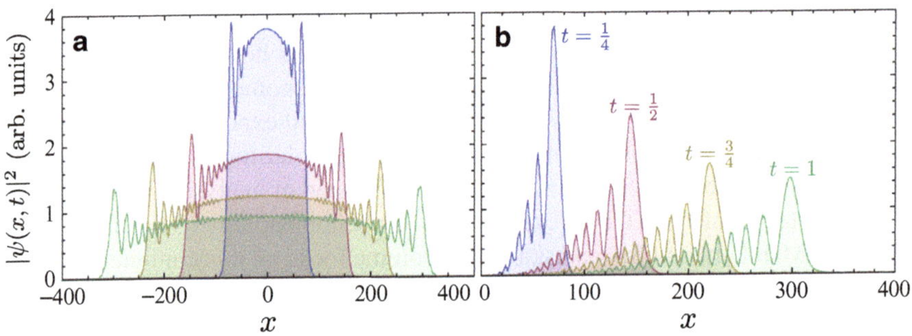

Fig. 1.11 Free propagation in real space of two polariton packets with a broad distribution of k: in (**a**), the average k is zero, and the condensate expands as shown as a function of time, with a uniform central part that pushes away its boundaries which develop oscillations. In (**b**), the average k is non-zero and the propagation proceeds with the characteristic shape of a wavefront that leaves an oscillating tail behind

for the points $k_x = 0$ (no average momentum), $k_x = k_/$ (where the dispersion is locally linear) and $\pm 50\%$ around this point, where the dispersion has a curvature. Superimposing them, one can reconstruct the polariton dispersion. The initial state has been set to be the lower polariton, so that the upper one is not, or very little, populated. One can reconstruct the whole polariton branch with a single packet which has a large spread of momentum. In Fig. 1.11, two narrow packets in real space—and therefore broad packets in reciprocal space—are propagated, one (a) with $k_0 = 0$, the other with $k_0 = k_{/}$. As they have a large spread in k space, the packets echo in real space the peculiar polariton dispersion, namely, they display an undulatory wavefront, not unlike the Airy beams of Berry and Balazs [121]. The diffusion of the $k = 0$ condensate is particularly interesting: the central part provides a uniform background, pushing away its boundaries that propagate with the characteristic shape of the polaritons of Fig. 1.11b.

In Fig. 1.12, we propagate packets with a narrower distribution of k, resulting in approximately Gaussian shapes (or smaller undulatory tails). In the upper row, we show the cavity ψ_C and the exciton ψ_X components of the wave function (1.3). They are plotted side by side for clarity. As the system is in strong coupling, they both behave qualitatively similarly, with different intensities as the exciton or photon character is more marked depending on where the polariton lies on the dispersion (at the negative detuning that we have chosen, the lower polariton is photon-like at $k = 0$). This situation may change in other circumstances. When the coupling is weaker, for instance, two beams propagate instead, dragging each other. One can observe how the packet diffuses and propagates as ruled by the dispersion with, indeed, a shape preserving propagation at $k_{/}$ where the dispersion is locally linear. In Fig. 1.13, we repeat the numerical experiment but interposing a potential in the middle of the trajectory. We have scaled the potential in the three cases so that it is corrected for the different energies at the three points in the branch. At low momentum, the potential destroys the bullet, one part is reflected, the other part is smeared over as it continues along its initial trajectory. At $k_{/}$, the packet interacts

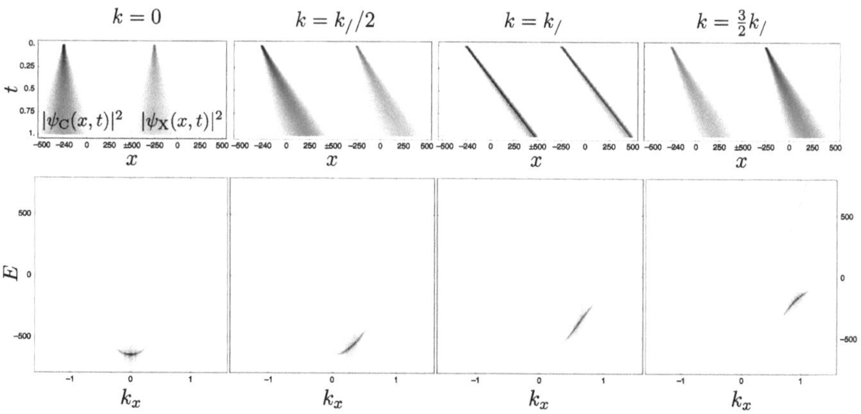

Fig. 1.12 Free propagation, in real and reciprocal space, for the points shown in Fig. 1.10. The cavity and the exciton fraction are both shown side by side. The packets expand quickly but for the case $k_{/}$ where the dispersion becomes linear, allowing to define a polariton packet and track it over long times. Above and below this point, the curvature has changed sign (from *concavity to convexity*), resulting in similar patterns of diffusion and propagation. Below, the trace in the energy–momentum space, reconstructing locally the polariton branch. The relatively broad set of k components which has been chosen result in a small expansion of the polariton in the case $k = k_{/}$. In the spectral image, one can see a small deviation from linearity at the tips of the segment

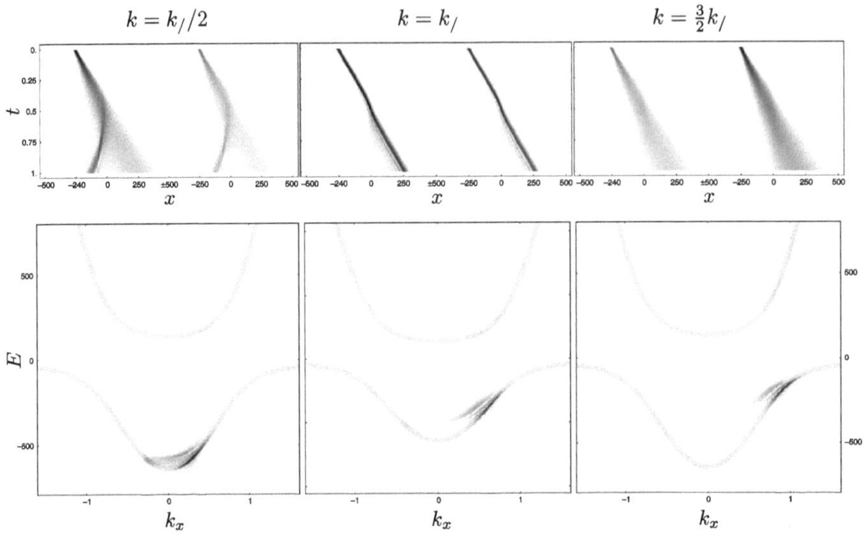

Fig. 1.13 Same as in Fig. 1.12 but with an obstacle interceded in the middle of the propagation (at $x = 0$). At low momentum, the system has the $-k$ states close enough to escape there, resulting in reflection of the packet, which is however much altered in the process, particularly the part which tunnels through. At $k = k_/$, the packet is slowed down as it tries to scatter off, but as it finds no final state, it eventually passes through the obstacle. It is slightly deformed in the process and propagates in its shape-preserving motion thereafter. At large momentum, the system also does not find where to scatter off and so continue its course, but diffuses profusely because of the large curvature of the dispersion at this point. The potential has been scaled to match the increasing energy when going up the polariton branch

with the potential, resulting in the appearance of a spread in the spectral trace as the packet is trying to scatter off, but as it finds no final state, it continues its course unaffected, in a way similar to that of the linear dispersion case, Fig. 1.9. The case at higher momenta is intermediate between the two previous ones. The state is also isolated in energy from possible scatterer and therefore carries on its course. However, as it also diffuses quickly because of the curvature, its existence as a wave packet is also limited in this sense. Only the case $k_/$ is spectacular in this respect: the bullet propagates essentially unaffected through the potential.

This is the simplest case of polariton propagation, and only remotely approaches the experimental situation that involves non-linearity and joint, or locked, propagation of the signal and idler, pinned by the pump. This nevertheless illustrates the new light that polariton physics can shed on problems as old as quantum mechanics itself.

1.6 Summary and Outlooks

Exciton–polaritons bring together antagonist properties of light and matter. In their condensed phase, they furthermore bring these peculiar properties to the macroscopic world. For these reasons, they are systems of rich fundamental interest and great prospects for possible technological applications.

We have focused on the dynamical aspect of polariton condensates, which they are particularly suited to address since they are open systems. The short photon lifetime in the cavity allows, for instance, a continuous monitoring of the quantum dynamics realised within the microcavity.

In a first part, we discussed the problem of the formation of the condensed phase, from either a thermal gas or from various regimes of strong coupling in the spontaneous emission regime. In a second part, we discussed the space–time propagation of wave packets, an emerging framework in microcavity physics pioneered by a new class of experiments that can set condensates in motion.

The formation of condensation is a fundamental problem with practical applications since in most systems, particles have a lifetime, even in atomic system where they can evaporate from their trap. The semi-classical Boltzmann equation has proved successful in describing condensation in the peculiar polariton geometry (with its characteristic dispersion and its specific microscopic details, such as strength of interaction and timescale of lifetimes). To access the quantum properties, one needs to go beyond the semi-classical description. We have presented an exact solution of quantum Boltzmann master equations for relaxation from the Rayleigh circle (with M states degenerate in energy) that spells out how coherence grows; Bose–Einstein condensation arises when a single quantum state—the ground state—copies properties of the system as a whole, such as peaked fluctuations of its number of particles. This is inherited from the macroscopic system, thanks to the central limit theorem, that states that a large sum (for the whole system) of random variables (populations) that fluctuate thermally with a small average (each state in isolation being classical) results in a normal distribution (a Gaussian that the ground state can copy). This is the quantum version of Einstein's picture of condensation of the ideal gas. It can be easily upgraded with pumping, lifetime and other specifics of a polariton system which, as numerical solutions show, provide coherence with a build-up as fast as, and steeper than, the population. As this mechanism relies on a large number of states and we assumed degeneracy in energy to keep the solution exact but compact, we regard this case as emblematic of the 2D realisation of polariton physics. In this case, one state only out of a large collection of a classical system becomes quantum. In the opposite limit of a fully quantised system, the regime of cavity QED is realised. The reduced simplicity brought by the fewer degrees of freedom allows an exact solution that keeps track of all quantum correlators. The condensation in this case appears as a striking transition from the quantum to the classical regime. In the former case, a small number of microscopic variables, namely, quantum correlators associated to transitions between well-defined quantum states (the polaritons), describe the dynamics. As pumping is increased and coherence grows, a breakdown of quantisation occurs with an emerging continuous field, in the form of an increasing number of vanishing quantum correlators. The condensation of polaritons (or dressed states) is accompanied by a complex display of crossings and level repulsions. When quantisation has been smoothen out completely, a classical description is adequate rather than the exceedingly complicated one accounting for all the quantum correlators, that loose meaning of their own. The classical limit, on the other hand, recovers a simple description in terms of a few macroscopic variables, namely, intensity, population inversion and the coherence degree. Spectrally, a counterpart of the Mollow triplet is observed with a lasing line sitting on the fluorescence spectrum. In this coherence build-up process, we outlined the case of a thresholdless laser where perfect coherence (factoring out of quantum correlators to all orders) is present throughout the entire process of the quantum to classical transition. In this case, however, formation of the condensate results in an intermediate stage with disruption of coherence and universal fluctuations in particle numbers, as the system shifts paradigm from strong coupling to condensation. Finally, we overviewed some prospects of polaritons to study the quantum dynamics of propagating wave packets, following recent breakthrough experiments that could set condensates in motion. We have shown how and explained why a linear dispersion allows a perfect tunneling of a shape-preserving packet through any obstacle. Polariton packets with a small spread in momentum space along the polariton dispersion exhibit a shape-preserving propagation at points where the dispersion is linear or where its curvature flattens. The collision of wave packets with a potential reproduces most of the phenomenology of the linear dispersion, letting it pass through essentially unaffected. At other points of the dispersion, the bullet is destroyed by the potential through its subsequent fast diffusion or scattering. As we kept only one variable of space (and momentum), we cast this part as 1D, but the results are essentially the same in 2D, from which we took a slice for simplicity.

We have kept in this chapter the description of these important questions of microcavity physics as simple as possible. We regard the term "simple" here as an important and noble one, to be understood as a synonymous of "*fundamental*," "*essential*" and "*elegant*," and nothing like "oversimplified" or "trivial." In fact, one can see that even our simple approach is considerably complex in the mundane sense of the term. The quantum-to-classical transition in 0D, for instance, required days of computer time on a cluster. Of course, this does not mean the picture is complete. Most importantly, we have left away the non-linearities (except when intrinsic like in 0D) which, in the spirit above, would have taken us too far astray. Non-linearities make the system considerably more complex, in both senses of more rich and in its traditional meaning. Other chapters in this volume explore some of this fascinating physics.

Acknowledgements I am indebted to Elena del Valle for constant exchanges and collaboration on most of the topics addressed here. Many of the results presented, and some of the most beautiful, are her own. I thank Daniele Sanvitto for his support with my contribution to this volume and for keeping this topic constantly exciting with his fresh experimental approach. I am grateful to

Alexey Kavokin for having introduced me to the physics of microcavity polaritons which he also kept exciting, with his own peculiar approach. I also thank C. Tejedor, G. Malpuech, M. Glazov, Yu. Rubo, I. A. Shelykh, A. Gonzalez-Tudela, E. Cancellieri, A. Laucht, J. J. Finley and many other colleagues for discussions on many parts of this work, to which they often contributed to large extents. Support from the Marie Curie IEF "SQOD" is acknowledged.

References

1. A. Kavokin, J.J. Baumberg, G. Malpuech, F.P. Laussy, *Microcavities*, 2 edn. (Oxford University Press, Oxford, 2011)
2. J.J. Hopfield, Theory of the contribution of excitons to the complex dielectric constant of crystals. Phys. Rev. **112**, 1555 (1958)
3. L.V. Keldysh, A.N. Kozlov, Collective properties of excitons in semiconductors. Sov. Phys. JETP **27**, 521 (1968)
4. Z.I. Alferov, Nobel lecture: The double heterostructure concept and its applications in physics, electronics, and technology. Rev. Mod. Phys. **73**, 767 (2001)
5. L.V. Butov, A.L. Ivanov, A. İmamoğlu, P.B. Littlewood, A.A. Shashkin, V.T. Dolgopolov, K.L. Campman, A.C. Gossard, Stimulated scattering of indirect excitons in coupled quantum wells: Signature of a degenerate Bose-gas of excitons. Phys. Rev. Lett. **86**, 5608 (2001)
6. C. Weisbuch, M. Nishioka, A. Ishikawa, Y. Arakawa, Observation of the coupled exciton-photon mode splitting in a semiconductor quantum microcavity. Phys. Rev. Lett. **69**, 3314 (1992)
7. M.G. Raizen, R.J. Thompson, R.J. Brecha, H.J. Kimble, H.J. Carmichael. Normal-mode splitting and linewidth averaging for two-state atoms in an optical cavity. Phys. Rev. Lett. **63**, 240 (1989)
8. R.J. Thompson, G. Rempe, H.J. Kimble, Observation of normal-mode splitting for an atom in an optical cavity. Phys. Rev. Lett. **68**, 1132 (1992)
9. K. Hennessy, A. Badolato, M. Winger, D. Gerace, M. Atature, S. Gulde, S. Fălt, E. L. Hu, A. İmamoğlu. Quantum nature of a strongly coupled single quantum dot–cavity system. Nature **445**, 896 (2007)
10. D. Press, S. Götzinger, S. Reitzenstein, C. Hofmann, A. Löffler, M. Kamp, A. Forchel, Y. Yamamoto. Photon antibunching from a single quantum dot-microcavity system in the strong coupling regime. Phys. Rev. Lett. **98**, 117402 (2007)
11. R. Houdré, C. Weisbuch, R.P. Stanley, U. Oesterle, P. Pellandin, M. Ilegems, Measurement of cavity-polariton dispersion curve from angle-resolved photoluminescence experiments. Phys. Rev. Lett. **73**, 2043 (1994)
12. P.G. Savvidis, J.J. Baumberg, R.M. Stevenson, M.S. Skolnick, D.M. Whittaker, J.S. Roberts, Angle-resonant stimulated polariton amplifier. Phys. Rev. Lett. **84**, 1547 (2000)
13. J.J. Baumberg, P.G. Savvidis, R.M. Stevenson, A.I. Tartakovskii, M.S. Skolnick, D.M. Whittaker, J.S. Roberts, Parametric oscillation in a vertical microcavity: A polariton condensate or micro-optical parametric oscillation. Phys. Rev. B **62**, R16247 (2000)
14. A. İmamoğlu, R.J. Ram, S. Pau, Y. Yamamoto, Nonequilibrium condensates and lasers without inversion: Exciton-polariton lasers. Phys. Rev. A **53**, 4250 (1996)
15. M. Kira, F. Jahnke, S.W. Koch, J.D. Berger, D.V. Wick, T.R. Nelson Jr., G. Khitrova, H.M. Gibbs, Quantum theory of nonlinear semiconductor microcavity luminescence explaining "Boser" experiments. Phys. Rev. Lett. **79**, 5170 (1997)
16. L.V. Butov. A polariton laser. Nature **447**, 540 (2007)
17. H. Deng, G. Weihs, C. Santori, J. Bloch, Y. Yamamoto, Condensation of semiconductor microcavity exciton polaritons. Science **298**, 199 (2002)
18. J. Bloch, B. Sermage, M. Perrin, P. Senellart, R. André, Le Si Dang. Monitoring the dynamics of a coherent cavity polariton population. Phys. Rev. B **71**, 155311 (2005)

19. J. Kasprzak, M. Richard, S. Kundermann, A. Baas, P. Jeambrun, J.M.J. Keeling, F.M. Marchetti, M.H. Szymanska, R. André, J.L. Staehli, V. Savona, P.B. Littlewood, B. Deveaud, Le Si Dang, Bose–Einstein condensation of exciton polaritons. Nature **443**, 409 (2006)
20. A. Kavokin, G. Malpuech, F.P. Laussy, Polariton laser and polariton superfluidity in microcavities. Phys. Lett. A **306**, 187 (2003)
21. I. Carusotto, C. Ciuti, Probing microcavity polariton superfluidity through resonant Rayleigh scattering. Phys. Rev. Lett. **93**, 166401 (2004)
22. K.G. Lagoudakis, M. Wouters, M. Richard, A. Baas, I. Carusotto, R. André, Le Si Dang, B. Deveaud-Plédran, Quantized vortices in an exciton-polariton condensate. Nat. Phys. **4**, 706 (2008)
23. A. Amo, D. Sanvitto, F.P. Laussy, D. Ballarini, E. del Valle, M. D. Martin, A. Lemaître, J. Bloch, D.N. Krizhanovskii, M.S. Skolnick, C. Tejedor, L. Viña, Collective fluid dynamics of a polariton condensate in a semiconductor microcavity. Nature **457**, 291 (2009)
24. M.D. Fraser, G. Roumpos, Y. Yamamoto, Vortex–antivortex pair dynamics in an exciton–polariton condensate. New J. Phys. **11**, 113048 (2009)
25. A. Amo, J. Lefrère, S. Pigeon, C. Adrados, C. Ciuti, I. Carusotto, R. Houdré, E. Giacobino, A. Bramati, Superfluidity of polaritons in semiconductor microcavities. Nat. Phys. **5**, 805 (2009)
26. K.G. Lagoudakis, T. Ostatnický, A.V. Kavokin, Y.G. Rubo, R. André, B. Deveaud-Plédran, Observation of half-quantum vortices in an exciton-polariton condensate. Science **326**, 974 (2009)
27. D. Sanvitto, F.M. Marchetti, M.H. Szymańska, G. Tosi, M. Baudisch, F.P. Laussy, D.N. Krizhanovskii, M.S. Skolnick, L. Marrucci, A. Lemaître, J. Bloch, C. Tejedor, L. Viña, Persistent currents and quantized vortices in a polariton superfluid. Nat. Phys. **6**, 527 (2010)
28. G. Roumpos, M.D. Fraser, A. Löffler, S. Höfling, A. Forchel, Y. Yamamoto, Single vortex-antivortex pair in an exciton-polariton condensate. Nat. Phys. **7**, 129 (2011)
29. H.T.C. Stoof, Nucleation of Bose–Einstein condensation. Phys. Rev. A **45**, 8398 (1992)
30. D.V. Semikoz, I.I. Tkachev, Kinetics of Bose condensation. Phys. Rev. Lett. **74**, 3093 (1995)
31. Yu. Kagan, B.V. Svistunov. Evolution of correlation properties and appearance of broken symmetry in the process of Bose–Einstein condensation. Phys. Rev. Lett. **79**, 3331 (1997)
32. C.W. Gardiner, P. Zoller, Quantum kinetic theory: A quantum kinetic master equation for condensation of a weakly interacting Bose gas without a trapping potential. Phys. Rev. A **55**, 2902 (1997)
33. L. Banyai, P. Gartner, Real-time Bose–Einstein condensation in a finite volume with a discrete spectrum. Phys. Rev. Lett. **88**, 210404 (2002)
34. F. Tassone, C. Piermarocchi, V. Savona, A. Quattropani, P. Schwendimann, Bottleneck effects in the relaxation and photoluminescence of microcavity polaritons. Phys. Rev. B **56**, 7554 (1997)
35. D. Porras, C. Ciuti, J.J. Baumberg, C. Tejedor, Polariton dynamics and Bose–Einstein condensation in semiconductor microcavities. Phys. Rev. B **66**, 085304 (2002)
36. G. Malpuech, A. Di Carlo, A. Kavokin, J.J. Baumberg, M. Zamfirescu, P. Lugli, Room-temperature polariton lasers based on GaN microcavities. Appl. Phys. Lett. **81**, 412 (2002)
37. Yu. G. Rubo, F.P. Laussy, G. Malpuech, A. Kavokin, P. Bigenwald, Dynamical theory of polariton amplifiers. Phys. Rev. Lett. **91**, 156403 (2003)
38. F.P. Laussy, G. Malpuech, A. Kavokin, Spontaneous coherence buildup in a polariton laser. Phys. Stat. Sol. C **1**, 1339 (2004)
39. H.T. Cao, T.D. Doan, D.B. Tran Thoai, H. Haug, Condensation kinetics of cavity polaritons interacting with a thermal phonon bath. Phys. Rev. B **69**, 245325 (2004)
40. T.D. Doan, H. Thien Cao, D.B. Tran Thoai, H. Haug, Coherence of condensed microcavity polaritons calculated within Boltzmann-master equations. Phys. Rev. B **78**, 205306 (2008)
41. M. Wouters, V. Savona, Stochastic classical field model for polariton condensates. Phys. Rev. B **79**, 165302 (2009)
42. I.G. Savenko, E.B. Magnusson, I.A. Shelykh, Density-matrix approach for an interacting polariton system. Phys. Rev. B **83**, 165316 (2011)

43. J. Keeling, P.R. Eastham, M.H. Szymanska, P.B. Littlewood, Polariton condensation with localized excitons and propagating photons. Phys. Rev. Lett. **93**, 226403 (2004)
44. J. Keeling, P.R. Eastham, M.H. Szymanska, P.B. Littlewood, BCS–BEC crossover in a system of microcavity polaritons. Phys. Rev. B **72**, 115320 (2005)
45. M.H. Szymanska, J. Keeling, P.B. Littlewood, Nonequilibrium quantum condensation in an incoherently pumped dissipative system. Phys. Rev. Lett. **96**, 230602 (2006)
46. F.M. Marchetti, J. Keeling, M.H. Szymanska, P.B. Littlewood, Thermodynamics and excitations of condensed polaritons in disordered microcavities. Phys. Rev. Lett. **96**, 066405 (2006)
47. E.T. Jaynes, F.W. Cummings, Comparison of quantum and semiclassical radiation theory with application to the beam maser. Proc. IEEE **51**, 89 (1963)
48. Y. Zhu, D.J. Gauthier, S.E. Morin, Q. Wu, H.J. Carmichael, T.W. Mossberg, Vacuum Rabi splitting as a feature of linear-dispersion theory: Analysis and experimental observations. Phys. Rev. Lett. **64**, 2499 (1990)
49. G. Khitrova, H.M. Gibbs, F. Jahnke, M. Kira, S.W. Koch, Nonlinear optics of normal-mode-coupling semiconductor microcavities. Rev. Mod. Phys **71**, 1591 (1999)
50. C.N. Cohen-Tannoudji, Manipulating atoms with photons. Rev. Mod. Phys. **70**, 707 (1998)
51. R.H. Dicke, Coherence in spontaneous radiation processes. Phys. Rev. **93**, 99 (1954)
52. J.P. Reithmaier, G. Sek, A. Löffler, C. Hofmann, S. Kuhn, S. Reitzenstein, L.V. Keldysh, V.D. Kulakovskii, T.L. Reinecker, A. Forchel, Strong coupling in a single quantum dot–semiconductor microcavity system. Nature, **432**, 197 (2004)
53. T. Yoshie, A. Scherer, J. Heindrickson, G. Khitrova, H.M. Gibbs, G. Rupper, C. Ell, O.B. Shchekin, D.G. Deppe, Vacuum Rabi splitting with a single quantum dot in a photonic crystal nanocavity. Nature **432**, 200 (2004)
54. E. Peter, P. Senellart, D. Martrou, A. Lemaître, J. Hours, J.M. Gérard, J. Bloch, Exciton-photon strong-coupling regime for a single quantum dot embedded in a microcavity. Phys. Rev. Lett. **95**, 067401 (2005)
55. E. del Valle, F.P. Laussy, C. Tejedor, Luminescence spectra of quantum dots in microcavities. II. Fermions. Phys. Rev. B **79**, 235326 (2009)
56. E. del Valle, F.P. Laussy, Mollow triplet under incoherent pumping. Phys. Rev. Lett. **105**, 233601 (2010)
57. E. del Valle, F.P. Laussy, Regimes of strong light-matter coupling under incoherent excitation. Phys. Rev. A **84**, 043816 (2011)
58. D. Sanvitto, A. Amo, F.P. Laussy, A. Lemaître, J. Bloch, C. Tejedor, L. Viña, Polariton condensates put in motion. Nanotechnology **21**, 134025 (2010)
59. E. Wertz, L. Ferrier, D.D. Solnyshkov, R. Johne, D. Sanvitto, A. Lemaître, I. Sagnes, R. Grousson, A.V. Kavokin, P. Senellart, G. Malpuech an J. Bloch, Spontaneous formation and optical manipulation of extended polariton condensates. Nat. Phys. **6**, 860 (2010)
60. F.P. Laussy, M.M. Glazov, A. Kavokin, D.M. Whittaker, G. Malpuech, Statistics of excitons in quantum dots and their effect on the optical emission spectra of microcavities. Phys. Rev. B **73**, 115343 (2006)
61. E. del Valle, *Microcavity Quantum Electrodynamics*. (VDM Verlag, 2010)
62. V. Savona, Z. Hradil, A. Quattropani, P. Schwendimann, Quantum theory of quantum-well polaritons in semiconductor microcavities. Phys. Rev. B **49**, 8774 (1994)
63. D.W. Snoke, The quantum boltzmann equation in semiconductor physics. Annalen der Physik **523**, 87 (2010)
64. E.A. Uehling, G.E. Uhlenbeck, Transport phenomena in Einstein–Bose and Fermi–Dirac gases. I. Phys. Rev. **43**, 552 (1933)
65. C. Ciuti, V. Savona, C. Piermarocchi, A. Quattropani, and P. Schwendimann. Role of the exchange of carriers in elastic exciton-exciton scattering in quantum wells. Phys. Rev. B **58**, 7926 (1998)
66. G. Malpuech, A. Kavokin, A. Di Carlo, J.J. Baumberg, Polariton lasing by exciton-electron scattering in semiconductor microcavities. Phys. Rev. B **65**, 153310 (2002)

67. J. Kasprzak, D.D. Solnyshkov, R. André, Le Si Dang, G. Malpuech, Formation of an exciton polariton condensate: Thermodynamic versus kinetic regimes. Phys. Rev. Lett. **101**, 146404 (2008)
68. V.E. Hartwell, D.W. Snoke, Numerical simulations of the polariton kinetic energy distribution in GaAs quantum-well microcavity structures. Phys. Rev. B **82**, 075307 (2010)
69. R.J. Glauber, Coherent and incoherent states of the radiation field. Phys. Rev. **131**, 2766 (1963)
70. R. Hanbury Brown, R.Q. Twiss, A test of a new type of stellar interferometer on Sirius. Nature **178**, 1046 (1956)
71. L.D. Landau, Zur Theorie der Phasenumwandlungen I, II. Zh. Ekspr. Teoret. Fiz **19**, 627 (1937)
72. F.P. Laussy, Y.G. Rubo, G. Malpuech, A. Kavokin, P. Bigenwald, Dissipative quantum theory of polariton lasers. Phys. Stat. Sol. C **0**, 1476 (2003)
73. Yu. G. Rubo, Kinetics of the polariton condensate formation in a microcavity. Phys. Stat. Sol. A **201**, 641 (2004)
74. D. Sarchi, V. Savona, Long-range order in the Bose–Einstein condensation of polaritons. Phys. Rev. B **75**, 115326 (2007)
75. F.P. Laussy, G. Malpuech, A. Kavokin, P. Bigenwald, Spontaneous coherence buildup in a polariton laser. Phys. Rev. Lett. **93**, 016402 (2004)
76. D. Jaksch, C.W. Gardiner, P. Zoller, Quantum kinetic theory. II. Simulation of the quantum Boltzmann master equation. Phys. Rev. A **56**, 575 (1997)
77. C.W. Gardiner, P. Zoller, Quantum kinetic theory. III. quantum kinetic master equation for strongly condensed trapped systems. Phys. Rev. A **58**, 536 (1998)
78. D. Jaksch, C.W. Gardiner, K.M. Gheri, P. Zoller, Quantum kinetic theory. IV. intensity and amplitude fluctuations of a Bose–Einstein condensate at finite temperature including trap loss. Phys. Rev. A **58**, 1450 (1998)
79. C.W. Gardiner, P. Zoller, Quantum kinetic theory. V. quantum kinetic master equation for mutual interaction of condensate and noncondensate. Phys. Rev. A **61**, 033601 (2000)
80. M.D. Lee, C.W. Gardiner, Quantum kinetic theory. VI. the growth of a Bose–Einstein condensate. Phys. Rev. A **62**, 033606 (2000)
81. M.J. Davis, C.W. Gardiner, R.J. Ballagh, Quantum kinetic theory. VII. the influence of vapor dynamics on condensate growth. Phys. Rev. A **62**, 063608 (2000)
82. A. Einstein, Quantentheorie des einatomigen idealen Gases. Sitzungsberichte der Preussischen Akademie der Wissenschaften. **1**, 3–14 (1925)
83. R.L. Graham, D.E. Knuth, O. Patashnik, *Concrete Mathematics*. 2nd edn. (Addison-Wesley Longman, Amsterdam, 1994)
84. F.P. Laussy, I.A. Shelykh, G. Malpuech, A. Kavokin, Effects of Bose–Einstein condensation of exciton polaritons in microcavities on the polarization of emitted light. Phys. Rev. B **73**, 035315 (2006)
85. E. del Valle, D. Sanvitto, A. Amo, F.P. Laussy, R. André, C. Tejedor, L. Viña. Dynamics of the formation and decay of coherence in a polariton condensate. Phys. Rev. Lett. **103**, 096404 (2009)
86. E. del Valle, S. Zippilli, F.P. Laussy, A. Gonzalez-Tudela, G. Morigi, C. Tejedor. Two-photon lasing by a single quantum dot in a high-Q microcavity. Phys. Rev. B **81**, 035302 (2010)
87. F.P. Laussy, E. del Valle, C. Tejedor, Luminescence spectra of quantum dots in microcavities. I. Bosons. Phys. Rev. B **79**, 235325 (2009)
88. F.P. Laussy, E. del Valle, J.J. Finley, Lasing in strong coupling (2011). http://arxiv.org/pdf/1106.0509.pdf
89. Y. Mu, C.M. Savage, One-atom lasers. Phys. Rev. A **46**, 5944 (1992)
90. P. Gartner, Two-level laser: Analytical results and the laser transition, Phys. Rev. **A 84**, 053804 (2011)
91. S. Strauf, K. Hennessy, M.T. Rakher, Y.S. Choi, A. Badolato, L.C. Andreani, E.L. Hu, P.M. Petroff, and D. Bouwmeester, Self-tuned quantum dot gain in photonic crystal lasers. Phys. Rev. Lett. **96**, 127404 (2006)

92. Z. G. Xie, S. Götzinger, W. Fang, H. Cao, G.S. Solomon, Influence of a single quantum dot state on the characteristics of a microdisk laser. Phys. Rev. Lett. **98**, 117401 (2007)
93. S.M. Ulrich, C. Gies, S. Ates, J. Wiersig, S. Reitzenstein, C. Hofmann, A. Löffler, A. Forchel, F. Jahnke, P. Michler, Photon statistics of semiconductor microcavity lasers. Phys. Rev. Lett. **98**, 043906 (2007)
94. M. Witzany, R. Roßbach, W.-M. Schulz, M. Jetter, P. Michler, T.-L. Liu, E. Hu, J. Wiersig, and F. Jahnke, Lasing properties of InP/($Ga_{0.51}In_{0.49}P$) quantum dots in microdisk cavities. Phys. Rev. B **83**, 205305 (2011)
95. F.P. Laussy, A. Laucht, E. del Valle, J.J. Finley, J.M. Villas-Bôas, Luminescence spectra of quantum dots in microcavities. III. Multiple quantum dots, Phys. Rev. **B 84**, 195313 (2011)
96. F.P. Laussy, E. del Valle, Optical spectra of the Jaynes-Cummings ladder. AIP Conf. Proc. **1147**, 46 (2009)
97. B.R. Mollow, Power spectrum of light scattered by two-level systems. Phys. Rev. **188**, 1969 (1969)
98. M. Löffler, G.M. Meyer, H. Walther, Spectral properties of the one-atom laser. Phys. Rev. A **55**, 3923 (1997)
99. A.N. Poddubny, M.M. Glazov, N.S. Averkiev, Nonlinear emission spectra of quantum dots strongly coupled to a photonic mode. Phys. Rev. B **82**, 205330 (2010)
100. E. del Valle, Strong and weak coupling of two coupled qubits. Phys. Rev. A **81**, 053811, (2010)
101. L.D. Landau, Theory of the superfluidity of helium II. Phys. Rev. **60**, 356 (1941)
102. N.N. Bogoliubov, Theory of the weakly interacting Bose gas. J. Phys. (Moscow) **11**, 23 (1947)
103. R.P. Feynman, Application of quantum mechanics to liquid helium. Progr. Low Temp. Phys. **1**, 17 (1955)
104. E. Cancellieri, F.M. Marchetti, M.H. Szymanska, C. Tejedor, Superflow of resonantly driven polaritons against a defect. Phys. Rev. B **82**, 224512 (2010)
105. S. Ianeselli, C. Menotti, A. Smerzi, Beyond the Landau criterion for superfluidity. J. phys. B.: At. Mol. Phys. **39**, S135 (2006)
106. W.F. Wreszinski, On translational superfluidity and the Landau criterion for Bose gases in the Gross–Pitaevski limit. J. phys. A.: Math. Gen. **41**, 392006 (2008)
107. R.H. Stuewer, Resource letter Sol-1: Solitons. Am. J. Phys. **66**, 486 200 (1998)
108. S. Burger, K. Bongs, S. Dettmer, W. Ertmer, K. Sengstock, A. Sanpera, G.V. Shlyapnikov, M. Lewenstein, Dark solitons in Bose–Einstein condensates. Phys. Rev. Lett. **83**, 5198 (1999)
109. K.E. Strecker, G.B. Partridge, A.G. Truscott, R.G. Hulet, Formation and propagation of matter-wave soliton trains. Nature **417**, 150 (2002)
110. O.A. Egorov, D.V. Skryabin, A.V. Yulin, F. Lederer, Bright cavity polariton solitons. Phys. Rev. Lett. **102**, 153904 (2009)
111. A. Amo, S. Pigeon, D. Sanvitto, V.G. Sala, R. Hivet, I. Carusotto, F. Pisanello, G. Leménager, R. Houdré, E. Giacobino, C. Ciuti, A. Bramati, Polariton superfluids reveal quantum hydrodynamic solitons. Science **332**, 1167 (2011)
112. J. Keeling, N.G. Berloff, Going with the flow. Nature **457**, 273 (2009)
113. E. Schrödinger. Der stetige übergang von der Mikro- zur Makromechanik. Naturwissenschaften **14**, 664 (1926)
114. C.G. Darwin, Free motion in the wave mechanics. Proc. Roy. Soc **A117**, 258 (1928)
115. J.R. Klein, Do free quantum-mechanical wave packets always spread? Am. J. Phys. **48**, 1035 (1980)
116. M. Berry. Quantum physics on the edge of chaos. New Scientist **116**(1587), 44 (1987)
117. J. Scott Russell, Report on waves. *Fourteenth meeting of the British Association for the Advancement of Science*, (1844)
118. D.J. Korteweg, G. de Vries, On the change of form of long waves advancing in a rectangular canal, and on a new type of long stationary waves. Philosophical Mag. **39**, 422 (1895)
119. C.C. Yan, Soliton like solutions of the schrödinger equation for simple harmonic oscillator. Am. J. Phys. **62**, 147 (1994)

120. D.F. Walls, Squeezed states of light. Nature **306**, 141 (1983)
121. M.V. Berry, N.L. Balazs, Nonspreading wave packets. Am. J. Phys. **47**, 264 (1979)
122. F.P. Laussy, Propagation of polariton wavepackets. ICSCE4 conference, Cambridge, http://www.tcm.phy.cam.ac.uk/BIG/icsce4/talks/laussy.pdf, 2008
123. I.A. Shelykh, A.V. Kavokin, Yu.G. Rubo, T.C.H. Liew, G. Malpuech, Polariton polarization-sensitive phenomena in planar semiconductor microcavities. Semicond. Sci. Technol. **25**, 013001 (2010)
124. T.S. Raju, C. Nagaraja Kumar, P.K. Panigrahi, On exact solitary wave solutions of the nonlinear Schrödinger equation with a source. J. phys. A.: Math. Gen. **38**, L271 (2005)
125. V.M. Vyas, T.S. Raju, C.N. Kumar, P.K. Panigrahi, Soliton solutions of driven nonlinear schrödinger equation. J. phys. A.: Math. Gen. **39**, 9151 (2006)
126. C. Ciuti, I. Carusotto, Quantum fluid effects and parametric instabilities in microcavities. Phys. Stat. Sol. B **242**, 2224 (2005)
127. M.H. Szymanska, F.M. Marchetti, D. Sanvitto, Propagating wave packets and quantized currents in coherently driven polariton superfluids. Phys. Rev. Lett. **105**, 236402 (2010)
128. C. Ciuti, P. Schwendimann, B. Deveaud, A. Quattropani, Theory of the angle-resonant polariton amplifier. Phys. Rev. B **62**, R4825 (2000)
129. C. Ciuti, P. Schwendimann, A. Quattropani, Parametric luminescence of microcavity polaritons. Phys. Rev. B **63**, 041303 (2001)
130. D.M. Whittaker, Effects of polariton-energy renormalization in the microcavity optical parametric oscillator. Phys. Rev. B **71**, 115301 (2005)
131. A.E. Garriz, A. Sztrajman, D. Mitnik, Running into trouble with the time-dependent propagation of a wavepacket. Eur. J. Phys. **31**, 785 (2010)
132. H. Shao, Z. Wang, Numerical solutions of the time-dependent Schrödinger equation: Reduction of the error due to space discretization. Phys. Rev. E **79**, 056705 (2009)

Chapter 2
Polariton Nonlinear Dynamics: Theory and Experiments

Vladimir D. Kulakovskii, Sergei S. Gavrilov, Sergei G. Tikhodeev, and Nikolay A. Gippius

Abstract The results of experimental studies are presented of the polariton system in a semiconductor microcavity excited resonantly at various wave vectors by ns-long pulse laser with various light polarizations along with a theoretical description of the nonlinear effects in the polariton system. The interplay between the parametric scattering and self-instability of the driven mode results in a rich variety of scattering scenarios (or cavity dynamics) sensitive to variation of both the intensity and polarization state of the external pump. The observed instabilities and hysteresis effects in a scalar polariton system excited with circularly polarized pulses can be qualitatively reproduced within the semiclassical model of dynamically self-organized optical parametric oscillator (OPO), based on the resonant approximation of cavity electrodynamics and the Ginzburg-Landau-Gross-Pitaevskii-type equation for coherent excitonic interband polarization. However, this model fails to explain the polarization instabilities in the effectively spinor system excited with elliptically polarized pulses. The dynamics of such a system is strongly affected by the long-lived exciton reservoir (excited due to polariton scattering) which brings about additional blueshift of both components of bright excitons and results in the qualitative changes in the development of the polarization instabilities in the driven mode and in the OPO signal. Those transitions are phenomenologically introduced into the modified semiclassical model. In spite of some limitations, this model provides a self-consistent approach to description of intracavity field dynamics under both pulse and continuous wave excitation conditions and gives a

V.D. Kulakovskii (✉) · S.S. Gavrilov
Institute of Solid State Physics, RAS, Chernogolovka 142432, Russia
e-mail: kulakovs@issp.ac.ru

S.G. Tikhodeev · N.A. Gippius
A. M. Prokhorov General Physics Institute, RAS, Moscow 119991, Russia and LASMEA, UMR 6602 CNRS, Université Blaise Pascal, 63177 Aubière, France
e-mail: na.gippius@gmail.com

D. Sanvitto and V. Timofeev (eds.), *Exciton Polaritons in Microcavities*,
Springer Series in Solid-State Sciences 172, DOI 10.1007/978-3-642-24186-4_2,

good qualitative description of the observed polarization instabilities and hysteresis effects in the dynamics of both the driven mode and OPO signal.

2.1 Introduction

Exciton-polaritons in planar semiconductor microcavities (MCs) are unique bosonic excitations occurring in an active layer of the MCs due to the strong coupling of the exciton and MC photon modes [1–3]. They are characterized by a very small effective mass of the order 10^{-5} of the electron mass in vacuum. The photoexcited system of exciton-polaritons demonstrates a large number of interesting physical phenomena due to the combination of several specific properties such as quasi- (2D), Bose statistics, and unique dispersion law with a very small effective mass and an inflection point in the range of small (light) wave vectors. The most striking of them are the giant stimulated polariton–polariton scattering under resonant excitation near the inflection point of the LP dispersion curve [4–8], polarization multistability in highly excited MC [9–12], and nonequilibrium Bose–Einstein condensation of exciton-polaritons under nonresonant excitation, resulting in formation of a macropopulated state at the low polariton (LP) band bottom and, as a consequence, in a drastic change of the polariton properties [13].

This chapter is organized as follows: First, we describe the experiments with a cw pump which demonstrate the polariton–polariton parametric scattering with a low threshold intensity. Then a quasiclassical nonlinear model of the interacting cavity and exciton modes is introduced, and the role of nonlinearities is discussed. It is shown that the marked peculiarities of the parametric LP–LP scattering are due to the competition between the bistability of the pumped LP mode and its parametric instability. Next, the experiments with ns circularly polarized pump are discussed, and the hysteresis effects in the kinetics of LP–LP scattering are demonstrated. The last part is devoted to the experiments with elliptically polarized pump. It is shown that in order to explain these experiments, fast scattering of LPs into a long-living incoherent exciton reservoir has to be added to the theoretical model.

2.2 Giant Stimulated LP–LP Scattering Under CW Excitation

Giant stimulated LP–LP scattering in the optical response of planar MCs was first observed in GaAs-based MCs with InGaAs quantum wells (QWs) in the active layer under cw excitation at wave vector $\mathbf{k}_p$ slightly above the inflection point of LP branch $E_{LP}(\mathbf{k}_{infl})$, when the scattering exhibits an unusually low (smaller than 400 W/cm^2) threshold of parametric scattering [4, 5, 14] (see the scheme of the experiment in Fig. 2.1a). Specifically, such excitation results in a strong parametric scattering into states positioned approximately on the LP dispersion curve $E_{LP}(\mathbf{k})$ with $\mathbf{k}_s = 0$ and $\mathbf{k}_i = 2\mathbf{k}_p$, called *signal* and *idler*, respectively.

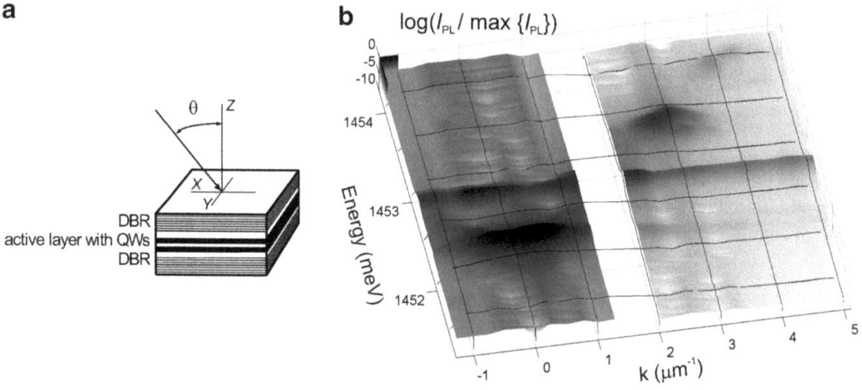

Fig. 2.1 The experimental geometry (**a**). The measured energy — wave vector spectra of the LP emission under a resonant excitation at $E_{\mathrm{p}} = 1452.7\,\mathrm{meV}$ and $\vartheta = 12.5°$ with intensity above the scattering threshold (**b**)

That is illustrated in Fig. 2.1b showing the distribution of the scattered light vs energy $E = \hbar\omega$ and in-plane wave vector $k = \omega \sin\vartheta/c$. The emission spectrum is recorded from GaAs/AlAs MC containing six 10-nm-thick InGaAs quantum wells in a $3/2\lambda$ GaAs cavity with a MC mode–QW exciton detuning of $-0.5\,\mathrm{meV}$. The excitation occurs at $E_{\mathrm{p}} = E_{\mathrm{LP}}(k_{\mathrm{p}}) + 0.5\,\mathrm{meV} = 1452.7\,\mathrm{meV}$, $k_{\mathrm{p}} = 1.6\,\mu\mathrm{m}^{-1} \sim k_{\mathrm{infl}}$ and intensity 880 W/cm^2, which is above the threshold of stimulated parametric scattering.

A simple explanation of such LP–LP scattering could be a four-wave mixing or parametric scattering of two pump polaritons into the signal and idler polaritons with conservation of energy and in-plane wave vector $2E_{\mathrm{p}} = E_{\mathrm{s}} + E_{\mathrm{i}}$, $2\mathbf{k}_{\mathrm{p}} = \mathbf{k}_{\mathrm{s}} + \mathbf{k}_{\mathrm{i}}$. However, assuming that scattered polaritons belong to the LP branch, one finds that the idler and the signal have to be on the intersections of $E_{\mathrm{LP}}(\mathbf{k})$ and $2E - E_{\mathrm{LP}}(2\mathbf{k}_{\mathrm{p}} - \mathbf{k})$. Thus, it is expected that the signal and idler have to shift along the LP curve with the shift of the pump from the inflection point on E_{LP}, as illustrated in Fig. 2.2a. However, further studies [6, 7, 15] showed that it did not occur: the signal and idler continue to stick to $\mathbf{k}_{\mathrm{s}} = 0$ and $\mathbf{k}_{\mathrm{i}} = 2\mathbf{k}_{\mathrm{p}}$, as shown in Fig. 2.2b. The energy — wave vector conservation is ensured by the strong blue shift of both the signal and idler.

The early theoretical descriptions of LP–LP scattering were based on the OPO model in the simplified polariton basis, when the intensity of the pumped mode is assumed to be fixed by external excitation [16–19]. Practically, this OPO-based model is often reduced [16–18] to consideration of three macrooccupied modes, i.e., the pump, signal, and idler modes, which provides a qualitative explanation of the threshold behavior of stimulated scattering. With increasing pump intensity, the signal and idler branches become renormalized via the parametric coupling (Fig. 2.3a). Simultaneously, *dampings*, i.e., the imaginary parts of the signal and idler eigenenergies approach zero (see Fig. 2.3b). At some critical pump intensity,

Fig. 2.2 The prediction of a simple four-wave mixing model: with a shift of the pump from the inflection point, the signal and idler are expected to shift along the LP dispersion curve (**a**). In reality, the signal and idler are found always at 0 and $2k_p$ (**b**)

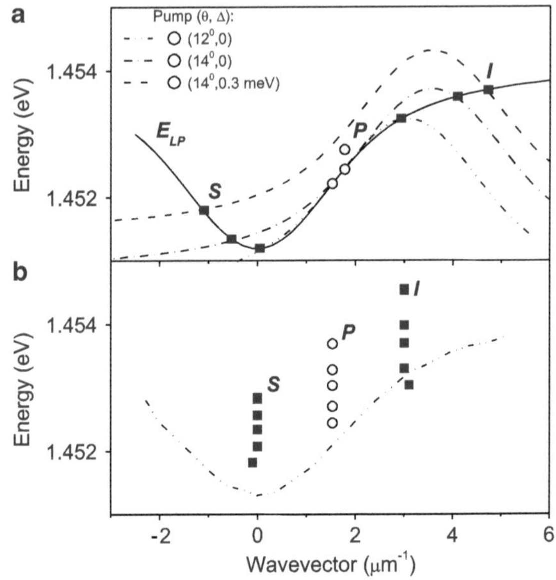

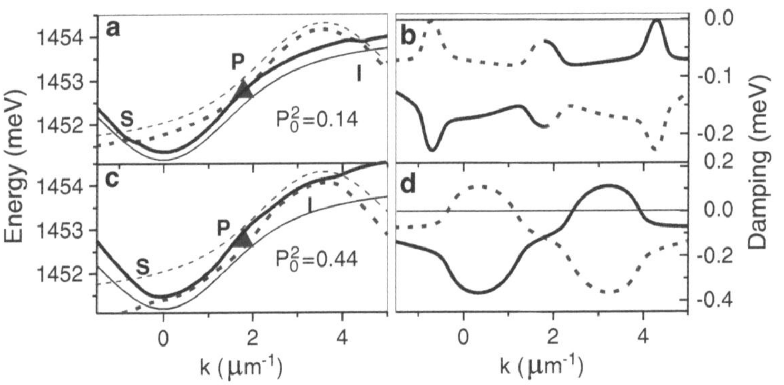

Fig. 2.3 The calculated real (*left side*) and imaginary (*right side*) parts of the signal and idler LP branches (*thick solid and dashed lines*, respectively, for pump intensity $|\mathscr{P}_p|^2 = 0.14$ a,b and 0.44 c,d. Left (**a**, **c**) and right (**b**, **c**) panels show the corresponding real and imaginary eigenenergies. Thin lines in panels (**a**, **c**) are signal and idler at $|\mathscr{P}_p|^2 = 0$ from Fig. 2.2a. The pump (*triangle*) is at $(\vartheta, \Delta) = (14°, 0.3\,\text{meV})$

the damping changes sign (becomes *gain*), which points to the instability and indicates the threshold of the parametric stimulated scattering.

However, the OPO model also predicts that the signal and idler at the threshold pump intensity for parametric scattering have to shift along the renormalized LP curves into regions $k_s < 0$, $k_i > 2k_p$ with the shift of the pump from the LP inflection point (as shown in Fig. 2.3a), whereas in the experiments, the signal and idler maxima are always at $\mathbf{k}_s \approx 0$, $\mathbf{k}_i \approx 2\mathbf{k}_p$.

2.3 Semiclassical Theory of Microcavity Nonlinearities

The main peculiarity of four-wave mixing in planar semiconductor MCs is its 2D character. This makes a strong difference from the bulk (3D) case, when the driven mode is excited on the 3D polariton dispersion curve, although renormalized by the nonlinear interactions in the system. In a planar MC, in contrast, external monochromatic pumping at given frequency ω_p and in-plane wave vector $\mathbf{k}_p$ excites the polaritonic wave with the same frequency and in-plane wave vector. ω_p can be set away from the polariton dispersion, which allows scanning large areas of $(\omega_p, \mathbf{k}_p)$ space at various pump intensities.

The semiconductor MC is a nonlinear optical system with a complicated spatial structure that, at first sight, makes the nonlinear optical approach impractical because of severe computational difficulties. However, in fact, all the nonlinearities in this problem arise from the QW exciton states located in specific positions inside the structure. The empty cavity (without QW resonances) is well treated by the linear approach and described within the resonant approximation [20].

Within this approach, the response function of the empty cavity can be approximated in the frequency domain by an oscillator with the coupling coefficients being the residuals of the exactly calculated response function. As to the excitonic polarization, owing to the exciton–exciton interaction, it can be described by a nonlinear equation of the Gross-Pitaevskii type. Within the semiclassical approach, it is convenient [9, 15] (see also [21] and the references therein) to describe the exciton-cavity photon dynamics by coupled equations for the in-plane components of the electric field $\mathbf{E}$ and the excitonic part of electric polarization $\mathbf{P}$ in the cavity active layer. The basis of linear xy or circular $\pm$ polarizations can be used, $\mathbf{E} = (\mathscr{E}_x, \mathscr{E}_y)$, $\mathbf{P} = (\mathscr{P}_x, \mathscr{P}_y)$, or $\mathbf{E} = (\mathscr{E}_+, \mathscr{E}_-)$, and $\mathbf{P} = (\mathscr{P}_+, \mathscr{P}_-)$, respectively. In what follows, we will call $\mathbf{P}$ the *excitonic field*, in order to distinguish it from the polarization states of the involved waves (elliptic, linear, circular, etc.). These equations can be written as:

$$\mathrm{i}\dot{\mathbf{E}}_\mathbf{k} = \omega_c(\mathbf{k})\,\mathbf{E}_\mathbf{k} + \beta(\mathbf{k})\mathbf{P}_\mathbf{k} + \alpha(\mathbf{k})\mathbf{F}_\mathbf{k}, \tag{2.1}$$

$$\mathrm{i}\dot{\mathbf{P}}_\mathbf{k} = A\mathbf{E}_\mathbf{k} + [\omega_x + \delta\omega_x(\mathbf{P},\ldots)]\,\mathbf{P}_\mathbf{k} + \xi_\mathbf{k}\,. \tag{2.2}$$

Here, $\mathbf{F}$ stands for the incident electric field that is usually regarded as a plane wave, $\mathbf{F}_\mathbf{k} \propto e^{-\mathrm{i}\omega_p t}\delta(\mathbf{k}-\mathbf{k}_p)$; $\xi_\mathbf{k}$ is the stochastic Langevin force; $\omega_{c,x} = \Re(\omega_{c,x}) - \mathrm{i}\gamma_{c,x}$ are the eigenfrequencies and decay rates of the intracavity photon and exciton modes; α and β are the cavity response coefficients and A is excitonic polarizability (so that $2\sqrt{A\beta}$ equals the Rabi splitting between the lower and upper polariton dispersion branches). $omega_{c,x}$ are complex-valued eigenfrequencies of the intracavity photon and exciton modes (with imaginary parts corresponding to the decay rates. The matrix over the exciton polarization (or spin) states $\hbar\delta\omega_x(\mathbf{P},\ldots)$ is composed of the energy shifts of the exciton states that depend on the excitonic field and, in general, on some other parameters.

The most important thing is that the exciton energy shift in (2.2) $\hbar\delta\omega_x(\mathbf{P},\ldots)$ depends on excitonic field **P**. This makes the whole problem nonlinear and its physics really exciting. For example, even in the case of stationary pump, the solution of nonlinear (2.1), (2.2) for the electric and excitonic fields of pumped mode $\mathbf{E}_p$, $\mathbf{P}_p$ as a function of the external pump **F** becomes bistable or even multistable (if the vectorial nature of the fields is taken into account). As a result, the one-to-one dependence of the polariton population on the external pump disappears; the evolution of the system exhibits the hysteresis, memory, and self-organization effects.

The use of the exciton-photon base (instead of a simplified polariton base) brings the advantage of taking into account the TE-TM splitting effects, the angular dependence of the polariton interaction constants, and the influence of the incoherent exciton reservoir.

In the simplest case of circularly polarized external pump the (2.1), (2.2) can be written for scalar amplitudes of excitonic and electric fields $\mathscr{P}$ and $\mathscr{E}$ with the same circular polarization. This corresponds to the experimental fact that in the case of circular polarized pump, the polariton scattering into states with opposite circular polarization is small and can be neglected [22]. Then the exciton energy shift in (2.2) can be written as a simple scalar blue shift, proportional to exciton intensity, $\delta\omega_x(\mathscr{P},\ldots) = V|\mathscr{P}|^2$. As a result, (2.2) takes the form of a nonlinear Shrödinger or Gross-Pitaevskii equation for exciton field [15].

The numerical solution of (2.1), (2.2) in the scalar approximation for long (≈ 1 ns) nearly rectangular pump pulses (shown as the dash-dotted line in Fig. 2.4a) demonstrates a sharp threshold-like transition from the picture which agrees with the above simple model of four-wave mixing to a completely different one (see Fig. 2.5). Note that the difference in the pump intensities between the panels (a) and (b) is $\approx 1\%$ only. The integrated intensity of the scattered polaritons increases above transition by several orders of magnitude (note the logarithmic vertical scale in Fig. 2.4).

Figure 2.4 explains the time kinetics of the scattering above the threshold pump intensity. The shape of excitation pulse $|\mathscr{F}|^2$ (dash-dotted line in Fig. 2.4a) was chosen so that the threshold intensity is approached slowly during pulse duration. Actually two sharp transitions occur (at $t \approx 600$ and 700 ps). Both transitions are characterized by jumps of driven mode amplitude $\mathscr{P}_p$ itself (see the solid line in Fig. 2.4a).

In order to understand these transitions, one has to investigate the stability of solutions of (2.1), (2.2) in the case of stationary external field $\mathscr{F}_p(t) =$ const with only one macroscopically filled pumped mode, i.e., of the form $\mathscr{P}(k,t) = \tilde{\mathscr{P}}(k,t) + \delta_{k,k_p}\mathscr{P}_p e^{-i\omega_p t}$, $\mathscr{E}(k,t) = \tilde{\mathscr{E}}(k,t) + \delta_{k,k_p}\mathscr{E}_p e^{-i\omega_p t}$, where $|\tilde{\mathscr{P}}/\mathscr{P}_p|$, $|\tilde{\mathscr{E}}/\mathscr{E}_p| \ll 1$. In the zero order, we get the following cubic equation for the amplitude of the driven mode $\mathscr{P}_p$,

$$\left[(\omega_p - \omega_c)(\omega_p - \omega_x - V|\mathscr{P}_p|^2) - A\beta\right]\mathscr{P}_p = A\alpha\mathscr{F}_p. \tag{2.3}$$

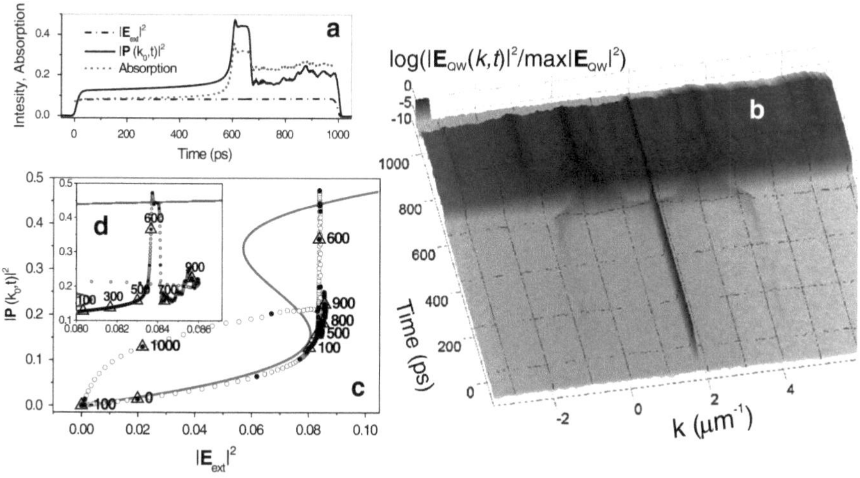

Fig. 2.4 The calculated time dependences of input excitation pulse (*dash-dotted line*), of excitonic field (*solid line*) and absorption coefficient (*dots*) (**a**), and of angular spectra of the scattered light (**b**). Excitonic field vs pump intensity during the excitation pulse (**c**); inset shows the magnified part of the region of transitions. The time delay in panel (**c**) between empty, black, and triangled dots is 1, 10, and 100 ps, respectively; the triangles are labeled by time in ps. Solid S-shaped curve is the solution of (2.3)

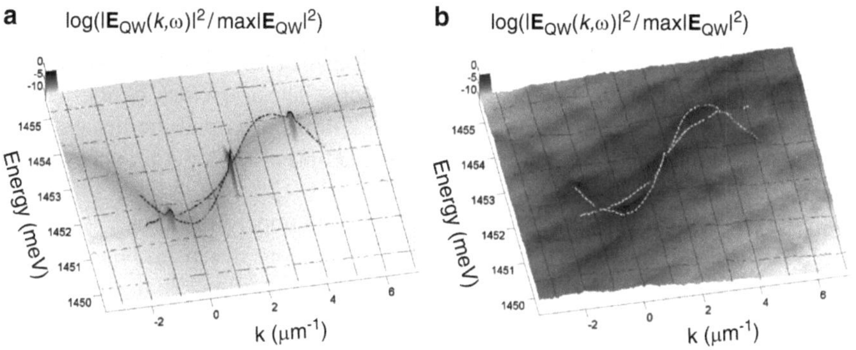

Fig. 2.5 The calculated time-integrated energy-wave vector spectra of the scattered light from MC when the pump pulse amplitude is slightly below (**a**) and above (**b**) the threshold. *Dotted lines* are the signal and idler branches from Fig. 2.3a calculated at the parametric scattering threshold $|\mathscr{P}_p|^2 = 0.14$

The solution of (2.3) is shown as the solid S-shaped line in Fig. 2.4c. Such S-shape is well known in the theory of interacting excitons [23] and nonlinear cavities [24, 25], it brings *bistability* into the behavior of MC polaritons that was first demonstrated in the MC reflectivity under normal light incidence [26]. The bistability occurs under the resonant pumping of the polariton mode at $\mathbf{k} = \mathbf{k}_p$ and changes significantly the scenario of the polariton–polariton scattering.

In order to understand these changes, one has to analyze the stability conditions, linearizing (2.1), (2.2) near the solutions with one macrooccupied mode. This is described [15, 27] by the linear eigenproblem for parametrically coupled signal $\tilde{\mathscr{E}}, \tilde{\mathscr{P}}(k,t) = \tilde{\mathscr{E}}, \tilde{\mathscr{P}}(k)\mathrm{e}^{-\mathrm{i}\omega t}$ and idler $\tilde{\mathscr{E}}^*, \tilde{\mathscr{P}}^*(\bar{k},t)\mathrm{e}^{-2\mathrm{i}\omega_{\mathrm{p}}t} = \bar{\tilde{\mathscr{E}}}, \bar{\tilde{\mathscr{P}}}(\bar{k})\mathrm{e}^{-\mathrm{i}\omega t}$ (where $\bar{k} = 2k_{\mathrm{p}} - k$),

$$\hbar\omega \begin{pmatrix} \tilde{\mathscr{E}}(k) \\ \tilde{\mathscr{P}}(k) \\ \bar{\tilde{\mathscr{E}}}(\bar{k}) \\ \bar{\tilde{\mathscr{P}}}(\bar{k}) \end{pmatrix} = \hat{H}_{\mathrm{eff}} \begin{pmatrix} \tilde{\mathscr{E}}(k) \\ \tilde{\mathscr{P}}(k) \\ \bar{\tilde{\mathscr{E}}}(\bar{k}) \\ \bar{\tilde{\mathscr{P}}}(\bar{k}) \end{pmatrix}, \tag{2.4}$$

$$\hat{H}_{\mathrm{eff}} = \hbar \begin{pmatrix} \omega_{\mathrm{c}}(k) & \beta & 0 & 0 \\ A & \omega_{\mathrm{x}} + 2V|\mathscr{P}_{\mathrm{p}}|^2 & 0 & V\mathscr{P}_{\mathrm{p}}^2 \\ 0 & 0 & 2\omega_{\mathrm{p}} - \omega_{\mathrm{c}}^*(\bar{k}) & -\beta^* \\ 0 & -(V\mathscr{P}_{\mathrm{p}}^2)^* & -A^* & 2\omega_{\mathrm{p}} - \omega_{\mathrm{p}}^* - 2V^*|\mathscr{P}_{\mathrm{p}}|^2 \end{pmatrix}.$$

The change of the sign (from the negative to positive one) of the imaginary parts of the eigenenergies of (2.4) indicates instability of the solution with a single macro-occupied mode. The idler and signal branches discussed above (see Figs. 2.5a,b, 2.3) are the eigenvalues of (2.4) calculated with $\mathscr{P}_{\mathrm{p}}^2 = 0.14$, the threshold value for stimulated scattering into $k_{\mathrm{s}} < 0$, $k_{\mathrm{i}} > 2k_{\mathrm{p}}$. Indeed, Fig. 2.4b shows that at $t < 600\,\mathrm{ps}$, the scattered intensity is maximum at predicted angles $k_{\mathrm{s}} < 0$, $k_{\mathrm{i}} > 2k_{\mathrm{p}}$. However, at $t \approx 600\,\mathrm{ps}$, instead of stimulated scattering into these modes, another type of instability develops. The trajectory of the system on the $\left[|\mathscr{P}(k_{\mathrm{p}},t)|^2, |\mathscr{F}(k_{\mathrm{p}},t|^2\right]$ plane (see Fig. 2.4c), shows that the instability at $t \approx 600\,\mathrm{ps}$ is the jump of the k_{p} mode between the lower and upper branches of the S-shaped curve.

In the empty cavity with quadratic dispersion, the upper branch of S is usually stable [24, 25]. In an MC with inflective LP dispersion, it can become unstable against parametric scattering. This can be seen from, e.g., Fig. 2.3c,d, showing the signal and idler branches calculated on the upper part of the S curve at $|\mathscr{P}_{\mathrm{p}}|^2 = 0.44$. Instead of damping, a large gain is realized for polariton modes in fairly wide ranges around $k_{\mathrm{s}} \gtrapprox 0$, $k_{\mathrm{i}} \lessapprox 2k_{\mathrm{p}}$. This results in fast developing stimulated scattering into these modes (see Fig. 2.4b). It is enhanced by increased absorption (see the dashed line in Fig. 2.4a). But because of the fast scattering into these modes, the pumped mode becomes unstable and eventually jumps back into a lower position (at $t \approx 700\,\mathrm{ps}$). Now, owing to the well-developed scattered noise in the system, this lower position becomes more or less stable, although subject to noisy fluctuations because of the already developed scattered states and the competition between them and the pumped mode.

Thus, the unusual behavior of the signal and idler in stimulated LP–LP scattering results from the interplay between two instabilities in the resonantly excited MC: the *single-mode* bistability of the pumped LP mode intensity with respect to the external pump, and the parametric instability *of the highly excited LP mode* with respect to the decay into scattered LPs in a *wide* range of $\mathbf{k}$ [15, 27–30]. The signal at $\mathbf{k}_s \approx 0$ appears as a result of *dynamic self-organization* of the scattered polaritons into a pair of new macrooccupied signal-idler modes $\mathbf{k}_s \approx 0,\ \mathbf{k}_i \approx 2\mathbf{k}_p$.

2.4 Kinetics of Stimulated Polariton Scattering: Hysteresis Behavior of LP–LP Scattering and Experimental Evidence for Dynamic Self-Organization

Direct evidence of a multimode self-organized nature of scattering was obtained from the studies of LP system dynamics under ns-long pump pulses with a spectral width of 0.7 meV [31, 32]. Due to spectral broadening of excitation, the shifts in polariton energy are reflected by temporal variations in the transmission energy spectrum. Thus, the technique makes it possible to visualize the temporal correlations between the resonance energy and intensity of the intra-cavity field. Such experiments allow to investigate the formation of signal at $k \approx 0$; strong hysteresis effects were discovered in the kinetics of the optical response of pumped as well as of scattered LPs [31, 32].

Figures 2.6a and b show the transmission spectra recorded at $\mathbf{k} = \mathbf{k}_p$ with a time interval of 100–200 ps at peak excitation intensity P smaller and larger P_{thr}, respectively. It is clearly seen that both the spectral position and the FWHM of the transmission signal recorded at the low excitation density change very weakly

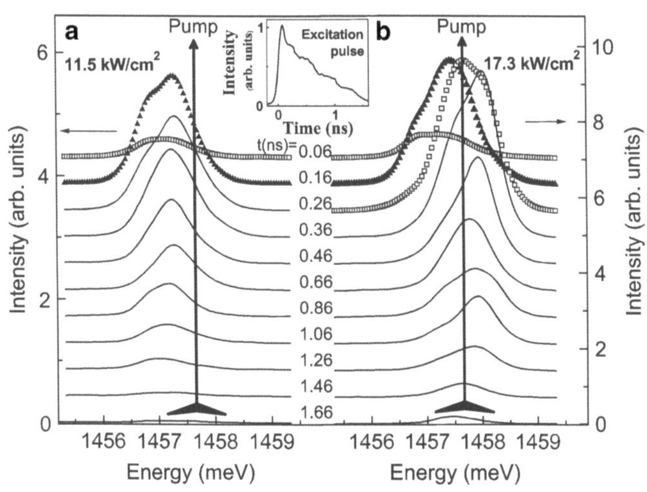

Fig. 2.6 Transmission spectra recorded at $\mathbf{k} = \mathbf{k}_p$ with time delays between 0.06 and 1.66 ns at the peak excitation densities $P = 11.3$ (**a**) and 17.3 kW/cm^2 (**b**). The spectral position of the excitation pulse is shown by black vertical arrows labeled as Pump. The pump pulse profile $I_p(t)$ is shown in the inset

during pulse duration. Meanwhile, the spectra recorded at high P exhibit a well-pronounced broadening and a strong blue shift of the signal. The former behavior is characteristic of a driven mode in a nearly linear regime, while the latter is the fingerprint of a highly nonlinear regime.

The typical time dependence of excitation pulse intensity $I_p(t)$ used in [31, 32] is shown in the inset to Fig. 2.6 (see also the solid line in Fig. 2.7a); the pump polarization is σ^+, and pump frequency is approximately 0.5 meV above the LP dispersion branch at $\mathbf{k}_p = (k_{px}, k_{py}) = (1.96, 0)\ \mu\text{m}^{-1}$. Figure 2.7 summarizes the measured scattering dynamics.

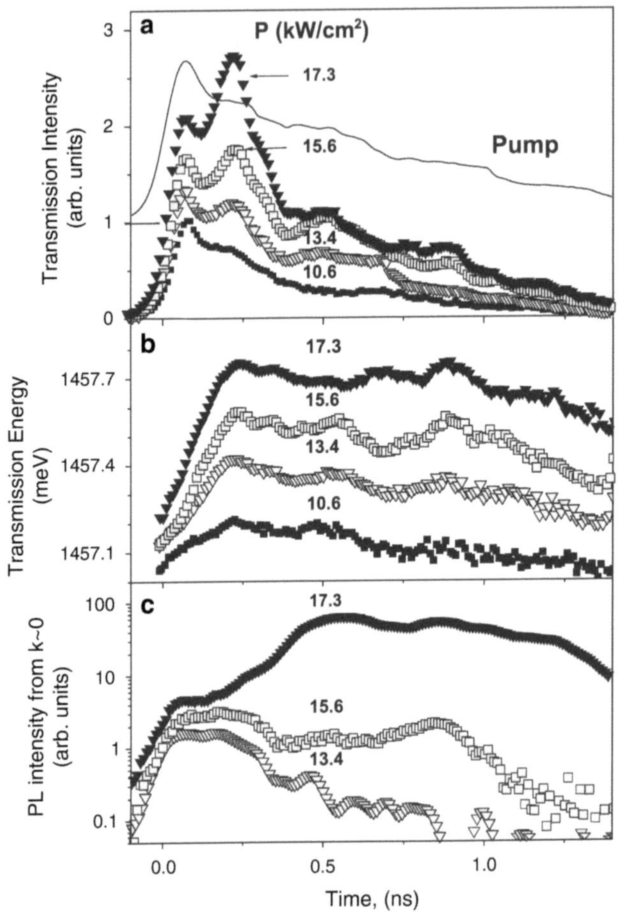

Fig. 2.7 Time dependences of the MC transmission intensity (**a**), spectral position of the transmitted signal (**b**), and the emission intensity at $\mathbf{k} = 0$ (**c**) recorded for several peak excitation intensities P (shown as *numbers near curves*) below and above the threshold of the stimulated parametric LP–LP scattering. Pump profile $I_p(t)$ is shown in panel (**a**) by a solid line labeled as Pump

The population dynamics of the driven LP mode (with $\mathbf{k} = \mathbf{k}_p$) was obtained from the MC transmission measurements. The active region in the cavity is separated from the detector by a Bragg mirror that does not introduce any nonlinearity and/or spectral selectivity. Because of that, the spectral intensity of the pump pulse, transmitted through the MC, $I_{tr}(\omega, t)$, is proportional to the squared magnitude of intracavity electric field $|\mathcal{E}(\omega, \mathbf{k}_p, t)|^2$ and, therefore, provides a direct information on its spectral and time dependences. Measuring of the spectra at different time delays with averaging over time intervals of ~ 70 ps enables one to obtain information both on the time dependence of $|\mathcal{E}(\omega, \mathbf{k}_p|^2$ and its blueshift due to the interparticle interaction in the photoexcited LP system: The time dependence of a spectrally integrated intensity $\int I_{tr}(\omega, t)\mathrm{d}\omega = I_{tr}(t)$ is proportional to $|\mathcal{E}(\mathbf{k}_p, t)|^2$, whereas the time dependence of the first momentum

$$\tilde{E}_{LP}(t) = \left[\int I_{tr}(\omega, t)\mathrm{d}\omega\right]^{-1} \int \hbar\omega I_{tr}(\omega, t)\mathrm{d}\omega, \tag{2.5}$$

provides information on the dynamics of the average energy of the pumped LP mode.

The time dependence of MC transmission signal $I_{tr}(\mathbf{k}_p, t) \propto |\mathcal{E}(\mathbf{k}_p, t)|^2$ is shown in Fig. 2.7 along with the pump profile $I_p(t)$ for various peak excitation intensities $P = \max I_p(t)$. At small $P \lesssim 10.7\,\mathrm{kW/cm^2}$ $|\mathcal{E}(\mathbf{k}_p, t)|^2$ is a superlinear function of I_p on both the ascending and descending parts of the excitation pulse. However, the monotonous dependence $I_{tr}(I_P)$ is violated with increasing P, namely, I_{tr} starts showing a very narrow peak in the range of nearly constant exciting field at $t \sim 0.2$ ns indicating the instabilities in $|\mathcal{E}(\mathbf{k}_p, t)|^2$. The peak grows fast with P and shifts slightly towards the pulse onset. Measurements of the spectral position of the maximum of the transmission signal showed that it changes very weakly during the pulse duration at $P < 10.7\,\mathrm{kW/cm^2}$, whereas a further increase in P results in its strong blueshift (Fig. 2.7b). Such a behavior is indicative of a nearly linear regime of the driven mode at low P and a highly nonlinear regime at high P.

It is revealing to redraw the experimental relationship between $I_{tr}(t)$ and $I_p(t)$ as the implicit function $I_{tr}(I_p)$ (equivalent to the dependence $|\mathcal{E}(\mathbf{k}_p)|^2$ as a function of $|\mathcal{F}|^2$) (see Fig. 2.8a). It can be seen that already at $P = 10.7$ kW/cm^2, I_{tr} as a function of I_p demonstrates a weak hysteresis. The hysteresis magnitude grows quickly with increasing P, which indicates a critical transformation of the driven LP mode, related to its bistability. The well-pronounced hysteresis behavior at high P also appears in the dependence of the blueshift vs I_p, as shown in Fig. 2.8b. The vertical width of the hysteresis loop is less than 0.1 meV at $P = 10.7\,\mathrm{kW/cm^2}$ increasing to ~ 0.6 meV at $P = 17.3\,\mathrm{kW/cm^2}$. Note, however, the significant difference in the hysteresis loops in $I_{tr}(I_p)$ and $\tilde{E}_{LP}(t)(I_p)$. The sharp decrease in $I_{tr}(I_p)$ is followed by a rather weak change in $\tilde{E}_{LP}$. That indicates the appearance of a strong contribution to $\tilde{E}_{LP}$ made by polaritons scattered from the driven mode during highly efficient stimulated LP–LP scattering.

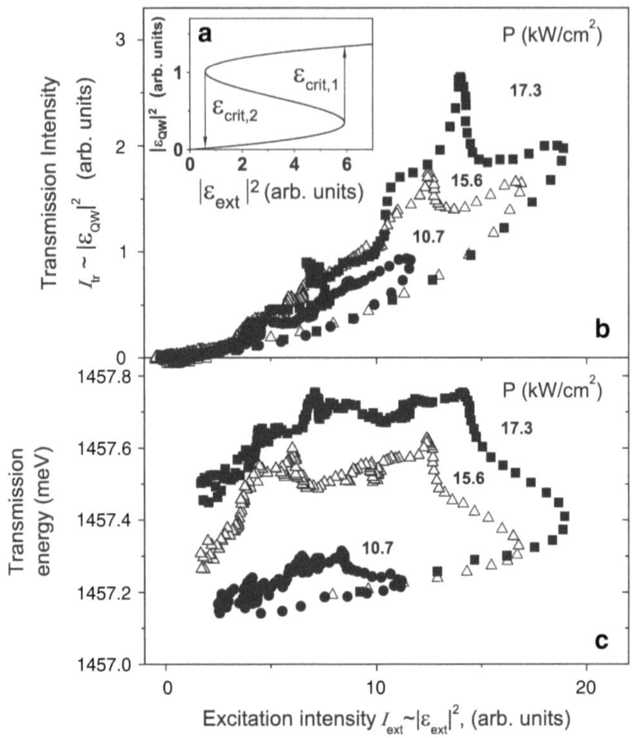

Fig. 2.8 Measured dependences of intensity (**a**) and spectral position (**b**) of the MC transmission signal as a function of the external field intensity for excitation densities below, close to, and above the threshold for the stimulated parametric LP–LP scattering. The inset in (**a**) shows a typical single-mode response of the intracavity field on the external pump, calculated for a nonlinear exciton oscillator

In case of single driven mode instability, the hysteresis in the dependence of the driven mode intensity vs pump is expected to be maximum, i.e., between the upper and lower borders of the unstable part of S-curve $I_{\rm crit,1}$ and $I_{\rm crit,2}$, respectively (see the inset in Fig. 2.8a). The significant shrinking of this hysteresis (see Fig. 2.4), may be caused by the fast parametric scattering after the bistability transition of the driven mode.

The measured dynamics of the scattered $\mathbf{k} = 0$ harmonic of the electric field in QW *inside* the MC, found from the time dependences of the MC emission normal to its plane $I_{\rm s}(\mathbf{k} = 0, t) \propto |\mathscr{E}(\mathbf{k} = 0, t)|^2 \propto I_{\rm s}(0, t)$, is shown in Fig. 2.7c. It is clearly seen that $I_{\rm s}(0, t)$ and, hence, $|\mathscr{E}(0, t)|^2$ differs significantly from exciting pulse shape $I_{\rm p}(t)$. At low $P < 13.4\,\rm kW/cm^2$, the signal reaches its peak value at $t = 0.2{-}0.3$ ns, i.e., only somewhat later than the pumping pulse and then decreases fast, by an order of magnitude at $t \sim 0.35$ ns, at $I_{\rm p} \sim 0.5P$. At $P > 15.5\,\rm kW/cm^2$, the signal behavior changes drastically. After the marked decrease of the signal at $t = 0.2 - 0.35$ ns, it starts to grow and reaches its second peak at $t \sim 0.85$ ns

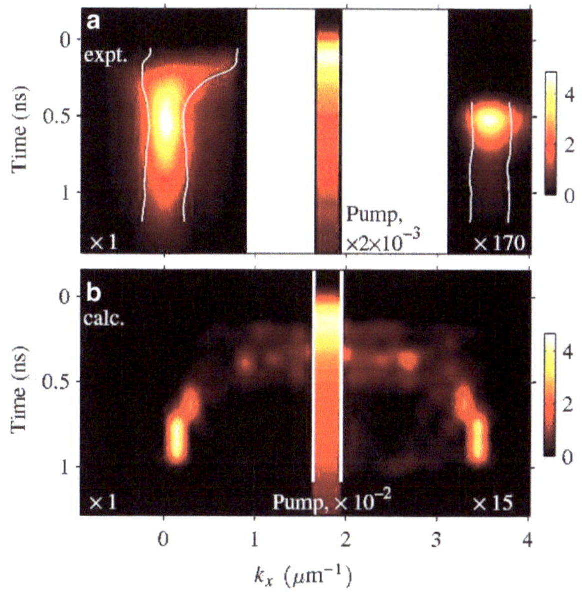

Fig. 2.9 Experimental (**a**) and calculated (**b**) time dependences of the k_x distribution of LP emission at $k_y = 0$ and pump intensity $P > P_{\mathrm{thr}}$. The pulse profile is shown at the excitation wave vector $k_x = 1.8\ \mu\mathrm{m}^{-1}$. The scale (in *arbitrary units*) is explained in color bars. k resolution is equal to 0.06 $\mu\mathrm{m}^{-1}$ at $k_x < 1\ \mu\mathrm{m}^{-1}$ and 0.2 $\mu\mathrm{m}^{-1}$ at $k_x > 3\ \mu\mathrm{m}^{-1}$. Time dependences of k_{HM} corresponding to $1/2 \max_{k_x} I_{\mathrm{PL}}(k_x, t)$ are displayed in the upper panels by white lines (i.e., the FWHM of the measured "signal" and "idler" k_x distributions can be seen between these lines)

when $I_{\mathrm{p}}(t)$ decreases already by more than twice. The intensity of this second peak increases threshold-like, by more than two orders of magnitude in the range of P between 14.9 and 17.2 kW/cm^2.

The measured dynamics of the wave vector distribution of the scattered LPs in a wide range of k is shown in Fig. 2.9a. The time evolution of the k-distribution at $k_y = 0$ for k_x from –0.9 to 0.9 $\mu\mathrm{m}^{-1}$ and from 3.1 to 4 $\mu\mathrm{m}^{-1}$ is recorded at $P = 1.4P_{\mathrm{thr}}$. The strong LP emission connected with LP–LP scattering appears only at $t > 0.35$ ns, just after the abrupt increase in $\mathscr{E}(\mathbf{k}_{\mathrm{p}})$ induced by the self-instability of the driven LP mode.

The scattered signal appears in the range of k_x between -0.4 and 0.8 $\mu\mathrm{m}^{-1}$ (with a maximum at $k_x \sim +0.7\ \mu\mathrm{m}^{-1}$) and then its maximum shifts fast, during 0.15 ns, to $k \sim 0$. The growth of the signal at $k \sim 0$ on $t \sim 0.2 - 0.5$ ns is followed by a strong narrowing and the appearance of a narrow idler at $\mathbf{k} \sim 2\mathbf{k}_{\mathrm{p}}$. Note that the signal states from $\mathbf{k} = -0.7$ to $0.8\,\mu\mathrm{m}^{-1}$ at $t \sim 0.3$ ns are already in the range of stable states of the three-mode OPO [33]. Nevertheless, the signal shifts with time towards $\mathbf{k} = 0$. Hence, the simplified three-mode OPO model is not sufficient to describe the scattering evolution: the development of the signal at an initially less populated state $k \sim 0$ compared to $k_x \sim 0.7\ \mu\mathrm{m}^{-1}$ indicates that the selection of the final OPO states is determined by multiple scattering of LPs into the whole range of populated k-states. The stimulation of the scattering into the signal and idler states by their macrooccupation leads to a strong hysteresis in their population vs that of the driven mode [31]. The decay of the stimulated signal at $\mathbf{k} = 0$ starts only at $t > 1.2$ ns.

Numerical simulations using the time dependence of the excitation pulse taken from the experiment show that this behavior can be qualitatively described within the framework of the semiclassical model, (2.1), (2.2) [15, 27, 30]. For example, Fig. 2.9b shows the calculated time evolution of the k distribution of LPs which reproduces qualitatively the experimental panel (a): the simulations predict (1) the filling of a wide range of k-states with a maximum at positive k just after the jump of the internal electromagnetic field at $\mathbf{k}_\mathrm{p}$ and (2) subsequent formation of OPO states with a dominating signal and idler at $\mathbf{k} \sim 0$ and $2\mathbf{k}_\mathrm{p}$.

Comparison of the experiment and numerical simulations elucidates the nature of the observed jumps and hysteresis behavior in the dynamics of the excited LP system. The hysteresis of I_tr vs I_p is initiated by the self-instability of the driven LP mode that leads to a fast increase of $\mathscr{E}(\mathbf{k}_\mathrm{p})$ and blueshift of the driven mode towards the pump frequency. Even with decreasing I_P, the system retains a considerable population of scattered LPs. This keeps the driven mode close to the resonance with the pump frequency, which provides its strong excitation.

The reason for the unusual "hard" regime of the onset of stimulated polariton scattering is a stepwise increase of $\varepsilon_\mathrm{QW}(\mathbf{k}_\mathrm{p})$ to values well above the parametric scattering threshold, which occurs due to the bistability of the driven polariton mode. The strong overshoot of P_thr in MCs becomes possible due to the presence of the inflection point in the LP dispersion. It leads to a very high rate of the LP–LP scattering which provides an explosive growth of LP population in a wide k range and, consequently, a rapid decay of the driven mode. The latter causes the strong modification of the hysteresis loop in I_tr vs I_P and gives rise to the hysteresis of I_S vs I_tr observed in the same experiment.

The *eventual* quasi-three-mode state with $\mathbf{k} = \sim 0, \mathbf{k}_\mathrm{p}, \sim 2\mathbf{k}_\mathrm{p}$ occurs as an essentially collective formation. The OPO state forms as a result of a dynamical self-organization in the multimode scattering including many other weakly populated LP states along with the three dominating macrooccupied states. Thus, the final OPO state differs from the "three-mode" OPO solution discussed in [15]. It can be numerically shown that the actual stability of the three-mode pattern is maintained by the presence of numerous weak above-condensate modes so that the whole system appears to be highly correlated; i.e., it demonstrates a new—dynamically self-organized—type of OPO.

However, a careful examination of the measured hysteresis effects in Figs. 2.8a,c shows [32] a significantly more pronounced hysteresis of the blueshift as compared to that of the transmission intensity (panels (b) and (a) in Fig. 2.8, respectively). This difference cannot be explained within the model [15, 27, 30], based on (2.1), (2.2) and, hence, dealing only with coherent polariton scattering into optically active states. The scattering into incoherent exciton states (localized and/or with large wavevectors), not taken into account in this model, seems to play an important role.

Filling of such relatively long-living optically inactive states may explain the fast shrinking of the $I_{tr}(I_p)$ hysteresis loop, while the blueshift which integrates all the occupied exciton states is still demonstrating a well-pronounced hysteresis (see next section).

2.5 Polariton Multistability in a "Spinor" Polariton System

In the previous, it has been found that the "scalar" approximation of (2.1), (2.2) neglecting the polariton spin degrees of freedom describes satisfactorily the instabilities and hysteresis effects in the transmission and scattering signal intensities in MCs excited with a circularly polarized light. It is obvious that the response of the optically excited polariton system becomes more complicated in the case of arbitrary optical polarization when the exciton spin degrees of freedom are allowed. In a general case, one can expect that the system can have several stable states under a given cw pump, whereas the state actually chosen by the system is determined by the history of the excitation process [9,21,34,35]. In this section, for simplicity, we consider the case of pumping at $\mathbf{k}_p = 0$ (a single-mode case).

Figure 2.10a–e compare the measured temporal dependences of transmission of MC with detuning of 1 meV under excitation with linearly, elliptically, and circularly polarized pulses with degrees of circular polarization (DCP) 0, 0.2, 0.6, 1 at $P = 28.5\,\mathrm{kW/cm^2} > P_{thr}$ normal to the QW plane ($\mathbf{k} = 0$). The temporal delay of the dominating σ^+ and π_x polarization components of the signal with respect to

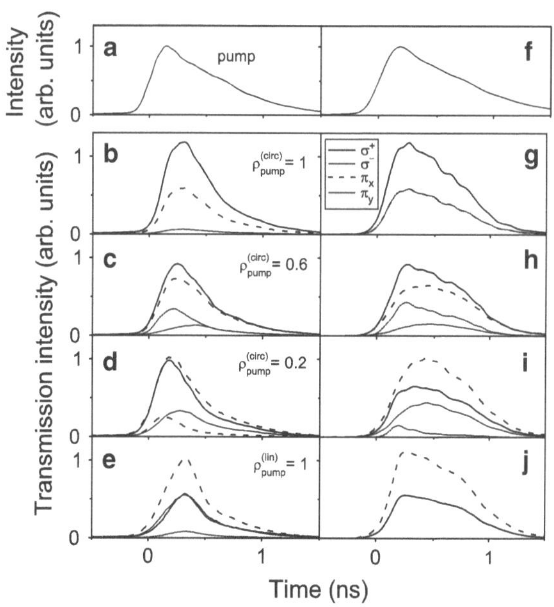

Fig. 2.10 Time dependences of the cavity transmission intensity, in circular and linear polarization components ($\sigma^{\pm}$ and $\pi_{x,y}$, respectively) for different pump polarizations $\rho_p^{(circ)}$. The uppermost panels show the time shape of exciting pulse. Left and right panels represent experimental and calculated data, respectively. In the experiment $P = 28.5\,\mathrm{kW/cm^2}$

the pump peak is seen to be in the range of 150–200 ps for all DCPs. The increase in I_{tr} in the range of decreasing pump power (after the pump peak is already passed) is provided by the blue-shift of effective polariton energy.

Figures 2.10b and e show that the transmission signal retains both the circular and linear pump polarizations; in such cases, only a weak depolarization of the transmitted light is observed. Retaining of circular polarization agrees with the angular momentum conservation law. On the other hand, under elliptically polarized excitation $0 < \rho_p^{(circ)} < 1$, the ratio between both circularly and linearly polarized components of the signal intensity ($\sigma^{\pm}$ and $\pi_{x,y}$, respectively) varies with time (Fig. 2.10c,d). The dominant σ^+ component increases relatively to σ^-, whereas the dominant π_y component decreases relatively to π_x at the pulse onset. However, these changes are not stable: reverse changes are observed after several tens of picoseconds, and the ratios of the components in both circular and linear polarizations return eventually to the initial values close to their values in the pump pulse. The changes in $\rho_{circ}^{tr}(t)$ and $\rho_{lin}^{tr}(t)$ are in antiphase for any $\rho_p^{(circ)}$ and excitation densities (Fig. 2.11).

The increase in $\rho_{lin}^{tr}(t)$ of the transmitted pulse at the pulse onset is easily explained by the model based on vectorial (2.1), (2.2). This model suggests that

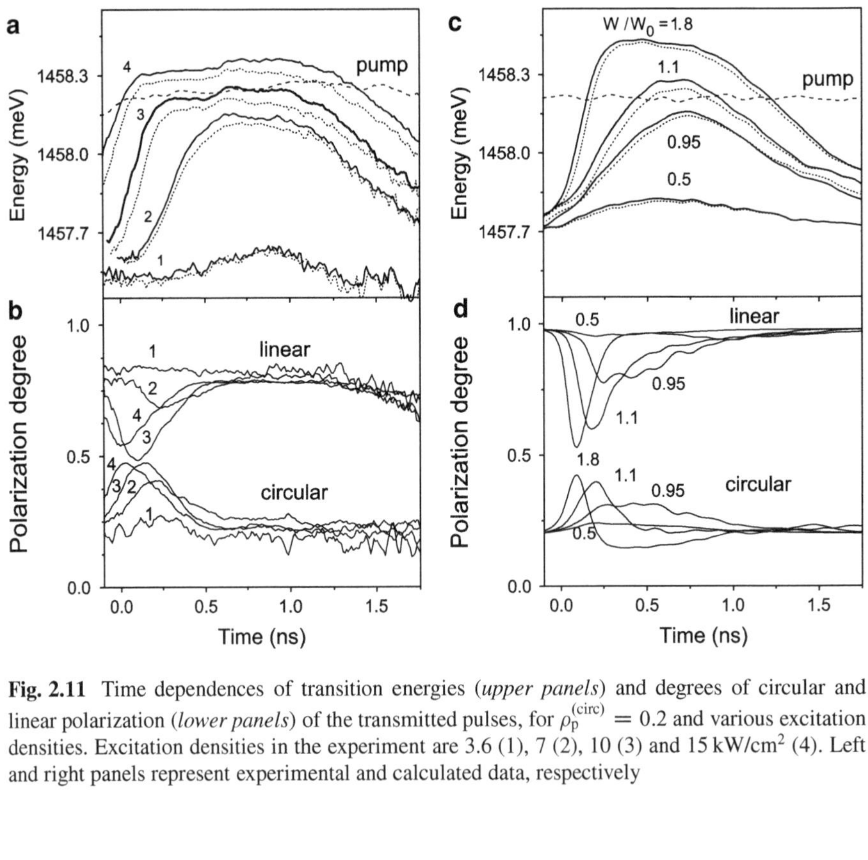

Fig. 2.11 Time dependences of transition energies (*upper panels*) and degrees of circular and linear polarization (*lower panels*) of the transmitted pulses, for $\rho_p^{(circ)} = 0.2$ and various excitation densities. Excitation densities in the experiment are 3.6 (1), 7 (2), 10 (3) and 15 kW/cm^2 (4). Left and right panels represent experimental and calculated data, respectively

the nonlinear blue shift of polaritons is described by the coherent exciton field only. It can be characterized by the contact interaction constants between excitons with same (V_1) and opposite (V_2) circular polarizations. Taking this into account, one immediately arrives to a Gross-Pitaevskii equation for spinor two-component (σ^+,σ^-) exciton polarization with two constants V_1 and V_2 instead of one $V = V_1$ used above for description of the scalar system. For instance, the blue shift of the $\mathscr{P}_+$ component in this approximation reads as $\delta\omega_x^+ = V_1|\mathscr{P}_+|^2 + V_2|\mathscr{P}_-|^2$. In the polariton system in GaAs-based MCs, V_2 is negative (the excitons with opposite spins are attracted to each other) and $|V_2/V_1| \ll 1$ [22, 36]. More studies on the polariton interaction constants can be found in [37–41].

As $|V_2/V_1| \ll 1$, the repulsive interaction between polaritons with the same J_z values is larger than with the opposite ones, and hence, an increase in the pump density is accompanied by a higher increase in the effective frequency of the polariton resonance (see (2.5)) $\tilde{E}^+_{\mathrm{LP}}$ in the dominant σ^+ polarization. This increase is well seen in Fig. 2.11a. The positive feedback between $\tilde{E}^+_{\mathrm{LP}}$ and the magnitude of electric field in σ^+ polarization explains the instability of the system and the fast increase in $\rho^{\mathrm{tr}}_{\mathrm{circ}}$ in the pumped mode. The decrease in $\rho^{(\mathrm{circ})}_{\mathrm{p}}$ observed just after its jump up correlates with the sharp blueshift of the spectral maximum in the minor σ^- polarization. Indeed, Fig. 2.11a shows that $\tilde{E}^-_{\mathrm{LP}}$ increases only with a small delay compared to $\tilde{E}^+_{\mathrm{LP}}$ and rapidly reaches values close to values of $\tilde{E}^+_{\mathrm{LP}}$. As a result, the delaying σ^- component of the electric field in the QW increases along with decreasing degree of circular polarization of the passed pulse. However, the increase in $\tilde{E}^-_{\mathrm{LP}}$ cannot be explained by the model based on (2.1), (2.2), where only the coherent field mode on the quantum well is taken into account. Indeed, at $V_2 \lesssim 0$ an increase in $\rho^{\mathrm{tr}}_{\mathrm{circ}}$ should lead to a decrease rather than an increase in $\tilde{E}^-_{\mathrm{LP}}$.

Figure 2.11a demonstrates another feature of the time dependence of the spectral position of the transmission maximum that cannot be explained within the framework of the coherent model based on (2.1), (2.2). That is, its very weak dependence on t at $t = 0.5 - 1$ ns, where the transmitted signal and, hence, $|\mathscr{E}(\omega, \mathbf{k}_{\mathrm{p}}, t)|^2$ decreases by almost an order of magnitude. The characteristic time of the relaxation of $\tilde{E}^{\pm}_{\mathrm{LP}}$ to the initial (unperturbed) value is about 0.3 ns. Such time is characteristic of long-lived exciton states in the exciton reservoir rather than of polaritons near the LP band bottom which is by two orders of magnitude shorter. Thus, the exciton reservoir gives a dominant rather than negligible contribution to the blueshift in spite of the fact that the pump energy is slightly below the exciton level, and hence, the reservoir states have to be taken into account for description of the resonantly excited polariton system.

Recently, there have been developed models that introduce phenomenologically transitions of optically driven excitons into the incoherent exciton reservoir state in which the overall pseudospin is relaxed so that the reservoir provides equal blueshifts for both polarization components in the coherent state [11, 42]. The equations for intracavity electric field $\mathscr{E}$ and exciton polarization $\mathscr{P}$ in the circularly polarized basis coupled with the integral population of reservoir $\mathscr{N}$ are written as:

$$i\dot{\mathscr{E}}_+ = (\omega_c - i\gamma_c)\,\mathscr{E}_+ + \alpha\mathscr{F}_+ + \beta\mathscr{P}_+, \tag{2.6}$$

$$i\dot{\mathscr{E}}_- = (\omega_c - i\gamma_c)\,\mathscr{E}_- + \alpha\mathscr{F}_- + \beta\mathscr{P}_-, \tag{2.7}$$

$$\begin{aligned} i\dot{\mathscr{P}}_+ = {} & \big[\omega_x + V_1|\mathscr{P}_+|^2 + V_2|\mathscr{P}_-|^2 + (V_1 + V_2)\,\mathscr{N}/2 \\ & - i\left(\gamma_x + \gamma_{xr} + V_r|\mathscr{P}_-|^2\right)\big]\mathscr{P}_+ + A\mathscr{E}_+, \end{aligned} \tag{2.8}$$

$$\begin{aligned} i\dot{\mathscr{P}}_- = {} & \big[\omega_x + V_1|\mathscr{P}_-|^2 + V_2|\mathscr{P}_+|^2 + (V_1 + V_2)\,\mathscr{N}/2 \\ & - i\left(\gamma_x + \gamma_{xr} + V_r|\mathscr{P}_+|^2\right)\big]\mathscr{P}_- + A\mathscr{E}_-, \end{aligned} \tag{2.9}$$

$$\begin{aligned} \dot{\mathscr{N}} = {} & -\gamma_r\mathscr{N} + 2\gamma_{xr}\left(|\mathscr{P}_+|^2 + |\mathscr{P}_-|^2\right) \\ & + 4V_r|\mathscr{P}_+|^2|\mathscr{P}_-|^2. \end{aligned} \tag{2.10}$$

This model is a generalization of (2.1), (2.2). Here, $omega_{c,x}$ are real-valued and $\gamma_{c,x}$ are decay rates of the intracavity photon and exciton, γ_{xr} is an additional decay rate of excitons that corresponds to the linear mechanism of light absorption, providing the term $2\gamma_{xr}(|\mathscr{P}_+|^2 + |\mathscr{P}_-|^2)$ in (2.10); V_r stands for the rate of the nonlinear interaction between $\mathscr{P}_\pm$ which provides additional occupation of the reservoir ($4V_r|\mathscr{P}_+|^2|\mathscr{P}_-|^2$ per unit time) due to the mixture of excitons with opposite polarizations; and γ_r stands for the reservoir own decay rate.

Microscopically, the nonlinear absorption of cross circularly polarized excitons occurs due to the scattering of a pair of bright excitons with opposite spins ($J_z = -1$ and $J_z = +1$) into dark excitons ($J_z = -2$ and $J_z = +2$), which is closely related to biexciton creation (see [38] for details). The model does not consider reverse transitions of incoherent (reservoir) excitons into the driven mode (induced scattering). Thereby, this approximation is valid as long as the occupation of the reservoir states is relatively small.

Equations (2.6–2.10) can be easily generalized to the case of a multimode system with exactly the same nonlinear interaction terms (corresponding to the contact exciton–exciton interaction in real space), its linear part in the k-space being the same as in [21]. In the many-mode calculations represented in Figs. 9–11, the cavity dispersion $E_c^{TE,TM}(k)$ and response coefficients $\alpha_{TE,TM}(k)$, $\beta_{TE,TM}(k)$ for the TE and TM cavity modes are properly taken into account using the transfer matrix technique [43].

The numerical solutions of (2.8–2.10) carried out with negative constant V_2 of the contact interaction between excitons with opposite circular polarizations are shown in the right panels of Figs. 2.10–2.11. It is seen that they reproduce qualitatively all the experimental features discussed above. The parameters used in the simulations, $V_2/V_1 = -0.1$, $V_r/V_1 = 7\times10^{-3}$, $\hbar\gamma_r = 2\times10^{-3}$ meV, and $\gamma_{xr} = 3\times10^{-3}$ meV, are chosen to meet the following experimentally established conditions: (1) the overall occupation of the reservoir is comparable to that of the driven (optically active) polariton mode, so that $P_{thr}^{(lin)} \lesssim P_{thr}^{(circ)}$, (2) the temporal peak of the signal can be delayed by one or two hundreds of picoseconds with respect to the excitation peak, and (3) the decay time of the reservoir states is of the order of $\sim 300\,\text{ps}$, which is characteristic of excitons localized due to fluctuations of the QW potential and/or free excitons with large lateral wave numbers.

Comparison of the calculated dependences of the LP mode blueshift and DCP in Fig. 2.11c and d shows clearly that the increase in ρ_p^{circ} is due to the instability leading to the jump in the effective resonance frequency of polaritons in the dominant σ^+ polarization. The subsequent return of the system into the original polarization state is due to the occupation of the exciton reservoir; in view of the fast spin relaxation of incoherent excitons, this occupation leads to equal shifts for both circularly polarized components of the macro-occupied mode of the field in the quantum well.

Finally, let us consider the polarization properties of the signal of parametric scattering under pumping at the magic angle ($k_p \approx 1.8\,\mu\text{m}^{-1}$). Here, two points deserve a discussion. First, in the general case of elliptically polarized, excitation the signal is expected to reflect variations in the DCP of the driven mode that are

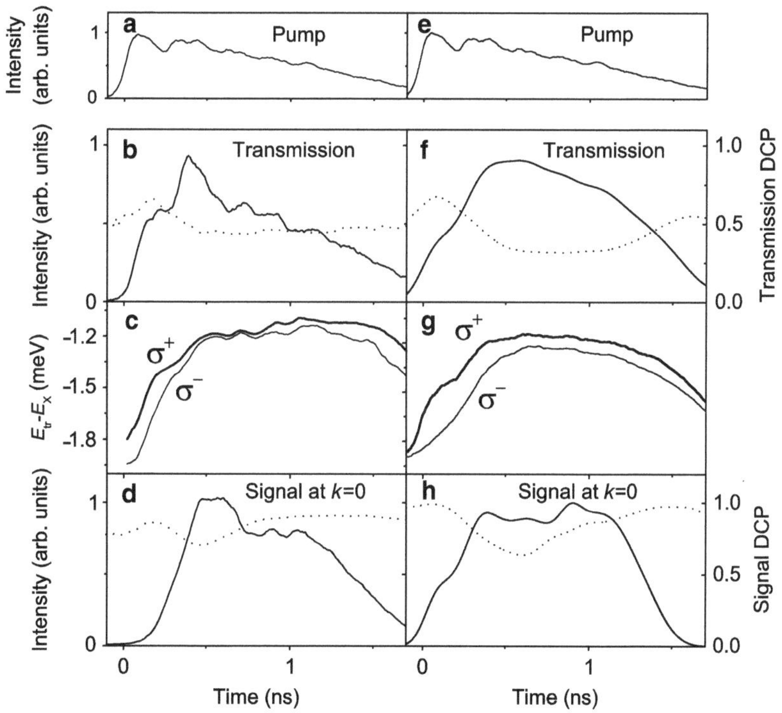

Fig. 2.12 Time dependences of the pump pulse intensity (**a, e**), transmission intensity (*solid line*), and DCP (*dotted line*) (**b, f**), weighted mean values of the transmission energy $\bar{E}_{\pm}$ in two polarization components σ^+ and σ^- (*thick and thin lines*, resp.) (**c, g**), OPO signal intensity (*solid line*), and DCP (*dotted line*)(**d, h**), for $P > P_{thr}$. Straight lines in panels (**c, i**) indicate the mean value of pump energy. Left and right panels represent experimental and calculated data, respectively. In the experiment $P = 18\,\text{kW/cm}^2$

strongly affected by the reservoir. Second, the signal polarization axis under the linearly polarized pump is rotated by 90° [44]. Thus, one should check if (2.6–2.10) can reproduce this effect in spite of the reservoir-induced interaction between cross-circularly polarized excitons.

The scattering is investigated at negative photon-exciton detuning ($\Delta \approx -1.5$ meV) with a view to make the pump frequency close to that used for normal incidence pumping considered above.

Figure 2.12 represents measured and calculated dynamics of intra-cavity field at $k = k_p$ (which corresponds to the cavity transmission) and $k = 0$ (OPO signal) under elliptically polarized excitation with DCP $= 0.5$ for $P > P_{thr}$. The behavior of the driven mode is similar to that under normal incidence pumping. Indeed, Fig. 2.12b shows a sharp increase in the transmission intensity at $t = 0.1$–0.4 ns. The growth of the field amplitude in the driven mode is first followed by an increase in its DCP (Fig. 2.12c), which is followed by a fast reverse change. This occurs as soon as the system enters the above-threshold region where the energies of $\sigma^{\pm}$ components of intracavity field (Fig. 2.12d) are partially leveled due to the reservoir filling.

Figure 2.12 shows that the OPO signal develops with a delay of about 0.1 ns compared to that in the driven mode. The domination of the leading polarization component (σ^+) is strongly enhanced in this signal, its DCP reaching ~ 0.85 at $t \approx 0.2$ ns. The lowering of the DCP of the driven mode at $t > 0.2$ ns makes the OPO signal lose its polarization significantly; however, the signal restores a high DCP during further evolution accompanied by the decrease in transmission intensity.

The simulation of the time dependences of the driven and signal mode intensities, DCPs, and blue shifts is shown in the right panels of Fig. 2.12. It is seen that the calculated dependences reproduce qualitatively the measured instability peculiarities observed when studying both of the driven and signal modes.

Let us now turn to the case of an exactly linear polarization of the excitation at $k_p > 0$ parallel to that of the TM cavity mode. The dynamics of the OPO signal at $P > P_{thr}$ is shown in Fig. 2.13. In the range of high signal intensity, the DLP

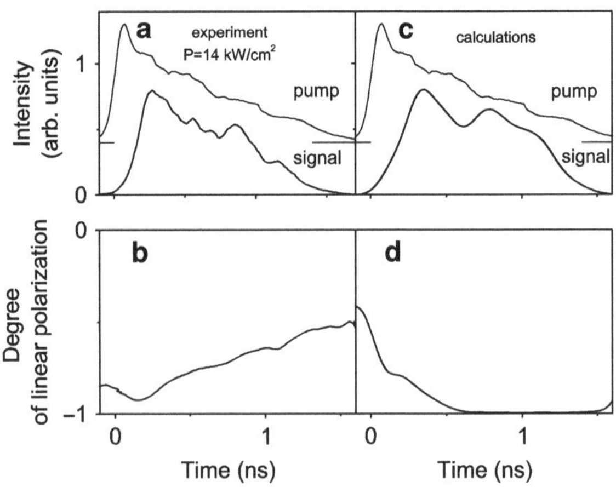

Fig. 2.13 Time dependences of the OPO signal intensity (**a**, **c**) and degree of linear polarization (**b**, **d**) under the pump $\vec{\mathscr{F}}(t)$ polarized linearly along the TM cavity mode ($\vec{\mathscr{F}} \parallel \vec{\mathscr{E}}_{TM}$), for $P > P_{thr}$. Left and right panels represent experimental and calculated data, respectively. Time dependences of the pump pulse intensity are shown in (**a**, **c**) by thin lines

of the signal reaches 90%, and the polarization axis is 90° rotated with respect to the pump (which corresponds to negative DLP values). Figure 2.13c,d show that the calculations using (2.5–2.9) reproduce both the signal dynamics and its high linear polarization.

Thus, the dynamics of the OPO signal confirms the important role of the reservoir. Under elliptic excitation, the $\sigma^{\pm}$ polarization components tend to be levelled by the reservoir. Note, that this effect by itself could be assigned to repulsion between cross circularly polarized excitons, $V_2 > 0$. On the other hand, the axis of signal polarization is found to be inverted with respect to the pump, which shows that the sign of V_2 is negative. Reproduction of these effects within (2.6–2.10) proves that they provide a self-consistent qualitative explanation for the correlation of the time dependences of the energies and amplitudes of the polarized field components in the QW and explains the coincidence of the instability thresholds for the cases of linearly and circularly polarized pump waves. At the same time, such a simplified model is insufficient for an exact description of the kinetics of photoexcited polaritons. For example, the calculation does not describe the partial depolarization of the signal observed in the experiment with linearly polarized pump light. Comparison of Fig. 2.11a and b indicates that the blue shift at the rear front of the pump pulse (at $t > 0.2$ ns) remains stabilized for a much longer time than that predicted by the model. Such discrepancies are expected, since (2.8–2.10) do not take into account the nonzero spin relaxation time in the reservoir and the possibility of reverse transition of excitons from the reservoir to the coherent state.

2.6 Conclusion

The paper addresses non-equilibrium transitions in a multistable system of exciton-polaritons in planar MCs under ns-long resonant (by frequency and angle) excitation at a normal to the MC plane and at the magic angle at $k = k_{\text{infl}}$. Excitation with spectrally broadened pulses allowed visualizing the temporal correlations between the effective resonant energy, intensity, and optical polarization of the intra-cavity field at $k = 0$ and k_{infl}, all of them undergoing the strong changes on reaching the threshold pump power. The widely used theoretical model based on the Gross-Pitaevskii equations written for purely coherent macrooccupied polariton states has been shown to explain qualitatively the instabilities only in a scalar polariton system excited resonantly with circularly polarized pulses. The hysteresis effects are due to multimode scattering when the self-instability of the nonlinear pumped polariton mode is accompanied by the explosive growth of the scattered polaritons population due to parametric instability of the highly excited polariton mode. The subsequent self-organization process in the nonlinear polariton system results in a new—dynamically self-organized—type of OPO.

However, this model fails to provide even a qualitative explanation of the polarization instabilities and hysteresis effects in the effectively spinor polariton system excited with linearly and/or elliptically polarized pulses. Most important is that the observed phenomena in principle are not described within a model considering only the two exciton–exciton interaction constants ($V_{1,2}$). The dynamics of the polariton system above P_{thr} has been found to be strongly affected by the long-lived exciton reservoir (excited by polariton scattering) which influences the blue shifts of the LP modes. This results in the qualitative changes in the development of the polarization instabilities in the driven mode and the OPO signal.

The proposed modified model based on (2.6–2.10) for describing the intra-cavity field dynamics accounts phenomenologically for the coupling of the macro-occupied polariton states with an exciton reservoir. In spite of some limitations, this model provides a self-consistent approach to the intra-cavity field dynamics under both pulse and continuous wave excitation conditions and a good qualitative description of the observed polarization instabilities and hysteresis effects both in the driven mode and OPO signal.

Authors thank A. Forchel, L.V. Keldysh, Yu.V. Kopaev, M.S. Skolnick, V.B. Timofeev, and D.M. Whittaker for fruitful discussions, and A.S. Brichkin, A.A. Demenev, A.A. Dorodnyi, A.V. Kavokin, D.N. Krizhanovskii, G. Malpuech, I.A. Shelykh, D.D. Solnyshkov, and A.I. Tartakovskii for collaboration. This work was supported by the Russian Foundation for Basic Research, the Russian Academy of Sciences, and the Russian Ministry of Science and Education.

References

1. C. Weisbuch, M. Nishioka, A. Ishikawa, Y. Arakawa, Phys. Rev. Lett. **69**(23), 3314 (1992).
2. Y. Yamamoto, T. Tassone, H. Cao, *Semiconductor Cavity Quantum Electrodynamics*, (Springer, Berlin, Heidelberg, 2000) pp. 4958
3. A.V. Kavokin, G. Malpuech, *Cavity Polaritons*, (Elsevier, Amsterdam, 2003)
4. J.J. Baumberg, P.G. Savvidis, R.M. Stevenson, A.I. Tartakovskii, M.S. Skolnick, D.M. Whittaker, J.S. Roberts, Phys. Rev. B **62**(24), R16247 (2000).
5. R.M. Stevenson, V.N. Astratov, M.S. Skolnick, D.M. Whittaker, M. Emam-Ismail, A.I. Tartakovskii, P.G. Savvidis, J.J. Baumberg, J.S. Roberts, Phys. Rev. Lett. **85**(17), 3680 (2000)
6. V.D. Kulakovskii, A.I. Tartakovskii, D.N. Krizhanovskii, N.A. Gippius, M.S. Skolnick, J.S. Roberts, Nanotechnology **12**, 475 (2001)
7. R. Butte, M.S. Skolnick, D.M. Whittaker, D. Bajoni, J.S. Roberts, Phys. Rev. B **68**(11), 115325 (2003)
8. D.N. Krizhanovskii, S.S. Gavrilov, A.P.D. Love, D. Sanvitto, N.A. Gippius, S.G. Tikhodeev, V.D. Kulakovskii, D.M. Whittaker, M.S. Skolnick, J.S. Roberts, Phys. Rev. B **77**(11), 115336 (2008).
9. N.A. Gippius, I.A. Shelykh, D.D. Solnyshkov, S.S. Gavrilov, Y.G. Rubo, A.V. Kavokin, S.G. Tikhodeev, G. Malpuech, Phys. Rev. Lett. **98**(23), 236401 (2007).
10. T.K. Paraïso, M. Wouters, Y. Léger, F. Morier-Genoud, B. Deveaud-Plédran, Nat. Mater **9**(8), 655 (2010).
11. D. Sarkar, S.S. Gavrilov, M. Sich, J.H. Quilter, R.A. Bradley, N.A. Gippius, K. Guda, V.D. Kulakovskii, M.S. Skolnick, D.N. Krizhanovskii, Phys. Rev. Lett. **105**(21), 216402 (2010).
12. C. Adrados, A. Amo, T.C.H. Liew, R. Hivet, R. Houdré, E. Giacobino, A.V. Kavokin, A. Bramati, Phys. Rev. Lett. **105**(21), 216403 (2010).
13. J. Kasprzak, M. Richard, S. Kundermann, A. Baas, P. Jeambrun, J.M.J. Keeling, F.M. Marchetti, M.H. Szymanska, R. Andre, J.L. Staehli, V. Savona, P.B. Littlewood, B. Deveaud, L.S. Dang, Nature **443**(7110), 409 (2006).

14. A.I. Tartakovskii, D.N. Krizhanovskii, V.D. Kulakovskii, Phys. Rev. B **62**(20), R13298 (2000)
15. N.A. Gippius, S.G. Tikhodeev, V.D. Kulakovskii, D.N. Krizhanovskii, A.I. Tartakovskii, Europhys. Lett. **67**(6), 997 (2004).
16. C. Ciuti, P. Schwendimann, B. Deveaud, A. Quattropani, Phys. Rev. B **62**(8), R4825 (2000)
17. C. Ciuti, P. Schwendimann, A. Quattropani, Phys. Rev. B **63**(4), 041303 (2001)
18. D.M. Whittaker, Phys. Rev. B **63**(19), 193305 (2001).
19. P.G. Savvidis, C. Ciuti, J.J. Baumberg, D.M. Whittaker, M.S. Skolnick, J.S. Roberts, Phys. Rev. B **64**(7), 075311 (2001).
20. N.A. Gippius, S.G. Tikhodeev, V.D. Kulakovskii, Phys. Stat. Sol. (c) **2**(2), 744 (2005)
21. S.S. Gavrilov, N.A. Gippius, S.G. Tikhodeev, V.D. Kulakovskii, JETP **110**(5), 825 (2010).
22. P. Renucci, T. Amand, X. Marie, P. Senellart, J. Bloch, B. Sermage, K.V. Kavokin, Phys. Rev. B **72**(7), 075317 (2005).
23. V.F. Elesin, Y.V. Kopaev, Sov. Phys. JETP **36**(4), 767 (1973)
24. W.J. Firth, A.J. Scroggie, Phys. Rev. Lett. **76**(10), 1623 (1996)
25. R. Kuszelewicz, I. Ganne, I. Sagnes, G. Slekys, M. Brambilla, Phys. Rev. Lett. **84**(26), 6006 (2000)
26. A. Baas, J.P. Karr, H. Eleuch, E. Giacobino, Phys. Rev. A **69**(2), 023809 (2004)
27. N.A. Gippius, S.G. Tikhodeev, J. Phys.: Condens. Matter **16**, S3653 (2004).
28. V.D. Kulakovskii, D.N. Krizhanovskii, A.I. Tartakovskii, N.A. Gippius, S.G. Tikhodeev, Phys. Uspekhi **46**(9), 967 (2003). [Uspekhi Fiz. Nauk **173**, 995 (2003)]
29. N.A. Gippius, S.G. Tikhodeev, L.V. Keldysh, V.D. Kulakovskii, Phys. Uspekhi **48**(3), 306 (2005). [Uspekhi Fiz. Nauk **175**, 327 (2005)]
30. S.S. Gavrilov, N.A. Gippius, V.D. Kulakovskii, S.G. Tikhodeev, Zh. Eksp. Teor. Fiz. **131**, 819 (2007). [JETP **104**, 715–723 (2007)]
31. A.A. Demenev, A.A. Shchekin, A.V. Larionov, S.S. Gavrilov, V.D. Kulakovskii, N.A. Gippius, S.G. Tikhodeev, Phys. Rev. Lett. **101**(13), 136401 (2008).
32. A.A. Demenev, A.A. Shchekin, A.V. Larionov, S.S. Gavrilov, V.D. Kulakovskii, Phys. Rev. B **79**(16), 165308 (2009).
33. D.M. Whittaker, Phys. Rev. B **71**(11), 115301 (2005).
34. I.A. Shelykh, T.C.H. Liew, A.V. Kavokin, Phys. Rev. Lett. **100**(11), 116401 (2008).
35. T.C.H. Liew, A.V. Kavokin, I.A. Shelykh, Phys. Rev. Lett. **101**(1), 016402 (2008).
36. K.V. Kavokin, P. Renucci, T. Amand, X. Marie, P. Senellart, J. Bloch, B. Sermage, Phys. Stat. Sol. (c) **2**(763) (2005)
37. M. Vladimirova, S. Cronenberger, D. Scalbert, K.V. Kavokin, A. Miard, A. Lemaître, J. Bloch, D. Solnyshkov, G. Malpuech, A.V. Kavokin, Phys. Rev. B **82**(7), 075301 (2010).
38. J. ichi Inoue, T. Brandes, A. Shimizu, Phys. Rev. B **61**(4), 2863 (2000).
39. M. Kuwata-Gonokami, S. Inouye, H. Suzuura, M. Shirane, R. Shimano, T. Someya, H. Sakaki, Phys. Rev. Lett. **79**(7), 1341 (1997).
40. D.N. Krizhanovskii, D. Sanvitto, I.A. Shelykh, M.M. Glazov, G. Malpuech, D.D. Solnyshkov, A. Kavokin, S. Ceccarelli, M.S. Skolnick, J.S. Roberts, Phys. Rev. B **73**(7), 073303 (2006).
41. M. Wouters, Phys. Rev. B **76**(4), 045319 (2007).
42. S.S. Gavrilov, A.S. Brichkin, A.A. Dorodnyi, S.G. Tikhodeev, N.A. Gippius, V.D. Kulakovskii, JETP Lett. **92**, 171 (2010).
43. S.G. Tikhodeev, A.L. Yablonskii, E.A. Muljarov, N.A. Gippius, T. Ishihara, Phys. Rev. B **66**(4), 045102 (2002)
44. D.N. Krizhanovskii, D. Sanvitto, A.P.D. Love, M.S. Skolnick, D.M. Whittaker, J.S. Roberts, Phys. Rev. Lett. **97**(9), 097402 (2006).

Chapter 3
Vortices in Spontaneous Bose–Einstein Condensates of Exciton–Polaritons

Benoit Deveaud-Plédran and Konstantinos G. Lagoudakis

Abstract One of the most striking quantum effects in an interacting Bose gas at low temperature is superfluidity. First observed in liquid ^{4}He, this phenomenon has been intensively studied in a variety of systems for its remarkable features such as the persistence of superflows and the proliferation of quantized vortices. The achievement of Bose–Einstein condensation in dilute atomic gases provided the opportunity to observe and study superfluidity in an extremely clean and well-controlled environment. In the solid state, Bose–Einstein condensation of exciton polaritons now allows to plan for the observation of similar phenomenology. Polaritons are interacting light–matter quasiparticles that occur naturally in semiconductor microcavities in the strong coupling regime and constitute an interesting example of composite bosons. Here, we report the observation of spontaneous formation of pinned quantized vortices in the Bose-condensed phase of a polariton fluid. Theoretical insight into the possible origin of such vortices is presented in terms of a generalized Gross–Pitaevskii equation. In the second part of the chapter, we provide the clear observation of half vortices, special to spinor condensates. We then go no, in the last part of this chapter, to study the dynamics of spontaneously created vortices. We show that their path is determined by the disorder landscape towards their final stable position.

3.1 Introduction

Exciton polaritons have been cornered, almost by chance, by a careful reflectivity study of a vertical cavity surface emitting laser (VCSEL) by Claude Weisbuch in 1992 [1]. In this seminal paper, the normal mode splitting was observed for the first time and attributed to the strong coupling between the excitons confined in the

B. Deveaud-Plédran (✉) · K.G. Lagoudakis
Ecole Polytechnique Fédérale de Lausanne (EPFL), 1015 Lausanne, Switzerland
e-mail: benoit.deveaud-pledran@epfl.ch

D. Sanvitto and V. Timofeev (eds.), *Exciton Polaritons in Microcavities*,
Springer Series in Solid-State Sciences 172, DOI 10.1007/978-3-642-24186-4_3,

quantum wells placed in the spacer layer and the photons confined in the Fabry–Pérot cavity. Since this initial demonstration, a few years were needed to get to our present understanding that polaritons can be considered, in a wide density range, as good bosons. A considerable amount of time has also been necessary to get convinced that these half-matter/half-light quasiparticles could behave in a way that is really different from what is observed for laser photons in VCSELs.

The outline of this chapter will then be the following: in the first part, we will rephrase some on the important properties that allow polaritons to become such interesting quasiparticles. In the second part, we will show that these particles are good bosons and that they can Bose Einstein condense, although as a highly non equilibrium and strongly interacting type of condensate. In the third part, we will describe our observation of pinned vortices in condensates. In the fourth part, we will give evidence for the existence of half vortices due to the fact that polaritons do have a spin. We will then terminate by some observations of the dynamics of vortices.

3.2 Basics of Polaritons

As is shown in a very schematic way on Fig. 3.1, polaritons are the result of the coupling between quantum well excitons and cavity photons (dashed lines in Fig. 3.1). On the scale of the figure, only ten inverse microns, the parabolic dispersion of excitons appears to be flat. When the quality of the mirrors is sufficient, strong coupling occurs between the photon and the exciton modes, leading to coupled states called polaritons, as proposed initially by Hopfield [2]. Here, we call the new quasi-particles microcavity polaritons, and we will study the properties given by their dispersion relation (continuous lines in Fig. 3.1). The strength of the coupling gives the splitting between the two modes, and is usually labeled Ω_R.

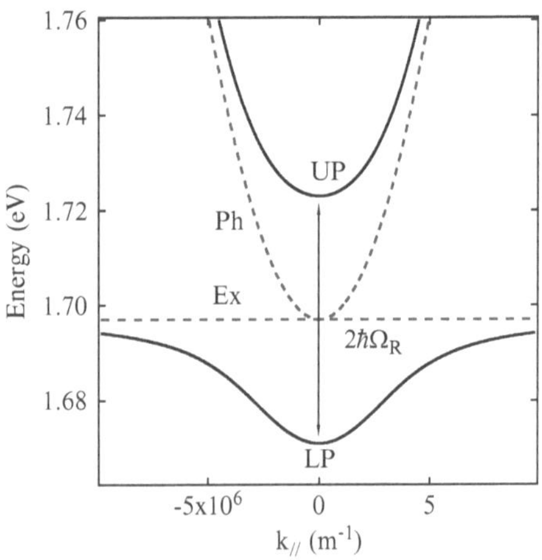

Fig. 3.1 Dispersion curve for microcavity polaritons. The *dashed red curves* show the dispersion of the uncoupled exciton (Ex) and photon (Ph) modes. We assume zero detuning, i.e., the same energy at $k = 0$ The modes resulting from the coupling are the two polariton modes shown by the *black continuous lines*: the lower polariton (LP) and the upper polariton (UP)

Polaritons possess quite a number of very interesting properties directly given by the shape of their dispersion. Importantly enough, it is worth noting that the window shown in Fig. 3.1 represents only a very small window of the Brillouin zone. The two properties that we will use extensively in the present work are the fact that polaritons have a very small mass coupled to the fact that they may be considered as good bosons in a very extended temperature and density range. One important feature of polaritons is that, despite the quality of the cavity achievable these days, with Q factors up to 10^5, the lifetime of the photon component of the polariton is very short. The photon part therefore leaks out of the cavity within 1–10 ps, carrying out all the information on the polariton states. First, the wave vector of the polariton is directly related to the angle of emission of the photon [3] (10° corresponding roughly to one inverse micron). Second, the intensity, polarization, noise, and coherence of the emitted photons exactly map out the properties of the polaritons inside the cavity [4].

The mass of polaritons comes mainly from their photon part and is smaller than 10^{-4} times the mass of a free electron. This extremely light effective mass has been recognized quite soon as a key feature for the Bose–Einstein condensation of polaritons at relatively high temperatures [5]. In fact, the idea of using quasiparticles, such as, for example, excitons, in solids was even put forward much earlier by Blatt [6] and by Moskalenko [7] with the same kind of idea in mind.

Different experiments have been carried out following this idea, in particular experiments using luminescence under high excitation density, in order to try obtaining evidence for Bose–Einstein condensation. The first experiments were rather deceiving as, when reaching the saturation density, the oscillator strength of excitons vanishes and the strong coupling regime is lost [8]. Improving on the sample quality and optimizing the number of quantum wells included in the spacer layer (at the antinodes of the electromagnetic field), it has been possible to observe highly non-linear effects. The first observations were wrongly attributed to polariton lasing [9] and are now considered as transition to the standard VCSEL regime. High-quality samples allowed 3 years later to observe similar non-linear effects, this time in the strong coupling regime [10, 11] At the time, the authors did claim for stimulated scattering only, when they apparently got the first real signature of Bose–Einstein condensation in a solid.

Nevertheless, the experiments carried out at the time, and in particular the experiments on parametric amplification with polaritons, showed that polaritons do behave as good bosons up to quite high densities [12, 13]. In such experiments, the non-parabolicity of the lower polariton branch is used that allows getting proper energy and momentum conservation at the same time in the parametric scattering process. The efficiency of such processes is very large as it corresponds to real states.

3.3 Bose–Einstein Condensation of Exciton–Polaritons

The constant improvement of the quality of the samples, and of the experimental skills, allowed after a series of preliminary papers by the group of Dang in

Grenoble [14], to come to a clear demonstration of Bose–Einstein condensation, first in a CdTe-based microcavity [15]. All experiments that will be described in this chapter have been realized on the very same sample that already allowed the group of Dang in Grenoble to realize very promising steps on the way to the demonstration of Bose–Einstein condensation [16, 17].

The demonstration took a long series of steps that basically consist in: first finding conditions where, under non resonant excitation of the system at energy such that the coherence imprinted by the exciting laser is not passed onto the polariton gas, the polaritons show a well-behaved Boltzmann distribution below threshold. Second, the distribution of polaritons that we can directly probe-over the whole bottom polariton branch, turns from Boltzmann to Bose like. Third, a very strong narrowing at threshold is observed, that corresponds to the sudden increase of the intensity emitted at $k = 0$, as well as of the $g^{(1)}$ of polaritons. Fourth, this threshold, with an intensity increase by more than a factor of 10, exactly corresponds to our estimation of approximately one polariton per state, i.e., the onset of stimulated scattering.

Last, but not least, the sudden appearance of long-range order above threshold has been probed with the use of a stabilized Michelson interferometer. As this set-up will be an essential part in our experiments on vortices in polariton condensates, it is worth describing the set-up as is done in Fig. 3.2. The essential part of this set-up, for the demonstration of the existence of Bose–Einstein condensation of polaritons, is the stabilized Michelson interferometer in the mirror-retroreflector configuration. This interferometer provides interference between the image of the condensate and its centro-symmetric replica, the center of which is the so-called autocorrelation point.

This system has allowed us to demonstrate, for the first time, the build-up of long-range order in a polariton condensate when driven above threshold [18]. The typical result is highlighted in Fig. 3.3. Below threshold, the interferogram only shows high contrast fringes (and thus a high degree of spatial coherence $g^{(1)}$) around the autocorrelation point, over a size that is of the order of less than one micron, typically the de Broglie wavelength of the polaritons at the estimated temperature of the gas. Above threshold, clear interference fringes appear that basically extend up to the edges of the excitation spot, and consequently, the measured spatial coherence has significant values over the whole condensate. Strong fluctuations of the intensity and of the contrast originate in the disorder inherent to the sample itself, and in particular to the disorder of the Bragg mirrors.

Bose condensation of polaritons has been subsequently confirmed by other groups on different samples with, for example, the work of the group of David Snoke [19], using a needle to trap the condensate, the group of Yamamoto [20], the group of Jacqueline Bloch in particular with the work on 1D condensates [21] and eventually in the group of Nicolas Grandjean, at room temperature, in GaN microcavities [22].

One of the main reasons allowing the polaritons to Bose condense, despite the importance of disorder clearly visible in the images, has been found to be related to mode synchronization [23]. Such a mechanism, which is very general and gas

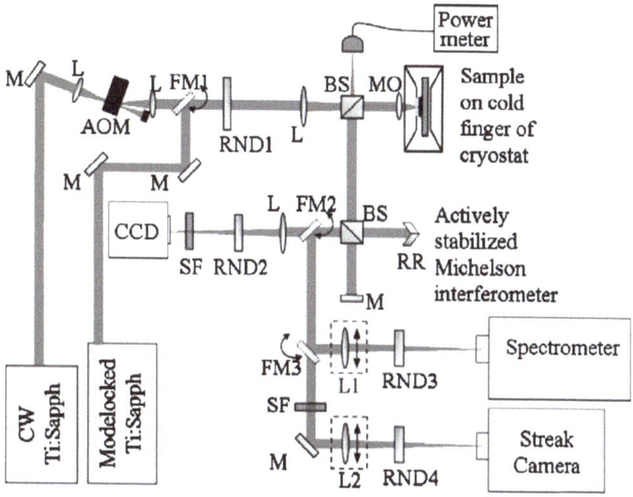

Fig. 3.2 Experimental setup. *M* Mirror, *RND1–4* (variable) rotational neutral density filters, *L*, *L*1−2 Lenses, *RR* Retroreflector, *BS* Beam splitter, *MO* Microscope objective, *SF* Spectral filter, *FM1–3* Flipping mirrors. The flipping mirror FM1 allows switching between CW or pulsed excitation. FM2 allows for the direction of the luminescence either to the CCD at the output of the interferometer or to the streak camera if the mirror is in place. Finally, if FM3 is in the indicated place, the luminescence is directed to the spectrometer entranced slit. The lenses L1 and L2 are mounted on electronically controlled translation stages which, depending on the position, allow for the displacement of the real-space image on the entrance slit of the spectrometer or the of the streak camera. RND1 is the neutral density which regulates the excitation intensity (monitored by the power meter), and RND2–4 are used to prevent saturation of the detection devices. The acquisition of images and the translation stage control are computerized, which allows for the tomographic decomposition experiments to be performed

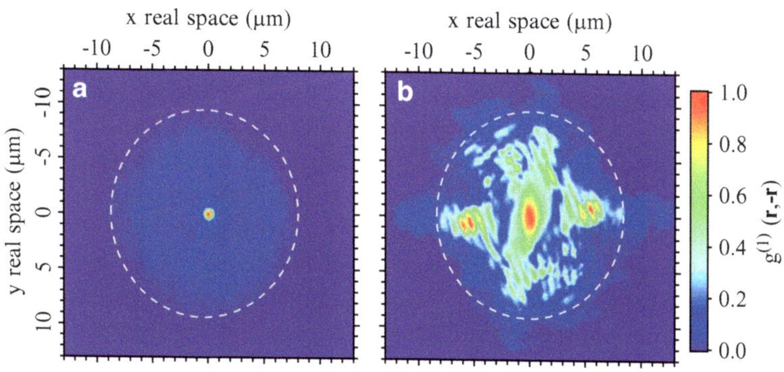

Fig. 3.3 *Left*, spatial coherence of the polariton cloud below threshold, coherence is only observed at the autocorrelation point, over a size that corresponds to the de Broglie wavelength of the polaritons. *Right*, above threshold, sizable coherence is observed over the whole of the excitation spot. This coherence is strongly modulated by the disorder

way beyond Bose–Einstein condensates, allows different oscillators, initially with slightly different frequencies, to come to the same common frequency through minimal coupling strength. In our case, mode synchronization is achieved through the polariton interactions, which are coming mostly from their excitonic fraction.

3.4 Vortices in Polariton Condensates

In nature, any flow of fluid with closed streamlines can be considered as vortex flow. Everyone is familiar with vortices like the ones forming in a water container that is quickly emptied through a hole in the bottom. Classical vortices are characterized by a rotational flow of the fluid around the vortex core, which is easily traceable through the density minimum. The "fluid rotation" is better described by a quantity called circulation that can take any value.

In quantum fluids, vortices are similar objects (the fluid rotates around the vortex core, and the core is a density minimum), but here, the circulation is quantized. The quantization of the circulation is directly imprinted in the phase of the quantum fluid which is linearly changing from 0 to multiple integers of 2π when going around the vortex core. The integer number of 2π phase jumps is the so-called vortex charge.

One of the prominent observations, in superfluids such as Bose–Einstein condensates, is the existence of such quantized vortices. In the case of atom condensates and liquid ^{4}He, beautiful experiments have been realized through stirring of the fluid with lasers or through rotation of the magnetic trap (the fluid container), and vortices have been observed in several forms (single vortex or vortex lattices).

In the case of polaritons, the disorder in the sample and the non-equilibrium nature of the fluid concur to the spontaneous creation and pinning of vortices. As all properties of the polariton quantum fluid are carried on the emitted photons, by studying the luminescence image, one should be able to see both the phase winding and the density minimum of possible vortices. The observation of the density minimum is straightforward, but the phase winding is not readily available. In order to access the phase winding property, it is necessary to perform an interferometric measurement identical to the one mentioned above for the estimation of the $g^{(1)}$ function. As it is known, the interference of two oblique constant phase wavefronts will result in a series of straight interference fringes. On the contrary, when interfering a constant phase wavefront with a field carrying a vortex (a 2π singularity), the resulting interference pattern will be similar to the constant phase interference, the only difference being that at the center of the singularity; the interference fringes will show a forklike dislocation (the number of fringes changes by one from above to below the singularity). A vortex in a polariton condensate is thus evidenced in a very direct way by the presence of a fork-like dislocation in the interference pattern (Fig. 3.4).

In principle, one would expect that, for polaritons as it is the case for atom condensates, the observation of quantized vortices should be obtained upon setting the condensate in rotation through the appropriate means [24]. However, the

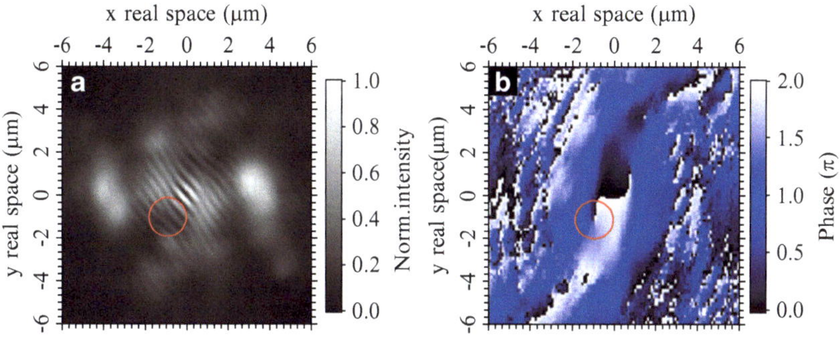

Fig. 3.4 (**a**) Interference pattern at one given point of the sample featuring a forklike dislocation. Fringes give evidence for the long range order. The forklike dislocation highlighted by a *circle* corresponds to a quantized vortex. (**b**) The fourier transform of the pattern allows to demonstrate a 2π phase shift around the core of the vortex

initial observation of vortices in polariton condensates has been obtained without any rotation [25]. In this work vortices appear spontaneously (without setting the polariton condensate in rotation) and are evidenced to be pinned at specific locations. Two issues are thus necessary to address: the origin of the vortices and the pinning mechanism.

There exist a number of scenarios for the vortex generation. Vortices may be created during the first moments of the condensate formation when "islands" of particles with different phases merge to create a large condensate with a common phase. Another possibility is related to the fact that polariton condensates are far out of equilibrium and are governed by gain-loss mechanisms. It has been shown that nucleation and trapping of vortices may occur, thanks to the interplay of losses and gain [26]. Finally, hydrodynamic nucleation of vortices is another possible mechanism when a polariton condensate flows with high velocity against an obstacle [27] (flow of particles down a steep potential wall against a small-size local potential maximum). Pinning of the created vortices can then occur at locations where the flows are such that a vortex is energetically favored. This can also explain the necessary pinning of the sign of the vortices that are observed here.

3.5 Half Quantum Vortices in Polariton Condensates

Up to now, the spin of polaritons has been systematically ignored because in the experiments that were presented above, the polarization of the condensate was either highly linear or a combination of non degenerate linear polarization states along the crystallographic axes. In the case of polaritons with low enough disorder, as will be shown, it is necessary to take into account the spin of exciton–polaritons. In the circular polarization basis, the linear polarization is a superposition of the two circular components (the two spins) with equal amplitude and phase. In this case, both spin components are forced to behave precisely the same way, thus suppressing the spin degree of freedom. This type of quantum fluid, that only bears the phase degree of freedom, is the so-called scalar fluid. On the contrary, if the polarization degree is not highly linear and the different polarization states (e.g., x and y) are degenerate, then the two spin components are not necessarily identical thus allowing for an independent spin contribution. In this case, there is another degree of freedom coming into play: the spin. This type of quantum fluid, having not only the phase but also the spin degree of freedom, is the commonly known spinor quantum fluid. Quantized vorticity is a property of quantum fluids that has been widely studied in various types of superfluids either with or without spin.

Superfluids with a two-component (spinor) order parameter are characterized by a different type of vortices than those found in conventional scalar superfluids. This new type of vortices is the so-called half-quantum vortices (HQV). They have lower energy with respect to full vortices and constitute the elementary excitations of spinor superfluids [28, 29]. When circumventing their core, the phase and the spin (or polarization) vectors experience $a \pm \pi$ rotation. In this sense, HQV can be understood as a half-phase vortex combined with a half-polarization vortex [30]. In ^{4}He superfluids, the HQV cannot be formed as the spin degree of freedom is absent. However, in ^{3}He or in triplet superconductors, the order parameter has two or more components, so that the formation of HQV is possible. So far, experiments have not presented any direct evidence for HQV in ^{3}He [31], while more reliable indications of existence of HQV have been reported in cuprate superconductors [32]. Recently, HQVs have been proposed as a smoking gun for the superfluid of exciton–polaritons in semiconductor microcavities [33]. Indeed, the spatially inhomogeneous polarization splitting that is present in CdTe microcavities, makes polariton condensates on different areas of the sample to have a different topology, either that of a scalar or that of a spinor quantum fluid. Taking advantage of the spinor topology, we will provide evidence of HQVs in polariton condensates.

In order to fully characterize a vortex in a polariton condensate, one needs two winding numbers, (k, m), one for the polarization angle and one for the phase. One can express the order parameter of the condensate in the linear xy basis as

$$\phi_{\text{lin}} = \sqrt{ne^{i\theta(r)}} \begin{pmatrix} \cos\eta(r) \\ \sin\eta(r) \end{pmatrix},$$

where $\eta(\boldsymbol{r})$ is the polar angle that characterizes the orientation of the polarization and $\theta(\boldsymbol{r})$ is the phase of the coherent polariton fluid. Any kind of vortices (fractional or full) are described in this notation by rotation of the phase and the polarization following $\theta(r) \rightarrow \theta(r) + 2\pi m$ and $\eta(r) \rightarrow \eta(r) + 2\pi k$ where the two winding numbers k and m can each take either integer or half-integer values provided their sum is an integer. Four types of half vortices are described by these two winding numbers being equal to $\pm 1/2$. In order to reveal the specific phenomenology of HQVs with respect to the integer vortices, it is convenient to analyze the circularly polarized components of the order parameter, which can be expressed as

$$\phi_{\text{lin}}(r) = \frac{\sqrt{n}}{2}\left[e^{i(\theta(r)-\eta(r))}\begin{pmatrix}1\\ i\end{pmatrix} + e^{i(\theta(r)+\eta(r))}\begin{pmatrix}1\\ -i\end{pmatrix}\right]$$

Two spin components, the phase and the polarization rotations, are now conveniently mixed, and for the case of half-quantum vortex where $\theta(r) \to \theta(r) + \pi$ and $\eta(r) \to \eta(r) + \pi$, it is straightforward that a zero rotation takes place for one circular polarization, and a full 2π rotation is achieved for the other circular polarization. This means that if one were to detect a half vortex, it would be easiest when looking in σ^+ and $\sigma-$ polarizations simultaneously. Then HQV would be observed as a full vortex in one polarization and no vortex in the other one.

Note that in the case of full phase vortices, the forklike dislocations are expected to be seen in the same place in both circular polarizations, whereas in the case of HQV the fork should only appear in one of the circular polarizations. This is indeed what we have performed by changing the configuration of our interferometer so as to allow for the separate imaging of the two circular polarizations [34].

The result of the experiment is shown on Fig. 3.5 on a location of the sample where the polarization splitting is almost negligible (spinor topology). At this location, a HQV was observed, and indeed, as expected from the simple theoretical

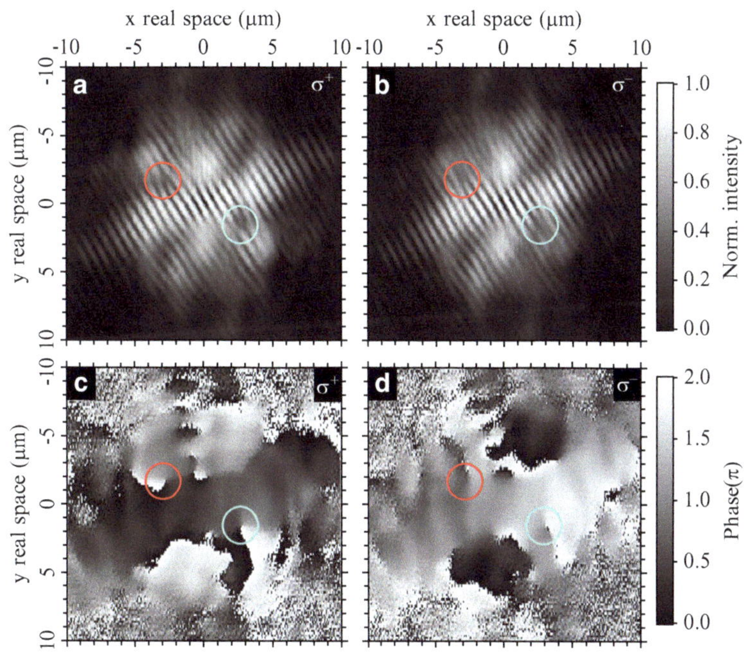

Fig. 3.5 (**a**) Circular polarization resolved interferograms of the polariton condensate. On the left of the image is shown the σ^+ polarization, and the *red circle* shows a well resolved fork-like dislocation. (**b**) σ^- polarization where the same position does not show any dislocation. This is exactly what is expected for a half quantized vortex. (**c, d**) phase extracted from the respective interferograms showing the existence (or not) of the phase singularities

comments given above, a forklike dislocation is observed in the right circular polarization interference when no dislocation is observed on the counter circular polarization.

In order for the claim of the observation of the half vortices to be solid and unquestionable, it is necessary to rule out a possibility that has not been taken into account up to now and might, if true, have devastating consequences: how can we be sure that we are not observing two independent condensates, one with spin-up polaritons and a vortex and one in spin down and no vortex? If this were to be true, it is impossible to claim that we have evidence of HQVs. One hand-waving argument is that there are many occurrences where a half vortex coexists with full vortices in the same excitation spot (different topologies depending on the position), and thus, there has to be phase coherence between the two spin components.

Although solid enough as an argument, it does not apply everywhere because in most of the cases, HQVs are seen alone. For this purpose, a new experiment is needed to be designed in order for the phase coherence between the two opposite spins to be unquestionably demonstrated in the presence of a half-quantum vortex. A polarization mixing Mach–Zehnder interferometer in the mirror-retroreflector configuration was used, where the first splitter is a polarizing beam splitter that separates the two spin components and sends them in the two different paths: one with a mirror and one with a retroreflector. Note that the polarization in the two paths is now orthogonal, and so no interference is possible after recombining the two spin components. A half-wave plate was thus used in one of the two arms to get the two polarizations collinear which then got recombined at the second beam splitter and were sent on the camera. The resulting interference pattern can be found in Fig. 3.6 and corresponds to the same HQV as in Fig. 3.5. The mere existence of interference fringes provides direct evidence that there is excellent coherence between the two spin components. Moreover the interference pattern observed has

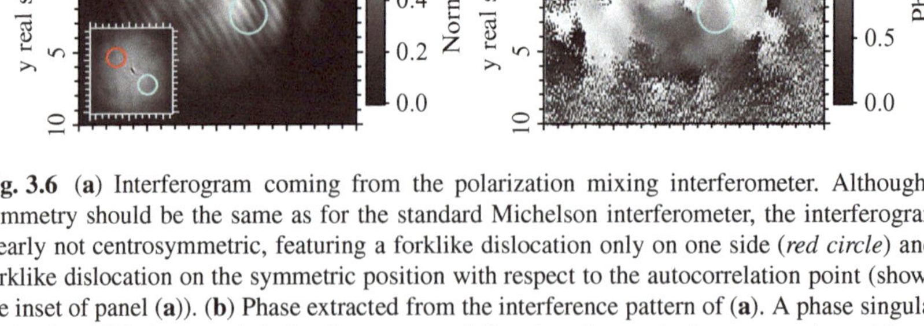

Fig. 3.6 (**a**) Interferogram coming from the polarization mixing interferometer. Although the symmetry should be the same as for the standard Michelson interferometer, the interferogram is clearly not centrosymmetric, featuring a forklike dislocation only on one side (*red circle*) and no forklike dislocation on the symmetric position with respect to the autocorrelation point (shown in the inset of panel (**a**)). (**b**) Phase extracted from the interference pattern of (**a**). A phase singularity is clearly visible in the *red circle* whereas a quasi-flat phase is seen in the symmetric position

a surprising feature: It is not symmetric with respect to the autocorrelation point (shown in the inset of Fig. 3.6a) even though there is a retroreflector in one of the two arms of the interferometer. This is another indication that we are dealing with a half vortex. By definition, a half vortex can be conceived as a full vortex in one spin and no vortex in the other spin. The interference between a condensate with a vortex and a condensate without a vortex will only show *one* forklike dislocation at the location of the vortex.

It is worth noting here that half vortices have not been predicted or observed in VCSELs. No arguments exist though on whether this is impossible or not, and thus, it is a subject worth more investigation.

3.6 Dynamics of Vortices in Polariton Condensates

In all the experiments described above, the condensate is created non resonantly through excitation with a quasi-CW laser having an energy much higher than the polariton resonance. At such energies, the laser photons create unbound electron–hole pairs, which then relax to the bottom of their respective band, then bind into excitons in the non radiative part of the Brillouin zone, the so-called "reservoir". Excitons therefore accumulate in the reservoir, and, when their density is large enough, exciton–exciton interactions allow them to scatter to the bottom of the lower polariton branch. At sufficient excitation density, the occupation of the lower branch may exceed one, and the stimulated scattering then populates this lower branch very efficiently. Such a density exactly corresponds to the threshold for polariton condensation.

A very interesting question is then, how long does the condensate need to build up the long-range order that we have detected and that in fact constitutes the smoking gun of the demonstration that condensation exists. We have therefore decided to carry out time-resolved experiments on the build-up of long-range order in polariton condensates. For such experiments [35], we use non resonant femtosecond-pulsed excitation.

Preliminary measurements had shown that, in such a case, the population if the exciton reservoir indeed builds up for tens of ps, until the threshold for stimulated scattering is reached [36, 37]. The larger the excitation density, the shorter is the time to reach this threshold. In Fig. 3.7, we show the streak image after non-resonant excitation of the condensate, of a cross-section of the interferogram of the condensate. The images are recorded along a narrow line that does not pass through the center (the autocorrelation point) of the interferogram but is shifted below by a few microns. Therefore, below condensation threshold (panel (a) of Fig. 3.7), no coherence is observed.

When we increase the excitation density so as to reach the stimulation threshold, the population in the lower polariton branch raises with a delay of about 45 ps, and we do not observe any delay in the build up of coherence (panel (b) of Fig. 3.7). Rising further the excitation power allows to reach the stimulation threshold much

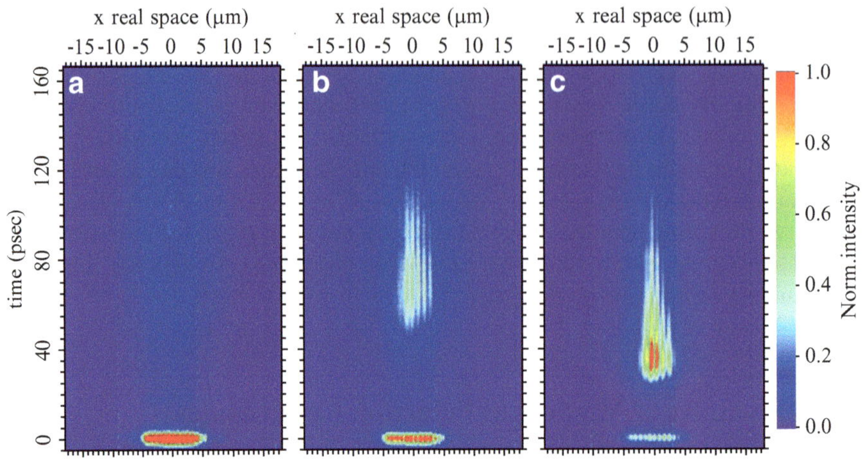

Fig. 3.7 Streak images corresponding to cuts of the interferogram of the condensate off the autocorrelation point. Below threshold, *right*, no fringes are observed. At threshold, *center*, the condensate builds up within about 40 ps, and the fringes appear simultaneously with the population. At large densities, *right panel*, the same phenomenology is observed with even shorter delays

sooner (see panel (c) of Fig. 3.7). Once again, the interference pattern appears with negligible delay with the build up of the population in the polariton ground state.

The next question is therefore to be able to understand how fast are vortices created in the disordered landscape of the sample: do they appear directly at the position where we observe them on time-averaged images, or are they created elsewhere and then move and get trapped at their final pinning position?

Such an experiment requires the time resolution over the full image of the interferogram. To this aim, we have to send the interferogram to the entrance slits of a streak camera where it is "*decomposed*" into multiple slices with real-space and temporal resolution throughout a tomographic decomposition. This allowed for a full reconstruction of the two-dimensional interferogram from 0 to 155 ps with a 3 ps resolution given by the streak camera. The results of the experiment are shown in Fig. 3.8.

The forklike dislocation observed in the time-integrated interferogram is initially not pinned. It exhibits a smooth motion along a well-defined path towards the center of the condensate. The most interesting migration dynamics is found within the first 35 ps of the condensate life.

The vortex appears close to the side of the excitation spot simultaneously with the formation of the condensate at about 20 ps after the excitation pulse arrival. The subsequent smooth vortex motion is then abruptly interrupted at 55 ps when the vortex encounters a potential trap created by the stationary disorder that acts as a pinning center. Four snapshot interferograms taken during this time interval are shown in Fig. 3.8. For visibility purposes the constant component of the 2D interferograms is removed, and the color scale of the remaining interference pattern

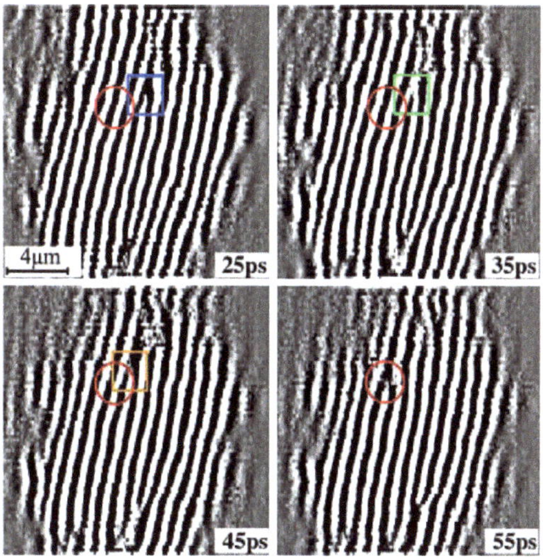

Fig. 3.8 Dynamics of the migration of the vortex. Each image corresponds to a subsequent time frame (25, 35, 45, and 55 ps). The *red circle* shows the final vortex location and is present in each time frame, whereas the *colored squares* show the current vortex location. The CW part of the interferogram has been here removed and the contrast is saturated for readability purposes

is saturated, giving a very high contrast. At the sample location we studied here, the condensate of exciton–polaritons was linearly polarized with a polarization pinned to one of the crystal axis, suppressing the existence of half-quantum vortices detected at different places on the same sample.

We have performed numerical modeling of our experiment using the stochastic generalized Gross–Pitaevskii equation. We have introduced into the model a set of quite reasonable assumptions. The non-resonant excitation creates hot electron–hole plasma that further relaxes to form high energy excitons, which we separate in two subsets:

1. The "inactive" excitons that do not fulfill the required energy and momentum conservation conditions to scatter directly into the condensate state, described by the density $n_j(r,t)$.
2. The "active" excitons that do fulfill those conditions and act as the condensate source, described by the density $n_A(r,t)$.

One has to distinguish between active and inactive excitons in such a way because, once stimulated scattering into the condensate takes place, the reservoir of active excitons is immediately depleted.

The initial inactive exciton concentration $n_I(r,t)$ is created by the external pump $P(r,t)$. These excitons disappear by non-radiative recombination, which is described by the decay rate γ_I, or they can turn into the active excitons at a rate $1/\tau_R$. Those excitons may further relax into the condensate. The dynamics of optically injected excitons is given by a first rate equation:

$$\frac{\partial n_I(r,t)}{\partial t} = -\gamma_I n_I(r,t) - \frac{1}{\tau_R} n_I(r,t) + P(r,t)$$

The active exciton population may decay radiatively (with a rate γ_A) or be scattered to the condensate of exciton–polaritons. This scattering is stimulated by the population of the condensate. The rate equation for the reservoir of active excitons reads

$$\frac{\partial n_A(r,t)}{\partial t} = -\left(\gamma_A + R_R|\psi(r,t)|^2\right) n_A(r,t) - \frac{1}{\tau_R} n_I(r,t),$$

where R_R is the rate at which excitons scatter into the condensate. $\psi(r,t)$ is the mean scalar polariton field. The Gross–Pitaevskii equation then writes:

$$i\frac{\partial \psi(r,t)}{\partial t} = \left\{ -\frac{\hbar \nabla^2}{2m} + g|\psi(r,t)|^2 + V(r) + g_R n_A(r,t) + g_R n_I(r,t) \right.$$
$$\left. -\frac{i}{2}[\gamma_C + R_R n_A(r,t)] \right\} \psi(r,t).$$

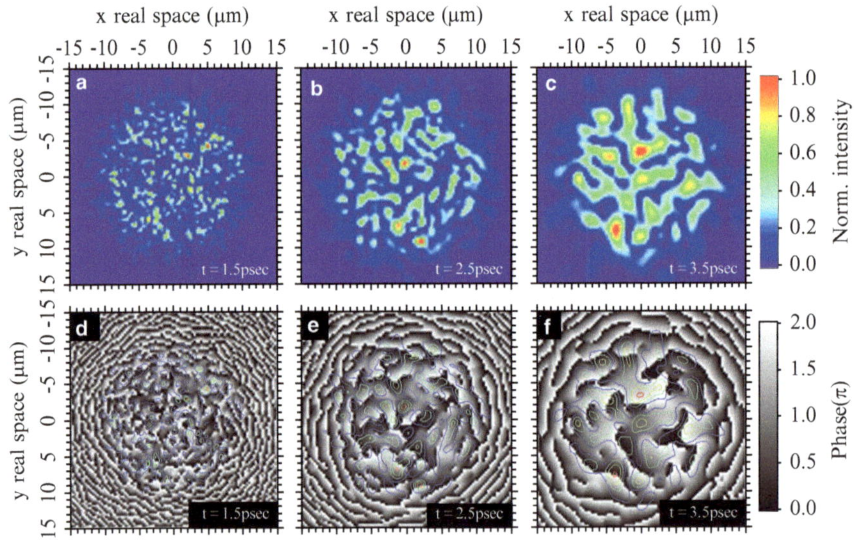

Fig. 3.9 (**a–c**) Amplitude of the condensate wavefunction during the early moments of the condensate formation. When stimulated relaxation of particles from the excitonic reservoir to the condensate takes place, one can observe the formation of particle clusters with a well defined phase but random phase from cluster to cluster. As the clusters grow and start merging together, vortices start to form and either leave the condensate or annihilate via vortex-antivortex binding. In later moments, only few vortices remain that might get eventually pinned. (**d–f**) Corresponding phase of the condensate wavefunction. The *colored contours* depict the wavefunction amplitude of panels (**a–c**). At 3.5 ps (panel (**f**)), one can see that the 2π phase singularities correspond to density minima, representative of the vortex cores

The interactions between condensed polaritons in the mean-field approximation are characterized by the constant g, whereas interactions between incoherent excitons and polaritons are accounted for by g_R. The stationary disorder $V(r)$ has been chosen as a randomly generated Gauss correlated disorder [38]. Polaritons decay from the condensate at a rate γ_C.

The stochastic element in our model is given by the initial condition:

$$\psi(r,t) = [\theta(r)/2]\mathrm{d}r,$$

where $\theta(r)$ is a stochastic noise, drawn from a Gauss distribution and characterized by the correlators:

$$\theta^*(r), \theta(r') = 2\delta(r - r'), \langle\theta(r), \theta(r')\rangle = 0.$$

This initial condition implies an average mode occupancy of 1/2 and random phase, which samples the Wigner distribution of a vacuum state and physically represents the quantum noise of the polariton vacuum.

The relaxation of excitons from the reservoir to the condensate is triggered by the initial noise and amplifies the density creating small isolated clusters with well-defined phase as shown in Fig. 3.9 for a single realization. The growth and merging of these clusters with different relative phase at the transition towards a state with macroscopic phase coherence leaves initially many topological defects in the phase. This spontaneous vortex generation process is a manifestation of the Kibble–Zurek mechanism [39, 40] and is also considered as a vortex generation mechanism for atomic condensate systems [41].

The above equations can be solved numerically, and each realization of the initial noise term corresponds to the arrival of a different pulse in the experiment. Averaging over multiple realizations of the noise is equivalent to the experimental procedure where the images registered by the streak camera result from averaging over multiple pulses. Different initial conditions result in an altered distribution of vortices. We show some typical results on Fig. 3.10.

The kinetics of vortices is defined by the interplay between the local disorder potential and the non-equilibrium character that induces flows of polaritons. Vortices appear in a spontaneous way within the excitation spot. They are then dragged by the existent flows resulting in a vortex migration. The observation of a stable vortex is the result of a dynamical equilibrium in this disordered system. Recently, a similar experiment has been realized by Daniele Sanvitto and coworkers whereby they resonantly imprinted a vortex in the condensate [42] and they observed the kinetics due to the residual momentum of the beam.

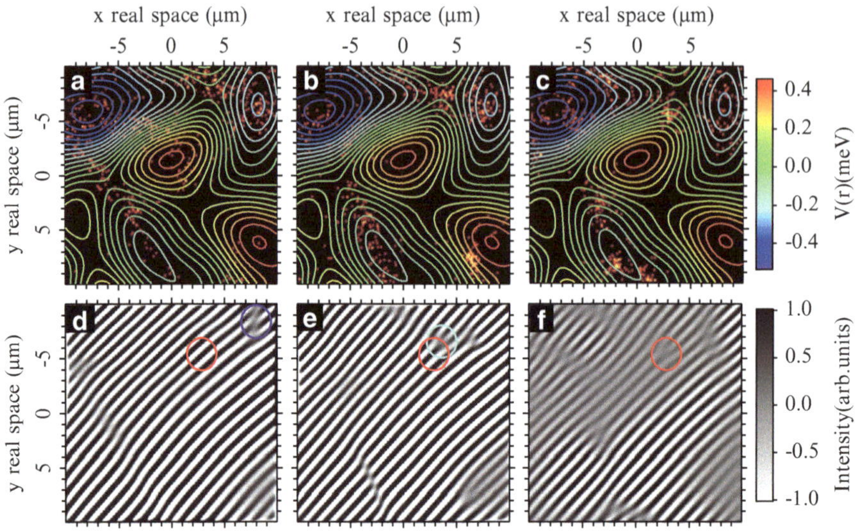

Fig. 3.10 (**a–c**) Simulation of real-space vorticity map for subsequent time frames (40, 50, 90) where the *orange* pixels correspond to the appearance of vortices under different initial condition realizations. The *colored contours* depict the local disorder potential. (**d–f**) Realization averaged interference patterns for the same time frames as in (**a–c**). Many forklike dislocations migrate in time to different real-space locations. The *red circle* shows the final location of a pinned vortex. The *colored squares* show the current position of the specific vortex for each time frames

3.7 Conclusions and Outlook

In this review chapter, we have detailed the different observations that were made on a single CdTe microcavity sample, excited by an non-resonant laser so as not to imprint the phase or polarization of the laser onto the polariton population. The different realizations of the random disorder on the surface of the sample allowed us to study different effects linked with the local configuration of disorder, exactly as if we had many samples available.

In such a way, we have been able, after the clear demonstration of the occurrence of Bose–Einstein condensation in particular through the build-up of long-range order, to evidence the occurrence of spontaneously formed vortices. Further on, we have demonstrated, as predicted by theory, the occurrence of half vortices due to the spinor nature of the polaritons. Eventually, we have been able to follow dynamically the formation of vortices, although their initial position is random. Many novel experiments are coming along that line of research, with more samples [43].

It is important to emphasize that interesting quantum fluid properties of polaritons can also been obtained in a non spontaneous fashion. One of the major interests of polaritons indeed lies in the ease with which they can be optically generated. This possibility has first been raised by theoreticians, which ideas have soon be implemented by experimentalists. For example, Iacoppo Carusotto and

Cristiano Ciuti proposed to evidence superfluid properties of resonantly created polaritons through Rayleigh scattering [44], which proved to be a very clever idea. This has allowed, for example, very clear demonstration of the superfluid properties of polaritons in different experimental configurations [45–47]. In another interesting design, Michiel Wouters has proposed to study, through four-wave mixing experiments [48], the ghost branch of the Bogoliubov superfluid dispersion. Such an observation has indeed been recently obtained with the proposed scheme [49]. Other theoretical prediction does pave the way for future experiments [50, 51].

The hydrodynamics of a superfluid on polaritons in the presence of an obstacle is one of the research lines that, although studied only very recently, have already shown results of major interest [52, 53].

The ability to realize, with the available techniques at hand, confined environments for polaritons [54, 55], has proven and will prove again to be a very promising avenue for the finding of novel exciting effects.

Acknowledgements This work is inspired from the PhD work of Konstantinos Lagoudakis who has carried most of the experiments reported here. It has been performed with the strong support of a group of very talented PhD student and postdocs, that I wish to congratulate warmly for their outstanding work: Stefan Kundermann, Francesco Manni, Verena Kohule, Barbara Pietka, Maxime Richard, Augustin Baas, and Yoan Léger. I also would like to convey my most sincere thanks to the theoreticians who allowed us to understand our results, by spending enough time with us and putting into simple words the results of their equations. In particular, Michiel Wouters, Vincenzo Savona, Alexei Kavokin and Yuri Rubo deserve very special acknowledgements. The work has been carried out within the framework of the Quantum Photonics National Center of Competence in research financed by the Swiss National Science Foundation. Complementary funding for conferences and meetings has been obtained through the Latsis Foundation, the Polatom network of the European Science Foundation and a King Saud University associate professorship.

References

1. C. Weisbuch et al., Phys. Rev. Lett. (1992)
2. J.J. Hopfield, Phys. Rev. (1958)
3. R.H. Houdre et al., Phys. Rev. Lett. **73**, 2043 (1994)
4. V. Savona et al., Phys. Rev. B **49**, 8774 (1994)
5. A. Imamoglu et al., Phys. Rev. A **53**, 4250 (1996)
6. J.M. Blatt, Phys. Rev. **126**, 1691 (1962)
7. S.A. Moskalenko, Soviet Phys. Solid State **4**, 199 (1962)
8. R.H. Houdré et al., Phys. Rev. B **52**, 7810 (1995)
9. S. Pau et al., Phys. Rev. **A54**, R1789 (1996)
10. L.S. Dang et al., Phys. Rev. Lett. **81**, 3920 (1998)
11. P. Senellart et al., Phys. Rev. Lett. **82**, 1233 (1999)
12. P.G. Savvidis et al., Phys. Rev. Lett. **84**, 1547 (2000)
13. M. Saba et al., Nature **414**, 6865 (2001)
14. M. Richard et al., Phys. Rev. Lett. **94**, 187401 (2005)
15. J. Kasprzak et al., Nature **443**, 7110 (2006)
16. M. Richard, J. Kasprzak, R. André et al., Phys. Rev. B **72**, 201301 (2005)
17. J. Kasprzak, R. André et al., Phys. Rev. B **75**, 045326 (2007)

18. J. Kasprzak et al., Nature **443**, 409 (2006)
19. Balili et al., Science **316**, 5827 (2007)
20. C.W. Lai et al., Nature **450**, 5289 (2007)
21. E. Wertz et al., Nat. Phys. **6**, 860 (2010)
22. G. Christmann et al., Appl. Phys. Lett. **93**, 051102 (2008)
23. A. Baas et al., Phys. Rev. Lett. **100**, 170401 (2008)
24. J.R. Abo-Shaeer, C. Raman, J.M. Vogels et al., Science **292**, 476 (2001)
25. K.G. Lagoudakis et al., Nat. Phys. **4**, 706 (2008)
26. J. Keeling, N. Berloff, Phys. Rev. Lett. **100**, 250401 (2008)
27. G. Nardin et al., Nat. Phys. doi:10.1038/nphys1959
28. M.C. Cross, W.F. Brinkmann, J. Low Temp. Phys. **5–6**, 683 (1977)
29. G.E. Volovik, V.P. Mineev, JETP Lett. **24**, 561–563 (1976)
30. M.M. Salomaa, G.E. Volovik, Rev. Mod. Phys. **59**, 533 (1987)
31. M. Yamashita et al., Phys. Rev. Lett. **101**, 025302 (2008)
32. J.R. Kirtley et al., Phys. Rev. Lett. **76**, 1336 (1996)
33. Y.G. Rubo, Phys. Rev. Lett. **99**, 106401 (2007)
34. K.G. Lagoudakis et al., Science **326**, 974 (2009)
35. G. Nardin et al., Phys. Rev. Lett. **103**, 256402 (2009)
36. P. Senellart, J. Bloch, B. Sermage et al., Phys. Rev. B. **62**, 16263 (2000)
37. M. Muller, J. Bleuse, R. Andre, Phys. Rev. B. **62**, 16886 (2000)
38. V. Savona, W. Langbein, Phys. Rev. B. **74**, 75 311 (2006)
39. T.W.B. Kibble, J. Phys. A. **9**, 1387 (1976)
40. W.H. Zurek, Nature (London) **317**, 505 (1985)
41. C.N. Weiler et al., Nature **455**, 948–951 (2008)
42. D. Sanvitto, F.M. Marchetti, et al., Nat. Phys. **6**, 527 (2010)
43. E. Wertz, L. Ferrier, D.D. Solnyshkov et al., Nat. Phys. **6**, 860, (2010)
44. I. Carusotto, C. Ciuti, Phys. Rev. Lett. **93**, 166401 (2004)
45. A. Amo, D. Sanvitto, F.P. Laussy et al., Nature **457**, 291 (2009)
46. A. Amo, J. Lefrere et al., Nat. Phys. **5**, 805 (2009)
47. D. Sanvitto, F.M. Marchetti, M.H. Szymanska et al., Nat. Phys. **6**, 527 (2010)
48. M. Wouters, I. Carusotto, Phys. Rev. Lett. **105**, 020602 (2010)
49. V. Kohule et al., Phys. Rev. Lett, **106**, 255302 (2011)
50. G. Malpuech, D.D. Solnyshkov, H. Ouerdane et al., Phys. Rev. Lett. **98**, 206402 (2007)
51. M. Wouters, V. Savona, Phys. Rev. B. **81**, 054508 (2010)
52. D. Sanvitto, A. Amo, F.P. Laussy et al., Nanotechnology **21**, 134025 (2010)
53. G. Nardin et al., Nat. Phys. doi:10.1038/nphys1959 (2011)
54. L. Ferrier, E. Wertz, R. Johne et al., Phys. Rev. Lett. **106**, 126401 (2010)
55. O. El Daif et al., Appl. Phys. Lett. **88**, 061105 (2006)

Chapter 4
The Berezinskii–Kosterlitz–Thouless Phase Transition in Exciton–Polariton Condensates

Georgios Roumpos and Yoshihisa Yamamoto

Abstract In a homogeneous two-dimensional system at nonzero temperature, although there can be no ordering of infinite range, a superfluid phase is expected to occur for a Bose particle system. Theory predicts that, in this phase, the correlation function decays with distance as a power law, and quantum vortices are bound to antivortices to form molecular-like pairs. We study the relevance of this theory to microcavity exciton polaritons. These are two-dimensional bosonic quasiparticles formed as a superposition of a microcavity photon and a semiconductor quantum well exciton and have been shown to condense at high enough densities. Because of the short lifetime, full equilibrium is not established, but we instead probe the steady state of the system, in which particles are continuously injected from a pumping reservoir.

We create a large exciton–polariton condensate and employ a Michelson interferometer setup to characterize the short- and long-distance behavior of the first-order spatial correlation function. Our experimental results show distinct features of the two-dimensional and nonequilibrium characters of the condensate. We find that the Gaussian short-distance decay is followed by a power-law decay at longer distances, as expected for a two-dimensional condensate. The exponent of the power law is measured in the range 0.9–1.2, larger than is possible in equilibrium. We compare the experimental results to a theoretical model to understand the features and to clarify the influence of external noise on spatial coherence in nonequilibrium phase

G. Roumpos (✉)
E. L. Ginzton Laboratory, Stanford University, Stanford, California, USA

Current address: JILA, National Institute of Standards and Technology, and the University of Colorado, Boulder, Colorado, USA
e-mail: georgios.roumpos@jila.colorado.edu

Y. Yamamoto
E. L. Ginzton Laboratory, Stanford University, Stanford, California, USA

National Institute of Informatics, Hitotsubashi 2-1-2, Chiyoda-ku, Tokyo, Japan

D. Sanvitto and V. Timofeev (eds.), *Exciton Polaritons in Microcavities*,
Springer Series in Solid-State Sciences 172, DOI 10.1007/978-3-642-24186-4_4,

transitions. Our results indicate that the Berezinskii–Kosterlitz–Thouless (BKT)-like phase order survives in open dissipative systems.

We also present our observation of a single vortex–antivortex pair in a condensate of the appropriate size. Pairs are generated due to pump noise and are formed sequentially at the same point due to the inhomogeneous pumping spot profile. They are revealed in the time-integrated phase maps acquired using Michelson interferometry. Our results suggest that vortex–antivortex pairs can be created in a two-dimensional condensate without rotation or stirring. The observed correlated motion of a vortex and antivortex imply that vortex–antivortex pairs do not dissociate, which is consistent with the BKT theoretical prediction as well as with our observation of a power-law decay of the spatial correlation function.

4.1 Introduction

Lower dimensionality is known to enhance the role of fluctuations and interactions in the formation of a spontaneous coherence in a macroscopic level, which often leads to the appearance of exotic phases. Such low-dimensional ordered systems are currently the subject of intense study.

Bose–Einstein condensation (BEC) started as a theoretical peculiarity [1] but was unearthed when F. London realized its connection to superfluidity of liquid helium [2]. The subject became popular recently, thanks to experimental breakthrough in atomic gases [3, 4]. Experiments that probe many-body effects in atoms have now reached a high level of sophistication (see, e.g., [5] and references therein), so the limitations of the system start to appear. In particular, matter-wave interference and measurement of particle statistics are challenging in atomic condensates. It has not been possible yet to isolate truly lower dimensional gases, and engineered lattice potentials have been limited to very periodic structures.

Solid state implementations hold promise to solve all of the above problems. Optical studies using emitted photons naturally enable interference and particle statistics experiments, perfect confinement in two or one dimensions is achieved in quantum well or quantum wire structures, and it is not unreasonable to believe that, once a suitable experimental system has been identified, any lattice structure will be realized through electron beam lithography and nanofabrication techniques.

To understand the experimental challenges involved in this approach, we compare the required particle density in solids with that in atomic gases. In three dimensions, the BEC transition can be crossed by increasing the following phase-space density to a value of order one

$$n\lambda_{\mathrm{T}}^{3} = n\left(\frac{2\pi\hbar^{2}}{m^{*}k_{\mathrm{B}}T}\right)^{3/2} \sim 1, \tag{4.1}$$

where n is the particle density, m^* is the particle mass, T is the temperature, and λ_{T} is the thermal de Broglie wavelength. Atoms are heavy particles, and atomic gases cannot reach high densities. BEC was possible after the development of trapping and cooling techniques that allowed the isolation of low-density cold gases. Excitons in semiconductors have masses comparable to the free electron mass m_0, so they are lighter than atoms. Unfortunately, current technology does not allow to cool them down to very low temperatures (less than 0.1 K). Therefore, the threshold (4.1) must be reached by increasing the particle density. In most semicondutor systems, however, spurious nonlinear effects take place at high densities and obscure the BEC transitions one wishes to study.

The approach we have taken is to study new quasiparticles of much smaller effective mass. Microcavity exciton polaritons are superpositions of a quantum well exciton and a microcavity photon. Because of their photonic component, their mass is very small, on the order of $10^{-4}m_0$, so that the condensation threshold can be reached with a reduced particle density. The unambiguous evidence for exciton–polariton condensation was obtained in a relatively homogeneous GaAs quantum well (QW) microcavity [6], highly disordered CdTe QW microcavity [7], an artificial trap [8], and lattice structures [9]. However, photons are very difficult to confine so that the polariton lifetime is short, and full thermal equilibrium cannot be reached. In our experiments, we continuously pump a reservoir exciton state which feeds the system with fresh quasiparticles at the same rate as their decay rate, so that a constant particle density is maintained. It is instructive to compare experiments on exciton polaritons to experiments on atomic systems that show extended phase coherence, since the exciton–polariton condensates are intrinsically open-dissipative systems at nonequilibrium conditions (see, e.g., [10] and references therein). Indeed, signatures of nonequilibrium physics are evident in our data, but this feature can be considered as an advantage. Namely, this unique feature of the system enables experimental studies that would be otherwise impossible and raises new questions, such as what superfluidity means in a nonequilibrium context.

The exciton–polariton system is two dimensional. It is known that Bose–Einstein condensation cannot occur in an infinite two-dimensional gas [11], but a superfluid transition of a universal character is expected to take place at low temperatures [12]. Powerful spectroscopic and interference techniques enable us to probe the many-body configuration and extract details that are impossible to observe in other settings.

Section 4.2 presents the equilibrium theory of two-dimensional superfluidity. In Sect. 4.3, we review the experimental methods for the initial characterization of the sample and the condensation phenomenon, including real-space and momentum-space luminescence spectroscopy. In Sect. 4.4, we detail on the measurement of the spatial correlation function, and Sect. 4.5 addresses the formation and observation of vortex–antivortex pairs. The dynamics of vortices in OPO polariton superfluids are theoretically studied in the chapter by F.M. Marchetti and M.H. Szymańska.

4.2 Two-Dimensional Bose Gas and Superfluidity

The goal of this section is to describe equilibrium two-dimensional superfluidity. The theory developed here serves as a background for the experiments described later in this chapter, which probe superfluid-like behavior in microcavity exciton–

polariton condensates. In this section, thermodynamic equilibrium is assumed, so the reader should be careful, as the theory cannot be applied directly to nonequilibrium condensation studied in this chapter. The theory is still useful, however, as a guide to theoretical ideas and for comparisons to experiments in different systems, e.g., atomic gases.

4.2.1 Order Parameter, Spontaneous Symmetry Breaking, and Long-Range Order

In order to describe a phase transition, it is always useful to define an *order parameter* [13]. If $\psi_0(\mathbf{r})$ is the lowest energy single particle state with population N_0, we define the order parameter in a BEC transition of noninteracting Bosons as

$$\Psi(\mathbf{r}) \equiv \sqrt{N_0}\psi_0(\mathbf{r}). \tag{4.2}$$

Since $\psi_0(\mathbf{r})$ is normalized to unity, the normalization for $\Psi(\mathbf{r})$ is

$$\int |\Psi(\mathbf{r})|^2 \, \mathrm{d}\mathbf{r} = N_0. \tag{4.3}$$

An interacting system cannot be described as a collection of particles, each one occupying a single-particle state. Instead, one uses a many-body state $\psi_n(\mathbf{r}_1, \mathbf{r}_1, \ldots, \mathbf{r}_N)$, where N is the total number of particles, $\mathbf{r}_i$ is the position vector of the i-th particle, and ψ_n is symmetric under exchange of any pair of indices. In the general case, the system is a statistical mixture of mutually orthogonal states m with weights p_m. We then define the *one-body density matrix* as [14, 15]

$$\begin{aligned} n^{(1)}(\mathbf{r}, \mathbf{r}') &\equiv N \sum_m p_m \int \mathrm{d}\mathbf{r}_2 \, \mathrm{d}\mathbf{r}_3 \ldots \mathrm{d}\mathbf{r}_N \, \psi_m^*(\mathbf{r}, \mathbf{r}_2, \ldots, \mathbf{r}_N)\, \psi_m(\mathbf{r}', \mathbf{r}_2, \ldots, \mathbf{r}_N) \\ &\equiv \left\langle \hat{\psi}^\dagger(\mathbf{r}) \hat{\psi}(\mathbf{r}') \right\rangle. \end{aligned} \tag{4.4}$$

Since $n^{(1)}(\mathbf{r}, \mathbf{r}')$ is Hermitian, it can be diagonalized as

$$n^{(1)}(\mathbf{r}, \mathbf{r}') = \sum_i n_i \chi_i^*(\mathbf{r}) \chi_i(\mathbf{r}'). \tag{4.5}$$

BEC occurs when one and only one of the eigenvalues n_i is of order N, while all the others are of order 1. We label this macroscopically occupied state as $i \equiv 0$ and its occupation number $n_{i=0} \equiv N_0$ and redefine the order parameter instead of (4.2) as

$$\Psi(\mathbf{r}) \equiv \sqrt{N_0}\chi_0(\mathbf{r}). \tag{4.6}$$

The diagonal ($\mathbf{r} = \mathbf{r}'$) part of the one-body density matrix gives the particle density. The off-diagonal part ($\mathbf{r} \neq \mathbf{r}'$) describes the coherence established between distant points. To understand its behavior, consider first the momentum distribution

$$n(\mathbf{p}) = \left\langle \hat{\psi}^{\dagger}(\mathbf{p})\hat{\psi}(\mathbf{p}) \right\rangle, \tag{4.7}$$

where $\hat{\psi}(\mathbf{p}) = (2\pi\hbar)^{-3/2} \int \mathrm{d}\mathbf{r}\, \exp(\mathrm{i}\mathbf{p}\cdot\mathbf{r}/\hbar)\hat{\psi}(\mathbf{r})$ is the field operator in momentum space. In a homogeneous system, $n^{(1)}(\mathbf{r}, \mathbf{r}')$ only depends on the relative distance $\mathbf{s} \equiv \mathbf{r} - \mathbf{r}'$, and $n^{(1)}(\mathbf{s})$ ($= n^{(1)}(\mathbf{r} + \mathbf{s}, \mathbf{r})$) is the Fourier transform of the momentum distribution

$$n^{(1)}(\mathbf{s}) = \frac{1}{V} \int \mathrm{d}\mathbf{p}\, n(\mathbf{p}) \mathrm{e}^{-\mathrm{i}\mathbf{p}\cdot\mathbf{s}/\hbar}, \tag{4.8}$$

where V is the total volume of the system. In a noninteracting BEC state, the lowest energy state with momentum zero is macroscopically occupied, so the density distribution shows the singular behavior

$$n(\mathbf{p}) = N_0 \delta(\mathbf{p}) + \tilde{n}(\mathbf{p}), \tag{4.9}$$

where $\tilde{n}(\mathbf{p})$ is a smooth function. The long-distance behavior of $n^{(1)}(\mathbf{s})$ is then dominated by the delta function in the above equation. Specifically, it approaches a nonzero value as $s \to \infty$

$$n^{(1)}(\mathbf{s})_{s\to\infty} \to \frac{N_0}{V}. \tag{4.10}$$

This behavior is referred to as *off-diagonal long-range order.*

$n^{(1)}(\mathbf{s})$ is an important function that gives information about the thermodynamics of the system and the ordering mechanism. In the exciton–polariton system, $n^{(1)}(\mathbf{s})$ is a measurable quantity. In an interference experiment between two points at distance $\mathbf{s}$, similar to a Young's double slit setup, $n^{(1)}(\mathbf{s})$ is proportional to the fringe visibility.

For convenience, we also define the *first-order correlation function* $g^{(1)}(\mathbf{s})$, which is just $n^{(1)}(\mathbf{s})$ normalized by the local particle density

$$g^{(1)}(\mathbf{r}) = \frac{\left\langle \hat{\psi}^{\dagger}(\mathbf{r}+\mathbf{s})\hat{\psi}(\mathbf{r}) \right\rangle}{\sqrt{\left\langle \hat{\psi}^{\dagger}(\mathbf{r}+\mathbf{s})\hat{\psi}(\mathbf{r}+\mathbf{s}) \right\rangle \left\langle \hat{\psi}^{\dagger}(\mathbf{r})\hat{\psi}(\mathbf{r}) \right\rangle}} = \frac{n^{(1)}(\mathbf{r}+\mathbf{s}, \mathbf{r})}{\sqrt{n(\mathbf{r}+\mathbf{s})n(\mathbf{r})}}. \tag{4.11}$$

To quantitatively describe an interacting BEC state, we use the (time dependent) *Gross–Pitaevskii equation* (GPE)

$$\mathrm{i}\hbar \frac{\partial \Psi(\mathbf{r}, t)}{\partial t} = \left[-\frac{\hbar^2}{2m^*}\nabla^2 + V_{\mathrm{ext}}(\mathbf{r}, t) + g\left|\Psi(\mathbf{r}, t)\right|^2 \right] \Psi(\mathbf{r}, t), \tag{4.12}$$

where m^* is the single-particle mass, $V_{\text{ext}}(\mathbf{r}, t)$ is an externally applied potential, and g is the interaction parameter. The above equation is a good approximation as long as the interaction is sufficiently weak and short-ranged. In particular, it works for a weak interraction of the form

$$U(\mathbf{r}) = g\delta(\mathbf{r}). \tag{4.13}$$

If the external potential $V_{\text{ext}}(\mathbf{r}, t)$ is time independent, then (4.12) has solutions of the form $\Psi(\mathbf{r}) \exp(-\mathrm{i}\mu t/\hbar)$, where $\Psi(\mathbf{r})$ satisfies the time-independent GPE

$$\mu\Psi(\mathbf{r}) = \left[-\frac{\hbar^2}{2m^*}\nabla^2 + V_{\text{ext}}(\mathbf{r}) + g\,|\Psi(\mathbf{r})|^2 \right] \Psi(\mathbf{r}). \tag{4.14}$$

In a region of space with nearly constant density $n_0(\mathbf{r}) = |\Psi(\mathbf{r})|^2$, we can define a characteristic length

$$\xi = \frac{\hbar}{\sqrt{2m^* n_0 g}}. \tag{4.15}$$

ξ is called the *healing length*, and it is the lengthscale over which $\Psi(\mathbf{r})$ can change significantly. To make this point more explicit, consider a situation in which a condensate is confined in a box, so that the order parameter vanishes on the wall, which is defined by the plane $z = 0$. Then, as $z \to 0^+$, $\Psi(z)$ falls off to zero as $\tanh\left[z/(\sqrt{2}\xi)\right]$. Therefore, we can think of the healing length as some kind of screening length, since obstacles (e.g., walls) are not visible at distances longer than ξ. In the subsequent section, we will see that ξ is also the size of the vortex core.

4.2.2 *Superfluidity*

The most famous criterion for superfluidity is the Landau criterion. Landau [16] showed that if $\epsilon(p)$ is the excitation spectrum of a liquid, then a particle moving inside the liquid with velocity smaller than the critical value

$$v_L = \left[\frac{\epsilon(p)}{p} \right]_{\text{min}} \tag{4.16}$$

does not experience any dissipation.

As is now understood, the Landau criterion is not a sufficient condition for superfluidity. Indeed, superfluidity is a collective phenomenon. When a superfluid moves, all particles coherently participate in this motion. Therefore, phase coherence needs to be established across the whole volume. The idea of phase coherence and long-range order is the basis of the modern understanding of superfluidity [17].

Furthermore, experiments designed to directly test the Landau criterion are difficult to analyze. In particular, when a large object moves inside a superfluid, or when a superfluid enters a small tube, turbulent flow occurs and vortices can be excited. Creation and movement of vortices causes dissipation, so the critical velocity measured in this type of experiments is almost always smaller than what (4.16) implies.

Conceptually simpler manifestations of superfluidity occur when a superfluid is put into rotation [17]. We start from the assumption that the superfluid component can be described by an order parameter. But now $|\Psi(\mathbf{r})|$ is going to represent the superfluid density instead of the condensate density. The theoretical foundation of this assumption is not rigorous. There is strong evidence that BEC of the type such as He II has a condensate fraction $\sim$10% at $T = 0$. But as we shall see later, superfluidity can occur without BEC. So, one can only argue that, based on the current experimental facts, the phase coherence established across 3D superfluids allows the description of the superfluid component by a complex order parameter $\Psi(\mathbf{r})$.

$\Psi(\mathbf{r})$ is characterized by its amplitude $|\Psi(\mathbf{r})|$ and phase $\phi(\mathbf{r})$

$$\Psi(\mathbf{r}) = |\Psi(\mathbf{r})| \exp\left[\mathrm{i}\phi(\mathbf{r})\right]. \tag{4.17}$$

Similar to the quantum mechanical definition of probability density and current density, we can define the superfluid density $\rho_s(\mathbf{r})$ and current $\mathbf{j}_s(\mathbf{r})$

$$\rho_s(\mathbf{r}) = |\Psi(\mathbf{r})|^2, \tag{4.18}$$

$$\mathbf{j}_s(\mathbf{r}) = -\frac{\mathrm{i}\hbar}{2m^*}\Psi^*(\mathbf{r})\nabla\Psi(\mathbf{r}) + \text{c.c.} = |\Psi(\mathbf{r})|^2 \frac{\hbar}{m^*}\nabla\phi(\mathbf{r}). \tag{4.19}$$

The ratio $\mathbf{j}_s(\mathbf{r})/\rho_s(\mathbf{r})$ has units of velocity. We call it the *superfluid velocity*, and it depends only on the superfluid phase

$$\mathbf{v}_s \equiv \frac{\hbar}{m^*}\nabla\phi(\mathbf{r}). \tag{4.20}$$

An immediate consequence of the definition (4.20) is that wherever $\Psi(\mathbf{r})$ is nonzero (so that $\phi(\mathbf{r})$ is well defined), we obtain

$$\nabla \times \mathbf{v}_s(\mathbf{r}) = 0. \tag{4.21}$$

That is, the velocity field is irrotational. This means that the integral of $\mathbf{v}_s$ over a closed contour will be zero unless the contour encloses one or more singularities, namely, points where $\Psi(\mathbf{r})$ vanishes. Since $\phi(\mathbf{r})$ is defined modulo 2π, the change of phase $\Delta\phi$ around a contour can only be an integer multiple of 2π

$$\Delta\phi = \oint \nabla\phi \cdot \mathrm{d}\mathbf{l} = 2\pi l, \quad l = 0, \pm 1, \pm 2, \ldots \tag{4.22}$$

So, the *circulation* Γ around a close contour is quantized

$$\Gamma \equiv \oint v_s \cdot \mathrm{d}\mathbf{l} = l\frac{h}{m^*}. \tag{4.23}$$

Equation (4.23) is the *Onsager–Feynman quantization condition*.

4.2.3 Quantized Vortices

One of the most unusual features of superfluids is the appearance of quantized vortices [18]. As remarked in the discussion before (4.23), the superfluid circulation can be nonzero if a singularity (a point in which the order parameter $\Psi(\mathbf{r})$ vanishes) exists in the fluid. In a 3D geometry, such singularities form lines that either terminate at the boundaries of the fluid or close upon themselves. In the former case, they are called *vortex lines*, while in the latter, they are called *vortex rings*.

The order parameter around a vortex can be calculated using the time-independent GPE (4.14) [19]. Consider an infinite superfluid with a vortex line along the z-axis. In a loop around the z-axis, the phase changes by an integer multiple of 2π, thus in cylindrical coordinates, the order parameter can be written as

$$\Psi(\mathbf{r}) = f(r)\mathrm{e}^{\mathrm{i}l\phi}. \tag{4.24}$$

We assume translational symmetry along the z-axis, so $\Psi(\mathbf{r})$ does not depend on z. Inserting this form into (4.14) and setting the external potential to zero, we find

$$-\frac{\hbar^2}{2m^*}\frac{1}{r}\frac{\partial}{\partial r}\left(r\frac{\partial f}{\partial r}\right) + \frac{\hbar^2}{2m^* r^2}l^2 f + g f^3 = \mu f. \tag{4.25}$$

Far from the vortex, the radial derivative $\frac{\partial f}{\partial r}$ and the term proportional to $1/r^2$ can be neglected, so f approaches the value

$$f(r)_{r\to\infty} = \sqrt{\mu/g} \equiv f_0. \tag{4.26}$$

Equation (4.25) can be transformed to a dimensionless equation by dividing r with the healing length ξ defined (4.15) and f with f_0. Then, it can be solved numerically. A good approximation to the solution of singly quantized vortex ($l = 1$) is

$$f(r) = f_0\frac{r}{\sqrt{2\xi^2 + r^2}} \Rightarrow \Psi(r,\phi) = f_0\frac{r}{\sqrt{2\xi^2 + r^2}}\mathrm{e}^{\mathrm{i}\phi}. \tag{4.27}$$

Let us look at the qualitative features of solution (4.27). The order parameter drops to zero on the z-axis $r = 0$, namely, along the vortex line. The phase ϕ is undefined along the z-axis, so this is a line of singularities. The amplitude of the order parameter Ψ is perturbed in a cylindrical region or radius $\sim \xi$ around the vortex line. This region is called *vortex core*. The phase of Ψ, however, is perturbed everywhere in space. In a contour around the vortex line, the change of phase will always be 2π even if the contour is infinitely far from the vortex core. These qualitative features are always true, even in a strongly interacting superfluid like He II, in which GPE is not a good approximation.

A vortex line is a metastable entity. Obviously, the ground state of a superfluid does not contain any vortices. But because the phase is perturbed everywhere around a vortex line, disappearance of a vortex from inside the bulk of the fluid is statistically highly improbable. A vortex can only disappear by reaching the boundary or by annihilating with an antivortex of opposite circulation.

The energy per unit length of a vortex line can be calculated by appropriate integration of the solution (4.27). A simpler and physically more transparent approach [18] is to note that from (4.23), the velocity field around the vortex line is

$$v(r) = \frac{\Gamma}{2\pi r}. \tag{4.28}$$

Neglecting the core of the vortex and assuming that the fluid extends up to radius R, we find that the kinetic energy per unit length associated with a single vortex is

$$K = \int_{\xi}^{R} \frac{1}{2}\rho_s \left(\frac{\Gamma}{2\pi r}\right)^2 2\pi r \mathrm{d}r = \frac{\rho_s \Gamma^2}{4\pi} \ln\left(\frac{R}{\xi}\right). \tag{4.29}$$

Γ is the total circulation lh/m^*, where $l = 0, \pm 1, \pm 2, \ldots$. In a 3D geometry, the total energy of the vortex line is proportional to its length, so it increases linearly with the system size. On the contrary, in a 2D geometry, the vortex energy increases as the logarithm of the system size; therefore, it is energetically easier to excite. This logarithmic functional dependence has profound implications, which will be discussed later.

Now consider a multiply quantized vortex ($|l| > 1$). Equation (4.29) predicts that such a singularity will be unstable, and it will eventually dissociate into multiple singly quantized vortices. For example, the kinetic energy of a doubly quantized vortex ($l = 2$) is four times the energy of a singly quantized vortex. Thus, it is energetically preferable for it to split into two singly quantized vortices.

The reason why vortices are important in superfluidity is that they modify the flow properties of the fluid. For a review of the theory of quantized vortices in atomic gases, see [20, 21], and for a guide to all relevant experiments, see [22].

Consider a system of one vortex and one antivortex line and assume that the lines are straight and parallel to each other at distance x. Following (4.29), a good approximation for the energy of this system is

$$K = 2\int_{\xi}^{x} \frac{1}{2}\rho_s \left(\frac{\Gamma}{2\pi r}\right)^2 2\pi r \mathrm{d}r = \frac{\rho_s \Gamma^2}{2\pi} \ln\left(\frac{x}{\xi}\right). \tag{4.30}$$

Therefore, the vortex and antivortex attract each other with a force

$$F(x) = -\frac{\mathrm{d}K}{\mathrm{d}x} = -\frac{\rho_s \Gamma^2}{2\pi x}. \tag{4.31}$$

Because this is a Coulomb's law of attraction, the problem of an ensemble vortices and antivortices in a plane can be reduced to the problem of an ensemble of dipole charges. This idea was employed by Kosterlitz and Thouless, as will be outlined below.

Around a vortex, there is a rotating superfluid flow. Thus, in a system of a vortex and antivortex, the flow around the vortex drags the antivortex and vice versa. In the absence of any other forces, the pair moves along the direction perpendicular to the line connecting the vortex and antivortex, as shown in Fig. 4.1a with velocity $v = \Gamma/2\pi x$. In the case of two vortices, on the other hand, the system rotates with angular velocity $\omega = \Gamma/\pi x^2$ (Fig. 4.1b).

4.2.4 Low Dimensions and the Hohenberg–Mermin–Wagner Theorem

Dimensionality of a system profoundly influences its density of states and excitation spectrum and, consequently, its thermodynamics and phase diagram. We illustrate this point by a simple example. We consider the one-dimensional (1D) Ising model with Hamiltonian

$$\hat{\mathcal{H}} = -J \sum_{<i,j>} S_i S_j, \quad S_i = \pm\frac{1}{2}. \tag{4.32}$$

The index $< i, j >$ refers to pairs of nearest neighbors. We assume that the coupling constant J is positive, so that the ground state consists of all spins pointing along the same direction, as shown in Fig. 4.2a. An excited state is shown in Fig. 4.2b. In the latter state, a number of neighboring spins, namely, a domain inside the spin chain, have flipped. The domain wall consists of just two points, so the energy of this state

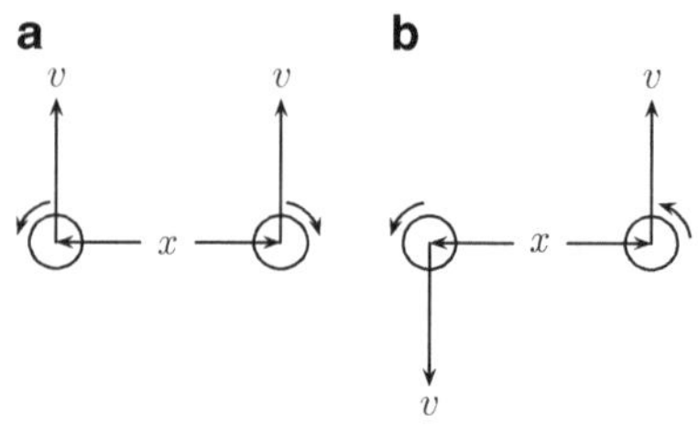

Fig. 4.1 Motion of vortex pairs. (**a**) In a vortex–antivortex pair, the antivortex is dragged by the velocity field around the vortex and vice versa. As a result, the pair undergoes linear motion. (**b**) A pair of two vortices rotates

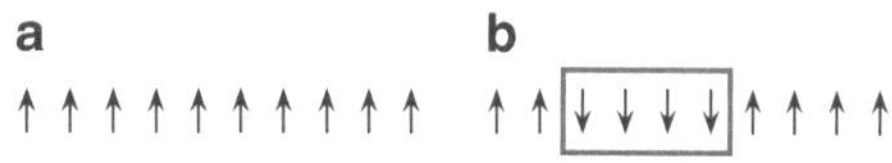

Fig. 4.2 One-dimensional Ising model. (**a**) Ground state. (**b**) Excited state

relative to the ground state is $\Delta E = J$, independent of the size of either the domain or the system itself. If m is the number of spins that have flipped relative to the ground state and N is the total number of spins, then there are in total $N - m - 1$ states with the same energy and total magnetization. Therefore, when the system is in a statistical mixture of all these states, it has entropy $\mathcal{S} = k_B \ln(N - m - 1)$. At nonzero temperature, the free energy $\mathcal{F} = E - T\mathcal{S}$ is minimized in equilibrium. In the above statistical mixture state, the change in free energy is

$$\Delta\mathcal{F} \equiv \Delta E - T\Delta\mathcal{S} = J - k_B T \ln(N - m - 1). \tag{4.33}$$

For an infinitely long chain ($N \to \infty$), $\Delta\mathcal{F}$ is always negative. Thus, even at infinitesimally small temperature, the system chooses to be in a mixture of excited states, and no long-range order can be established.

Another example relevant to the experiments described in this chapter is the case of a quantized vortex in a 2D superfluid. The energy of this state was calculated in (4.29) and is proportional to the logarithm of the system size R. In three dimensions, we need to further multiply by the length of the vortex line, so in that case, the vortex energy is proportional to the system size. Consider a 2D superfluid of size R^2 containing one vortex. There are R^2/ξ^2 possible positions for the vortex, so the free energy relative to the state with no vortex is

$$\Delta\mathcal{F} = \frac{\rho_s \Gamma^2}{4\pi} \ln\left(\frac{R}{\xi}\right) - k_B T \ln\left[\left(\frac{R}{\xi}\right)^2\right] = \left(\frac{\rho_s \Gamma^2}{4\pi} - 2k_B T\right) \ln\left(\frac{R}{\xi}\right), \tag{4.34}$$

where $\Gamma = h/m^*$ for singly quantized vortices. Depending on the temperature T, free vortices may or may not spontaneously occur in the system. When they do occur, they perturb the phase at infinite distances around them, so no long-range order (LRO) can occur. This is the essential idea behind the Kosterlitz–Thouless transition.

Hohenberg [11] has rigorously proven that the "assumption of long range order in Bose or Fermi liquids leads to a contradiction in one and two dimensions at finite temperature." His argument is of quite general applicability, and in particular, Mermin and Wagner [23] extended it to the one- and two-dimensional isotropic Heisenberg model with interactions of finite range and proved that this system can be "neither ferromagnetic nor antiferromagnetic at nonzero temperature." A special case of the Hohenberg argument is that the noninteracting 2D Bose gas does not show conventional BEC, in contrast to the 3D gas. In all those cases, LRO is destroyed by long-wavelength (low energy) thermal fluctuations. In the case of an interacting Bose gas, the long-wavelength modes are phonons, while in the Heisenberg model, they are spin waves. In both cases, because of the form of the density of states at low temperature, long-wavelength modes have a diverging contribution in thermal averages and destroy LRO. The above is true for an infinitely large system. In a finite geometry, however, considerable coherence can be established across the whole system.

4.2.5 *Two-Dimensional Bose Gas*

In this section, we present the theory of the two-dimensional (2D) Bose gas. We are mostly interested in the behavior of the correlation function, which we calculate for the homogeneous noninteracting and interacting gas. We also discuss the trapped noninteracting gas case, in which there is a crossover to a BEC-like state. We follow the pedagogical approach of [24] and [25]. See [26] for a review of more sophisticated theories and [5] for a review of some relevant experiments.

4.2.5.1 Uniform Noninteracting 2D Bose Gas

Consider a uniform noninteracting 2D Bose gas. The density of states for periodic boundary conditions is the constant $m^* L^2/(2\pi\hbar^2)$, where m^* is the particle mass and L is the linear dimension of the system. The total particle number is

$$N = N_0 + \frac{m^* L^2}{2\pi\hbar^2} \int_{\epsilon_0}^{\infty} \frac{\mathrm{d}\epsilon}{\mathrm{e}^{\beta(\epsilon-\mu)} - 1}. \tag{4.35}$$

When μ approaches the lowest energy ϵ_0, the integral in (4.35) diverges. This is easy to see from the Taylor expansion of the integrand

$$\left.\frac{1}{\mathrm{e}^{\beta(\epsilon-\epsilon_0)} - 1}\right|_{\epsilon\to\epsilon_0} \simeq \frac{1}{[1 + \beta(\epsilon - \epsilon_0)] - 1} = \frac{1}{\beta(\epsilon - \epsilon_0)}, \tag{4.36}$$

which shows a nonintegrable infrared (low energy) divergence. Therefore, the critical particle number N_c is infinite, so that always $N_0/N \ll 1$, and no BEC occurs.

We define the 2D *phase-space density* $D \equiv n\lambda_T^2$, where n is the particle density and λ_T is the thermal de Broglie wavelength $\lambda_T = h/\sqrt{2\pi m^* k_B T}$. In a semiclassical approach, λ_T is the extent of the wavepacket required to describe each particle at a temperature T. D is a dimensionless number, which is a measure of the interparticle separation relative to λ_T. When $D \ll 1$, the interparticle separation is much larger than λ_T, so the single-particle wavepackets do not overlap and the gas can be considered classical. Such a gas is also called *nondegenerate*. On the contrary, when $D > 1$ (*degenerate gas*), the single-particle wavepackets overlap and quantum effects arise. In the rest of this section, we use D as a free parameter, instead of the temperature T or density n. From (4.35), assuming $N_0 = 0$, $\epsilon_0 = 0$

$$D = \int_0^\infty \frac{\mathrm{d}x}{\mathrm{e}^{-\beta\mu}\mathrm{e}^x - 1} = \int_0^\infty \frac{\mathrm{e}^{\beta\mu}\mathrm{e}^{-x}}{1 - \mathrm{e}^{\beta\mu}\mathrm{e}^{-x}}\,\mathrm{d}x = -\ln\left(1 - \mathrm{e}^{\beta\mu}\right), \quad (4.37)$$

$$\Leftrightarrow \mu = k_{\mathrm{B}} T \ln\left(1 - \mathrm{e}^{n\lambda_{\mathrm{T}}^2}\right). \quad (4.38)$$

We now use (4.8) to calculate the off-diagonal part of the one-body density matrix $n^{(1)}(s)$. The momentum distribution function $n(\hbar\mathbf{k})$ is related to the occupation numbers $n_{\mathbf{k}}$ by

$$n(\hbar\mathbf{k}) = \left(\frac{L}{2\pi\hbar}\right)^2 n_{\mathbf{k}}; \quad (4.39)$$

therefore,

$$n^{(1)}(\mathbf{s}) = \frac{1}{(2\pi)^2}\int n_{\mathbf{k}}\mathrm{e}^{\mathrm{i}\mathbf{k}\cdot\mathbf{s}}\,\mathrm{d}\mathbf{k}. \quad (4.40)$$

For a nondegenerate gas ($D \ll 1$), (4.37) gives $\mathrm{e}^{\beta\mu} \simeq D$, and the momentum distribution approaches the Maxwell–Boltzmann distribution

$$n_k = n\lambda_{\mathrm{T}}^2 \mathrm{e}^{-k^2\lambda_{\mathrm{T}}^2/4\pi}. \quad (4.41)$$

From (4.40), we find that there are only short-range correlations and $n^{(1)}(s)$ has a Gaussian form

$$n^{(1)}(s) = n\mathrm{e}^{-\pi s^2/\lambda_{\mathrm{T}}^2}. \quad (4.42)$$

For a degenerate gas ($D > 1$), $\beta\mu \ll 1$. The high-energy states $\beta\epsilon_k \gg 1$ still follow the Maxwell–Boltzmann distribution and are responsible for the Gaussian decay of $n^{(1)}(s)$ (4.42) up to distances $\sim \lambda_{\mathrm{T}}$. The low-energy states ($\beta\epsilon_k \ll 1 \Leftrightarrow k^2 \ll 4\pi/\lambda_{\mathrm{T}}^2$), on the other hand, are strongly occupied

$$n_k = \frac{1}{\mathrm{e}^{\beta(\mu+\epsilon_k)} - 1} \simeq \frac{1}{\beta(\mu + \epsilon_k)} = \frac{4\pi}{\lambda_{\mathrm{T}}^2}\frac{1}{k^2 + k_{\mathrm{c}}^2}, \quad (4.43)$$

where $k_{\mathrm{c}} = \sqrt{2m|\mu|}/\hbar$. Therefore, for $s \gg \lambda_{\mathrm{T}}$, $n^{(1)}(s)$ falls exponentially

$$n^{(1)}(s) \simeq \frac{\sqrt{2\pi}}{\lambda_{\mathrm{T}}^2}\frac{\mathrm{e}^{-k_{\mathrm{c}}s}}{\sqrt{k_{\mathrm{c}}s}}, \qquad k_{\mathrm{c}}^{-1} = \lambda_{\mathrm{T}}\mathrm{e}^{n\lambda_{\mathrm{T}}^2/2}/\sqrt{4\pi}. \quad (4.44)$$

4.2.5.2 Trapped Noninteracting 2D Bose Gas

In the previous discussion, we showed that in a uniform noninteracting 2D Bose gas, the ground state population N_0 is always nonmacroscopic; therefore, no LRO can be established. However, the correlation length k_{c}^{-1} calculated in (4.44) grows exponentially as the temperature is reduced, or the density is increased. In a finite system, when this correlation length is of the same order as the system size, the ground state population N_0 can be a considerable fraction of the total particle number N and the system can show BEC-like features. In this section, we explicitly solve the problem of a trapped ideal 2D Bose gas.

Consider a gas trapped inside an infinitely deep quantum well of size L^2. The density of states is $2m^*L^2/\pi\hbar^2$, so the total number of particles is

$$N = N_0 - \frac{2m^*L^2}{\pi\hbar^2} k_B T \ln\left(1 - e^{\mu/k_B T}\right). \tag{4.45}$$

Using Bose–Einstein distribution, we arrive at the exact expression

$$N = N_0 + \frac{2m^*L^2}{\pi\hbar^2} k_B T \left[\ln(N_0) + \ln\left(1 + \frac{1}{N_0}\right)\right]. \tag{4.46}$$

Assuming $N \gg 1$ and that the temperature is small enough so that $N_0 \approx N$

$$N_0 = N\left(1 - \frac{T}{T_c}\right), \qquad k_B T_c = \frac{\pi\hbar^2 N}{2m^*L^2 \ln(N)}. \tag{4.47}$$

In Fig. 4.3, we plot the approximate expression (4.47) for the ground state fraction. N_0/N as a function of temperature T, as well as the solution to the exact expression (4.46) for several total particle numbers N. As $N \to \infty$, the exact solution approaches the approximation (4.47).

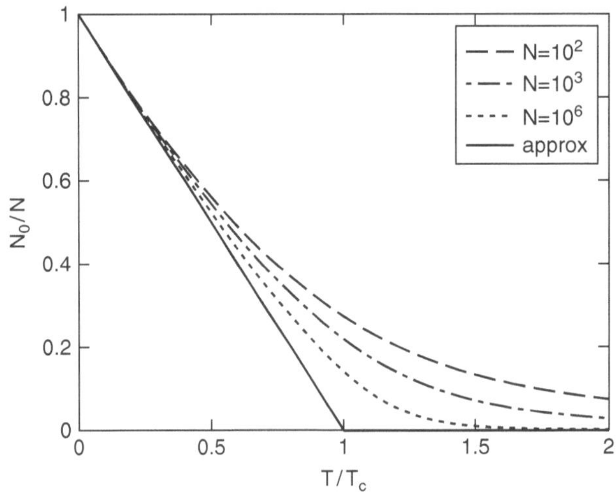

Fig. 4.3 2D trapped BEC. Ground state fraction N_0/N vs temperature T in a 2D Bose gas trapped in an infinitely deep quantum well. As the total number of atoms increases, solution of the exact expression (4.46) converges to (4.47), which is shown in continuous line

4.2.5.3 Uniform Interacting 2D Bose Gas

In (4.14), we wrote down the Gross–Pitaevskii equation, which depends on the interaction parameter g. We argued that (4.14) is valid when the interaction is weak. What "weak interaction" exactly means depends on the system dimensionality. In 2D geometry, we can define a dimensionless coupling constant $\tilde{g}$ from

$$g = \frac{\hbar^2}{m^*}\tilde{g}. \tag{4.48}$$

We can gain some intuition on what $\tilde{g}$ physically means from the definition of the healing length (4.15)

$$\xi = \frac{\hbar}{\sqrt{2m^* ng}} = \frac{1}{\sqrt{2\tilde{g}n}}. \tag{4.49}$$

In the discussion following (4.15), we argued that ξ is a lengthscale over which obstacles are screened from the bulk of the condensate. If interactions are strong, screening is effective, and ξ is small. From (4.49), we see that $\tilde{g}$ determines the ratio of ξ over the interparticle distance $1/\sqrt{n}$.

We can define the weak interaction limit as the values of $\tilde{g}$ for which the interaction energy of N particles is much smaller than the kinetic energy of N noninteracting particles equally distributed over the N lowest available states. Following this approach, the weak interaction limit is

$$\tilde{g} \ll 2\pi. \tag{4.50}$$

At zero temperature, a weakly interacting gas is condensed and is described by an order parameter $\Psi(\mathbf{r}) = \sqrt{n(\mathbf{r})}\mathrm{e}^{\mathrm{i}\phi(\mathbf{r})}$. At nonzero temperature, both the amplitude $\sqrt{n}$ and phase $\phi(\mathbf{r})$ of Ψ show thermal fluctuations. When the mean-field interaction energy gn is much larger than the thermal energy $k_{\mathrm{B}}T$, or equivalently when the phase-space density $D \gg 2\pi/\tilde{g}$, amplitude fluctuations are suppressed. The interaction energy is then a constant, and the only important contribution to the system energy is the kinetic energy of the superfluid component

$$H = \int \frac{1}{2}m^* n_{\mathrm{s}} \left|\mathbf{v}_s\right|^2 \mathrm{d}\mathbf{r} = \frac{\hbar^2}{2m^*}n_{\mathrm{s}} \int \left[\nabla\phi(\mathbf{r})\right]^2 \mathrm{d}\mathbf{r}, \tag{4.51}$$

where we have used (4.20) for the superfluid velocity v_{s}. Hamiltonian (4.51) completely neglects density fluctuations, which have a characteristic lengthscale of ξ. It also neglects higher energy modes, which involve both phase and density fluctuations. These modes determine the off-diagonal part of the one-particle density matrix $n^{(1)}(\mathbf{r})$ at distances $r \lesssim \lambda_{\mathrm{T}}$. Therefore, (4.51) describes long-range physics, namely, distances $r \gg \xi, \lambda_{\mathrm{T}}$.

We can then expand $\phi(\mathbf{r})$ in Fourier series, assume thermal occupation for each mode, and calculate

$$n^{(1)}(r) = \langle \Psi^*(\mathbf{r})\Psi(0)\rangle = n_s \left\langle e^{i[\phi(\mathbf{r})-\phi(0)]} \right\rangle. \tag{4.52}$$

The calculation uses the identity $\langle \exp(iu)\rangle = \exp\left(-\langle u^2\rangle/2\right)$ for a variable u which follows Gaussian statistics. The final result [24] is a power-law decay

$$n^{(1)}(\mathbf{r}) = n_s \left(\frac{\xi}{r}\right)^{1/\left(n_s \lambda_T^2\right)}. \tag{4.53}$$

This result suggests that even in infinitesimally small temperatures, $n^{(1)}(r)$ goes to zero as the distance r grows to infinity. So, no long-range order of the type defined in (4.10) can be established.

We have proven (4.53) in the limit $D \to \infty$. From the above discussion, it is not clear for which values exactly of the phase-space density Hamiltonian (4.51) is valid and, therefore, when we expect the power-law decay (4.53). Also, result (4.53) only proves that there can be no conventional LRO in the interacting Bose gas, but it leaves the question of whether there is another phase transition open. To find out the answers to these questions, we also need to take into account vortex excitations, which, as it turns out, dominate the thermodynamics at low temperature.

4.2.6 *The Berezinskii–Kosterlitz–Thouless Transition*

In the previous section, we saw that no conventional LRO can be established in an interacting Bose gas, since the correlation function $n^{(1)}(r)$ decays to zero as $r \to \infty$ at nonzero temperatures. But this result does not rule out the existence of a state with superfluid properties. In 1972, Kosterlitz and Thouless developed an elegant theory that explores the role played by quantized vortices [12,27]. The phase transition they discovered is now termed *Kosterlitz–Thouless phase transition* (KT phase transition) or BKT phase transition after Berezinskiǐ who put forward similar arguments at about the same time [28, 29].

At low enough temperature, the correlation function $n^{(1)}(r)$ decays slowly, so that superfluid properties can be observed locally. In particular, consider a small enough contour, so that the superfluid phase can be defined at all points on it. The phase change $\Delta\phi$ along the contour can only be a multiple of 2π. If $\Delta\phi \neq 0$, a superfluid current will flow, which decays when $\Delta\phi$ changes by 2π. This change can only occur when a quantized vortex moves across the contour, which is possible when free vortices exist in the system. Thus, a state with free vortices is nonsuperfluid. On the other hand, at very low temperatures, vortices and antivortices gather in clusters of zero total vorticity. When one cluster moves across the contour, the total phase change $\Delta\phi$ does not change, so supercurrents do not dissipate. This is the essential idea behind the BKT transition.

Equation (4.34) gives a first estimate of the BKT transition temperature

$$k_B T_{BKT} = \frac{\pi \hbar^2 n_s}{2m^*}. \tag{4.54}$$

When $T < T_{BKT}$ free vortices are not energetically allowed, whereas higher temperatures $T > T_{BKT}$ are favorable for the proliferation of free vortices. The BKT critical temperature (4.54) depends on the superfluid density n_s, which is not given beforehand. So this equation is not useful for computation of the critical temperature given the system parameters. However, the alternative form

$$n_s \lambda_{T_{BKT}}^2 = 4, \tag{4.55}$$

where $\lambda_{T_{BKT}}$ is the de Broglie wavelength at the critical temperature, predicts an interesting result. Since n_s can only increase when the temperature is decreased, we see that n_s cannot have any value smaller than $4/\lambda_T^2$ in the superfluid phase. Of course, $n_s = 0$ in the normal (nonsuperfluid) phase. Indeed, Nelson and Kosterlitz predicted [30] that the BKT transition is of first order, as there is a jump in the superfluid density n_s when the transition temperature is crossed. Another consequence of (4.55) is that the exponent in the power-law decay (4.53) is always smaller than $1/4$.

In (4.31), we showed that vortices of opposite sign attract each other with a Coulomb-like force. A similar energy argument can be used to show that vortices of the same sign repel each other according to the same law. Therefore, the problem of a 2D superfluid with an ensemble of vortices can be mapped to the problem of charged rods moving in the plane perpendicular to their axis [27]. The unbinding of vortices in the superfluid formally corresponds to the divergence of polarizability in the charged rod problem. Namely, a small electric field is capable of separating a positively charged rod from a negatively charged one at infinite distance. The mean square separation between pairs due to thermal fluctuations is

$$\langle r^2 \rangle = \frac{\int_\xi^\infty r^2 e^{-\beta K} 2\pi r dr}{\int_\xi^\infty e^{-\beta K} 2\pi r dr} = \frac{1}{2}\xi^2 \frac{k_B T - 2\frac{\pi \hbar^2 n_s}{2m^*}}{k_B T - \frac{\pi \hbar^2 n_s}{2m^*}}, \quad \text{for } T < \frac{\pi \hbar^2 n_s}{2m^*}. \tag{4.56}$$

The polarizability indeed diverges at the transition temperature (4.54).

For a more complete treatment, we need to consider how the interaction between vortices is influenced by the presence of other pairs. Namely, pairs of small size screen the interaction between vortices at long distances. In the problem of charged rods, this can be included as a distance-dependent dielectric constant $\epsilon(r)$ in the Coulomb law, so that the energy of a rod dipole is

$$U = \int_\xi^\infty \frac{q^2}{4\pi\epsilon(r)} dr - qEr\cos\theta. \tag{4.57}$$

In the above equation, q is the charge of each rod, E is the externally applied electric field, and θ is the angle of the dipole with respect to the electric field direction. The polarizability of the pair is then given by

$$p(r) \equiv q \frac{\partial}{\partial E} \langle r \cos\theta \rangle \Big|_{E\to 0}. \tag{4.58}$$

The polarization of small pairs (the charges at short distance) influences the energy of large pairs (the charges at long distance). We must account self-consistently for the short-distance behavior in order to calculate the long-distance response. The calculation [27] results in an improved estimate of the transition temperature T_{BKT}. As T_{BKT} is approached from below, the dielectric function at large distances diverges

$$\epsilon(r \to \infty)\big|_{T\to T_{\mathrm{BKT}}^-} \to \infty. \tag{4.59}$$

As T is further increased above T_{BKT} and $\epsilon(r)$ diverges for progressively shorter distances r. Physically, this means that pairs of infinite size dissociate first, and as the temperature is increased, pairs of smaller size become unbound, too.

As we discussed, (4.55) is not suitable for computation of the transition temperature given the system parameters. In particular, it does not predict any dependence of T_{BKT} on the interparticle interaction. Microscopic theories give the following result for the critical phase-space density [26, 31–33].

$$D_{\mathrm{BKT}} = n\lambda^2_{T_{\mathrm{BKT}}} = \ln\left(\frac{C}{\tilde{g}}\right), \qquad C = 380 \pm 3. \tag{4.60}$$

The Berezinskii–Kosterlitz–Thouless theory does not depend on the microscopic details of the system in study. As long as the geometry is two dimensional and the order parameter can be represented by an in-plane vector, the same arguments apply [34, 35], so that the theory can be considered to be universal. In a superfluid, the local order parameter is the superfluid wavefunction $\Psi(\mathbf{r}) = \sqrt{n_s(\mathbf{r})}\exp[i\phi(\mathbf{r})]$, but because at low temperature the amplitude fluctuations are frozen, the superfluid can be described solely by its phase $\phi(\mathbf{r})$ and the constant n_s. $\phi(\mathbf{r})$ indeed defines an in-plane vector. The 2D x-y model consists of a spin lattice, in which spins are free to move only on the lattice plane. So, every lattice cite is characterized by the angle $\phi(\mathbf{R})$ of its spin. A 2D crystal can be described by the displacement vector of the atom at each lattice cite. In all the above cases, when interactions are included, the low-energy Hamiltonian is of the form (4.51), and vortex dislocations are stable at low temperature.

In the x-y model, a vortex consists of a point around which the angle of spins changes by π. Because spins cannot precess out of the plane, a large energy barrier needs to be overcome in order for the vortex to be destroyed. This is shown schematically in Fig. 4.4. Considering the Hamiltonian

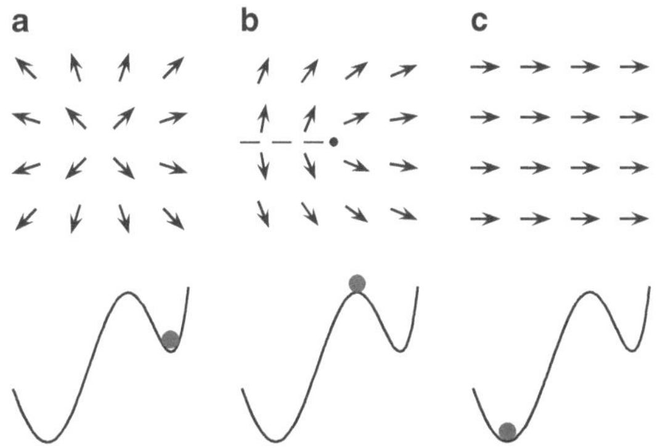

Fig. 4.4 Vortex stability in the 2D x-y model (*see text*). *Upper row*: schematic of (**a**) a vortex state, (**c**) the ground state of (4.61), and (**b**) a high-energy intermediate state between (**a**) and (**c**). *Bottom row*: schematic of the corresponding energy diagrams. The vertical axis is the energy, and the horizontal axis is an arbitrary parameter that interpolates between states (**a**) and (**c**)

$$\hat{\mathcal{H}} = -J \sum_{\langle i,j \rangle} \mathbf{S}_i \cdot \mathbf{S}_j, \tag{4.61}$$

the intermediate state (Fig. 4.4b) between a vortex state (Fig. 4.4a) and the ground state (Fig. 4.4c) shows a "cut" along one line. Thus, the energy of the state in Fig. 4.4b is proportional to $J \times L$, where L is the linear dimension of the system. This energy takes large values for large system sizes. As stated in [35], "The vortex state, even though it has a much higher energy than the ground state, is stable because there is no path to the ground state that is not energetically costly."

On the contrary, the BKT theory does not apply to the Heisenberg model, because of the dimensionality of the local order parameter. This model consists of a lattice of spins, which are free to rotate in all three directions. Consider a vortex of the type shown in Fig. 4.4a. We could slowly turn all spins toward the direction perpendicular to the lattice plane, so that the ground state could be reached without passing through an intermediate state of high energy.

Experimentally, the BKT theory has been studied in ^{4}He thin films [36, 37], Josephson junction arrays [38, 39], atomic hydrogen [40], and, more recently, in 2D atomic gas [41–44]. Although several of the predictions of the theory could be tested, in none of these experiments was it possible to directly measure the correlation function $n^{(1)}(r)$ and bound pair of vortex–antivortex. In this chapter, we present a direct measurement of the normalized correlation function $g^{(1)}(\mathbf{r})$ and the vortex–antivortex pair in a microcavity exciton–polariton condensate. Although the theory developed in the current section cannot be applied directly to such a nonequilibrium system, it can still be used as a guide to understand the experimental data and to point out the features arising from nonequilibrium physics.

4.3 Basic Characterisation of the Sample and Condensate

In this section, we present the initial characterization of the sample in order to estimate important parameters and associate the observed transition with exciton–polariton condensation.

4.3.1 *Major Scientific Instruments*

In the experiments described in this chapter, we used a Coherent Mira 900 Ti:Sapphire multimode laser. When used in modelocked mode, it generates ps-long pulses at 76 MHz repetition rate. In our experiments, however, we operate it in the continuous wave mode. It is pumped with a Verdi V-10 laser, operated at 8 W.

The input power is controlled by a variable metallic neutral density filter and a combination of a Meadowlark liquid crystal variable retarder and a polarizer. To create a flat-top pumping profile, we employ a refractive beam shaper (Newport GBS-AR16) designed for near infrared wavelength. The beam shaper transforms a collimated Gaussian beam of a particular diameter to a collimated flat-top beam. Since it uses low-dispersion refractive elements (lenses), its efficiency is much higher compared to diffractive beam shapers, and it operates over a wider wavelength range.

The sample is kept in vacuum inside a Janis ST-100 optical cryostat. Continuous flow of liquid helium cools the cold finger on which the sample is mounted using silver paint for good thermal conductivity. The cold finger is custom-designed so that the sample is as close as possible to the optical window. We collect luminescence through a long working distance objective lens (Mitutoyo BD Plan Apo 378-835-4). The numerical aperture (NA) is 0.55, the effective focal length is 4.0 mm, and the working distance is 13.0 mm. The long working distance provides us with enough flexibility in placement of the sample behind the cryostat window.

Spectroscopy is performed with an Acton Spectrapro 2,750 spectrometer of 0.75 m focal length coupled to a Princeton Instruments SPEC-10:256 Charge Couple Device (CCD) camera. The camera is cooled with liquid nitrogen and has an imaging array of $1,024 \times 256$ pixels of size $26 \times 26 \mu m^2$ each. Three different gratings are installed in the spectrometer. For high-resolution spectroscopy, we use the 1,800 g/mm holographic grating. To capture a large wavelength range, as in reflection measurements with white light, we use the 300 g/mm grating. The CCD camera in the Michelson interferometer setup is a high-sensitivity back-illuminated electron-multiplying CCD (EMCCD Andor iXon DV885KCS-VP). It has $1,004 \times 1,002$ pixels of size $8 \times 8 \mu\,\mathrm{m}^2$ each. Other CCD cameras used for recording real-space and momentum-space images are the Astrovid StellaCam II and the Watec 120N.

For time-resolved measurements, we use a Hamamatsu C5680 streak camera with 2 ps temporal resolution. The streak camera is placed at the output of a Chromex 500 is spectrometer. Finally, for motion control, besides standard translation stages, we employ a Newport ESP300 motion controller along with CMS series motorized actuators.

4.3.2 *Real-Space and Momentum-Space Spectroscopy*

Here, we describe the spectroscopy setup as is implemented in our lab [9]. We can measure energy-resolved luminescence in both real and momentum space. The real-space image is often referred to as *near field* (NF), while the momentum-space image is called *far field* (FF).

4.3.2.1 Real-Space Spectroscopy

Real-space luminescence is measured through the setup shown in Fig. 4.5a. A combination of two lenses is employed, and the real-space image is formed on the entrance slit of the spectrometer. A flip mirror in front of the spectrometer allows us to direct light to a CCD camera for imaging purposes (not shown in the figure). A sample real-space image of LP luminescence at low pumping power is shown in Fig. 4.5b. Closing the spectrometer slit, we can select just the central column from the 2D image and spectrally resolve it with the spectrometer. The result is a 2D

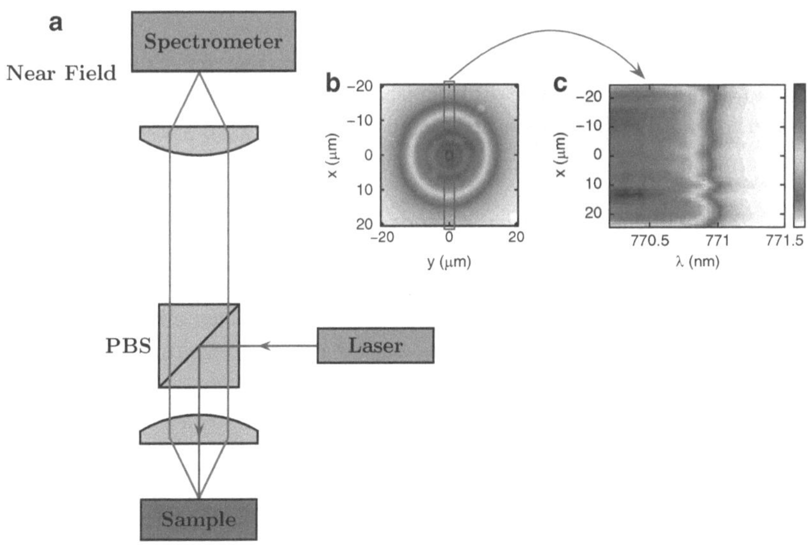

Fig. 4.5 Real-space (*near field*) spectroscopy (*see text*). (**a**) Schematic of the spectroscopy setup. (**b**) Example of a real-space image. (**c**) Example of a real-space spectrum

image in which the vertical axis corresponds to the vertical position on the sample and the horizontal axis corresponds to the energy (Fig. 4.5c). This is a real-space spectrum.

4.3.2.2 Sample Disorder Potential

Because of fluctuations in the cavity width, quantum well thickness and other inhomogeneities, such as local strain or charged impurities, the LP energy at zero momentum $k = 0$ is different from point to point. With our real-space spectroscopy setup, we can create a map of the disorder potential. We pump with a low laser power and large spot and measure the real-space spectrum (Fig. 4.5c). Translating the last lens before the spectrometer along the direction perpendicular to the slit and to the optical axis moves the real-space image, so we can acquire the real-space spectrum along a different column. Employing this technique, we can measure the spectrum of a two-dimensional grid of points with spatial resolution of 1 μm. The spectrum at each point contains contributions from all momenta. If we assume that the spectrum from every momentum component is Lorentzian, then the real-space spectrum should be a sum of Lorentzians centered at different energies. The lowest energy contribution should be from the $k = 0$ component, so the lowest (long-wavelength) part of the real-space spectrum should decay as a Lorentzian. We can fit this part of the spectrum with half of a Lorentzian and extract the local energy for every point (Fig. 4.6a).

We measure a striped pattern for the disorder potential (Fig. 4.6b), while the local energy follows a Gaussian distribution with standard deviation $\sigma = 71\,\mu$eV (Fig. 4.6c). This energy scale is much smaller than the repulsive LP–LP interaction energy, which is on the order of 0.1–1 meV at the pump rate above the condensation threshold. Therefore, the sample disorder is screened when the condensate is formed and should not affect the condensate dynamics appreciably.

4.3.2.3 Momentum-Space Spectroscopy

The in-plane translational symmetry enables us to perform momentum-space imaging. Each emitted photon has the same in-plane momentum as the polariton it originates from. Each photon carries a normalized momentum $k = P/\hbar = 2\pi/\lambda$ along the propagation direction, where λ is the wavelength. Consider photon traveling at an angle θ (inclination angle) with respect to the growth axis. The in-plane momentum is simply $k = 2\pi/\lambda \sin\theta$. Since all photons from the LP branch have approximately the same wavelength, we can use a common value of λ for the whole far field spectrum.

A traditional way to acquire a far-field image is to place a screen at infinity. The coordinates of this image would then correspond to the angle θ. A more compact setup is shown in Fig. 4.7a, and it includes an extra lens compared to Fig. 4.5a. The extra lens can be mounted on a kinematic base for easy switching between the two

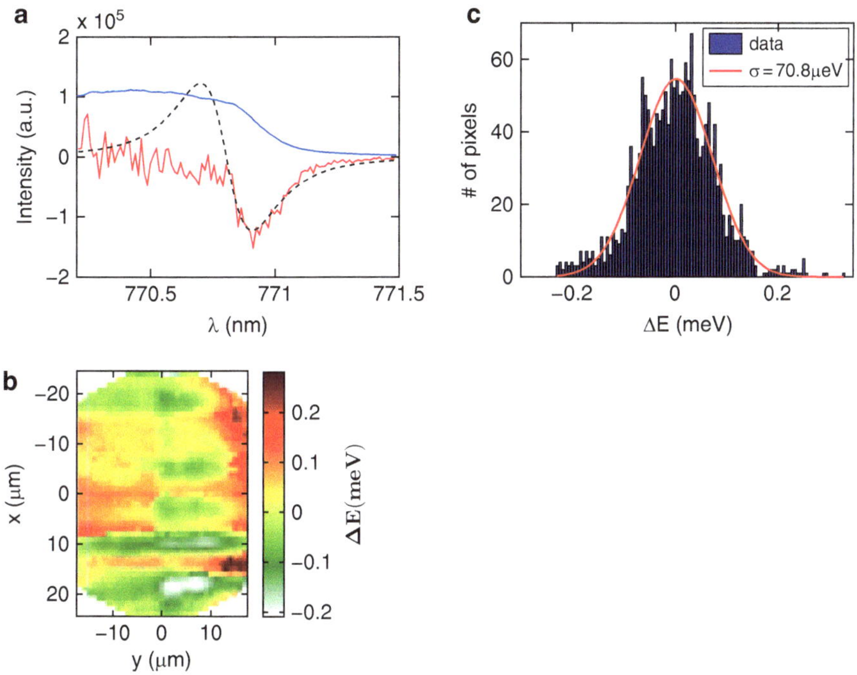

Fig. 4.6 Sample disorder potential. (**a**) The fitting method. *Blue*: spectrum selected from panel a along $x = 0\,\mu$m. *Red*: numerical derivative. *Black dashed line*: fitting of the long-wavelength part of the numerical derivative with the derivative of a Lorentzian. (**b**) Map of the disorder potential. White areas at the corners have low signal to noise ratio. (**c**) Histogram of the disorder potential shown in panel c, and fit with a Gaussian with $\sigma = 70.8\mu$ eV. From [45]

setups. The far-field image is created on the back focal plane of the last lens, since light along a given direction is focused at one point (Fig. 4.7b). The coordinates of the generated image correspond to the angle θ, which can be translated to the momentum. The resolution of our setup is 0.019 μm^{-1}, corresponding to one pixel on the spectrometer CCD camera. A sample FF image is shown in Fig. 4.7c, in which the excitation power is low, so no condensate is formed, and the momentum-space distribution is broad. In Fig. 4.7d we plot the corresponding FF spectrum. The parabolic form of the LP dispersion near zero momentum is evident, and the measured curvature gives the LP effective mass.

4.3.2.4 Reflection Measurement

To measure the reflection spectrum of the cavity, we replace the laser in Fig. 4.7 by a white-light source and the polarizing beam splitter (PBS) by a nonpolarizing beam splitter (NPBS). We can then measure a reflection spectrum in momentum space. An example of such a spectrum is shown in Fig. 4.8. The stop band extends from

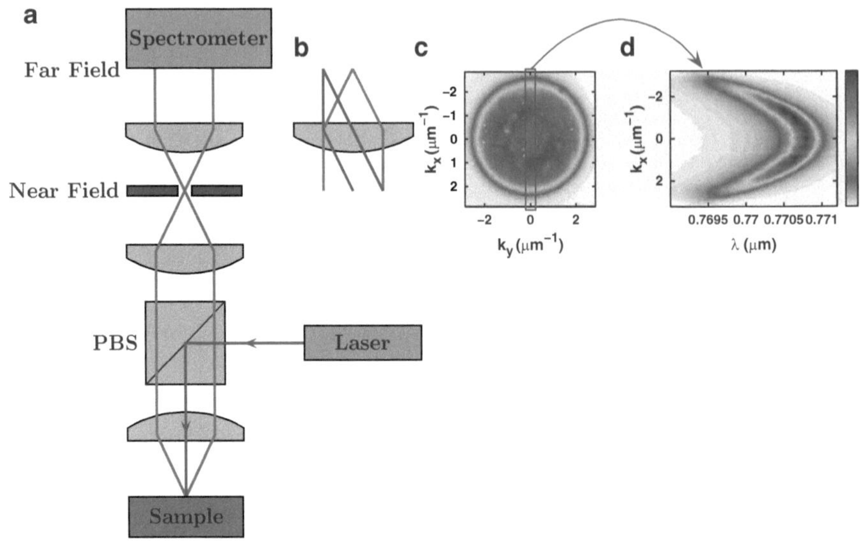

Fig. 4.7 Momentum-space (*far field*) spectroscopy (*see text*). (**a**) Schematic of the spectroscopy setup. (**b**) A far-field image is generated at the back focal plane of a lens. (**c**) Example of a momentum-space image. (**d**) Example of a momentum-space spectrum

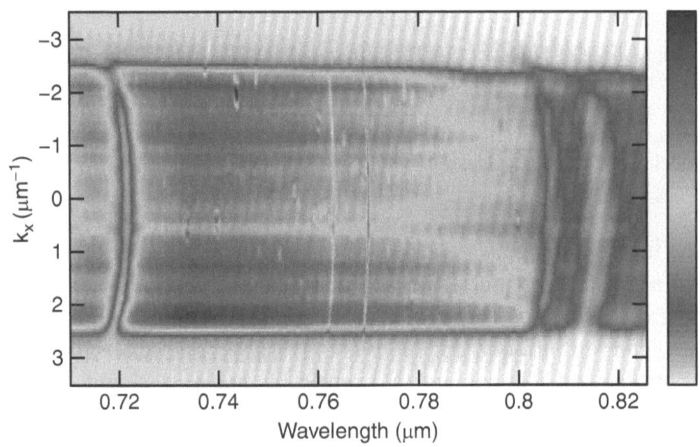

Fig. 4.8 Far-field reflection spectrum of our sample using a white-light source. The raw data is plotted. The stop band extends from 0.72 μm up to 0.8 μm. The upper polariton and lower polariton bands are visible in the interval 0.76–0.77 μm

0.72 μm up to 0.8 μm, and two narrow minima are visible in the interval 0.76–0.77 μm. The higher-energy minimum corresponds to the upper polariton, and the lower-energy minimum to the lower polariton.

For Fig. 4.8, we used an incandescent lamp as a white-light source. Such a lamp follows the black body spectrum, and its intensity is decreased as the wavelength is increased to the near infrared and beyond. This decay is visible in the reflection spectrum. If we wanted to be more precise, we should have divided the raw reflectance data by the reflectance spectrum of an ideal mirror.

4.3.3 *Condensation Characteristics*

4.3.3.1 Excitation Scheme

The sample is pumped at normal incidence at the first reflectivity minimum above the stop band, at around 723 nm (see Fig. 4.8). Transfer matrix calculation using complex refractive index for GaAs gives an absorption efficiency of ~25% at this wavelength for our sample design. In the experiments described later, we operate the laser in the continuous wave mode. The laser generates electron–hole pairs that subsequently relax and form LPs. Because LPs decay on a *ps*-timescale, the pump is continuously on to maintain a constant LP population. So we are probing the steady state of the system.

Since pumping occurs above band (~100 meV higher than the LP energy), the generated particles suffer multiple collisions with phonons, electrons and holes, excitons, and other LPs before reaching the condensate energy. Therefore, the phase of the pump wave cannot influence the condensate dynamics, and any coherence observed in the condensate cannot be attributed to laser-induced coherence.

The laser spot is created through the setup of Fig. 4.9a. The laser is first coupled to a polarization-maintaining single-model fiber, and a collimated Gaussian beam is created at the other end. The beam is then coupled to a commercial refractive beam shaper, which transforms a collimated Gaussian beam of a particular size to a top-hat profile. The objective lens needs to be focused with respect to the sample, since we image the system through it. We can use an extra lens just after the beam shaper, in order to move the focusing point of the laser beam away from the sample surface and have a large pumping spot on the sample. The focal length of this lens determines the size of the pumping spot.

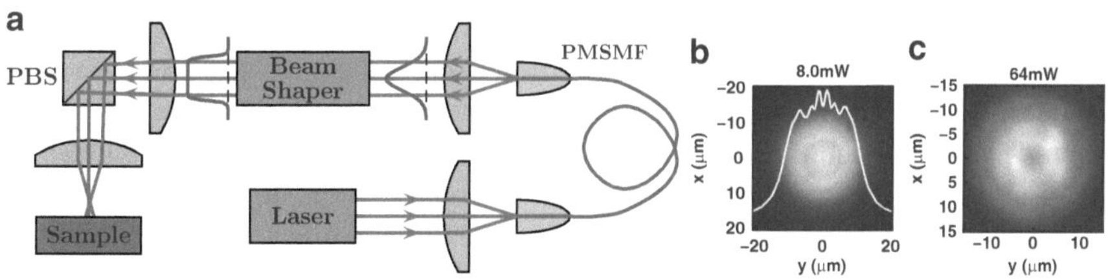

Fig. 4.9 Laser pumping setup. (**a**) Schematic of the laser pumping setup. (**b**) The pumping spot below the condensation threshold and its cross section along the horizontal axis. (**c**) Real-space image of the condensate above threshold

A flat-top spot is formed (Fig. 4.9b) with small loss of laser power. The most lossy part of the setup is coupling into the single-mode fiber, in which a diffraction-limited efficiency of $\sim$50% was achieved. Due to diffraction, the excitation spot on the sample cannot be perfectly flat, but instead, consists of closely spaced bright and dark rings. For very low pumping power, the polariton diffusion length is long, and thus, the rings are not visible in polariton luminescence. Above the condensation threshold, the condensate takes a doughnutlike shape (Fig. 4.9c). The physics behind the condensate shape is a repulsive interaction between the condensate and the reservoir, which will be discussed later.

It is theoretically predicted [46] that the condensate should be linearly polarized, with the direction of linear polarization chosen randomly. In experiments with highly disordered samples [7], there is a preferred direction of linear polarization. This happens because sample asymmetry (e.g., strain) introduces a splitting between orthogonal polarizations, so that an LP of a particular linear polarization has lower energy than an LP of the orthogonal linear polarization. This splitting can be a fraction of 1 meV. In samples with splittings < 0.1 meV, random selection of linear polarization occurs. This scenario was verified [47], in which the authors could detect the shot-to-shot variations in the linear polarization direction, while the time-integrated luminescence remained unpolarized. Time-integrated luminescence from our condensates is unpolarized as well (measured degree of linear polarization $< 20\%$), so we believe that the scenario of random selection of linear polarization direction is realized. Since the laser is focused on the sample through a polarizing beam splitter (Fig. 4.9a), the pump is linearly polarized. We have checked that the condensate luminescence remains unpolarized after we rotate the pump linear polarization by 90°, or when we pump with circularly polarized light.

4.3.3.2 Condensate Spectroscopy

To investigate the condensation characteristics, we perform momentum-space spectroscopy. In Fig. 4.10a, we plot the luminescence intensity measured around 0° collection angle versus the laser pump power [45]. The spectrometer discards the reflected laser light, and we select the integration angle by post-processing the far-field spectroscopy data. There is a threshold pump power at 20 mW, above which the signal intensity increases nonlinearly, similar to a lasing transition. In Fig. 4.10b, we plot the spectra as a function of pumping power. Across the threshold power, there is a blue shift of about 1 meV, which is mainly due to the repulsive interaction energy among LPs. A second threshold is also visible at $90mW$ with a larger blue shift. We can interpret the first threshold as polariton condensation, but the second one as a crossover to a new phase is not given a solid and unanimous interpretation right now. In [48], it has been described as photon lasing.

Figure 4.10c summarizes a set of measurements performed for varying photon–exciton detuning δ. The wedged sample structure allows us to change the detuning by simply moving to a different position. We can observe the anticrossing between the upper polariton (UP) and lower polariton (LP) resonances. Furthermore, when

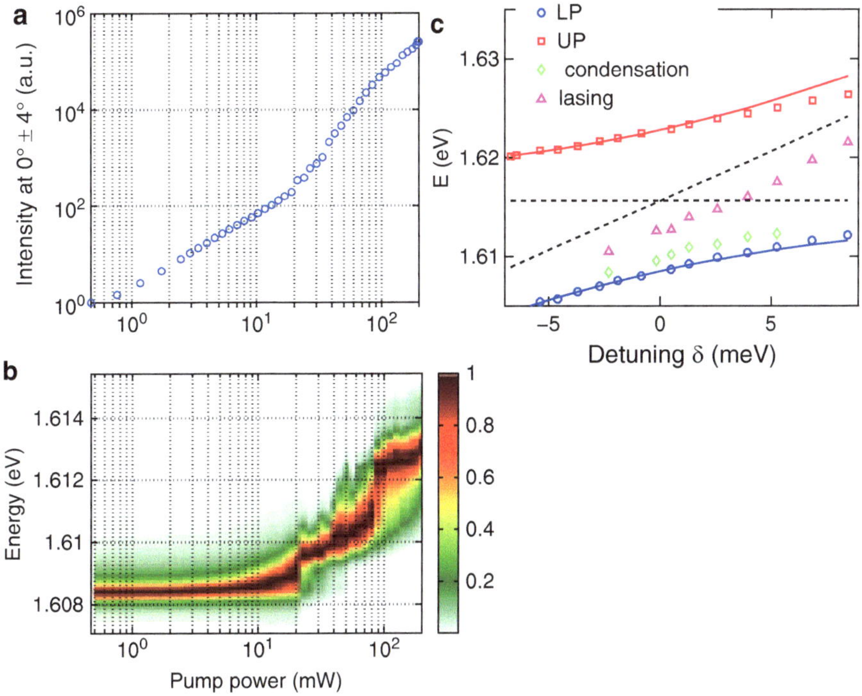

Fig. 4.10 Polariton condensation vs. photon lasing. (**a**) Signal intensity measured by the spectrometer at $0 \pm 4°$ ($0 \pm 0.57\,\mu\text{m}^{-1}$) as a function of pumping power. (**b**) Normalized spectrum at $0 \pm 4°$ as a function of pumping power. Two thresholds are visible. (**c**) LP (*blue circles*) and UP (*red squares*) energies as a function of photon–exciton detuning δ. Continuous lines are fits with Rabi splitting $2\hbar\Omega_R = 14$ meV. The flat and tilted dashed lines are the energies of the bare exciton and microcavity photon, respectively. The LP condensation threshold energy (*green diamonds*) follows the LP energies, while the photon lasing energy (*magenta triangles*) follows the photon energies. From [45]

we plot the energy of the first threshold (at 20 mW in Fig. 4.10b, where $\delta \sim 0$ meV), we see that it follows the LP resonance as shown by the green diamonds. On the other hand, the second threshold energy (at 80 mW in Fig. 4.10b) exists at a few meV below the photon resonance as shown by the red triangles. When we move to higher δ, so that LPs become more exciton-like, the first and second threshold windows shrink, and eventually disappear so that only second threshold is visible. This data is in accordance to measurements in different samples [48–50].

For the polariton gas to be considered dilute, the density must be smaller than the Mott density. We assume pumping efficiency of $\eta \sim 20\%$, based on a transfer matrix calculation and an LP lifetime of $\tau_{\text{LP}} \sim 4$ ps. The linewidth of the reflection spectra gives a lower bound of the lifetime, while the time-resolved measurements provide an upper bound, and 4 ps is a reasonable estimate. For the data of Fig. 4.10b, we used a circular pumping spot with radius $R \sim 12\,\mu$m and laser wavelength of $\lambda_{\text{las}} = 723$ nm. This results in an estimate of the threshold polariton density.

$$n_{\mathrm{th}} = a\frac{P_{\mathrm{th}}\tau_{\mathrm{LP}}}{\pi R^2 hc/\lambda_{\mathrm{las}}} \sim 10^2\mu\mathrm{m}^{-2} = 10^{10}\mathrm{cm}^{-2}. \quad (4.62)$$

Our sample contains twelve QWs, so the density per QW is twelve times smaller than (4.62). This density of $\sim 8\times 10^8\mathrm{cm}^{-2}$ is indeed two to three orders of magnitude smaller than the Mott density $1/\pi a_{\mathrm{B}}^{*2} \sim 3\times 10^{11}\mathrm{cm}^{-2}$.

From the blue shift of ~ 1 meV evidenced in Fig. 4.10b when the condensate threshold is crossed and (4.15), we can estimate the healing length $\xi \sim 0.7\,\mu$m, where we have used a typical effective mass of $7\times 10^{-5}m_{\mathrm{e}}$.

4.3.4 Comparison of Pumping Schemes

In our early experiments, we pumped the sample in the high-k part of the LP band and, in particular, at about 60° incidence angle. Because the cavity is highly blue detuned at high momenta, this part of the band is exciton-like. The absorption efficiency calculated with the transfer matrix method is only a few percent, but condensation can be reached easily with pulsed excitation because of faster relaxation. Namely, the excitons only need to loose an excess energy of a few meV before they reach the condensate energy. In this configuration, particles are created far from the so-called magic wavenumber [51,52], so they need to suffer multiple scattering events before reaching the condensate. Therefore, the condensate coherence cannot be inherited directly from the pump laser. In [53], we studied the polarization characteristics of the condensate, and we showed that the relaxation process for every LP involves multiple scatterings with phonons and, on the average, only one scattering with another LP.

The least efficient but conceptually cleaner pumping method is the one implemented in the experiments of this chapter, namely, pumping at normal incidence above band. In this way, electron–hole pairs of approximately zero momentum are generated, which then relax toward the LP branch and eventually form a condensate if the particle density is high enough. Because the laser energy is much higher than the condensate energy ($\sim$ 100 meV), there is no doubt that in this case, the condensate coherence is generated spontaneously. This pumping method, however, involves a large reservoir density, since the relaxation path is now longer. Consequently, electron–hole interaction is screened, so that the exciton oscillator strength is reduced and the microcavity can enter easily the weak coupling regime. From the data of Fig. 4.10, we conclude that in our sample, strong coupling is preserved at the pumping power required for condensation.

4.4 Power-Law Decay of the Spatial Correlation Function

The spatial correlation function quantifies the coherence properties of a system [35]. In a three-dimensional Bose-condensed gas, long-range order is observed, and the correlation function decays toward a plateau at large distances [14, 54].

In the homogeneous two-dimensional (2D) Bose gas [24], however, no long-range order can be established [11]. Instead, BKT theory of the equilibrium interacting gas predicts a transition to a low-temperature superfluid phase, which shows a power-law decay of the correlation function [12,29]. Unfortunately, it is frequently hard to directly measure this, and only very recently [41] was indication of the power-law decay of coherence seen in a two-dimensional atomic gas. It has been theoretically predicted [55,56] that power-law decay of coherence survives in the nonequilibrium problem, and it is this prediction that the current experiment sets out to test. Due to the nonequilibrium nature of polariton condensation, understanding its coherence properties is quite revealing regarding the different roles of fluctuations in the equilibrium and nonequilibrium problems. In particular, we show that laser pumping noise plays a role analogous to the role of temperature in an equilibrium situation.

Previous measurements on polariton condensates have demonstrated coherence at large distances, but were limited by large experimental uncertainties [57], or highly disordered samples [7,58], and the long-distance behavior could not be fully extracted. Recently, the correlation function at large distances was studied in one-dimensional condensates confined in a quantum wire [59] and in a valley of the disorder potential [60]. In both these experiments, the microcavity was excited by a single-mode laser with suppressed intensity noise. This resulted in highly coherent condensates, as was evident in the interferometric measurements. We are interested in the question of how the intensity noise of the external laser pump, namely, the laser, influences the coherence properties of the condensate state. For this purpose, we employ a multimode laser in the continuous wave operation. Because many laser cavity modes can lase simultaneously out of phase, the overall intensity fluctuates randomly on a ps timescale, as determined by the laser bandwidth.

With our setup, we can measure values of $g^{(1)}(r)$ as low as 0.02, so we can reliably extract the long-distance behavior. We find that, although true thermal equilibrium is not established, an effective thermal de Broglie wavelength can still be defined from the short-distance Gaussian decay of $g^{(1)}(r)$. Furthermore, $g^{(1)}(r)$ at long distances r decays according to a power law, in analogy to the equilibrium BKT superfluid phase. The exponent of the power-law decay is, however, higher than can be possible within the BKT theory. We apply a nonequilibrium theory [55, 56] to identify the source of the large exponent. We argue that, although the spectrum is modified due to dissipation, the exponent would still have the equilibrium value if the spectrum was thermally populated. If, on the other hand, a white noise source acts on the system, the exponent can have a large value, proportional to the noise strength. We therefore conclude that laser pumping noise can be responsible for the large value of the exponent.

4.4.1 Condensate in Real Space

The pumping spot setup is shown in Fig. 4.9a. We pump with a multimode laser that creates free electron–hole pairs at an excitation energy $\sim$100 meV above the lower polariton (LP) energy, and we probe LP luminescence in the steady state. The pump laser is continuously on and replaces LPs that leak out of the microcavity on a *ps* time scale. The laser is first coupled to a polarization-maintaining single-model fiber, and a collimated Gaussian beam is created at the other end. The beam then propagates through a commercial refractive beam shaper, which transforms a collimated Gaussian beam of a particular size to a collimated beam with top-hat profile. The objective lens needs to be focused with respect to the sample, since we image the system through it. We can use an extra lens just after the beam shaper, in order to move the focusing point of the laser beam away from the sample surface, and have a large pumping spot on the sample. The focal length of this lens determines the size of the pumping spot. In Fig. 4.11, we plot LP luminescence images for the four different spot sizes used in this experiment. The pumping power is ~10 mW, which is below the threshold power P_{th} for all four spots.

In Fig. 4.12, we plot the real-space images of LP luminescence for various laser pumping powers, both below and above threshold. We use a combination of two interference filters, one longpass at 750 nm and one bandpass at 770 ± 5 nm, which block the laser wavelength without distorting the LP luminescence spectrum. For very low pumping power, luminescence has a top-hat shape. Close to the threshold power of 55 mW, the diffusion length shortens, and luminescence takes the shape of the laser excitation spot. Airy-like patterns appear because of diffraction. Above threshold, the condensate progressively takes a doughnutlike shape. We studied this effect in [45], and concluded that the reservoir has a complementary profile with a density maximum at the center and that repulsive condensate-reservoir interactions render this distribution stable. Also, the condensate is smaller than the original spot

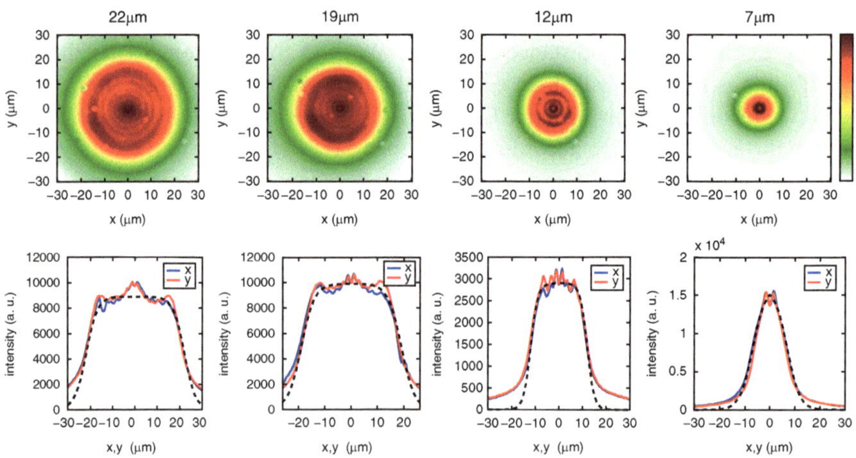

Fig. 4.11 Varying pumping spot size. By changing the lens just after the beam shaper in Fig. 4.9a, the pumping spot size can be varied. On the upper row, we plot the LP luminescence images below threshold for the four different spot sizes used in this experiment. The colorscale is linear, and the label on top of every figure is the spot radius. On the lower row, we plot the corresponding 1-D profiles along the *x*- and *y*-axes. The dashed line is an empirical fit to a Fermi–Dirac function, from which we determine the size of the pumping spot

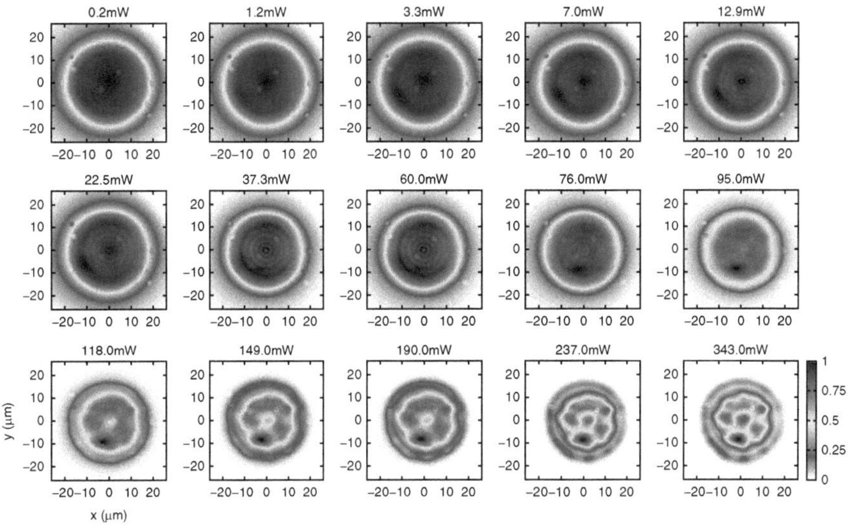

Fig. 4.12 Real space images of LP luminescence for increasing pumping power. Condensation threshold is at 55 mW

size because of repulsive LP–LP interactions. In particular, the large condensate density creates an antitrapping potential that pushes LPs away from the center. This effect only influences LPs close to the edge for our top-hat pumping spot, whereas it is prominent in cases of a Gaussian or very small pumping spot or samples with long LP lifetime [59, 61].

4.4.2 Michelson Interferometer Setup

The first-order spatial correlation function is defined as

$$g^{(1)}(\mathbf{r}t_1;\mathbf{r}t_2) = \frac{\left\langle \psi^\dagger(\mathbf{r}t_1)\,\psi(\mathbf{r}t_2)\right\rangle}{\sqrt{\left\langle \psi^\dagger(\mathbf{r}t_1)\,\psi(\mathbf{r}t_1)\right\rangle\left\langle \psi^\dagger(\mathbf{r}t_2)\,\psi(\mathbf{r}t_2)\right\rangle}}, \tag{4.63}$$

where $\psi^\dagger(\mathbf{r}t_i)$ and $\psi(\mathbf{r}t_i)$ are the creation and annihilation field operators at space-time point $(\mathbf{r}_i, t_i)$.

The intensity that a detector at space-time point $(\mathbf{r}, t)$ measures is $\left\langle \mathcal{E}^{(-)}(\mathbf{r}, t)\,\mathcal{E}^{(+)}(\mathbf{r}, t)\right\rangle$. Therefore, when light from point $(\mathbf{r}_1, t_1)$ interferes with light from point $(\mathbf{r}_2, t_2)$, the detector measures the quantity $\left\langle\left[\mathcal{E}^{(-)}(\mathbf{r}_1, t_1) + \mathcal{E}^{(-)}(\mathbf{r}_2, t_2)\right]\left[\mathcal{E}^{(+)}(\mathbf{r}_1, t_1) + \mathcal{E}^{(+)}(\mathbf{r}_2, t_2)\right]\right\rangle$. Hence, when a parameter such as the time difference $t_2 - t_1$ changes slowly, fringes are observed. Defining the fringe visibility as the ratio or the maximum minus the minimum intensity over the maximum plus the minimum intensity (4.63) gives the fringe visibility when the intensities at points $(\mathbf{r}_1, t_1)$ and

($\mathbf{r}_2, t_2$) are the same. If the intensities are different by a factor p, then the measured fringe visibility is equal to (4.63) times a factor $2\sqrt{p}/(1+p)$. Surprisingly, this factor remains close to 1, even for intensity ratios p much different than 1. For example, for $p = 2$, we find $2\sqrt{p}/(1+p) = 0.943$.

To measure this function, we built a Michelson interferometer setup. A schematic is shown in Fig. 4.13a. It includes a mirror in one arm and a right-angle prism in the other. We overlap the condensate real-space image with its reflected version, so that fringes similar to that of Fig. 4.13b are observed on the camera. By changing the length of one interferometer arm, as shown in Fig. 4.13a, the relative phase of the two beams is shifted. As a result, the intensity measured at one pixel point shows a sinusoidal modulation (Fig. 4.13c). From the data of Fig. 4.13c, we extract the phase

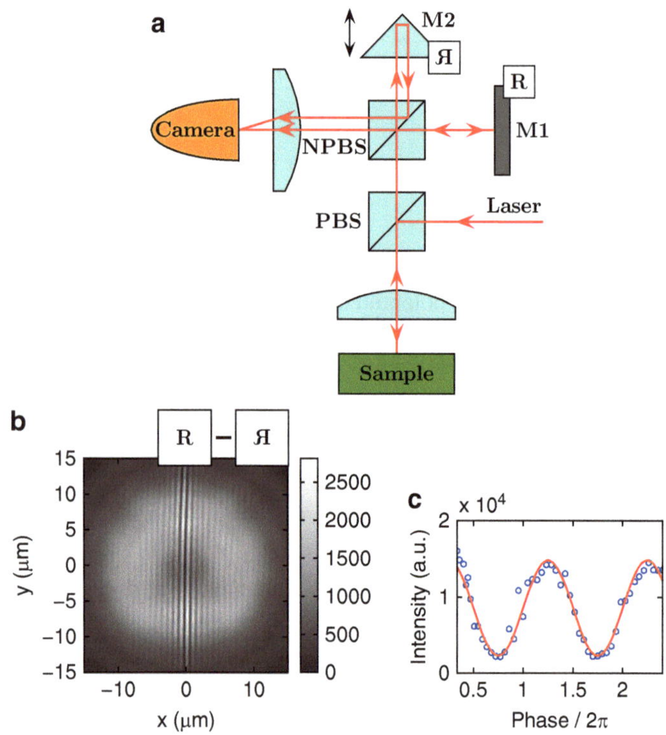

Fig. 4.13 Michelson interferometer. (**a**), Schematic of the setup for measurement of the correlation function. The laser is linearly polarized, and we record luminescence of the orthogonal linear polarization through a polarizing beam splitter (PBS). We then employ a 50–50 nonpolarizing beam splitter (NPBS), a mirror (M1), and a right-angle prism (M2). The latter creates the reflection of the original image along one axis, depending on the prism orientation. A two-lens microscope setup overlaps the two real-space images of the polariton condensate on the camera. (**b**), Typical interference pattern observed above the polariton condensation threshold along with a schematic showing the orientation of the two overlapping images. (**c**), Blue circles: Measured intensity on one pixel of the camera as a function of the prism (M2) position in normalized units. Red line: Fitting to a sine function

difference of the two images at a particular pixel point as well as the fringe visibility. The latter is proportional to the first-order correlation function, which is the physical quantity we are interested in this experiment.

The prism M2 in Fig. 4.13a forms the reflection of the condensate image along the prism axis. Therefore, point (x, y) overlaps with either $(-x, y)$, or $(x, -y)$ on the camera, depending on the orientation of the prism. This allows us to measure

$$g^{(1)}(x, -x; \tau) \equiv \left\langle g^{(1)}(x, y, t + \tau; -x, y, t)\right\rangle_t , \tag{4.64}$$

$$\text{or } g^{(1)}(y, -y; \tau) \equiv \left\langle g^{(1)}(x, y, t + \tau; x, -y, t)\right\rangle_t , \tag{4.65}$$

where $\langle\rangle_t$ denotes time average. In this experiment, we are mainly interested in interference at $\tau = 0$, so when the time argument is not mentioned explicitly, we imply $\tau = 0$.

We repeat the procedure explained in Fig. 4.13 for every pixel, so that we measure the phase difference between the two interfering images in addition to the correlation function across the whole spot. Representative data are shown in Fig. 4.14. Recording both these quantities allows us to identify useful signal from systematic or random noise. Because the prism displaces the beam that is incident on it, the images from the mirror and the prism are focused on the camera from different angles, so the two phase fronts are tilted with respect to each other. As a consequence, we expect to measure a constant phase tilt. This is the case in Fig. 4.14b, in which the laser power is above threshold and a condensate has formed. We conclude that our measurement of the correlation function in Fig. 4.14d is reliable over this whole area. On the other hand, at a pump rate below threshold, only short-range correlations exist. Figure 4.14a shows that in this case, the phase difference is measured correctly only over a small area around the center, ($|x| \leq 1\,\mu$m). So, the measured values of $g^{(1)}(x, -x)$ outside this area are not reliable and give an estimate of our measurement uncertainties. As is clear from Fig. 4.14c, the experimental error can be suppressed down to 0.01.

It is known that an ideal autocorrelation measurement with a Michelson interferometer provides the same information as an ideal measurement of the spectrum. In particular, $g^{(1)}(x, -x; t)$ is the Fourier transform of the power spectrum in momentum-space $S(\mathbf{k}, \omega)$[62]. However, systematic noise in measurement of $S(\mathbf{k}, \omega)$ currently makes the direct measurement of $g^{(1)}(x, -x; t)$ the only way to reliably extract λ_{eff} of Fig. 4.17b as well as the power-law decay at long distances to be explained later. The Fourier-transform relationship between $g^{(1)}(x, -x; t)$ and $S(\mathbf{k}, \omega)$ is illustrated in Fig. 4.15. The measured $g^{(1)}(x, -x; t)$ at very low pumping power is shown in Fig. 4.15a. At time delay $t = 0$, it has a Gaussian form as a function of x, but for increasing t, it broadens and acquires a multipeak structure. This unusual space-time dependence is reproduced by the numerical Fourier transform (Fig. 4.15c) of measured $S(\mathbf{k}, \omega)$ (Fig. 4.15b). Unfortunately, with the current experimental setup, measurement of the time dependence of $g^{(1)}(x, -x; t)$ is limited by inhomogeneous broadening due to time-integrated data, so it cannot provide an estimate of the homogeneous dephasing time.

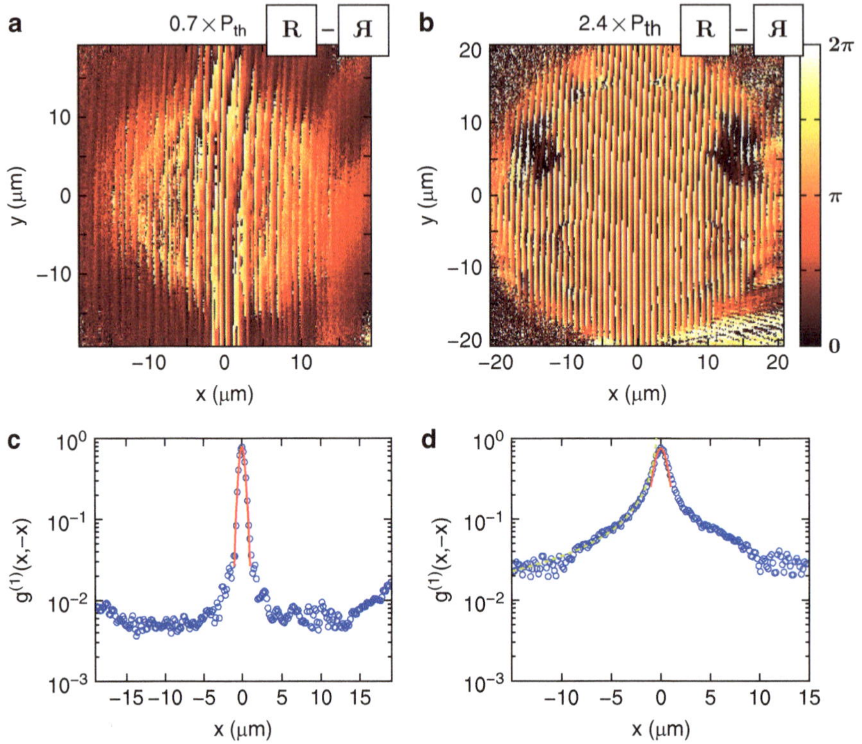

Fig. 4.14 Phase map measured for laser power (**a**) below and (**b**) above the threshold power P_{th}. The prism in the Michelson interferometer is oriented horizontally. The schematics on the top right of (**a**) and (**b**) show the orientation of the two interfering images. (**c**, **d**) Measured $g^{(1)}(x,-x)$ corresponding to (**a**) and (**b**), respectively, averaged over the y-axis inside the excitation spot area of 19 μm radius. Blue circles are experimental data. The continuous red and dashed yellow fitting lines are explained in Figs. 4.17 and 4.19, respectively

4.4.3 *Short-Distance Decay*

The results of the short-distance decay of the equal-time correlation function $g^{(1)}(r)$ can be understood by considering the simplest possible model, namely, the infinite two-dimensional noninteracting Bose gas. In (4.42), we proved that in the high-temperature or low-density limit, the decay is Gaussian with a width proportional to the thermal de Broglie wavelength. In Fig. 4.16a, we plot $g^{(1)}(r)$ by numerically calculating the Fourier transform of the momentum-space occupation function (4.40). We assume normal (quadratic) dispersion, Bose occupation function, and fixed temperature. The normalization density n_{c} is the density for which the phase-space density equals one: $n_{\mathrm{c}}\lambda_{\mathrm{T}}^2 = 1$. As the density is increased, the short-distance decay remains Gaussian with increasing width. In Fig. 4.16b, we plot the width of the Gaussian decay in units of the thermal de Broglie wavelength.

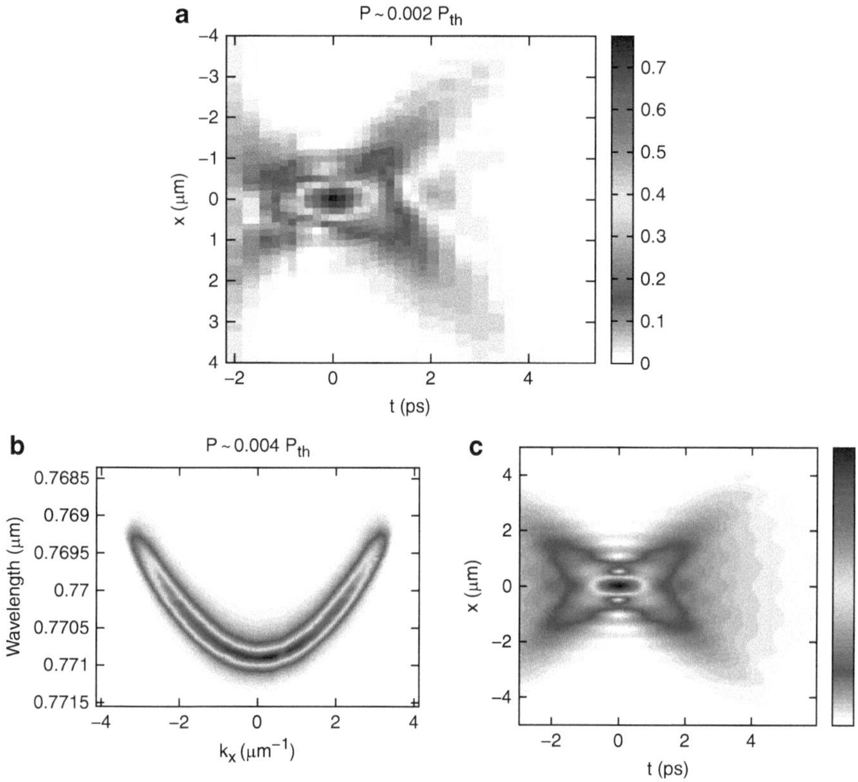

Fig. 4.15 (**a**) Measured $g^{(1)}(x, -x; t)$ for very low pumping power. (**b**) Measured momentum-space spectrum $S(k_x, \hbar\omega)$ for very low pumping power. As explained in the text, $g^{(1)}(x, -x; t)$ is the Fourier transform of $S(k_x, \hbar\omega)$. (**c**) Fourier transform of the experimental data shown in (**b**). The result indeed reproduces accurately (**a**). In (**b**) and (**c**), the data is plotted in linear color scale in arbitrary units

Figure 4.17a shows the measured short-distance dependence of $g^{(1)}(x, -x)$ for the same pumping power as in Fig. 4.14a,c. Because the prism of the Michelson interferometer is not completely horizontal, the axis of inversion is slightly tilted with respect to the columns of the charge-coupled device array. Every point in Fig. 4.17a corresponds to one pixel on the camera, and the x-axis is its distance from the axis of inversion. Data at distances $|x| > 1\ \mu$m is noise, since the measured phase in this area is random (Fig. 4.14a). At shorter distances, the correlation function clearly has a Gaussian form in analogy to a thermalized Bose gas.

We note that the temperature extracted from the short-distance behavior of $g^{(1)}(x, -x)$ is a measure of the occuparion of the higher energy part of the spectrum, i.e., the particle-like part of the spectrum. For an insufficiently thermalized system, it is quite possible that excitations in different energy ranges have different effective temperatures. Therefore, the temperature measured this way will not necessarily agree with other measures of temperature.

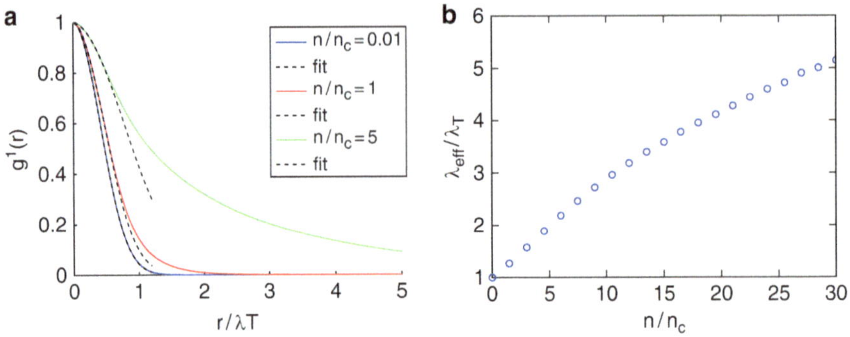

Fig. 4.16 $g^{(1)}(r)$ for a noninteracting 2D Bose gas. (**a**) Numerically calculated $g^{(1)}(r)$ for a noninteracting 2D Bose gas at increasing densities. We assume constant temperature and define $n_c = \lambda_T^{-2}$. Dashed lines are fits to a Gaussian function. (**b**) Width of the short-distance gaussian decay in units of the thermal de Broglie wavelength λ_T

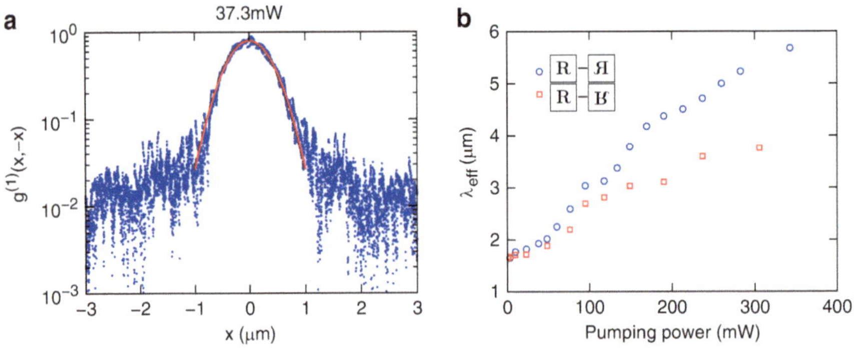

Fig. 4.17 Short-distance decay of the correlation function. (**a**) Measured $g^{(1)}(x, -x)$ at short distances. Blue dots are experimental data, the red line is a Gaussian fit. Data at $|x| > 1\,\mu$m is noise. (**b**) Effective thermal de Broglie wavelength λ_{eff} as a function of laser pumping power. λ_{eff} is extracted from the width of the Gaussian fit as shown in (**a**). Blue circles and red squares correspond to orthogonal orientations of the prism in the Michelson interferometer (*see text*). The condensation threshold is at ~55 mW

In Fig. 4.17b, we plot the effective wavelength λ_{eff} as a function of pumping power. If σ is the standard deviation of the Gaussian fit for $g^{(1)}(x, -x)$, $\lambda_{\text{eff}} = 2\sqrt{2\pi}\sigma$ in analogy to the thermal de Broglie wavelength. λ_{eff} shows a smooth increase for increasing pumping power with no obvious threshold, analogous to the theory of equilibrium noninteracting 2D Bose gas as the particle density is increased [24]. We performed the same experiment for two orthogonal prism orientations as shown in the legend of Fig. 4.17b, namely, horizontal and vertical orientations. In the former case, we measured $g^{(1)}(x, -x)$, whereas in the latter case, we measured $g^{(1)}(y, -y)$. We found that the effective temperature corresponding to λ_{eff} is larger for the vertical prism orientation. We attribute this difference to a small asymmetry of the laser pumping spot.

We also note that the resolution limit of our imaging setup is $\sim 1\,\mu$m, which influences the measurement of λ_{eff} at small pumping power. If the transfer function of our imaging system is Gaussian (namely, if a delta function is imaged to a Gaussian) with width λ_{res}, then the measured width of the Gaussian decay should be $\sqrt{\lambda_{\text{res}}^2 + \lambda_{\text{eff}}^2}$. Therefore, if the real λ_{eff} is similar to λ_{res}, then the measured λ_{eff} should be close to λ_{res}. The height of the Gaussian (unity for a perfect optical system) should also drop to a lower value, so that the total surface does not change. From the measured height of the Gaussian fits, namely, $g^{(1)}(0,0) \sim 0.7 - 0.8$ and the measured $\lambda_{\text{eff}} \sim 1.6 - 1.7\,\mu$m, we estimate a lower bound for the real $\lambda_{\text{eff}} \geq 1.3\,\mu$m at small pumping power.

4.4.4 Long-Distance Decay

In the equilibrium noninteracting 2D Bose gas, there is a crossover to an exponential decay of $g^{(1)}(r)$ when the phase-space density D increases to values higher than 1 (see (4.44)). This is true for the interacting gas, as well, although with a different value for the decay length k_c^{-1}. In the interacting case, however, there is a second transition at the Kosterlitz–Thouless transition temperature (4.60), below which $g^{(1)}(r)$ decays as a power law (4.53) for large distances. In this section, we investigate the functional form of the long-distance decay of the correlation function $g^{(1)}(r)$.

Below the condensation threshold, we can only measure nonzero $g^{(1)}$ up to distances $\Delta x = |2x| \sim 2\mu$m (see Figs. 4.14a and 4.17a). Above threshold, measurable phase coherence is observed across the whole condensate, and the decay is no longer Gaussian. We found that $g^{(1)}$ is influenced by the edge of the condensate. In Fig. 4.18, we plot the measured $g^{(1)}(\Delta x) = g^{(1)}(|2x|) \equiv g^{(1)}(x, -x)$ at pumping

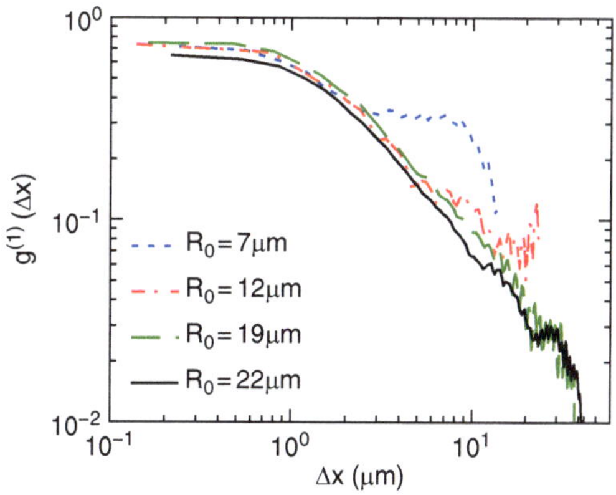

Fig. 4.18 Dependence on the condensate size. Measured $g^{(1)}(\Delta x) = g^{(1)}(|2x|) \equiv g^{(1)}(x, -x)$ vs Δx for various pumping spot radii R_0. All data is taken above threshold and is chosen such that $\lambda_{\text{eff}} \sim 4.1\,\mu$m. As the condensate size increases, $g^{(1)}(\Delta x)$ converges to a power-law decay

power $P \sim 3 \times P_{\text{th}}$ for increasing pumping spot radius. The measured $g^{(1)}(\Delta x)$ at long distances decreases as the spot size is increased and eventually converges toward a power-law decay for large condensates.

To determine whether the condensate size is large or small, one needs to compare the condensate radius R_0 with the healing length ξ defined (4.15). ξ is the length-scale at which boundaries are screened because of interparticle interactions, so when $R_0 \gg \xi$, boundaries do not influence the dynamics in the bulk of the condensate. Also, ξ is the size of the vortex core. Therefore, a condensate with $R_0 \gg \xi$ supports many vortices and vortex pairs, and the thermodynamics discussed in Sect. 4.2 are valid. More precisely, they would be valid if the system was in thermal equilibrium. The effective mass of LP's in our sample near zero photon–exciton detuning (where the current experiments were performed) is $m^* \sim 7 \times 10^{-5} m_{\text{e}}$ and the mean-field interaction energy $n_0 g \sim 1$ meV just above threshold, as follows from Fig. 4.10b. Plugging these numbers in (4.15), we find $\xi \sim 0.7\mu$m. Thus, a condensate with radius $R_0 = 19\mu$m can indeed be considered as infinitely large.

In the case of a large condensate, we should recover the limit of (infinitely large) homogeneous polariton gas. Therefore, we consider a pumping spot radius $R_0 = 19\mu$m. In Fig. 4.19a, we plot the correlation function $g^{(1)}(\Delta x)$ versus Δx as the pumping power is increased. Only short-range correlations exist for small pumping power, whereas above the condensation threshold of $\sim$55 mW, substantial phase coherence appears across the whole spot. The functional form of the long-distance decay is measured to be a power law over about one decade, as can be seen in Fig. 4.19b, in which we plot the data at one specific laser power. We fit the data to a function $g^{(1)}(\Delta x) = \left(\lambda_{\text{p}}/\Delta x\right)^{a_{\text{p}}}$ and plot the exponent a_{p} as a function of pumping power in Fig. 4.19c. It is found to be in the range 0.9–1.2.

In Fig. 4.20, we show log–log plots of the experimental data for $g^{(1)}(\Delta x)$ as a function of Δx for several pumping powers above threshold, along with the corresponding power-law fits. The data is the same as in Fig. 4.19a. As is evident from the figure, all our data can be fit equally well with a power-law function.

In Fig. 4.21, we plot the parameter λ_{p} of the $\left(\lambda_{\text{p}}/\Delta x\right)^{a_{\text{p}}}$ fit. Although λ_{p} has units of length, it is not a characteristic length of the system. In equilibrium BKT theory, the long-distance decay of $g^{(1)}(\Delta x)$ is of the form

$$g^{(1)}(\Delta x) = \frac{n_{\text{s}}}{n}\left(\frac{\xi}{\Delta x}\right)^{1/(n_{\text{s}}\lambda_{\text{T}}^2)}, \qquad (4.66)$$

where n_{s} is the superfluid density, n is the total density, ξ is the healing length, and λ_{T} is the thermal de Broglie wavelength. In this notation, $\lambda_{\text{p}} = \left(\frac{n_{\text{s}}}{n}\right)^{n_{\text{s}}\lambda^2} \xi$. For a thermalized gas, we expect that as the density is increased above the critical density, the superfluid fraction n_{s}/n increases, and λ_{p} increases accordingly approaching the healing length ξ. This behavior is qualitatively reproduced by our experimental results in Fig. 4.21.

In an equilibrium 2D Bose gas, the exponent of the coherence decay is limited to $a_{\text{p}} = 1/n_S\lambda^2 = mk_{\text{B}}T/2\pi\hbar^2 n_{\text{s}} \leq 1/4$, where n_{s} is the superfluid density and λ is

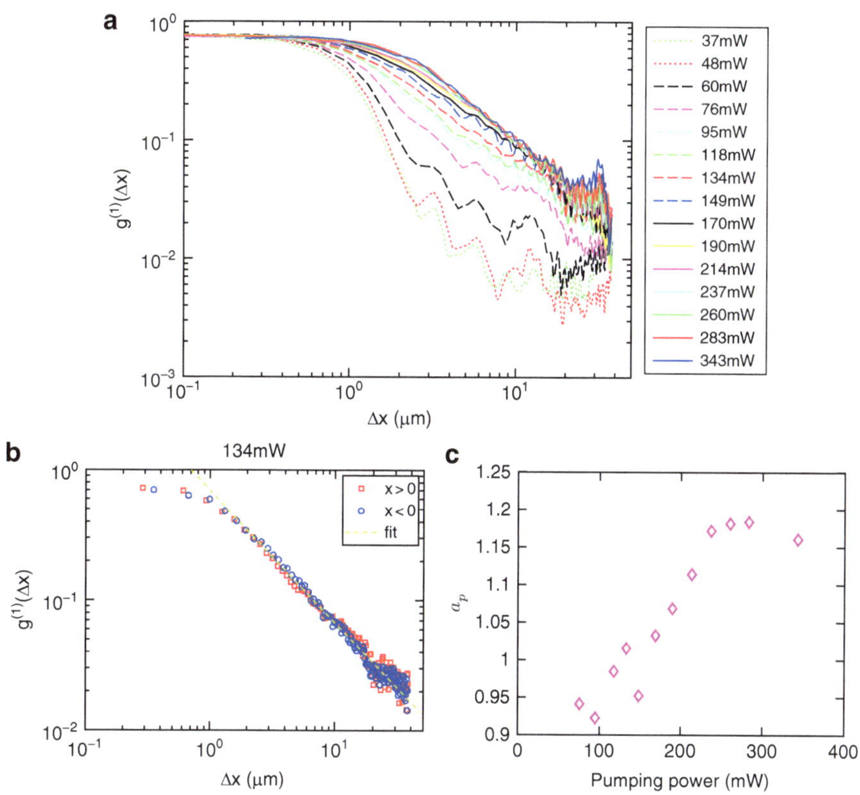

Fig. 4.19 Long-distance decay of the correlation function. (**a**) Measured $g^{(1)}(\Delta x) = g^{(1)}(|2x|) \equiv g^{(1)}(x, -x)$ vs Δx for increasing laser power. The laser pumping spot radius is $R_0 = 19\mu$m and the threshold power $P_{\mathrm{th}} = 55$ mW. (**b**) $g^{(1)}(\Delta x)$ vs Δx for one particular laser power and for x both positive (*blue circles*) and negative (*red squares*). Dashed line is a power-law fit. (**c**) Exponent a_{p} of the power-law decay as a function of laser power

the de Broglie wavelength. This restriction occurs because increasing temperature has two effects: it excites long wavelength phase fluctuations, which are responsible for the power-law decay, and it can also excite vortex pairs. Consider a 2D Bose gas at zero temperature. As the temperature is increased, thermal phase fluctuations cause $g^{(1)}(r)$ to decay as a power law with increasing exponent as the temperature is increased. When the BKT transition temperature T_{BKT} is reached, however, vortex pairs unbind. Because of the random motion of free vortices in the system, $g^{(1)}(r)$ decays exponentially. The maximum value of the exponent a_{p} occurs just below T_{BKT}. The observation here of a power-law $a_{\mathrm{p}} > 1/4$ implies that effects beyond thermal equilibrium are required to explain the data; i.e., there is noise which excites phase fluctuations without leading to vortex proliferation.

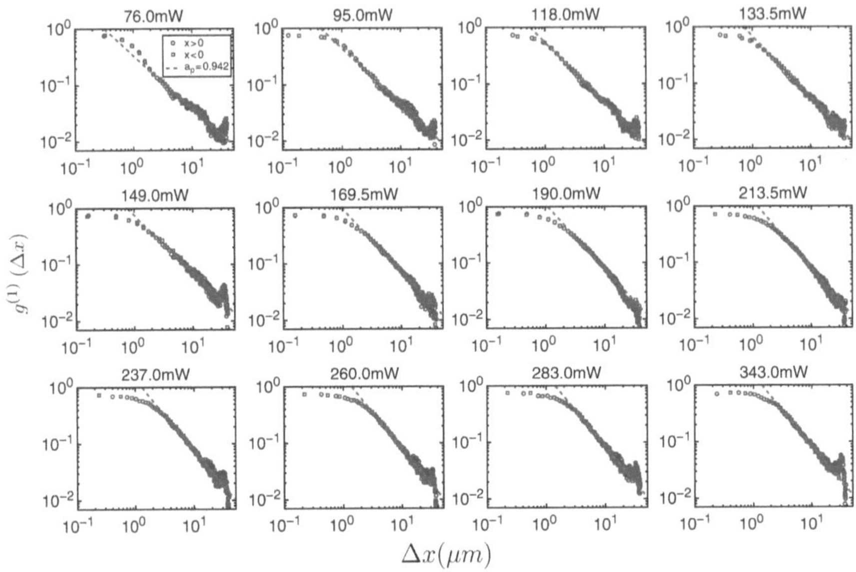

Fig. 4.20 Power-law fits of $g^{(1)}(\Delta x)$ for several pumping powers above threshold

Fig. 4.21 Parameter λ_p of the power-law fit of the correlation function $g^{(1)}(\Delta x) = (\lambda_p/\Delta x)^{a_p}$ at long distances Δx

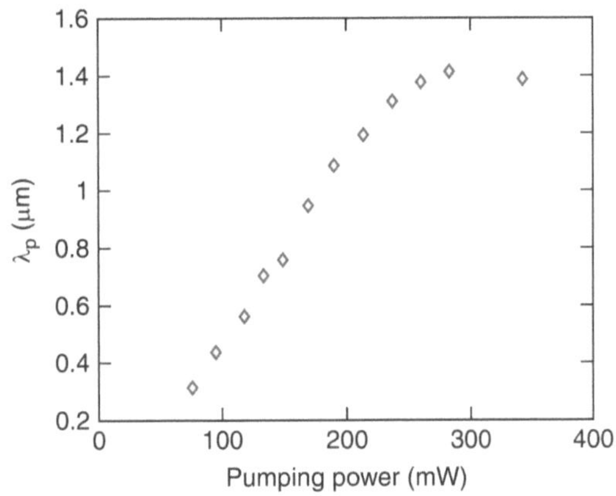

4.4.5 *Different Detunings and Orthogonal Prism Orientation*

We have repeated the same measurements at several photon–exciton detunings δ in the range $-3.7\,\text{meV} \le \delta \le 2.6\,\text{meV}$ (corresponding to effective mass $7.8 \times 10^{-5} \le \frac{m^*}{m_e} \le 11.5 \times 10^{-5}$). The results are summarized in Fig. 4.22. In Fig. 4.22a, we plot the experimental results for the effective thermal de Broglie wavelength λ_{eff} as a function of pumping power. The behavior is similar for all detunings. The measured λ_{eff} takes about the same values for all detunings at very low pumping power, as the measurement is resolution-limited in this region. In Fig. 4.22b, we plot the fitting

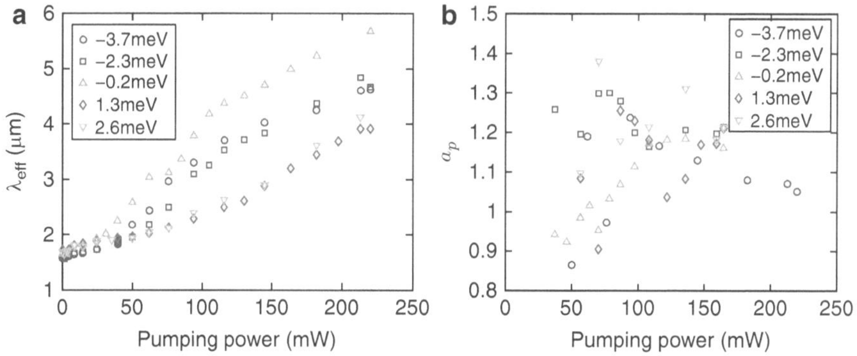

Fig. 4.22 Dependence on photon–exciton detuning. (**a**) Effective thermal de Broglie wavelength λ_{eff} as a function of laser pumping power for several photon–exciton detunings δ. The prism is in the horizontal orientation. (**b**) Exponent a_{p} of the power-law decay as a function of laser pumping power

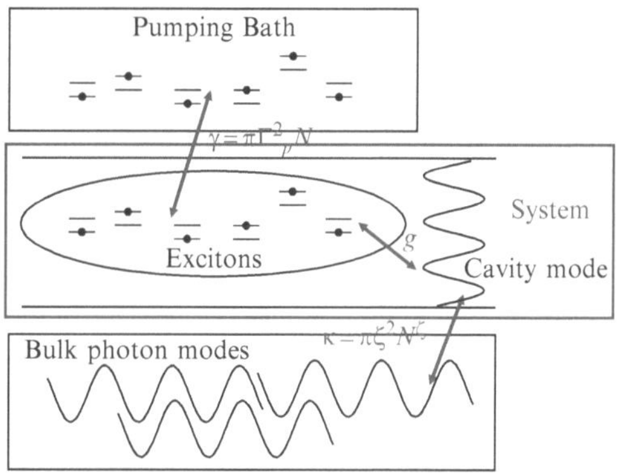

Fig. 4.23 Schematic of the nonequilibrium theoretical model. From [56]

results for the exponent a_{p} of the power-law decay of $g^{(1)}(\Delta x)$ at long distances. The long-distance decay was indeed found to be a power law, with an exponent in the range 0.9–1.3 when including different detunings.

4.4.6 Nonequilibrium Model

The nonequilibrium theory of polariton condensation that best explains our data was developed in [55,56] (see also [63] for a review of the essential results with minimal technical details). The model is schematically illustrated in Fig. 4.23. The main part of the system consists of cavity photon fields and localized two-level atoms randomly distributed in energy. The localized two-level atoms represent the excitons and do not interact with each other. Because of their random energy distribution, though, a nonlinear effect is present and models the interactions. Namely, as the exciton population increases, all low-energy atoms are occupied, so new excitons can only be accommodated in high-energy atoms, and the average exciton energy increases. Pumping is modeled through the coupling of excitons to a pumping bath, which also consists of localized two-level modes with random energy distribution. Cavity photons, on the other hand, can leak outside of the cavity, so they are coupled to external photon modes with the same momentum.

We start with the Hamiltonian

$$\hat{H} = \hat{H}_{\text{sys}} + \hat{H}_{\text{bath}} + \hat{H}_{\text{sys,bath}}, \tag{4.67}$$

where

$$\hat{H}_{\text{sys}} = \sum_{\alpha} \epsilon_{\alpha} \left(\hat{b}_{\alpha}^{\dagger}\hat{b}_{\alpha} - \hat{a}_{\alpha}^{\dagger}\hat{a}_{\alpha}\right) + \sum_{\mathbf{p}} \omega_{\mathbf{p}} \hat{\psi}_{\mathbf{p}}^{\dagger}\hat{\psi}_{\mathbf{p}} + \frac{1}{\sqrt{L^2}} \sum_{\alpha}\sum_{\mathbf{p}} \left(g_{\alpha,\mathbf{p}}\hat{\psi}_{\mathbf{p}}\hat{b}_{\alpha}^{\dagger}\hat{a}_{\alpha} + \text{H.c.}\right) \tag{4.68}$$

is the system Hamiltonian. The first term in 4.68 describes the excitonic modes (two-level atoms). ϵ_{α} is the energy of atom α, and $\hat{b}_{\alpha}$, $\hat{a}_{\alpha}$ are fermionic operators. $\hat{b}_{\alpha}^{\dagger}$ creates a particle on the upper level of the two-level atom α, and $\hat{b}_{\alpha}$ destroys a particle from the same level. Equivalently $\hat{a}_{\alpha}^{\dagger}$, and $\hat{a}_{\alpha}$ act in the lower level of mode α. Because only one of the two levels is occupied, it is necessary that $\hat{b}_{\alpha}^{\dagger}\hat{b}_{\alpha} + \hat{a}_{\alpha}^{\dagger}\hat{a}_{\alpha} = 1$ for each atom α. In the calculation, this constraint is imposed on average. The second term in (4.68) corresponds to the cavity photons. $\omega_{\mathbf{p}}$ is the energy, and $\hat{\psi}_{\mathbf{p}}^{\dagger}$, $\hat{\psi}_{\mathbf{p}}$ are the field operators of the cavity mode with momentum $\mathbf{p}$. The last term in (4.68) models the coupling between the exciton and cavity modes through the coupling matrix elements $g_{\alpha,\mathbf{p}}$. $g_{\alpha,\mathbf{p}}$ determine the Rabi splitting of the cavity.

The pumping bath is described by fermionic operators $\hat{A}_k$ and $\hat{B}_k$ that act on the lower and upper levels, respectively, of the two-level modes k. Correspondingly, external photons are described by $\hat{\Psi}_k^{\dagger}$ and $\hat{\Psi}_k$, and the bath Hamiltonian is given by

$$\hat{H}_{\text{bath}} = \sum_{k} \omega_k^{\Gamma^a} \hat{A}_k^{\dagger}\hat{A}_k + \sum_{k} \omega_k^{\Gamma^b} \hat{B}_k^{\dagger}\hat{B}_k + \sum_{k} \omega_k^{\zeta} \hat{\Psi}_k^{\dagger}\hat{\Psi}_k. \tag{4.69}$$

Finally, coupling between the system and the baths is described by

$$\hat{H}_{\text{sys,bath}} = \sum_{\alpha,k} \Gamma_{\alpha,k}^{a}(\hat{a}_{\alpha}^{\dagger}\hat{A}_k + \text{H.c.}) + \Gamma_{\alpha,k}^{b}(\hat{b}_{\alpha}^{\dagger}\hat{B}_k + \text{H.c.}) + \sum_{\mathbf{p},k} \zeta_{\mathbf{p},k}(\hat{\psi}_{\mathbf{p}}^{\dagger}\hat{\Psi}_k + \text{H.c.}). \tag{4.70}$$

The path integral formulation of nonequilibrium Keldysh field theory is used [64]. It is assumed that the baths are much larger than the system, so their behavior is not affected by the interaction with the system. So, we can integrate over the bath degrees of freedom and calculate an effective action expressed only in terms of the fields describing the system. Including phase fluctuations to all orders and density fluctuations to second order, the first-order correlation function is proven to follow the equation

$$g^{(1)}(\mathbf{r},t) = [1 + \mathcal{O}(1/\rho_0)] \exp[-f(\mathbf{r},t)], \tag{4.71}$$

where ρ_0 is the condensate density and $f(\mathbf{r},t)$ is a function to be determined.

As the long-distance and long-time decay of $g^{(1)}(\mathbf{r},t)$ depends on phase fluctuations, the function $f(\mathbf{r},t)$ depends both on the occupation function and on the spectral function of the excited phase modes. The spectral function determines the excitation spectrum of the system; thus, changes to the excitation spectrum can (but need not necessarily) change the behavior of $g^{(1)}(\mathbf{r},t)$. In equilibrium, the change of spectrum from normal to condensed state leads to the appearance of a power-law decay. In this case, $f(\mathbf{r},t)$ is

$$f(\mathbf{r},t) \simeq a_\mathrm{p} \ln\left(\frac{k_\mathrm{B}T\sqrt{c_\mathrm{s}^2 t^2 + r^2}}{\hbar c_\mathrm{s}}\right), \tag{4.72}$$

where the exponent $a_\mathrm{p} = 1/\rho_0\lambda_\mathrm{T}^2$. As expected, the equilibrium result (4.53) is recovered. c_s in (4.72) is the sound velocity, whose exact value is not important for the rest of the discussion.

Including pumping and decay further modifies the condensed spectrum, leading to diffusive behavior at long wavelengths [55,56,65]; however, whether this change of spectrum causes a change of power law depends on the occupation function. For thermal occupation of the excitations, the power law for the spatial decay is not affected by pumping and decay, so that $f(\mathbf{r},t = 0)$ is given by the equilibrium result (4.72). The temporal decay of $g^{(1)}(\mathbf{r} = 0,t)$ for large t is also a power law, but the exponent is different that the spatial decay.

For other occupation functions, the power law can be affected by dissipation. Given that spatial coherence extends further when the same sample is pumped by a single-mode laser [66], and the known role of pumping noise on temporal coherence discussed in [67, 68], it is likely that the current results are the manifestation of such pumping noise acting on the continuum of long wavelength modes. As a crude model of this, one could take the spectral function of the dissipative system and assume a flat occupation function (i.e., flat noise spectrum) with strength ζ. This predicts $a_\mathrm{p} \propto \zeta/n_\mathrm{s}$ for the spatial decay.

Unfortunately, this approach cannot lead to an expression of the pumping strength ζ in terms of experimentally measurable quantities. It does predict, though that in a nonequilibrium system, it is possible to observe a power-law decay of $g^{(1)}(\mathbf{r})$ with a large exponent if the noise is sufficiently strong.

4.4.7 Conclusion

In conclusion, the measured power-law decay of the correlation function suggests that some form of the BKT superfluid phase survives in nonequilibrium condensates, namely, phase fluctuations are excited but no vortices. The large value of the exponent implies that, in the current experiment, this ordered phase is more robust against external noise than would be expected in equilibrium, in which equipartition holds. We conjecture that the noise source is intensity fluctuations of the pumping laser and apply a nonequilibrium theory to show that pumping noise can induce a power-law decay with a large exponent. One may anticipate that sufficient noise could induce vortex proliferation and a transition to short-range coherence. This fascinating possibility remains an open question for future studies.

4.5 Single Vortex–Antivortex Pair

In Sect. 4.2, we argued that in a homogeneous two-dimensional system at nonzero temperature, although there can be no ordering of infinite range [11,23], a superfluid phase is predicted for a Bose fluid [12, 26, 29]. Bound vortex–antivortex pairs dominate the thermodynamics and phase coherence properties in this superfluid regime. It is believed that several different systems share this common behavior, when the order parameter describing their macroscopic state has two degrees of freedom [34] and the theory has been tested for some of them [37–41,44]. However, there has been no direct experimental observation of a quasicondensate which includes a bound pair of vortex and antivortex.

Here, we present an experimental result that can identify a single vortex–antivortex pair in a two-dimensional exciton–polariton condensate. We have created a pumping spot which generates a minimum of the condensate density at the center. A zero in density can be thought of as a superposition of a vortex and antivortex that can be separated by an external perturbation [69]. Thus, the central dip of the condensate acts as a source of vortex–antivortex pairs. By calculating the energies of vortex–antivortex pairs observed in the experiment, we conclude that thermal fluctuations cannot be responsible. We conjecture that this effect is due to pump laser intensity fluctuations.

We have found that for a particular condensate size, there is on the average one pair at any time. A Michelson interferometer is used to reconstruct the time-integrated phase map of the system. When the sample disorder potential is stronger than the blueshift induced by polariton–polariton interactions, pinned pairs appear at certain locations. When the sample disorder potential is weak, on the other hand, they are mobile. They appear along a fixed axis, because of a small asymmetry in the pumping spot, and are created with a random polarization, namely, the vortex can appear on the right side of the spot and the antivortex on the left, or vice versa. In the time-integrated measurement, two distinct characteristic phase defects appear in the above two cases of pinned pair and mobile pair.

We have applied an open-dissipative Gross–Pitaevskii (GP) equation model [70], which reproduces the observed density minimum at the condensate center. We next imprinted a vortex–antivortex pair in a condensate and numerically simulated its space-time evolution. We found that pair motion is significantly modified compared to atomic condensates [71] because of the dissipative nature of the polariton condensate and the repulsive interactions between condensate and reservoir particles. The vortex pair is found to migrate perpendicular to its dipole moment due to the velocity field and, in our experimental parameter space, recombines before reaching the condensate edge. Despite the short polariton lifetime, vortex pairs survive for a long enough time to be observed.

Our results suggest that vortex–antivortex pairs can be created in a 2D condensate without rotation or stirring. Our observation that the vortex and antivortex in each pair follow a correlated motion and do not separate is consistent with the power-law decay of the correlation function discussed in Sect. 4.4. If free vortices were present in the condensate, the correlation function would decay exponentially. Further measurements on larger condensates containing more vortex–antivortex pairs should clarify the influence of pair dynamics on phase coherence and polariton superfluidity [72–77].

4.5.1 Condensate Shape in Real Space

The excitation scheme is described already in Sect. 4.4. The pump laser at above bandgap energy creates free electron–hole pairs, which form excitons and finally relax toward the lower polariton (LP) branch, as shown in the schematic of Fig. 4.24a. When an LP decays, it emits one photon. In this experiment, we use a pumping spot radius of 12 μm. Figure 4.24b shows measured luminescence below threshold, which features Airy-function-like patterns because of diffraction effects. We assume that the shape is the same as the laser pumping spot, and use it as an input to our numerical calculation. Above the condensation threshold, a population dip develops at the condensate center, as shown in the measured luminescence image of Fig. 4.24d. Our numerical simulation reproduces the observed population dip (Fig. 4.24d) and suggests that the inhomogeneity of the pump spot profile, notably the ring near the center, generates this condensate shape. The condensate density minimum is stabilized by the repulsive reservoir-condensate interaction, since the reservoir shows a corresponding density maximum at the same point (Fig. 4.24c). This density minimum can be considered as an overlapping vortex and antivortex.

Phase fluctuations [78] of the condensate are responsible for the spontaneous vortex-pair generation [69]. In two dimensions, this is usually the result of interaction of the condensate population with thermal excitations. Knowing the energy E_v of a particular vortex-pair configuration, the probability of thermal nucleation of this vortex-pair configuration can be estimated. In three dimensions, the thermal excitation of a single vortex requires an energy proportional to the length of the vortex line and is thus highly improbable. In two dimensions (2D) though, the confinement along the axis of the vortex line results in a much smaller formation energy. In the present polariton condensate, the thermal energy is still insufficient to generate a single vortex whose energy we can estimate from (4.29)

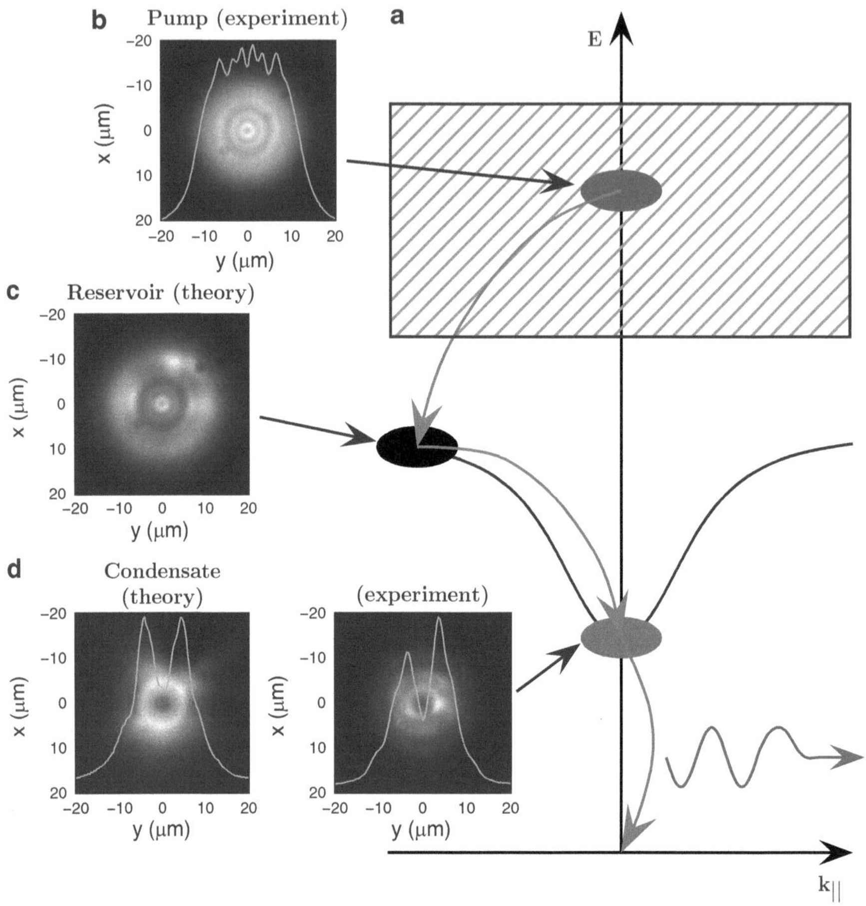

Fig. 4.24 Pump and condensate profile. (**a**) Schematic of the pumping mechanism (see text). (**b**) LP luminescence image at half the threshold power. The profile along $x = 0$ is shown in yellow line. We assume that it has the shape of the laser pumping spot, and use it as an input to our numerical simulation. (**c**) Numerically calculated reservoir density above threshold. (**d**) Calculated and measured condensate density images for pumping power three times above threshold.

$$E_{\mathrm{v}} \sim \frac{m_{\mathrm{LP}} n_{\mathrm{C}} \Gamma^2}{4\pi} \ln\left(\frac{R}{\xi}\right) \sim 10\,\mathrm{eV} \gg k_{\mathrm{B}} T. \tag{4.73}$$

In this equation, n_{C} is the condensate density, R is the condensate radius, and $\Gamma = h/m_{\mathrm{LP}}$. When dissipative effects are taken into account (through the dissipative GPE), the vortex energy is found to be lower than this estimate and nonlinear in n_{C} due to the reservoir presence, but is still of order eV. The energy of a 2D vortex–antivortex pair depends on its separation and thus can have a smaller energy for small separations:

$$E_p = \frac{m_{\mathrm{LP}} n_{\mathrm{C}} \Gamma^2}{2\pi} \ln\left(\frac{d_{\mathrm{v}}}{\xi}\right), \tag{4.74}$$

gives the energy for a uniform condensate, where d_{v} is the vortex–antivortex separation. Unless the vortex pair is given enough energy for the vortex cores to fully separate (core–core separation $d_{\mathrm{v}} \gtrsim 2\xi$), it will immediately recombine rapidly and will not be observed. The coefficient of the logarithm is large, though, and thermal fluctuations ($k_{\mathrm{B}}T \sim \mathrm{meV}$) cannot be solely responsible for formation of vortex pairs with a reasonably large separation $d_{\mathrm{v}} \geq \xi$.

Instead, density fluctuations in the reservoir, which has population density comparable to or higher than the condensate, introduce an alternative formation source. The mean-field interaction energy due to the reservoir is maximum at the center, where the reservoir density is maximum. Since the intensity fluctuations of our pump laser modulate the reservoir population strongly, this results in modulation of the local interaction energy and thus of the local phase of the condensate, which induces a separation of the vortex and antivortex. We confirmed that a vortex–antivortex pair is not regularly generated when a quiet single-mode pump laser is used [66].

There is no confining potential in this experiment, and the localization of polaritons near the pumping laser is a result of their finite lifetime, and hence, the limited distance they can travel before recombining. There are, however, subtleties regarding the interplay of this expansion of the condensate from the pump spot and the effects of interactions between the condensate polaritons and reservoir excitations created by the laser.

In particular, the repulsive interactions among condensate particles and between the condensate and reservoir influence the condensate shape, but whether this leads to spatial spreading of the condensate depends on the size and shape of the excitation spot. A small excitation spot creates a strong antitrapping potential, which leads to condensation in a high-momentum state, so that the condensate propagates away from the excitation spot and spreads in space. This was observed in [79] in a CdTe sample and reproduced in [59] in a GaAs sample. When the excitation spot is large, on the other hand, a smooth antitrapping potential is formed that cannot prevent condensation in the zero momentum state. In this case, the condensate is localized and does not propagate far from the excitation spot. This was observed in our experiment. Both cases of the small and large pumping spots were theoretically studied in [61], and the above picture was established. We theoretically studied the condensate using the model of [61] and confirmed the doughnut shape for our sample parameters.

4.5.2 Pinned Vortex–Antivortex Pair

The expected phase map for a pinned vortex–antivortex pair is shown in Fig. 4.25a. If we perform an interferometry measurement, in the way described in Sect. 4.4,

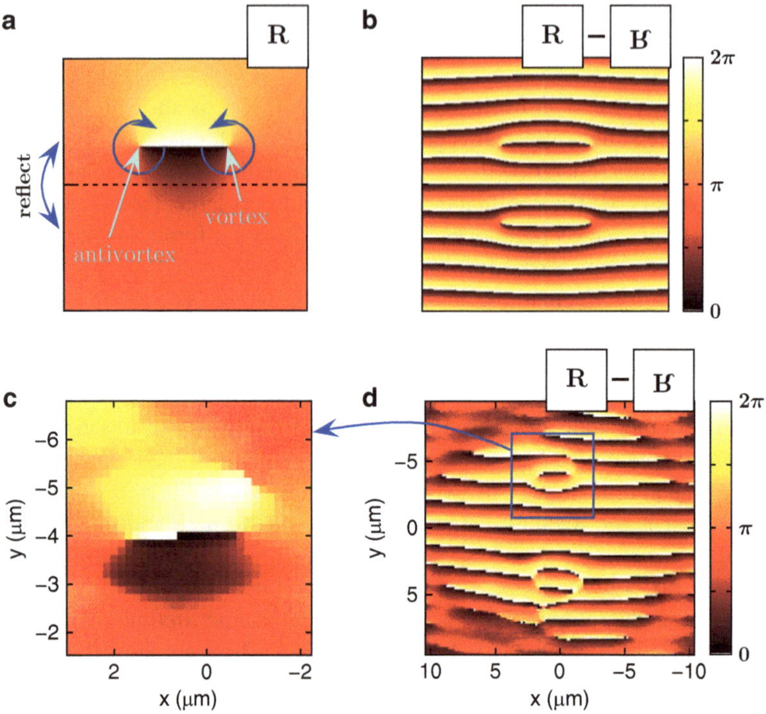

Fig. 4.25 Phase map of a pinned pair. (**a**) Expected phase map of a condensate including a single vortex–antivortex pair. Arrows show the direction of the phase increase around the vortex and antivortex. (**b**) Simulation of the experimentally measured phase map when panel a interferes with its reflection along the horizontal (*dashed*) line. A global phase slope along the vertical direction is added. (**c**) Experimentally measured phase map at 55 mW above the condensation threshold of 20 mW. The blue square marks the position of a double dislocation pattern. (**d**) Expanded view of the blue square in c, where the global slope along the horizontal direction is subtracted. c and d are rotated by 90° with respect to all other experimental data in this chapter

with this phase map as an input, the expected result is shown in Fig. 4.25b. The two double-dislocation patterns, one at the upper part and the other at the lower part of the figure, mark the position of the pair in the original and in the reflected images. Figure 4.25c shows a phase map measured experimentally. It features the double dislocation pattern characteristic of a pinned vortex–antivortex pair. In Fig. 4.25d, we have subtracted the global phase slope to reveal the actual phase map of a single vortex–antivortex pair, corresponding to the expected one (Fig. 4.25a).

The sample is mounted using silver paint on a copper sample holder, attached to the cold finger of a helium flow cryostat. Because of the different thermal expansion coefficients, when the system is cooled down, the sample experiences some amount of strain. Strain can induce a splitting ΔE_{gr} between emitted photons linearly polarized along the horizontal and vertical directions. Hence, ΔE_{gr} can be considered as a measure of local strain. Data in Fig. 4.25 are taken with

$\Delta E_{gr} = 150\,\mu$ eV. Due to the sample disorder potential, the pair has a pinned position and polarization, defined as the vector pointing from the antivortex to the vortex. This is similar to the experiment in [72], where a single vortex pinned by the disorder potential was observed.

However, when special care is taken to reduce strain during cooldown, by applying a homogeneous layer of silver paint between the sample and sample holder, ΔE_{gr} can be reduced below 50μ eV. In this case, the disorder potential is very weak, much smaller than the blueshift due to polariton–polariton interactions, so the disorder is screened when the condensate is formed. Therefore, there are only very few locations where pinned vortex–antivortex pairs can be observed.

4.5.3 Mobile Vortex–Antivortex Pair

From the previous discussion, we expect vortex–antivortex pairs in our second experiment to be mobile. We next numerically study the evolution of a mobile vortex pair imprinted along the horizontal axis, using a time-dependent open-dissipative GP equation [70]. In a harmonically trapped conservative condensate, a single vortex pair will undergo linear motion with velocity $v_{v-av} = \hbar/m^* d_v$, which is inversely proportional to the vortex–antivortex separation d_v (Fig. 4.1). Upon interaction with the boundary, it will wrap back upon itself with a cyclical motion [20, 71]. However, a polariton condensate with considerable repulsive condensate-reservoir interaction (due to large reservoir population present in the experiment) experiences drag forces strongly perturbing the vortex-pair motion [80], causing it to recombine after only a short travel distance. For these types of micromotion, despite the vortex-pair motion, a signature of the pair in time-integrated interferogram is preserved. There are two areas where the phase is shifted by π, one at the top and the other at the bottom of the figure, and they are surrounded by minima in the fringe visibility. The measured positions of the phase defects depend on the time averaged positions of the vortex and antivortex. The above scenario is illustrated in Fig. 4.26.

To better understand the characteristic pi-phase-shift interference pattern expected for a single vortex–antivortex pair with random polarization, consider Fig. 4.27. Figure 4.27a shows the interference fringes we expect to observe on the camera when there is a single pair at a fixed position along the horizontal axis, and the prism in the Michelson interferometer is oriented vertically. Two trident patterns appear. When the pair polarization is reversed (Fig. 4.27b), the trident patterns are also reversed. Figure 4.27c results from 50–50 statistical mixture between the two pair polarizations. There are two areas where the phase of the fringes is shifted by pi compared to outside, and around these areas, the fringes are blurred, namely, the fringe visibility has a minimum. This is because in the case of panel a, the fringes are bent one way, whereas in the case of panel b, the fringes are bent the other way. Figures 4.27d,e shows the phase and visibility maps corresponding to Fig. 4.27c, where the pi-phase-shift areas are clearly seen. Obviously, the shape of these areas depends on the position of the pair and should be different when the pair can move

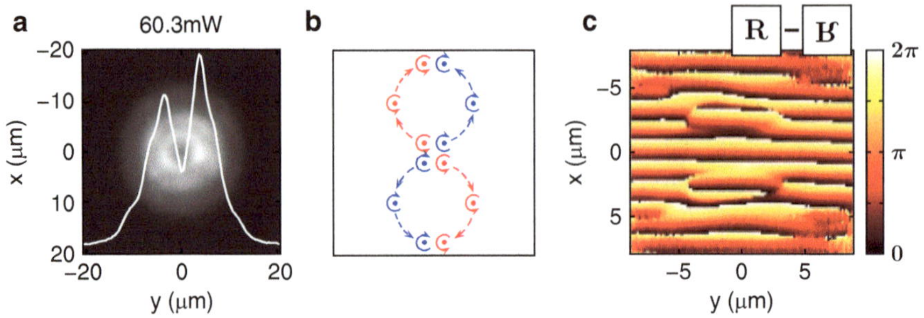

Fig. 4.26 Summary of the vortex–antivortex pair production, evolution, and detection. (**a**) Condensate shape and its cross section along the line $x = 0$. (**b**) Schematic of the vortex–antivortex pair trajectory. Depending on the pair polarization (whether the vortex is on the *left or right of the antivortex*), the pair moves in the upper or lower half of the condensate. (**c**) Measured phase map revealing the presence of a single vortex pair

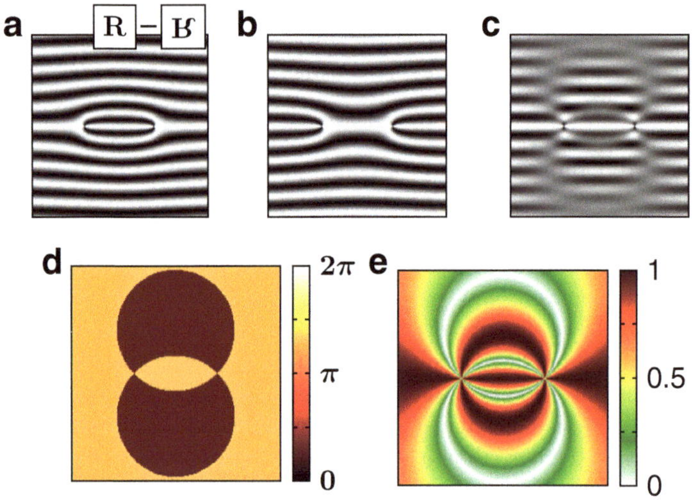

Fig. 4.27 Explanation of the interference pattern for a single pair with random polarization (*simulation results*). (**a**) Interference fringes observed in the camera for one vortex–antivortex pair polarization and for vertical orientation of the prism in the Michelson interferometer. It features two trident-like patterns. (**b**) Interference fringes for the opposite pair polarization. The trident-like patterns are now reversed. (**c**) Interference fringes for 50–50 statistical mixture of the two pair polarizations. (**d**) Phase map and (**e**) visibility map corresponding to c

inside the condensate without changing its orientation, as will be discussed later. We expect that this motion takes place in our experiments and is captured by our theoretical model.

Figure 4.28a,b shows the results of our numerical simulation for the time-integrated phase and visibility maps, respectively. We indeed observed such unique patterns experimentally when the sample strain was removed, as shown

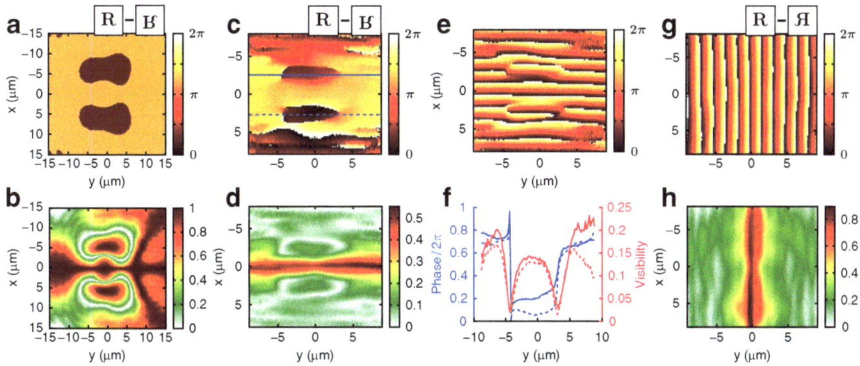

Fig. 4.28 Phase and fringe visibility map of a free pair. (**a**) Theoretical time-integrated phase map for a vortex–antivortex pair imprinted into a condensate and evolving according to the dissipative GP equation. (**b**) Corresponding fringe visibility map. (**c**) and (**d**) Measured phase and fringe visibility maps. (**e**) Same data as in c, but now the global phase slope is not subtracted. (**f**) *Blue*: phase cross section along the continuous and dashed lines in c. *Red*: fringe visibility cross section along the same lines. (**g**) Measured phase map when the prism is rotated by 90°, along with a schematic showing the orientation of the interfering images. (**h**) Corresponding fringe visibility map. Experimental data are taken at 55 mW, above the condensation threshold of 20 mW

in Fig. 4.28c,d, which illustrates typical experimental data. We found that there is no noticeable dependence of the measured phase maps on the laser pumping power, up to ~5 times the threshold power. In Fig. 4.28e, we have not subtracted the global phase slope, so that the direct experimental phase map is given. In Fig. 4.28f, we plot the cross section of the phase along two (continuous and dashed) lines passing through the two areas of π-phase shift, where this phase shift is shown quantitatively. We also plot the measured fringe visibility along the same lines and find that the visibility minima coincide with the phase jumps. This result demonstrates that the vortex–antivortex pairs are created with a random polarization after the removal of strain.

Moreover, when we rotate the prism by 90°, so that the reflected image is along the vertical axis, no phase defect is observed (Fig. 4.28g). The difference between Fig. 4.28e and g can be understood as follows (see Fig. 4.29): The vortex pair always remains on the horizontal axis, and we fold the reflected image along the vertical axis. In this case, the phase rotation around the vortex by 2π and that around the antivortex by -2π cancel out, thus there is no phase defect in the interference pattern. The corresponding visibility map without any defect is shown in Fig. 4.28h.

The free pair of Fig. 4.28 shows a distinct signature compared to the pinned pair of Fig. 4.25. Indeed, in the interferogram of Fig. 4.25d, there are two singularity points, one at the location of the vortex and the other at the location of the antivortex, and the measured phase is continuous everywhere else. The experimental data in Fig. 4.28 suggests that such vortex–antivortex pairs are not pinned and travel a distance comparable to the spot size in a process similar to the one described in [81]. When the pair can form with random polarization and also move, the measured

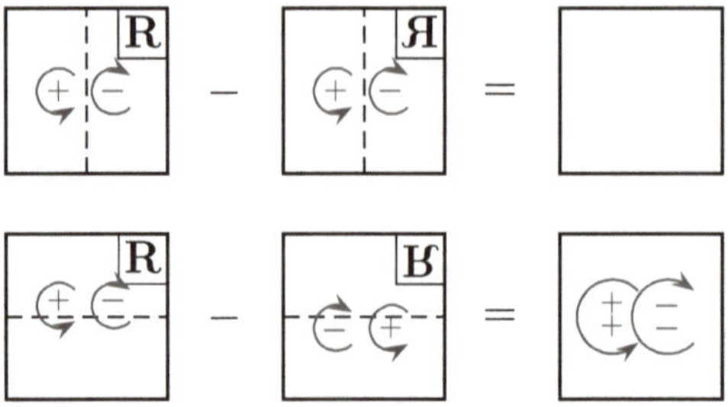

Fig. 4.29 Horizontal versus vertical prism orientation. (**a**) When the vortex–antivortex pair lies along the horizontal axis and the original image interferes with its reflection along the vertical axis, then the vortex and antivortex in the reflected image annihilate those in the original image. So the final image does not show any phase defects, corresponding to Fig. 4.28g. (**b**) When the prism creates the reflection of the original image along the horizontal axis, then the final image shows a double vortex and a double antivortex, corresponding to Fig. 4.28e. This is a qualitative change with respect to a.

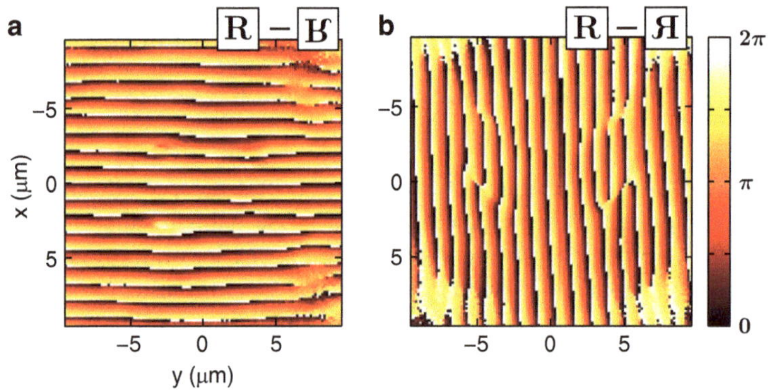

Fig. 4.30 Interference images with a pumping spot rotated by 90°. (**a**) Phase map with one prism orientation. (**b**) Phase map with the orthogonal prism orientation

time-integrated interferogram is a superposition of many interferograms of the same type as Fig. 4.25d with varying positions for the vortex and antivortex. This results in a pattern with two π-phase shift areas (Fig. 4.28c), which is consistent to our theory (Fig. 4.28a).

We consistently observe pairs along the same axis, even after we rotate the sample by 90°. However, when we rotate the pump laser spot by 90°, the polarization axis of the vortex pair is also rotated by 90°. We rotated the original pumping spot by using a Dove prism just before the PBS in Fig. 4.9a. The vortex–antivortex pair is also rotated by 90° in this case, as is clear from Fig. 4.30. The two panels can be directly compared with Fig. 4.28e,g. These results suggest that the pair polarization direction is determined by a small asymmetry of the pump laser spot, rather than by the disorder potential landscape in the sample. Although the pair orientation follows the rotation of the laser spot, the random switching of pair polarization confirms that the pair generation mechanism is nondeterministic.

4.5.4 Different Condensate Shapes

To make sure that the observed π-phase shift areas are not an experimental artifact, we perform the same Michelson interferometer measurement using a Gaussian pumping spot by removing the beam shaper from the setup of Fig. 4.9a. In this measurement, we switch to a single-mode Ti:Sapph ring laser with long coherence time. For both prism orientations, no phase defect is observed in the interference pattern directly seen on the camera (Fig. 4.31a,c) or in the phase maps (Fig. 4.31b,d). According to the generation mechanism of vortex–antivortex pairs discussed in the main text, the population dip in the condensate formed because of the maximum of the reservoir density at the center is essential. This mechanism is absent when the Gaussian pumping spot is used. This observation supports our argument that the minimum of the condensate density at the center of the spot acts as a source of vortex–antivortex pairs.

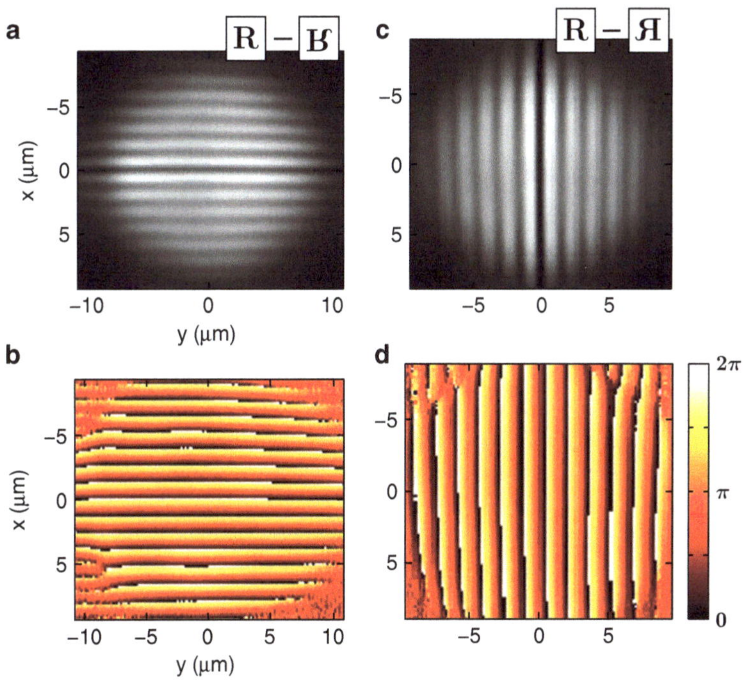

Fig. 4.31 Interference images with Gaussian pumping spot. (**a**) Interference fringes observed on the camera with one prism orientation. (**b**) Phase map corresponding to **a**. (**c**) Interference fringes observed on the camera with the orthogonal prism orientation. (**d**) Phase map corresponding to **c**

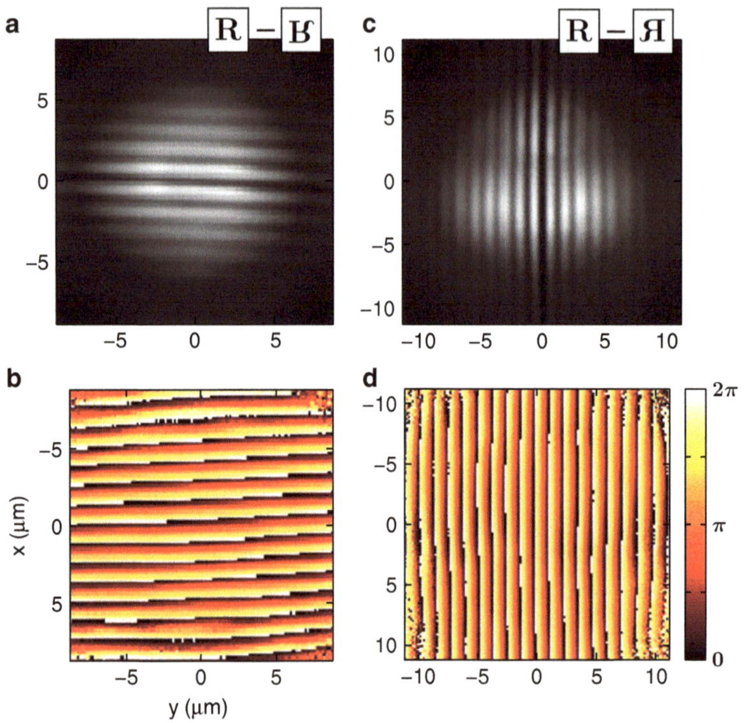

Fig. 4.32 Interference images with a small flat pumping spot. (**a**) Interference fringes observed on the camera with one prism orientation. (**b**) Phase map corresponding to **a**. (**c**) Interference fringes observed on the camera with the orthogonal prism orientation. (**d**) Phase map corresponding to **c**

Finally, we created a pumping spot with radius 7 μm, instead of 12 μm. No phase defects were observed. This is shown in Fig. 4.32, which features the measured interference patterns (Fig. 4.32a,c) and phase maps (Fig. 4.32b,d). We conclude that in this small condensate, pairs recombine quickly at the boundary and do not leave any trace in the interferogram. Therefore, the 12 μm-radius condensate is of just the right size for which on the average one pair is present at all times.

Using a prism in our Michelson interferometer (M2 in Fig. 4.13a) instead of a retroreflector has the advantage that a single vortex should always be observable, even if it is mobile. The idea is illustrated in Fig. 4.33a and is based on the fact that the sense of rotation changes when an image is reflected. Consider the case that the original image includes one vortex. The reflected image will include one antivortex. Because the interferometer measures the phase difference between the two images, the final image will feature two vortices. Figure 4.33b shows the simulated time-integrated interference pattern when a single vortex is moving randomly inside the red circle. Figure 4.33c shows the same interference pattern for the case of a mobile antivortex, while in Fig. 4.33d, we consider a 50–50 probability for either vortex or antivortex. None of these interference patterns is observed in our data, which

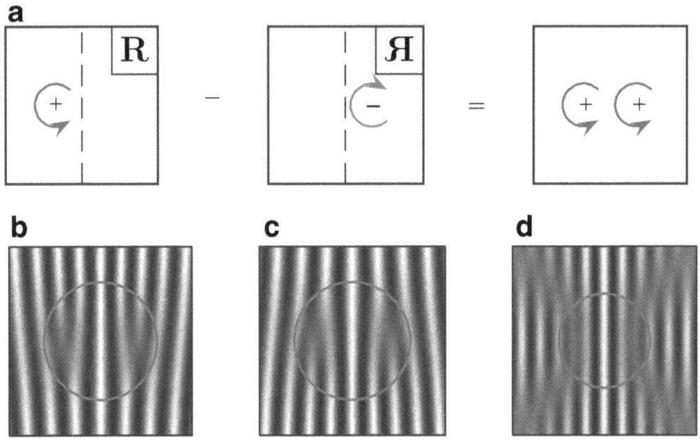

Fig. 4.33 Interference images for a single mobile vortex (*simulation results*). (**a**) Schematic of the Michelson interferometer measurement for the case that a single vortex is present in the initial image (*see text*). (**b, c**), Simulated time-integrated interference patterns for the case that a single vortex (**b**) or a single antivortex (**c**) are moving randomly inside the red circle. (**d**) The same as b and c, but now, there is a 50–50 probability for vortex or antivortex

suggests that there are no free vortices in the condensate, but only bound vortex–antivortex pairs.

4.5.5 Vortex-Pair Dynamics Described by Open-Dissipative Gross–Pitaevskii Equation

The open-dissipative Gross–Pitaevskii equation [70] consists of two coupled equations describing the time evolution of the condensate order parameter $\psi(\mathbf{r},t)$ and exciton reservoir density $n_{\mathrm{R}}(\mathbf{r},t)$

$$i\hbar\frac{\partial\psi(\mathbf{r},t)}{\partial t} = \left(-\frac{\hbar^2\nabla^2}{2m_{\mathrm{LP}}} - \frac{i\hbar}{2}\left[\gamma_{\mathrm{C}} - R\left((n_{\mathrm{R}}(\mathbf{r},t)\right)\right] + g_{\mathrm{C}}|\psi(\mathbf{r},t)|^2 + g_{\mathrm{R}}n_{\mathrm{R}}(\mathbf{r},t)\right)\psi(\mathbf{r},t), \tag{4.75}$$

$$\frac{\partial n_{\mathrm{R}}(\mathbf{r},t)}{\partial t} = P_{\mathrm{l}}(\mathbf{r},t) - \gamma_{\mathrm{R}}n_{\mathrm{R}}(\mathbf{r},t) - R(n_{\mathrm{R}}(\mathbf{r},t))|\psi(\mathbf{r},t)|^2. \tag{4.76}$$

$n_{\mathrm{R}}(\mathbf{r},t)$ is controlled by laser pumping gain $P_{\mathrm{l}}(\mathbf{r})$ and reservoir loss γ_{R}. Interaction between condensate and reservoir is represented by a coupling constant $g_{\mathrm{R}} = 2g_{\mathrm{C}}$, where the condensate–condensate interaction constant is $g_{\mathrm{C}} = 6 \times 10^{-3}\mathrm{meV}\mu\mathrm{m}^2$. Other parameters are the condensate polariton loss rate $\gamma_{\mathrm{C}} = 0.33\,\mathrm{ps}^{-1}$ and stimulated scattering rate $R(n_{\mathrm{R}}(\mathbf{r},t))$.

Though some of the parameters (particularly $n_R(\mathbf{r},t)$ and $R(n_R(\mathbf{r},t))$), cannot be measured directly, the strong influence of the reservoir population on the condensate spatial profile allows us to indirectly probe the reservoir population density and uniquely determine the parameter space. In particular, we obtain the best fits to experimental sequences of spatial data as a function of P_l for pumping threshold $P_{th} = 25$ mW with 24% pumping efficiency and $\gamma_R \approx 1.5\gamma_C$ and $R(n_R(\mathbf{r},t))/n_R(\mathbf{r},t) \approx \gamma_C\gamma_R/P_{th}$.

To study the vortex–antivortex pair dynamics, a vortex pair is imprinted in a steady-state condensate directly via a phase factor $e^{i\theta}$ for the vortex and $e^{-i\theta}$ for the antivortex and is subsequently left free to evolve in time. The temporal frames are stored and then cumulatively subjected to the same time-averaging mechanism as in the experiment.

We have analyzed the dynamics and stability of a vortex–antivortex pair in a polariton condensate in detail [80] and demonstrated significant deviations from the usual vortex-pair motion expected in the conservative atomic condensate. We found that the vortex pair will either recombine within the condensate or separate and dissipate from the condensate boundary. The choice between these two types of trajectory (in a largely homogeneous but confined condensate) is a competition between the force due to the radially directed superfluid flow of the unconfined repulsively interacting condensate and the drag forces due to the interaction with noncondensate (reservoir and thermal polaritons). The cross over between these regimes essentially depends only on the magnitude and profile of $n_R(\mathbf{r})$

The pumping profile used in this simulation is an experimentally measured top-hat profile (see Fig. 4.11), and the condensate polariton lifetime is $\tau_r = 3$ ps. The scattering rate $R(n_R) = R_{sc}n_R(\mathbf{r})$is assigned a linear dependence on $n_R(\mathbf{r})$, and a measurement of the threshold pumping power P_{th} permits us to estimate the scattering rate via $R_{sc} = \gamma_C\gamma_R/P_{th}$, where we assume γ_R is comparable to γ_C, necessary to study this experimental parameter space. Thus, with these parameters fixed to correspond to experiments, a variation in the normalized pumping power $\bar{P} = P_l/P_{th}$ notably alters the relative fraction of reservoir n_R to condensate n_C particles according to $\frac{n_R}{n_C} = \frac{\gamma_C}{\gamma_R}\frac{1}{\bar{P}-1}$, and thus, the pumping power should also control the choice between the two possible vortex-pair trajectories outlined in [78]. Additionally specific to these experiments, as the healing length of the condensate is not significantly smaller than the condensate size, the condensate boundaries are expected to also affect the vortex-pair motion. Depending on the vortex pair initial energy and the local condensate environment, the process of vortex pair either recombining within the condensate or splitting and leaving the condensate will happen on the order of the condensate polariton lifetime due to the small condensate size.

It is desirable to know in more detail how the dynamics and life cycle of a single vortex–antivortex pair are reflected in the time-integrated measurements and, specifically, why the defects in the fringes are not washed out by vortex motion. We assume that the vortex pair (imprinted in the calculation directly via a phase factor $e^{il\theta}$where$\theta = \tan^{-1}(\frac{y}{x \mp d_v/2})$ and $l = \pm 1$) is formed along the x-axis with core–core separation d_v and is subsequently free to evolve in time. Essentially, given the mirror symmetry of the problem across the y-axis, the interference fringes will be observed due to the correlated motion of the vortex and antivortex. The system topology restricts the pair motion to identical velocities in the y-direction and opposite velocities in the x-direction, so that the pair does not rotate.

In Fig. 4.34, we show time snapshots of the real-time evolution of an imprinted vortex pair, which subsequently moves along the $+y$-direction. The ratio of condensate density to reservoir density is defined approximated by

$$\frac{n_\mathrm{C}}{n_\mathrm{R}} \sim \frac{\gamma_\mathrm{R}}{\gamma_\mathrm{C}}(1 - \bar{P}(0)), \tag{4.77}$$

and the parameter space is adjusted such that the vortex-pair trajectory (which is largely determined by this ratio) is modified to produce the scenarios of (a) a radially splitting pair (indicative of small reservoir population and lack of confining potential) and (b) a recombining vortex pair (indicative of a strong dissipation with large reservoir population). A perfect top-hat pumping profile is used for simplicity. In (c), we show the simulation result for the system parameters most relevant to our experiment. The pumping profile is the experimentally measured one, and the parameters are the same as for the phase and visibility maps shown in Fig. 4.28, as well as the condensate shape images. The pair is found to recombine rapidly, as expected for a dissipative condensate with large reservoir population.

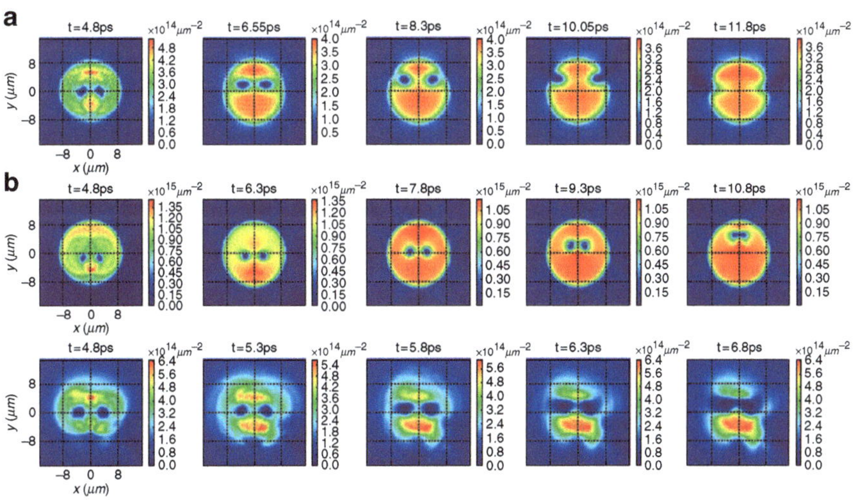

Fig. 4.34 Numerical simulations giving the time-dependent dynamics of (**a**) a radially splitting vortex pair and (**b**) a slowly recombining vortex pair and (**c**) a rapidly recombining vortex pair. The first two scenarios use a perfect top-hat pumping profile, while the third uses the measurement of the experimental profile. The parameter space of (**c**) is also chosen to closely match that of the experiment, notably that this parameter space dictates both the presence of a central density dip in the steady-state profile as well as a recombining vortex-pair trajectory

The dislocations in the interference measurement and visibility minima patterns are easily reproduced despite vortex-pair motion given that (a) the mirror symmetry of the vortex pair about its midpoint and (b) that the condensate is occupied by a vortex pair most of the time. The shape of these π-phase-shifted regions in the interference and fringe visibility experiments and simulations have a strong dependence on the vortex-pair dynamics, specifically a correspondence between the area of the π-phase-shifted region and area mapped by the vortex-pair trajectory. Small π-phase-shifted areas (as in experiments) are only reproduced with recombining vortex pairs, while separating vortex pairs generally give distinct and significantly larger areas.

Thus, two requirements are (a) strong modulation of condensate spatial profile by an inhomogeneous pump and (b) the vortex pair recombines rather than splits both imply a similar parameter space position to successfully explain the experimental results. In this parameter region, the condensate and vortex dynamics are strongly influenced by interaction with reservoir populations.

4.6 Future Directions

The results presented in this chapter provide a detailed microscopic picture of the steady-state condensate realized in two-dimensional exciton–polariton systems. We have presented the two characteristic features of the BKT phase transition in a homogeneous two-dimensional exciton–polariton system. One feature is a power-law decay of the spatial correlation function and the other is a bound pair of vortex and antivortex. In contrast to the BKT phase transition at thermal equilibrium, the exponent for the power-law decay of the spatial correlation function is larger than $1/4$ and the vortex–antivortex pair recombines to disappear before it is reflected from the condensate boundary. However, except for these details, the exciton–polariton condensate keeps the basic BKT-like behavior. The power-law decay of the spatial correlation function is consistent with our observation that no free vortices are spontaneously created in the cold enough condensate. According to the theoretical analysis, we conjecture that phase fluctuations are the main source of noise in our system. To test this hypothesis, two complementary directions can be followed.

The first one is to repeat the same measurements of the correlation function $g^{(1)}(r)$ at samples with longer lifetime. Although current technology cannot increase the lifetime to the point where thermal equilibrium is fully established, such a sample is expected to show weaker nonequilibrium effects. Therefore, the exponent of the power-law decay of $g^{(1)}(r)$ should be smaller.

The second path is to apply electrical pumping. Instead of creating electron–hole pairs by shining a noisy laser beam at above bandgap energy, carriers can be injected much more quietly by pn junctions. This way, we can control the intensity and phase noise of the system.

Acknowledgements We would like to acknowledge the contributions of our colleagues. M. Fraser performed numerical calculations using the open-dissipative Gross–Pitaevskii equation in Sect. 4.5. J. Keeling, M. Szymańska, and P. Littelwood developed the nonequilibrium theoretical model outlined in Sect. 4.4, and adapted it to the current experimental situation. Finally, A. Löffler, C. Schneider, S. Höfling, and A. Forchel provided the microcavity sample. This work is supported by the FIRST program of the JSPS.

References

1. A. Einstein, Quantentheorie des einatomigen idealen Gases. Sitzungber. Preuss. Akad. Wiss. **1**, 3–14 (1925)
2. F. London, The λ-phenomenon of Liquid Helium and the Bose-Einstein Degeneracy. Nature **141**(3571), 643 (1938)
3. M.H. Anderson, J.R. Ensher, M.R. Matthews, C.E. Wieman, E.A. Cornell, Observation of Bose-Einstein condensation in a dilute atomic vapor. Science **269**(5221), 198–201 (1995)
4. K.B. Davis, M.O. Mewes, M.R. Andrews, N.J. van Druten, D.S. Durfee, D.M. Kurn, W. Ketterle, Bose-Einstein condensation in a gas of sodium atoms. Phys. Rev. Lett. **75**(22), 3969–3973 (1995)
5. I. Bloch, J. Dalibard, W. Zwerger, Many-body physics with ultracold gases. Rev. Mod. Phys. **80**(3), 885–964 (2008)
6. H. Deng, G. Weihs, D. Snoke, J. Bloch, Y. Yamamoto, Polariton lasing vs. photon lasing in a semiconductor microcavity. PNAS **100**, 15318–15323 (2003)
7. J. Kasprzak, M. Richard, S. Kundermann, A. Baas, P. Jeambrun, J.M.J. Keeling, F.M. Marchetti, M.H. Szymańska, et al., Bose-Einstein condensation of exciton polaritons. Nature **443**, 409–414 (2006)
8. R. Balili, V. Hartwell, D. Snoke, L. Pfeiffer, K. West. Bose-Einstein condensation of microcavity polaritons in a trap. Science **316**, 1007–1010 (2007)
9. C.W. Lai, N.Y. Kim, S. Utsunomiya, G. Roumpos, H. Deng, M.D. Fraser, T. Byrnes, P. Recher, et al., Coherent zero-state and π-state in an exciton polariton condensate array. Nature **450**, 529–532 (2007)
10. H. Deng, H. Haug, Y. Yamamoto, Exciton-polariton Bose-Einstein condensation. Rev. Mod. Phys. **82**, 1489–1537 (2010)
11. P.C. Hohenberg, Existence of long-range order in one and two dimensions. Phys. Rev. **158**(2), 383–386 (1967)
12. J.M. Kosterlitz, D.J. Thouless, Ordering, metastability and phase transitions in two-dimensional systems. J. Phys. C: Solid State Phys. **6**(7), 1181–1203 (1973)
13. H.E. Stanley, *Introduction to Phase Transitions and Critical Phenomena* (Oxford University Press, Oxford, 1971)
14. L. Pitaevskii, S. Stringari, *Bose-Einstein Condensation* (Oxford University Press, 2003).
15. A.J. Leggett, *Quantum Liquids* (Oxford University Press, Oxford, 2006)
16. L.D. Landau, The theory of superfluidity of Helium II. J. Phys. USSR **5**, 71 (1941)
17. A.J. Leggett, Superfluidity. Rev. Mod. Phys. **71**(2), S318–S323 (1999)
18. R.J. Donnelly, *Quantized Vortices in Helium II* (Cambridge University press, Cambridge, 1991)
19. C.J. Pethick, H. Smith, *Bose-Einstein Condensation in Dilute Gases*. 2nd edn. (Cambridge University press, Cambridge, 2008)
20. A.L. Fetter, A.A. Svidzinsky, Vortices in a trapped dilute Bose-Einstein condensate. J. Phys.: Condens. Matter **13**(12), R135 (2001)
21. A.L. Fetter, Vortices and dynamics in trapped Bose-Einstein condensates. J. Low Temp. Phys. 1–15 (2010)
22. B.P. Anderson, Resource article: Experiments with vortices in superfluid atomic gases. J. Low Temp. Phys. 1–29 (2010)

23. N.D. Mermin, H. Wagner, Absence of ferromagnetism or antiferromagnetism in one- or two-dimensional isotropic Heisenberg models. Phys. Rev. Lett. **17**(22), 1133–1136 (1966)
24. Z. Hadzibabic, J. Dalibard, *Two-dimensional Bose fluids: An atomic physics perspective* (2009). Rivista del Nuovo Cimento, vol. 34, pp. 389–434, 2011
25. D. Petrov, D. Gangardt, G. Shlyapnikov, Low-dimensional trapped gases. J. Phys. IV France **116**, 5–44 (2004)
26. A. Posazhennikova, Colloquium: Weakly interacting, dilute Bose gases in 2D. Rev. Mod. Phys. **78**(4), 1111–1134 (2006)
27. J.M. Kosterlitz, D.J. Thouless, Long range order and metastability in two dimensional solids and superfluids. J. Phys. C: Solid State Phys. **5**(11), L124–L126 (1972)
28. V.L. Berezinskiǐ, Destruction of long-range order in one-dimensional and two-dimensional systems having a continuous symmetry group. I. Classical systems. Sov. Phys. JETP **32**(3), 493–500 (1971)
29. V.L. Berezinskiǐ, Destruction of long-range order in one-dimensional and two-dimensional systems having a continuous symmetry group. II. Quantum systems. Sov. Phys. JETP **34**(3), 610–616 (1972)
30. D.R. Nelson, J.M. Kosterlitz, Universal jump in the superfluid density of two-dimensional superfluids. Phys. Rev. Lett. **39**(19), 1201–1205 (1977)
31. D.S. Fisher, P.C. Hohenberg, Dilute Bose gas in two dimensions. Phys. Rev. B **37**(10), 4936–4943 (1988)
32. N. Prokof'ev, O. Ruebenacker, B. Svistunov, Critical point of a weakly interacting two-dimensional Bose gas. Phys. Rev. Lett. **87**(27), 270402 (2001)
33. N. Prokof'ev, B. Svistunov, Two-dimensional weakly interacting Bose gas in the fluctuation region. Phys. Rev. A **66**(4), 043608 (2002)
34. J. Kosterlitz, D. Thouless, *Chapter 5 Two-Dimensional Physics*, ed. by D. Brewer, *Prog. Low Temp. Phys.*, (Elsevier, Amsterdam, 1978). Vol. 7, Part 2, pp. 371–433
35. P.M. Chaikin, T.C. Lubensky, *Principles of Condensed Matter Physics* (Cambridge University Press, Cambridge, 2000)
36. D.J. Bishop, J.D. Reppy, Study of the superfluid transition in two-dimensional ^4he films. Phys. Rev. Lett. **40**(26), 1727–1730 (1978)
37. D.J. Bishop, J.D. Reppy, Study of the superfluid transition in two-dimensional ^4he films. Phys. Rev. B **22**(11), 5171–5185 (1980)
38. D.J. Resnick, J.C. Garland, J.T. Boyd, S. Shoemaker, R.S. Newrock, Kosterlitz-Thouless transition in proximity-coupled superconducting arrays. Phys. Rev. Lett. **47**(21), 1542–1545 (1981)
39. H.S.J. van der Zant, H.A. Rijken, J.E. Mooij, Phase transition of frustrated two-dimensional Josephson junction arrays. J. Low Temp. Phys. **82**(1/2), 67–92 (1991)
40. A.I. Safonov, S.A. Vasilyev, I.S. Yasnikov, I.I. Lukashevich, S. Jaakkola, Observation of quasicondensate in two-dimensional atomic hydrogen. Phys. Rev. Lett. **81**(21), 4545–4548 (1998)
41. Z. Hadzibabic, P. Krüger, M. Cheneau, B. Battelier, J. Dalibard, Berezinskii-Kosterlitz-Thouless crossover in a trapped atomic gas. Nature **441**, 1118–1121 (2006)
42. V. Schweikhard, S. Tung, E.A. Cornell, Vortex proliferation in the Berezinskii-Kosterlitz-Thouless regime on a two-dimensional lattice of Bose-Einstein condensates. Phys. Rev. Lett. **99**(3), 030401 (2007)
43. S. Tung, G. Lamporesi, D. Lobser, L. Xia, E.A. Cornell, Observation of the presuperfluid regime in a two-dimensional Bose gas. Phys. Rev. Lett. **105**(23), 230408 (2010)
44. P. Cladé, C. Ryu, A. Ramanathan, K. Helmerson, W.D. Phillips, Observation of a 2D Bose gas: From thermal to quasicondensate to superfluid. Phys. Rev. Lett. **102**(17), 170401 (2009)
45. G. Roumpos, M.D. Fraser, A. Löffler, S. Höfling, A. Forchel, Y. Yamamoto, Single vortex-antivortex pair in an exciton polariton condensate. Nature Phys. **7**, 129–133 (2011)
46. F.P. Laussy, I.A. Shelykh, G. Malpuech, A. Kavokin, Effects of Bose-Enstein condensation of exciton polaritons in microcavities on the polarization of emitted light. Phys. Rev. B **73**(3), 035315 (2006)

47. J.J. Baumberg, A.V. Kavokin, S. Christopoulos, A.J.D. Grundy, R. Butté, G. Christmann, D.D. Solnyshkov, G. Malpuech, et al., Spontaneous polarization buildup in a room-temperature polariton laser. Phys. Rev. Lett. **101**(13), 136409 (2008)
48. D. Bajoni, P. Senellart, A. Lemaître, J. Bloch, Photon lasing in GaAs microcavity: Similarities with a polariton condensate. Phys. Rev. B **76**(20), 201305 (2007)
49. D. Bajoni, P. Senellart, E. Wertz, I. Sagnes, A. Miard, A. Lemaître, J. Bloch, Polariton laser using single micropillar GaAs-GaAlAs semiconductor cavities. Phys. Rev. Lett. **100**(4), 047401 (2008)
50. R. Balili, B. Nelsen, D.W. Snoke, L. Pfeiffer, K. West, Role of the stress trap in the polariton quasiequilibrium condensation in gaas microcavities. Phys. Rev. B **79**(7), 075319 (2009)
51. P.G. Savvidis, J.J. Baumberg, R.M. Stevenson, M.S. Skolnick, D.M. Whittaker, J.S. Roberts, Angle-resonant stimulated polariton amplifier. Phys. Rev. Lett. **84**(7), 1547–1550 (2000)
52. C. Ciuti, P. Schwendimann, B. Deveaud, A. Quattropani, *Theory of the angle-resonant polariton amplifier*. Phys. Rev. B **62**(8), R4825–R4828 (2000)
53. G. Roumpos, C.-W. Lai, T.C.H. Liew, Y.G. Rubo, A.V. Kavokin, Y. Yamamoto, Signature of the microcavity exciton-polariton relaxation mechanism in the polarization of emitted light. Phys. Rev. B **79**(19), 195310 (2009)
54. I. Bloch, T.W. Hänsch, T. Esslinger, Measurement of the spatial coherence of a trapped Bose gas at the phase transition. Nature **403**, 166–170 (2000)
55. M.H. Szymańska, J. Keeling, P.B. Littlewood, Nonequilibrium quantum condensation in an incoherently pumped dissipative system. Phys. Rev. Lett. **96**(23), 230602 (2006)
56. M.H. Szymańska, J. Keeling, P.B. Littlewood, Mean-field theory and fluctuation spectrum of a pumped decaying Bose-Fermi system across the quantum condensation transition. Phys. Rev. B **75**(19), 195331 (2007)
57. H. Deng, G.S. Solomon, R. Hey, K.H. Ploog, Y. Yamamoto, Spatial coherence of a polariton condensate. Phys. Rev. Lett. **99**, 126403 (2007)
58. D.N. Krizhanovskii, K.G. Lagoudakis, M. Wouters, B. Pietka, R.A. Bradley, K. Guda, D.M. Whittaker, M.S. Skolnick, et al., Coexisting nonequilibrium condensates with long-range spatial coherence in semiconductor microcavities. Phys. Rev. B **80**(4), 045317 (2009)
59. E. Wertz, L. Ferrier, D. Solnyshkov, R. Johne, D. Sanvitto, A. Lemaître, I. Sagnes, R. Grousson, et al., Spontaneous formation and optical manipulation of extended polariton condensates. Nat. Phys. **6**(11), 860-864 (2010)
60. F. Manni, K.G. Lagoudakis, B. Pietka, L. Fontanesi, M. Wouters, V. Savona, R. André, B. Deveaud-Plédran, Polariton condensation in a one-dimensional disordered potential. Phys. Rev. Lett. **106**, 176401 (2011)
61. M. Wouters, I. Carusotto, C. Ciuti, Spatial and spectral shape of inhomogeneous nonequilibrium exciton-polariton condensates. Phys. Rev. B **77**(11), 115340 (2008)
62. A.M. Weiner, *Ultrafast Optics* (Wiley, New York, 2009)
63. M. Szymańska, F. Marchetti, J. Keeling, P. Littlewood, Coherence properties and luminescence spectra of condensed polaritons in CdTe microcavities. Solid State Comm. **144**(9), 364–370 (2007)
64. A. Kamenev, *Many-body theory of non-equilibrium systems*. ed. In H. Bouchiat, Y. Gefen, S. Guéron, G. Montambaux, J. Dalibard, (Elsevier, Amsterdam, 2005). *Nanophysics: Coherence and transport*, *Les Houches*, Vol. LXXXI, p. 177
65. M. Wouters, I. Carusotto, Absence of long-range coherence in the parametric emission of photonic wires. Phys. Rev. B, **74**, 245316 (2006)
66. G. Roumpos, W.H. Nitsche, S. Höfling, A. Forchel, Y. Yamamoto, Gain-induced trapping of microcavity exciton polariton condensates. Phys. Rev. Lett. **104**(12), 126403 (2010)
67. A.P.D. Love, D.N. Krizhanovskii, D.M. Whittaker, R. Bouchekioua, D. Sanvitto, S.A. Rizeiqi, R. Bradley, M.S. Skolnick, et al., Intrinsic decoherence mechanisms in the microcavity polariton condensate. Phys. Rev. Lett. **101**(6), 067404 (2008)
68. D.M. Whittaker, P.R. Eastham, *Coherence properties of the microcavity polariton condensate*. EPL **87**, 27002 (2009)

69. L. Giorgetti, I. Carusotto, Y. Castin, Semiclassical field method for the equilibrium Bose gas and application to thermal vortices in two dimensions. Phys. Rev. A **76**(1), 013613 (2007)
70. M. Wouters, I. Carusotto, Excitations in a nonequilibrium Bose-Einstein condensate of exciton polaritons. Phys. Rev. Lett. **99**, 140402 (2007)
71. T.W. Neely, E.C. Samson, A.S. Bradley, M.J. Davis, B.P. Anderson, Observation of vortex dipoles in an oblate Bose-Einstein condensate. Phys. Rev. Lett. **104**, 160401 (2010)
72. K.G. Lagoudakis, M. Wouters, M. Richard, A. Baas, I. Carusotto, R. André, L.S. Dang, B. Deveaud-Plédran, Quantized vortices in an exciton-polariton condensate. Nat. Phys. **4**(7), 706–710 (2008)
73. S. Utsunomiya, L. Tian, G. Roumpos, C.W. Lai, N. Kumada, T. Fujisawa, M. Kuwata-Gonokami, A. Löffler, et al., Observation of Bogoliubov excitations in exciton-polariton condensates. Nat. Phys. **4**(9), 700–705 (2008)
74. I. Carusotto, C. Ciuti, Probing microcavity polariton superfluidity through resonant rayleigh scattering. Phys. Rev. Lett. **93**(16), 166401 (2004)
75. A. Amo, J. Lefrère, S. Pigeon, C. Adrados, C. Ciuti, I. Carusotto, R. Houdré, E. Giacobino, et al., Superfluidity of polaritons in semiconductor microcavities. Nat. Phys. **5**, 805 (2009)
76. M. Wouters, V. Savona, Superfluidity of a nonequilibrium Bose-Einstein condensate of polaritons. Phys. Rev. B **81**(5), 054508 (2010)
77. D. Sanvitto, F.M. Marchetti, M.H. Szymańska, G. Tosi, M. Baudisch, F.P. Laussy, D.N. Krizhanovskii, M.S. Skolnick, et al., Persistent currents and quantized vortices in a polariton superfluid. Nat. Phys. **6**, 527–533 (2010)
78. D.S. Petrov, M. Holzmann, G.V. Shlyapnikov, Bose-Einstein condensation in quasi-2D trapped gases. Phys. Rev. Lett. **84**, 2551–2555 (2000)
79. M. Richard, J. Kasprzak, R. Romestain, R. André, L.S. Dang, Spontaneous coherent phase transition of polaritons in cdte microcavities. Phys. Rev. Lett. **94**(18), 187401 (2005)
80. M.D. Fraser, G. Roumpos, Y. Yamamoto, Vortex-antivortex pair dynamics in an exciton polariton condensate. New J. Phys. **11**, 113048 (2009)
81. S. Pigeon, I. Carusotto, C. Ciuti, Hydrodynamic nucleation of vortices and solitons in a resonantly excited polariton superfluid. Phys. Rev. B **83**, 144513 (2011)

Chapter 5
Coexisting Polariton Condensates and Their Temporal Coherence in Semiconductor Microcavities

D.N. Krizhanovskii, David M. Whittaker, M.S. Skolnick, K.G. Lagoudakis, and M. Wouters

Abstract In this chapter, we study macroscopically occupied condensates, which can be observed in semiconductor microcavities under conditions of resonant or non-resonant excitation. In the case of resonant excitation, polariton condensates form due to optical parametric oscillation (OPO) and are strongly non-equilibrium states. In case of non-resonantly incoherently pumped system, the distribution of the higher energy polaritons shows some thermalisation, but the resultant polariton condensates are also far from thermodynamic equilibrium due to finite polariton lifetime. In this chapter, we show that both systems have very similar properties. We reveal the effects of polariton–polariton interactions and non-equilibrium character on the condensate properties. Above threshold condensation into several polariton levels with different energies and k-vectors is observed, which arises from the non-equilibrium character of the polariton system. The specific k-vectors at which condensation is triggered are determined by the local disorder potential landscape. We also investigate the coherence of a single condensed mode by measuring the first ($g^{(1)}$)- and second ($g^{(2)}$)-order correlation functions. We find that the decay times of these functions are ~100–150 ps, much longer than the 1.5 ps polariton lifetime. Even though the polariton condensate is a non-equilibrium system, the strong slowing down of the decay allows coherence decay processes characteristic of an equilibrium, interacting BEC to be observed. The signature of the interactions is a Gaussian form for the $g^{(1)}$-function and a saturation of coherence time with increasing number of particles in the condensate, as observed experimentally and confirmed theoretically. Although predicted, these effects have not been observed for atom BECs.

D.N. Krizhanovskii (✉) · D.M. Whittaker · M.S. Skolnick
Department of Physics and Astronomy, University of Sheffield, Sheffield S3 7RH, UK
e-mail: d.krizhanovskii@sheffield.ac.uk; m.skolnick@sheffield.ac.uk

K.G. Lagoudakis · M. Wouters
Ecole Polytechnique Federale de Lausanne (EPFL), Station 3, CH-1015 Lausanne, Switzerland
e-mail: michiel.wouters@epfl.ch

D. Sanvitto and V. Timofeev (eds.), *Exciton Polaritons in Microcavities*,
Springer Series in Solid-State Sciences 172, DOI 10.1007/978-3-642-24186-4_5,

5.1 Coexisting Non-equilibrium Polariton Condensates

5.1.1 Introduction

As described elsewhere in this book, there is significant contemporary interest in the study of strongly coupled semiconductor microcavities, where mixed exciton-photon quasiparticles, two-dimensional (2D) polaritons with very small effective mass can be created. Such a small effective mass favours Bose–Einstein condensation (BEC) of polaritons at high temperatures and relatively small optical excitation densities. Most notably, the formation of macroscopically occupied states has been recently observed in CdTe [1,2], GaAs [3,4], and GaN [5,6] microcavities under conditions of non-resonant excitation. Polariton condensates are observed to exhibit long-range spatial coherence, which is a characteristic property of BEC. Although there is evidence for equilibrated polariton distributions [1–3, 7] and macroscopic occupation of the $k = 0$ ground state, polariton condensates are far from thermodynamic equilibrium: due to the rather small polariton lifetimes, these states originate from a dynamical balance of pumping and losses in the system. We also consider strongly non-equilibrium polariton condensates, which are formed due to optical parametric oscillation (OPO) in the case of resonant excitation into lower polariton branch [8]. Such condensates have properties very similar to those observed under no resonant excitation. Such similarity is discussed in this chapter.

It is important to note that the polariton condensate is a strongly interacting system due to the excitonic component in the polariton wave function. The interactions alter the dispersion of condensate excitations, which becomes nearly linear at high k-vectors with defined sound velocity [9]. They were shown to decrease the healing length of vortices, which is a fundamental concept of a condensed phase determining the length scale on which the locally perturbed condensate wave function returns to its unperturbed value [10]. There are several approaches based on both microscopic calculations and generalisation of the Gross–Pitaevskii equation, which describe the consequences of the non-equilibrium aspect of the interacting polariton BEC system [9, 11–13].

The imperfections of semiconductor microcavity result in a disordered polaritonic potential landscape, which plays an important role in the formation of 2D macroscopically occupied polariton states. The influence of polaritonic disorder was discussed for the case of the OPO, where condensed signal and idler modes arise from direct scattering of the resonantly excited pump polaritons. The polaritonic disorder strongly affects the real space distribution of the signal [14], which consists of spatially localised modes emitting at different energies [15]. Polariton condensates which are formed under condition of high-energy no resonant excitation are also observed to consist of several localised maxima [16], which occurs due to polariton trapping by the polaritonic potential disorder. It was shown that polariton condensates subject to a disorder potential may spontaneously exhibit quantised vortices due to interplay between the disorder potential, polariton flow, and decay of the non-equilibrium condensate [17].

In this chapter, we present spectrally resolved real and momentum space images of the multimode polariton condensate. We observe that above threshold condensation occurs into several polariton levels. This is a result of the non-equilibrium aspects of the system, which prevents thermodynamic equilibrium being reached for the states near the bottom of the lower polariton (LP) branch. The low-energy condensate states are found to be strongly localised within a deep potential minimum, whereas the higher energy states are spatially extended over a few potential minima. In momentum space, these delocalised states typically consist of a few bright maxima distributed on a ring. Similar to the phenomenon of coherent back scattering in disordered systems, a peak at a certain k is often accompanied by a weaker satellite at $-k$. Such observations became possible since the diode laser has reduced intensity noise on a timescale of 100–1,000 ps and results in reduced fluctuations of the number of excitons and polaritons, which otherwise lead to marked broadening of BEC spectra and prevent the observation of the coexisting condensates [18].

5.1.2 Experimental Technique and Sample

The sample employed here is the λ CdTe microcavity with 16 embedded quantum wells, which is one used in [1, 2] where a polariton condensate with extended spatial coherence was observed. The exciton–photon detuning at $k = 0$ is nearly zero, and the Rabi splitting is about 26 meV. The sample was cooled to 10 K. To avoid heating of the sample, quasi-CW non-resonant excitation was employed by means of CW diode lasers at 685 nm using a mechanical optical chopper with a frequency of 300 Hz. The size of the excitation spot was about 20 μm. High-resolution submicron imaging was achieved by a high numerical-aperture microscope objective (N.A. = 0.5). Spectrally and spatially resolved images were recorded using a double monochromator of about 40 μeV resolution and a CCD camera.

5.1.3 Real and Momentum Space Imaging of Coexisting Condensates

Figure 5.1a shows spectra from the bottom of the LP branch at excitation powers $P = 16$ mW well above the threshold ($P_{\text{th}} \sim 4$ mW) for polariton condensation. The spectra are recorded from a section of $\sim 1 \times 20\,\mu$m dimension across the 20 μm excitation spot in two perpendicular linear polarisations, X and Y parallel to the (110) principal crystal axes. The emission in each polarisation consists of six narrow peaks with energy separation 0.080–0.2 meV. A splitting of $\sim$0.1 meV between X- and Y-polarised narrow lines is observed for each of the modes. Such a splitting also exists for the LP mode below threshold and is probably due to intrinsic anisotropy in the microcavity mirrors, leading to a cavity birefringence.

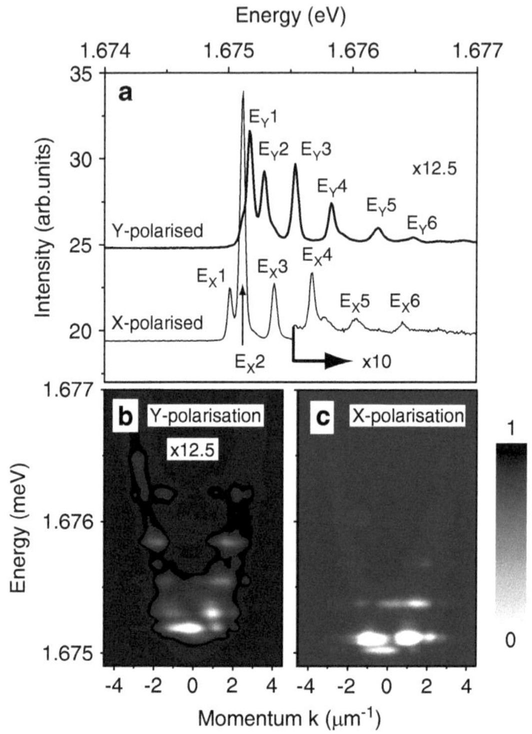

Fig. 5.1 (**a**) Spectra of polariton emission in X and Y polarisation above condensation threshold. (**b, c**) Images of polariton emission in energy-momentum space recorded in X (**c**) and Y (**b**) polarisations. The brighter regions in (**b**) and (**c**) indicate the higher intensities. The intensities are normalised by the factors displayed in each panel

In both polarisations, the peaks are quite narrow (FWHM $\sim$ 60 μeV, resolution limited). The strong spectral narrowing and the observed super linear increase of the intensity of each single mode (not shown) with excitation power [18] indicate the build-up of coexisting macroscopically occupied states (polariton condensates) and condensation into both no degenerate X- and Y-polarised states. The phenomena are only observable due to the significantly increased temporal coherence obtained using a noise-free laser.

Figure 5.2 compares spectra of incoherently pumped condensate emission recorded at power two times above threshold using a noise-free diode laser (a) and a multimode Ti:S laser (b). Obviously, the condensate emission is much broader in case of pumping using the "noisy" laser. The relevant difference between the Ti:S and the diode laser is the timescale of intensity fluctuations. For the Ti:S laser, the mode spacing is of the order of 250 MHz, resulting in intensity fluctuations on an ns timescale. The intensity fluctuations of the diode laser are much faster (ps timescale) due to their much larger mode spacing ($\sim$25 GHz). The importance of the intensity fluctuation timescale can be understood in the following way: slow laser intensity fluctuations are followed by the exciton reservoir (relaxation time of the order of ns), whereas fast fluctuations are averaged out. A fluctuating reservoir density results in a temporal variation of the blueshift of polariton condensates and hence a linewidth broadening. This explains why the condensate linewidth under

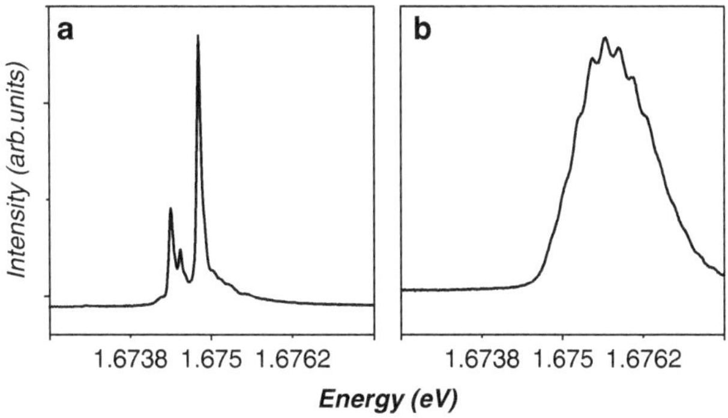

Fig. 5.2 Spectra of incoherently pumped condensate emission recorded at power two times above threshold using noise free diode laser (**a**) and multimode Ti:S laser (**b**)

Ti:Sapph excitation (hundreds of μeV) is much larger than the one under diode laser excitation (tens of μeV) (see Fig. 5.2) [16].

The multimode structure of the stimulated emission originates from the interaction of the non-equilibrium condensate with the fluctuations of the transverse photonic potential, which typically occurs on a length scale of 2–4 μm with amplitude of 0.5–0.7 meV in CdTe samples. To reveal the role of the potential on the formation of spectrally narrow coexisting condensates, we recorded spectrally and spatially resolved 2D images in real and momentum space. The images were obtained above threshold for each of narrow peaks in Fig. 5.1a. Figure 5.1b, c shows images of polariton emission recorded in energy-momentum space for the same excitation conditions and spot as in Fig. 5.1a. It is seen that the lowest energy state is localised at around $k = 0$ for both polarisations. By contrast, higher energy condensed states have well-defined k-vectors different from zero. Figure 5.3a–f shows 2D images recorded in momentum space for X-polarised modes. The first ground state mode emitting at 1.675005 eV is localised around $k = 0$ within $\pm 0.5\,\mu\text{m}^{-1}$. The second mode emitting at energy of 1.675115 meV already consists of two lobes at ($k_x = 0\mu\text{m}^{-1}$, $k_y = +1.2\,\mu\text{m}^{-1}$) and ($k_x = 0\,\mu\text{m}^{-1}, k_y = -1.2\,\mu\text{m}^{-1}$), respectively. The pattern of the higher energy modes 3–6 becomes more complicated: each mode has k-vectors distinct from zero, which corresponds to momentum of the propagating LP states at the same energy. We note that modes 3 and 6 consist of two strong maxima with opposite k-vectors at ($k_x = \pm 1.15\,\mu\text{m}^{-1}$, $k_y = 0$) and ($k_x \sim \pm 2.3\,\mu\text{m}^{-1}$, $k_y = 0$), respectively, suggesting efficient backscattering of the propagating polaritons in the condensed modes with finite k-vectors due to interaction with the disorder potential.

Although the Y-polarised modes are spectrally distinct from the corresponding X-polarised modes, they are observed to have very similar pattern (not shown). Such a similarity arises from the fact that photonic potential, which is given by the spatial fluctuations of optical thickness of the cavity and distributed Bragg reflectors, is identical for both X- and Y-polarised polariton modes despite the modes being split due to different refractive indexes along X and Y directions (birefringence).

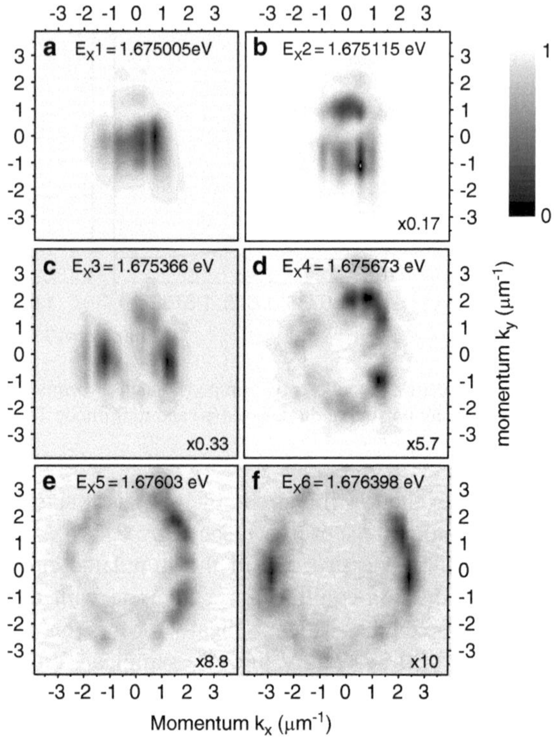

Fig. 5.3 (**a–f**) Images of polariton condensed modes in momentum space recorded in X polarisation at particular energy of each mode. Here the darker regions indicate the higher intensities. The emission intensities are normalised by the factors shown in each panels

The splitting between X and Y modes is of the order of 0.1–0.2 meV, which is smaller than the typical amplitude of the photonic potential disorder (0.5–0.7 meV).

Figure 5.4 shows spectrally resolved images in real space recorded for each of the narrow modes in X polarisation. The images of the condensed modes in real space recorded in two polarisations are almost identical, which is consistent with the similar X- and Y-polarised k-space images of these modes. Mode 1 is strongly localised in real space within an area of $\sim 5 \times 3\,\mu$m, whereas the wave functions of the higher energy polariton states are observed to consist of several maxima separated by $\sim$3–4 μm and are extended over a length scale 10–20 μm. Moreover, all the modes show strong spatial overlap, indicating coexistence of the condensed phases. We performed studies of the spatial coherence between the spatially separated regions of the spectrally distinct condensates using standard Michelson interferometry [18]. Above threshold an interference pattern is observed between the emissions from separated regions emitting at the same energy. Such observation indicates formation of a coherence length of the each high-density polariton mode comparable to its size ($\sim$10–20 μm). Similar coherence lengths were reported previously for polariton condensates excited with a noisy multimode Ti:Sapph laser [1,2]. The fact that the higher energy condensed modes have maxima in momentum space different from zero indicates the large phase difference between the polaritons at the different potential minima.

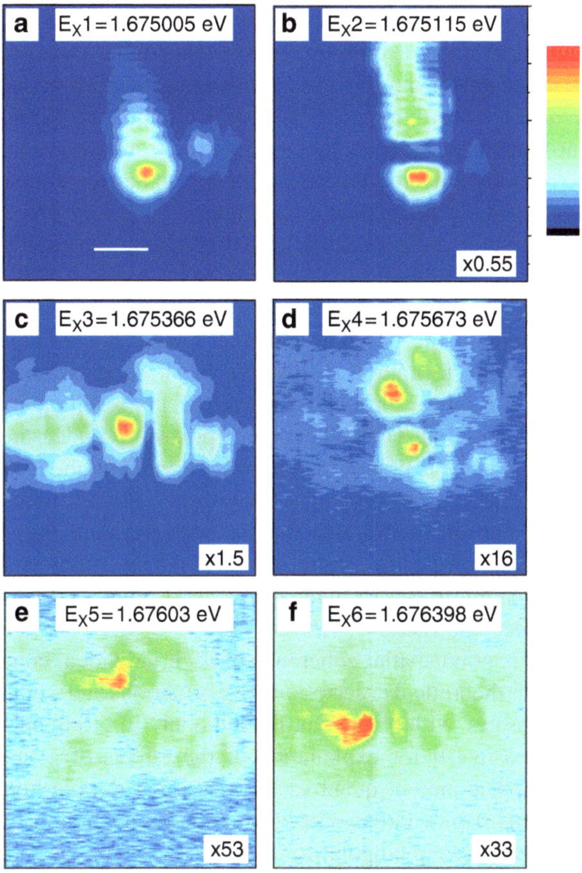

Fig. 5.4 Images of polariton condensed modes in real space recorded in X polarisation at particular energy of each mode. Here the emission intensity is indicated by colour. The emission intensities are normalised by the factors shown in each panels

5.1.4 Gross-Pitaevskii Formalism of Non-equilibrium Polariton Condensates

In a true equilibrium, BEC system condensation is expected to occur only into a single ground state with its resultant massive occupation below a critical temperature, whereas the population for the higher energy states should obey a Boltzmann distribution with a well-defined temperature. The polariton system is far from equilibrium. At powers much below threshold, polariton relaxation towards the ground state due to polariton–phonon scattering is very inefficient [19, 20], and hence, polaritons are not thermalised. By contrast, with increasing excitation power, polariton–exciton scattering from the reservoir of exciton states towards the LP states becomes dominant [20]. Such scattering enables simultaneous macroscopic occupation of several polariton modes with different k-vectors. The scattering between the condensed modes itself is an order of magnitude weaker than stimulated scattering from the reservoir into the modes since the reservoir population is much higher. Above threshold a dynamical balance is formed between losses in the condensed polariton modes and gain due to relaxation from the reservoir. Therefore, there is a close analogy between the phenomena of polariton condensation and a laser, with a significant difference that no population inversion is required to obtain a macroscopic occupation, because stimulated relaxation is ensured by the bosonic nature of polaritons.

The k-vectors at which polariton condensation is triggered are determined by the spatial fluctuations of the photonic potential. These fluctuations of the order of 0.5 meV result in the formation of discrete localised polariton levels. Above threshold condensation into each localised level occurs resulting in a multimode stimulated polariton emission. The higher energy polariton condensates, which have energies 0.5–1 meV above the polariton ground state and thus above the fluctuations in photonic potential, arise from delocalised polariton states as observed in Fig. 5.4. These condensed polariton modes also interact with the potential: particles with particular k-vectors determined by a pattern of local potential may undergo coherent backscattering, which minimises in-plane losses and triggers condensation into the states with these specific k-vectors.

Given that there is good spatial coherence of each individual condensate mode, we can neglect in our theoretical description the fluctuations and study the formation of multiple condensates within a mean-field model [9]. Here we will use the equations introduced in [9] for explanation of the effects of a finite size pump spot [21] and the appearance of quantised vortices [17] in polariton condensates. These equations consist of a Gross–Pitaevskii equation including losses and gain for the macroscopically occupied polariton field $\psi(r)$:

$$i\hbar\frac{\partial\psi(r)}{\partial t} = \Big\{ E_0 - \frac{\hbar^2}{2m}\nabla_r^2 + \frac{i\hbar}{2}\left[R\left[n_R(r)\right] - \gamma_c\right] + V_{\text{ext}}(r) + \hbar g\left|\psi(r)\right|^2 + V_R(r)\Big\}\,\psi(r) \tag{5.1}$$

where E_0 and m are respectively the minimum energy and the effective mass of the lower polariton branch, and $g > 0$ quantifies the strength of repulsive binary interactions between condensate polaritons. The term $V_{\text{ext}}(r)$ describes the external polaritonic potential, which originates from photonic potential disorder due to variation of the cavity optical thickness across the sample [14]. In addition, some long-range variation in the average exciton energy with a correlation length of a few microns may also contribute. At the simplest level, the corresponding gain rate $R[n_R]$ can be described by a monotonically growing function of the local density $n_R(r)$ of exciton reservoir in the so-called bottleneck region [19, 20]. At the same time, the reservoir produces a mean-field repulsive potential $V_R(r)$ that can be approximated by the linear expression $V_R(r) \sim g_{R\times} n_R(r) + G^* P(r)$, where $P(r)$

is the (spatially dependent) pumping rate and $g_R, G > 0$ are phenomenological coefficients. The GPE (5.1) for the condensate has then to be coupled to a rate equation for $n_R(r)$:

$$\frac{\partial n_R}{\partial t} = P(r) - \gamma_R n_R(r) - R\left[n_R(r)\right]|\psi(r)|^2 \tag{5.2}$$

where P describes the filling of the reservoir by the no resonant excitation. It is useful to consider the solutions of the (5.1) and (5.2) in the spatially uniform case $V_{\text{ext}} = 0$. For low excitation density P, a stable steady state is given by $\psi = 0$ and $n_R = P/\gamma_R$. The exciton density reaches the threshold for polariton condensation when the gain from stimulated scattering from the exciton reservoir into the lower polariton branch equals the polariton loss rate: $R(n_R) = \gamma_c$. For pump powers above the threshold $P > P_{th} = \gamma_R n_R(\text{th})$, the solution without a condensate $\psi = 0$ becomes dynamically unstable. The dynamically stable solution is then given by the condition that the gain is clamped to the losses $R(n_R) = \gamma_c$ and the condensate density is $|\psi|^2 = (P - P_{th})/\gamma$. The condensate wave function oscillates at the frequency $\hbar\omega = g|\psi|^2 + V_R$.

In the presence of an external potential, this simple picture breaks down. In particular, it is no longer guaranteed that only a single frequency appears above the condensation threshold. In the context of polariton condensation, multiple frequency solutions of the related complex Ginzburg–Landau equation have been found for certain types of regular external potentials [22].

The disorder potential that we used for the theoretical calculations is shown in Fig. 5.5a. The shape of the disorder potential is determined from the spatial distribution of the low pump intensity photoluminescence. The spatial variation of the lowest emission energy follows the disorder potential, albeit not perfectly because of the zero point kinetic energy that smooths out the variations of the potential energy. To compensate for the effect of the kinetic energy, we have multiplied the variations in the experimentally observed energy landscape by a heuristic factor of 3.5 so as to reproduce approximately the separation between the condensate frequencies. The solution of our model (5.1) and (5.2) is shown in Figs. 5.5b–5.7. As in the experiment, multiple frequencies are present in the spectrum. The observation of multiple frequencies is consistent with the parameters used in the simulations: the fluctuations in the disorder potential are larger than the blueshift due to condensate-condensate interactions $g|\psi|^2$ (its maximal value in the simulation is 0.3 meV). The coupling between the various modes at different energies is therefore too weak to ensure synchronisation [22]. The real and momentum space distributions of the peaks in Fig. 5.5b are shown in Figs. 5.7 and 5.6, respectively. First of all, we want to draw the attention to the strong qualitative similarities between the theoretical simulations in Figs. 5.6 and 5.7 and the experimental measurements in Figs. 5.3 and 5.4. In both cases, the lowest energy states are very localised in real space (extended over a few microns). In contrast, the states at higher energy are delocalised: they are extended over a much larger area, with a diameter of the order of 10 μm.

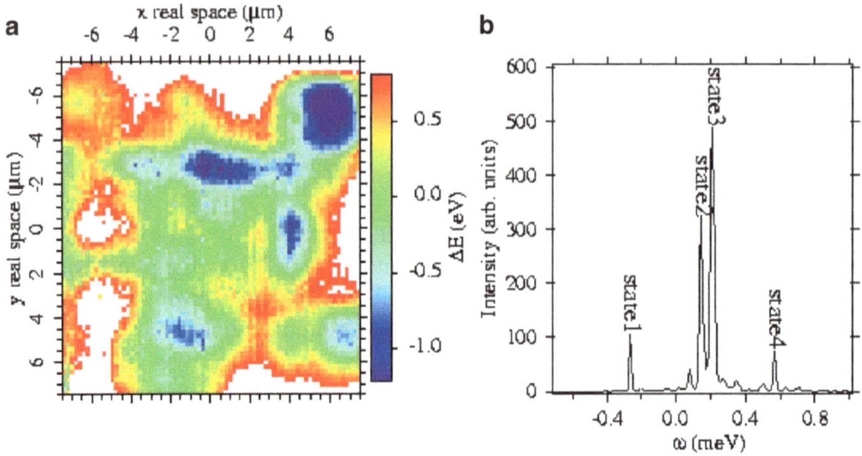

Fig. 5.5 (**a**) Disorder potential landscape used in the theoretical simulations and (**b**) energy spectrum from a simulation with the mean field model (1) and (2). Multiple condensates are found in the simulation. The zero of the energy is the same as in (**a**). Values of parameters used in the simulations: $\hbar g = 0.015\,\text{meV}/\mu\text{m}^2$, $\hbar y_c = 0.5\,\text{meV}$, $\hbar \gamma_R = 10\,\text{meV}$, $\hbar R[n_R] = (meV\mu m^2) \times n_R \hbar g_R = 0$, $G = 0$ and $P/P_{\text{th}} = 2$

The qualitative correspondence between the experimental and theoretical reciprocal space images is also excellent. The low-energy states form a disk around $k = 0$, while the states at highest energy lie on a ring in momentum space. Note that all the modes except the lowest one do not exhibit inversion symmetry in k-space. This lack of inversion symmetry is also present in the experimental momentum distributions above threshold: we have observed many peaks in the momentum distribution which have a satellite at the opposite wave vector. This is likely to be related to the phenomenon of backscattering in disordered systems as was mentioned above.

Time reversal of the linear Schrodinger equation requires that the eigenstates in the disorder potential are invariant under $k \to -k$. The condensate modes therefore do not coincide with the linear eigenstates of the microcavity [18]. An identification of the condensate states with the linear eigenstates can only be made for the lowest condensate state (it corresponds with the lowest eigenstate A). The higher energy states, on the other hand, cannot be identified with a linear eigenstate. Instead, the non-linear dynamics of pumping, dissipation, and polariton–polariton interactions define the new modes into which the polaritons condense.

5.1.5 Conclusion

We have presented a detailed experimental and theoretical analysis of a multimode polariton condensate. The real and reciprocal space densities of the different

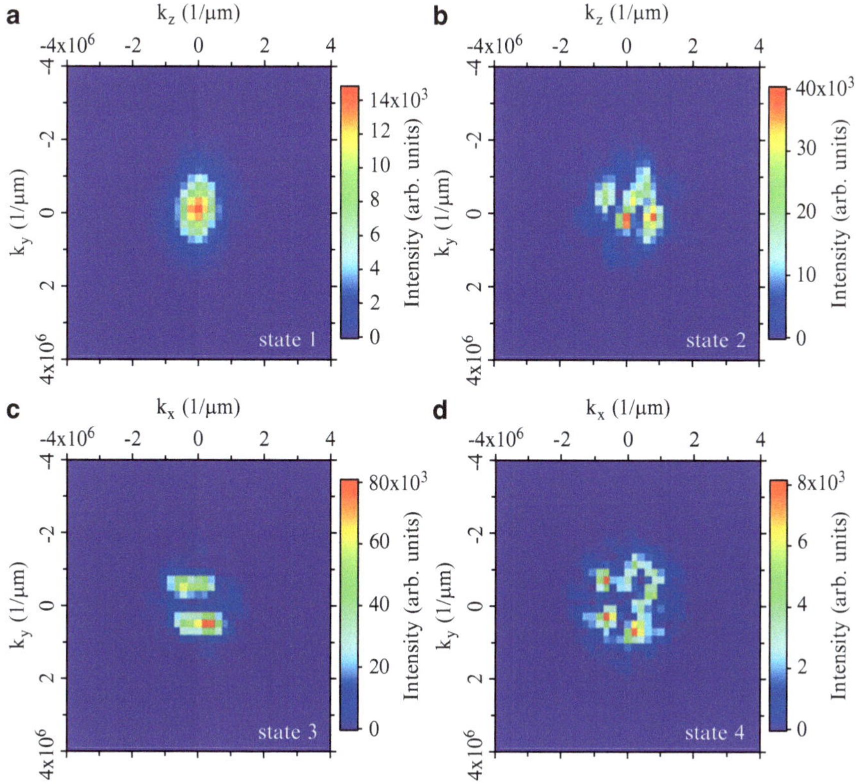

Fig. 5.6 Momentum space distribution of the polariton field at the corresponding mode frequencies in Fig. 5.5b

modes were recorded. It was shown that such a multimode structure of polariton condensation phenomena arises due to interplay between pump, polariton decay, and polariton propagation in a transverse potential disorder. Such a multimode structure is not expected to occur in a true non-interacting equilibrium BEC.

5.2 Effect of Interactions on Temporal Coherence of Polariton Condensates

5.2.1 Introduction

Temporal coherence is a fundamental property of system which undergoes transition into macroscopically occupied high-density states. The decay of the first-order correlation function $g^{(1)}$ defines the phase coherence time of a condensate, whereas

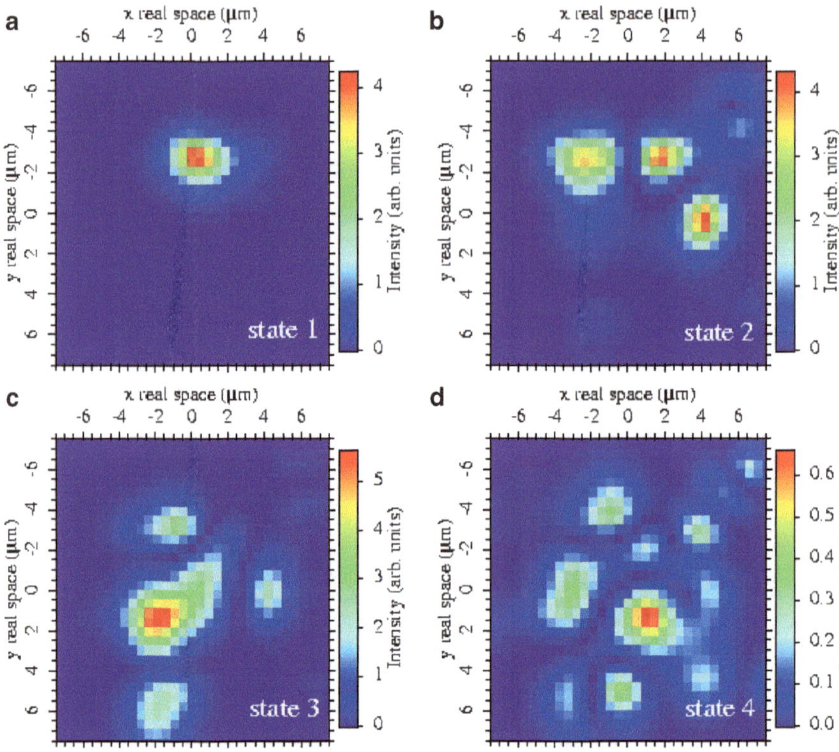

Fig. 5.7 Real space distribution of the polariton field at the corresponding peak frequencies in Fig. 5.4b

the second-order correlation function $g^{(2)}$ is determined by the statistics of the particle number in a non-equilibrium system with pump and decay [23].

The polariton condensate is a strongly interacting system due to its excitonic component. These interactions are expected to influence the condensate coherence as we show in this section. Here the phase coherent properties of a single non-equilibrium polariton condensed mode in a CdTe microcavity are studied. We observe coherence times of 100–150 ps for both the $g^{(1)}$ phase correlation function and the second-order intensity correlation function $g^{(2)}$. The $g^{(1)}$ decay is about 30 times longer than reports using a multimode excitation laser [1–3]. The coherence times of the condensate are nearly two orders of magnitude longer than the polariton lifetime below threshold ($\sim$1.5 ps), demonstrating a strong slowing of the decay in the coherent regime. Such slowing occurs because, as in a laser, stimulated scattering from the polariton reservoir into the condensate dominates over spontaneous processes.

The $g^{(1)}$ function is found to have a Gaussian shape and a decay time which saturates with increasing numbers of particles in the condensate. We show that these properties are well explained by a model which includes interactions between polaritons in the coherent state and the effects of gain and loss. The interactions play a central role in the decay of phase coherence because they cause spontaneous number fluctuations in the coherent state to be translated into random energy variations. Although this process has been predicted to occur in atomic BECs [24], it has yet to be observed experimentally. This is because the coherence of atom BECs is measured in atom-laser experiments [25, 26]. The atom laser has been realised only in pulsed mode; under these conditions, the coherence is limited by the lifetime of atoms in the trap (few ms). Although the coherence times of polaritons are several orders of magnitudes shorter than those of atom condensates, the interactions are stronger too. Combined with the slowing of the decay, this makes interactions the factor limiting the coherence of the polariton system.

5.2.2 *Measurement of the First-Order and the Second-Order Correlation Functions of Non-resonantly Pumped Condensates*

Having achieved spectral resolution of the localised condensed states in the previous section, we are able to investigate the coherence properties of the individual modes of polariton condensate. We employed a Mach–Zehnder interferometer to study the time dependence of the first-order coherence. Figure 5.8a shows a typical dependence of the first-order correlation function ($g^{(1)}(\tau)$) for an individual mode above threshold. It is found to have Gaussian time dependence with a coherence time τ_c of $\sim$100 ps. Figure 5.8b shows the variation of the measured coherence time of a single condensate mode as a function of the intensity of the polariton emission,[1] which in turn is proportional to the number of particles in the condensed state. The coherence time first increases rapidly with increasing intensity, from a value of $\sim$1.5 ps below threshold to a value of $\sim$150 ps at $\sim$3 times threshold, beyond which it then shows near saturation. This coherence time is about one order of magnitude longer than that reported previously for the polariton condensate [1–3].

The initial increase of coherence time with occupancy arises from the build of the macroscopic occupation as the ratio of stimulated to spontaneous processes increases [23]. The subsequent saturation value can be understood physically if the interactions between the coherent polaritons and the reservoir of polaritons in other modes are considered as we show below. These interactions also give rise to the Gaussian shape of the polariton emission.

Further insight into the quantum nature of the condensed polariton state is provided by the measurements of the second-order intensity correlation $g^{(2)}(\tau)$ using a standard Hanbury Brown-Twiss (HBT) setup [23].

[1]During the power dependence measurements of τ_c, the $g^{(1)}(\tau)$ is measured at the delay times τ of 5–150 ps and τ_c is extracted as $-\tau/\sqrt{\ln(g^{(1)}(\tau))}$.

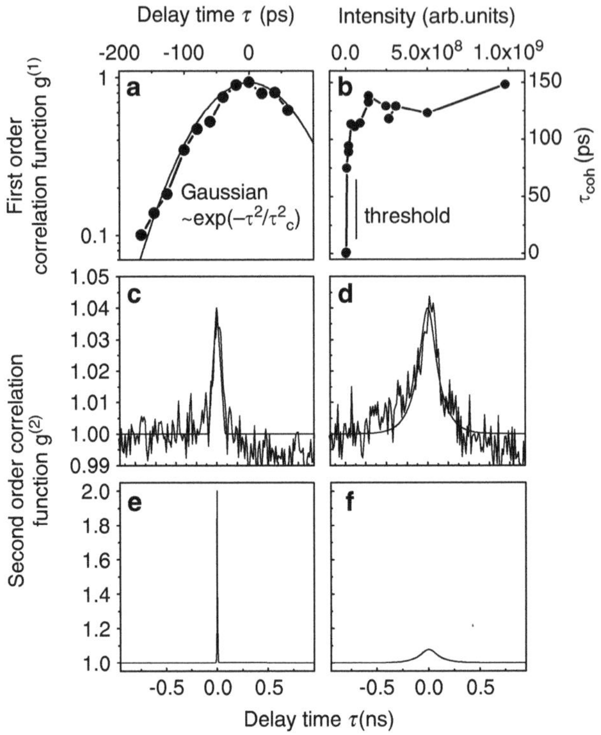

Fig. 5.8 (**a**) Variation of $g^{(1)}$ first order correlation function versus delay time for a single condensate mode above threshold. (**b**) Coherence times τ_c of condensate mode versus signal intensity I_s.(**c, d**) Typical variation of $g^{(2)}$-second order correlation function versus delay time for polariton emission below (**c**) and above (**d**) threshold. The true variation of the $g^{(2)}$—function versus delay time for polariton emission below (**e**) and above (**f**) threshold after correction for the efficiency and the resolution time of the detectors

The $g^{(2)}$ results below and above threshold are presented in Fig. 5.8c, d. For $P < P_{th}$, photon bunching is observed: the measured value of $g^{(2)}$, $g^{(2)}_{meas}$ has a maximum at $\tau = 0$ of 1.04 and then decays within ~40 ps, determined by the temporal resolution of the detectors (~40 ps). Below threshold, the state is expected to be thermal for which the actual $g^{(2)}(0) = 2$ and $g^2(\tau \gg \tau_c^{(1)}) = 1$, where $\tau_c^{(1)} \sim 1.5$ ps is the phase coherence time for the thermal polariton state given by the polariton lifetime. The small observed value of $g^{(2)}_{meas}$ 1.04 arises because of the finite efficiency of the detectors and because $\tau_c^{(1)}$ is very much less than the system time resolution of 40 ps. Above threshold in Fig. 5.8d, photon bunching is again observed: $g^{(2)}_{meas}$ has a maximum at $\tau = 0$ of ~1.04, but now decays on a longer timescale of 100 ps, twice as long as the resolution time. This indicates that the intensity fluctuations in the condensate are correlated on the same timescale as the phase coherence time (Fig. 5.8a).

By fitting the experimental data below threshold of Fig. 5.3c using the expression for the measured $g^{(2)}_{\rm meas}(\tau)$ in [23],

$$g^{(2)}_{\rm meas}(\tau) = 1 + (g^{(2)}(0) - 1)\mu^2 \int_0^{T_r} dt_1 \int_0^{T_r+\tau} dt_2 |\exp(-|t_1 - t_2|\tau_c^{(2)})|^2,$$

where T_r is the detector resolution time ($\sim$40 ps) and $\tau_c^{(2)}$ is the decay time of the second-order coherence which is given by the polariton lifetime below threshold, we obtain a value for the detector efficiency ($\sim$25%) in line with the manufacturer's specification. Using this value for the efficiency and the detector resolution time, we are then able to obtain the true value for $g^{(2)}$ of 1.1 ± 0.015 above threshold by fitting the experimental data of Fig. 5.8d. From the same fits the coherence time $\tau_c^{(2)}$ is obtained to be 100 ps above threshold.

5.2.3 Quantum Optical Treatment of Temporal Coherence in Non-equilibrium Polariton Condensates

Interparticle interactions and the fluctuations of the number of particles limit the coherence of polariton condensate. This can be understood qualitatively as follows: the energy blueshift of the polariton mode $E_{\rm shift}$ with respect of the unperturbed LP polariton branch is proportional to the number of particles N: $E_{\rm shift} \sim \kappa$N. Fluctuations in the number of particles δN cause fluctuations in the condensate energy $\delta E_{\rm shift} \sim \kappa\delta$N, thus establishing the upper limit for the condensate coherence time $\tau_c < \frac{\hbar}{\partial E_{\rm shift}}$.

We now discuss our detailed model for the first- and second-order coherence treating initially an isolated, equilibrium BEC and then introducing the non-equilibrium character. Consider an isolated state, with a Gaussian probability distribution for the number of polaritons, $P(n)$, characterised by its mean, $\bar{n}$, and variance, σ^2. This state evolves under a non-linear Hamiltonian $H = \kappa a^+ a a^+ a$, describing the polariton–polariton interaction. To find $g^{(1)}(\tau)$, we evaluate $< a^+(\tau)a(0) >$, which corresponds to removing a polariton at time $t = 0$ and putting it back at $t = \tau$. The phase change is τ times the difference in energy of states with n and $n-1$ polaritons, that is, $\kappa\tau(n^2 - (n-1)^2) \sim 2\kappa\tau n$. Averaging over the probability function $P(n)$ gives:

$$g^{(1)}(\tau) \sim \int dnP(n) \exp(2i\kappa\tau n) - \exp(-2\kappa^2\sigma^2\tau^2) \tag{5.3}$$

which has the Gaussian form observed experimentally in Fig. 5.3a.

We obtain the variance σ^2 from the second-order correlation function, $g^{(2)}(0) = 1 + (\sigma^2 - \bar{n})/\bar{n}^2$. Using the measured $g^{(2)}(0)$ of 1.1 and estimating $\bar{n} \sim 500$ from the emission intensity, we obtain $\sigma^2 = 25{,}500$. As shown below, κ is estimated to be $2 \times 10^{-5}\,\mathrm{ps}^{-1}$ from the blueshift (0.5 meV) of the mode from below to above threshold. This gives a decay time $\tau_c \sim 220\,\mathrm{ps}$, close to the experimental value of 150 ps.

The coherent mode is, of course, not isolated, as it is not in equilibrium; polaritons are lost by external emission at a rate determined by the cavity linewidth, γ, and are replaced from the reservoir of particles in other modes.[2] This disrupts the Hamiltonian evolution on a timescale τ_r, which depends on the loss rate, but is generally slower than γ^{-1}, because the stimulated component of the scattering into the mode exactly replaces the particle which is lost. This timescale is obtained most directly from the decay of $g^{(2)}(\tau)$, that is, $\tau_r \sim 100\,\mathrm{ps}$. It is this slowing down which allows us to see decoherence mechanism which occurs in equilibrium condensates, despite the short cavity lifetime.

To make these considerations more quantitative, we have solved a generalisation of a model for atom lasers with interactions [24]. Authors of [24] consider the regime far above threshold, where the gain is saturated, so no slowing of τ_r occurs. We extend this treatment to cover the case where the occupation of the mode may be less than or comparable to the saturation value, n_s. The population dynamics are given by a master equation of the form:

$$\dot{P}_n = \gamma n_c \left[\frac{n}{n+n_s} P_{n-1} - \frac{(n+1)}{(n+1)+n_s} P_n \right] + \gamma[(n+1)P_{n+1} - nP_n] \qquad (5.4)$$

where n_c is a measure of the pumping strength. Above threshold, the steady-state P_n has a Gaussian form with mean $\bar{n} = n_c - n_s$ and variance n_c. Solving this model, we find approximate expressions for the correlation functions, valid for $\bar{n} \geq 3\sqrt{n_c}$:

$$g^{(2)}(\tau) = 1 + \frac{n_s}{\bar{n}^2} \exp(-\bar{\gamma}\tau) \qquad (5.5)$$

$$|g^{(1)}(\tau)| \approx \exp(-2\kappa^2 n_c \tau^2) \exp(-\gamma\tau/2\bar{n}) (\bar{\gamma}\tau \leq 1) \qquad (5.6a)$$

$$\approx \exp(-4\kappa^2 n_c \tau/\bar{\gamma}) \exp(-\gamma\tau/4\bar{n}(\bar{\gamma}\tau \gg 1) \qquad (5.6b)$$

where the decay rate $\bar{\gamma} = \bar{n}\gamma/n_c$ is much slower than that of the bare cavity mode.

It is the slowed decay of $g^{(2)}$, with decay rate $\bar{\gamma} = \bar{n}\gamma/n_c$, which determines that we are in the early time regime $\bar{\gamma}\tau \leq 1$ for $g^{(1)}$ as given by (5.6a). In this regime, the first factor in $g^{(1)}$ is identical to the Gaussian expression for an equilibrium condensate of (5.3) since the variance $\sigma^2 = n_c$. The second factor, $\exp(-\gamma\tau/2\bar{n})$, corresponding to a Schawlow-Townes phase diffusion, is much slower. From the measured $g^{(2)}(0)$ of 1.1, we obtain $n_s = 25{,}000$ which predicts $\bar{\gamma}^{-1}$ to be $\sim$100 ps when $\bar{n} \sim 500$, in very good agreement with the experimental decay of $g^{(2)}(100\,\mathrm{ps})$. The deduced value of $\bar{\gamma}$ means that we are in the regime

[2]The reservoir consists of all the occupied modes in the system, predominantly exciton states at high wavevector.

$\bar{\gamma}\tau \leq 1$ for long enough to see almost the entire Gaussian decay of $g^{(1)}$ with the decay time of 220 ps, shown above to be in quantitative agreement with experiment. Using experimentally determined input parameters and our modelling, we are thus able to explain quantitatively the decay times of both $g^{(1)}$ and $g^{(2)}$, corresponding to very different physical mechanisms.

In addition to the exchange of particles above between the condensate and the reservoir, there is a direct interaction through the non-linear Hamiltonian, which means that noise in the reservoir will cause energy fluctuations and decoherence. This effect explains the short coherence times obtained previously using multimode laser excitation. For such a laser, we measure intensity fluctuations of ~15–20% on a nanosecond timescale; comparable variations in the reservoir population will be expected. The interaction with the reservoir population causes a blueshift of the mode energy to ~0.5 meV above the low power peak. The fluctuations will thus be 20% of this, ~0.1 meV, which translates into the coherence time ~20 ps, consistent with the experimental results under multimode excitation (~5–10 ps linewidths).

Even with a noise-free excitation laser, there will be thermal fluctuations in the reservoir population, which can cause decoherence. If the mean population is N_r, the variance will also be $\sim N_r$. The blueshift of the mode due to the interactions is $\sim 4\kappa N_r$.[3] From the blueshift and the variance, the coherence time is found to be $(8N_r\kappa^2)^{-1/2}$. Estimating $N_r \sim 10^4$ using the pump power, we get $\kappa \sim 2 \times 10^{-5}\,\mathrm{ps}^{-1}$ from the blueshift, and thus a coherence time ~180 ps.[4] We conclude that the decoherence due to the thermal fluctuations is unlikely to obscure the intrinsic effects of interactions between the polaritons in the mode.

5.2.4 *Temporal Coherence of a Polariton Condensate Excited in the Optical Parametric Oscillation Configuration*

5.2.4.1 Introduction

In the previous sections, we investigated polariton condensates which are formed under high-energy no resonant excitation in CdTe microcavity samples, when the photocreated carriers undergo multiple scattering before forming a polariton condensate. In GaAs microcavity samples, such non-resonant excitation is more problematic. Since the exciton Bohr radius in GaAs (~10 nm) is a factor of 3 larger than that in CdTe material (~3 nm) at high particle densities, exciton resonance

[3]The factor of 4 is a counting factor which arises when the interaction is between two different modes.

[4]Due to the small spatial overlap (~10%) between the pump mode and emitting spots, the relevant value of N_r may be a factor of 10 smaller than the total reservoir population, thus increasing the estimated coherence time due to interactions with the reservoir by up to a factor of ~3, further decreasing any contribution from thermal reservoir fluctuations.

can be bleached by particle–particle interactions resulting in collapse of polariton states before the condensation occurs. To overcome this effect, samples with a large number of quantum wells [27] or so-called "cold" excitation resonant with the upper polariton branch [4] or with high-momentum excitons have been employed [7].

An alternative way to create a polariton condensate in GaAs microcavities is to excite resonantly the lower polariton branch at k_P. In this case, with increasing pump power, polariton–polariton scattering enables efficient population of the lower polariton branch at $k \sim 0$ and condensation. This regime corresponds to the so-called OPO [11–15], where macroscopically occupied signal and idler are formed at $k \sim 0$ and $k = 2\,k_P$ due to polariton–polariton scattering from the pump. It is important that the phase of the signal or idler condensate is not determined by that of the pump but appears due to spontaneous U1 symmetry breaking as in the case of BEC [28]. As for non-resonantly pumped condensate, the signal condensate also exhibits long-range spatial coherence [29] and hence properties very similar to those in non-resonantly pumped system.

Here we show that temporal coherence of an OPO condensate shows coherence properties very similar to those observed for non-resonantly pumped polariton condensed phases studied above. We show that the signal consists of a number of narrow, spectrally distinct modes. Transverse spatial images show that these modes arise from regions with localisation size $\sim$10 µm and strongly overlap indicating coexistence of several condensates due to interactions with transverse photonic disorder. Using spatial and spectral filtering to isolate individual modes, we study the coherence of the OPO emission. As for incoherently pumped polariton condensates with increasing excitation power, the phase coherence times of OPO system exhibit saturation behaviour with state occupancy and achieve maximum values of $\sim$500 ps. The effect is explained by polariton–polariton interactions between macroscopically occupied signal, idler, and pump states. The coherence times of signal and idler are found to be very similar over the whole range of excitation powers, showing that the coherence is a property of the coupled OPO system, and that energy and phase fluctuations of signal and idler are anti-correlated. We also investigated the second-order intensity correlation function $g^{(2)}$ of an individual signal mode and observed its long decay of about 250 ps. Such observation is similar to that observed for non-resonantly pumped condensate reflecting similar physical process occurring in the two systems. However, up to now, no quantitative theoretical model exists to explain the observed effect.

The microcavity (MC) studied had very similar structure to that of [11–13]. Several regions were investigated, all with Rabi splitting $\Omega \sim 6$ meV and near zero detuning between exciton and cavity modes. The beam from a single-frequency diode laser (FWHM $\sim$10 MHz) was focused to $\sim$40 µm on the sample at an angle of incidence of $\Theta_p =\sim 12°–15°$ to achieve resonant excitation of the LP branch. Spatially and spectrally resolved images of the LP emission (collected in a solid angle $0 \pm 5°$) were recorded (see Fig. 5.8a). First-order correlation functions were measured using a Mach–Zehnder interferometer.

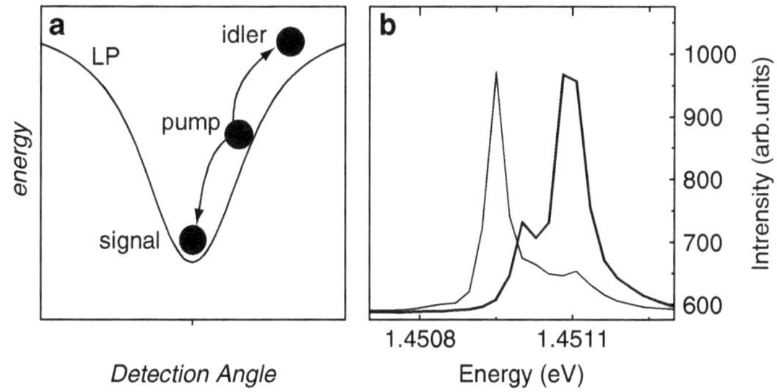

Fig. 5.9 (**a**) Schematic showing polariton–polariton scattering under resonant excitation resulting in condensate "signal" and "idler" formation. (**b**) Spectra of the signal emission recorded from different areas across the excitation spot revealing multimode condensate structure

5.2.4.2 The First-Order and the Second-Order Correlation Functions of OPO Condensates

Figure 5.9b shows spectra from the bottom of the LP branch (the signal emission at $\Theta_S =\sim 0^0$) at excitation powers $P = 31$ mW well above the threshold ($P_{th} \sim 3$ mW) for stimulated polariton–polariton scattering. The spectra are recorded from two spatially separated sections of $\sim 5 \times 40\,\mu$m dimension across the 40 μm excitation spot. The emission consists of a number of narrow features with energy separation and linewidth of 0.05–0.06 meV (FWHM, resolution limited). As seen in Fig. 5.9b, different peaks dominate depending on the detection area across the spot.

Spatially resolved images (5 μm resolution) of the signal are presented in Fig. 5.10a–d. Each individual image in Fig. 5.10a–d is recorded at a specific emission energy of an individual peak. Below threshold at a power $P = 1.5$ mW, the LP emission has a near-Gaussian distribution, determined by the excitation beam. By contrast at $P = 31$ mW above threshold (Fig. 5.10a–d), the signal consists of a number of spatially separated modes (spectrally resolved in Fig. 5.9b) with localisation size of ∼10–15 μm and with emission at different energies. It is notable that the different modes strongly overlap in space indicating again the formation of coexisting condensates.

The explanation for the multimode signal emission above threshold is similar to that presented for CdTe non-resonantly pumped condensates. There are fluctuations in the photonic potential of ∼0.2 meV possibly due to misfit dislocations in the Bragg mirrors [30], which contribute to a spatial variation of the signal energy. As a result, condensation occurs into different localised polariton modes above threshold. The detailed study with a theoretical modelling of OPO condensation in disorder potential are presented in [14].

Having achieved spectral and spatial resolution of individual signal modes, we are able to investigate the coherence properties of the OPO, without the influence of multimode phenomena which reduce the apparent coherence time and obscure the

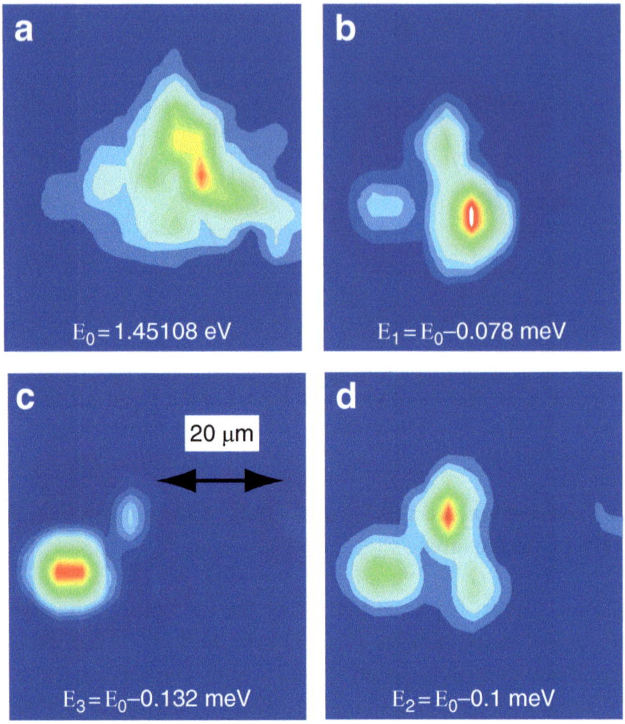

Fig. 5.10 (**a–d**) Two dimensional images of signal emission in real space recorded at each particular energy of the individual modes in Fig. 5.1

physics determining the coherence time. Figure 5.11a shows a typical dependence of the first-order correlation function $g^1(\tau_d)$ versus time for the single signal mode above threshold, decaying within a coherence time τ_{coh} of ~230 ps. To provide information on the factors determining the OPO coherence, τ_{coh} was studied as a function of excitation power for both signal and idler at detection angles $\Theta_{\mathrm{S}} \sim 0$ and $\Theta_{\mathrm{i}} \sim 24°$ (Fig. 5.8a), respectively. The ratio of the signal to idler intensities is measured to be ~50–100. Correcting for the photon fractions of the signal and idler states, N_{s} is deduced to be a factor of 5–10 larger than the population of the idler, N_{i}. The idler depletion mainly arises due to "idler–idler" and "idler-pump" scattering towards high k exciton states.

Surprisingly, despite the extra scattering channel out of the idler state, both the signal and idler have very similar coherence times. Moreover, as shown in Fig. 5.11b, very similar coherence times for the signal and idler are observed over the whole range of pump powers.[5] Coherence times (τ_{coh}) of ~100 ps are observed at powers just above threshold. As a function of signal intensity ($I \propto N_s$), a

[5]The idler intensity is too weak to permit study of its coherence close to threshold.

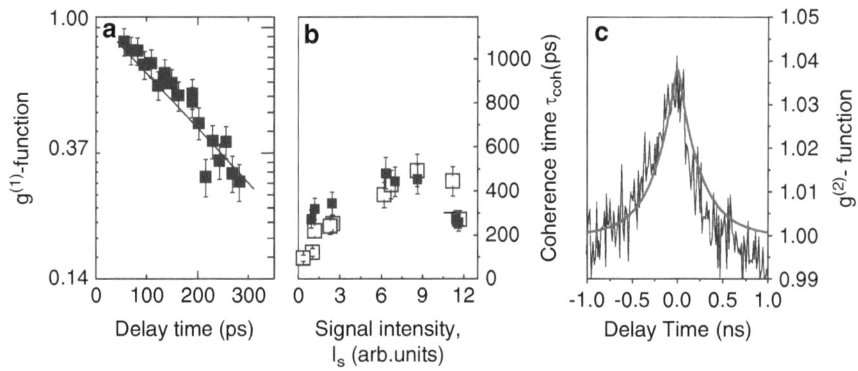

Fig. 5.11 (**a**) Decay of $g^{(1)}$ of an individual signal mode above threshold. (**b**) Dependence of coherence time of signal (*solid*) and the corresponding idler (*open*) modes as function of signal intensity. (**c**) Decay of $g^{(2)}$ of an individual signal mode above threshold

rapid increase is then seen, followed by saturation behaviour at values of ~500 ps, which is very similar to coherence time behaviour of the non-resonantly pumped condensates.

We have also investigated the decay of the second-order correlation function $g^{(2)}(\tau_d)$ of a signal mode at powers 2–3 times above threshold. As seen in Fig. 5.11c, photon bunching is again observed: $g^{(2)}$ has a maximum at $\tau = 0$ of ~1.04 and decays on a long timescale of 250 ps, with a decay time similar to that of $g^1(\tau_d)$. This indicates that the intensity fluctuations in the condensate are again correlated on the similar timescale as the phase coherence time (Fig. 5.11a). Taking into account the efficiency of detectors in our HBR setup at 850 nm and using similar procedure we described above, we deduced the true value of $g^{(2)}(0) \approx 1.6$, which indicates that the OPO condensate is not a fully coherent system. This value is larger than that observed for non-resonantly pumped condensates indicating stronger fluctuations of the number of particles in the system.

5.2.4.3 Semi-classical Qualitative Model of OPO Coherence

The observed behaviour of the first- and the second-order coherence of OPO condensate is very similar to that observed for non-resonantly pumped CdTe condensates. Nevertheless, one has to remember that the OPO system is much more complicated. It is a system of three coherently coupled macroscopically occupied states: pump, signal, and idler, which requires a much more complicated theoretical treatment. To explain the observed results, here we will only use qualitative explanation based on semi-classical treatment of OPO.

In a non-interacting stimulated system such as a laser [23],[6] the coherence time is proportional to the ratio of the number of stimulated to noise photons, due to spontaneous emission into the lasing mode. In the OPO, spontaneous scattering occurs within a time $\tau_{\rm sp} \sim \sqrt{\gamma_s \gamma_i} \sim 5\,$ps, determined by the polariton losses, where γ_s and γ_i are polariton linewidths below threshold. Therefore, making the analogy with the laser and assuming the limitation of the OPO coherence only by spontaneous scattering, the coherence time of the signal and idler is expected to vary linearly with polariton population $(N_S + N_i) \approx N_s$ $(N_S \gg N_i)$ and is given by $\tau_{\rm coh} = < N_s > \tau_{\rm sp}$. We note, despite the strong depletion of the idler population, the coherence times of signal and idler are expected to be the same within this picture since an equal polariton flux into these states due to polariton–polariton scattering implies the same fractional contribution of spontaneous noise polaritons for signal and idler.

The observed variation of the temporal coherence, which saturates with N_s, finds a natural explanation if polariton–polariton interactions are included [15,16,31]. The repulsive interaction leads to the blueshift of the condensate energy, determined by the total polariton population in the system [32]. Following [32] for a fixed energy of the pump, the energy shift of the signal, determined by the polariton–polariton interactions between pump, signal, and idler states, is given by

$$\Delta E_s \sim \chi^3 \left(\frac{|X_s|^4 \gamma_i}{(\gamma_i + \gamma_s)} \left(N_s - N_i \frac{\gamma_s |X_i|^4}{\gamma_i |X_s|^4} \right) + \frac{2|X_p|^4}{(\gamma_i + \gamma_s)} \left(\gamma_i |X_s|^2 - \gamma_s |X_i|^2 \right) N_p \right) \tag{5.7}$$

$N_{s,p,i}$ is the mean number of polaritons in the signal, pump, and idler respectively, $\gamma_s \sim 0.3\,$meV, $\gamma_i \sim 1\,$meV are the polariton linewidths below threshold and $|X_s|^2 \sim 0.5$, $|X_p|^2 \sim 0.7$ and $|X_i|^2 \sim 0.95$ are the excitonic contents of the signal, pump and idler states, respectively; x^3 is proportional to the third-order nonlinearity coefficient which determines the strengths of the scattering processes. The energy of the idler is given by energy conservation by $E_i = 2E_p - E_s$.

In macroscopically occupied states, there are always fluctuations in the number of particles. For coherent states, the minimum amplitude of the fluctuations is determined by Poissonian statistics with $\delta N_s = N_s^{0.5}$, $\delta N_i = N_i^{0.5}$, $\delta N_p = N_p^{0.5}$ [23]. These fluctuations will result in equal energy fluctuations for signal and idler $\delta E_s = -\delta E_i$ (and thus reduction of the coherence time) due to the population-dependent blueshift and hence equal coherence times in agreement with experiment. The coherence is thus a property of the whole OPO system, rather than of either signal or idler separately. The scattering out at the idler is stronger than at the signal due to its higher excitonic content and proximity in energy to uncoupled exciton states. Such scattering affects not only the coherence of the idler state but also that

[6]VCSEL Design, Fabrication, Characterization and Applications (Cambridge University Press, 1999), p. 240.

of the signal, which is coupled to the idler and the pump by polariton–polariton pair scattering.

Taking into account the broadening of the OPO emission due to spontaneous processes discussed above, and the broadening due to fluctuations in polariton number, derived from (5.1), the coherence times of the signal and idler are given by

$$1/\tau_{\rm coh} = 1/(\tau_{\rm sp} \times N_s) + \chi^3 \sqrt{\left(\frac{|X_s|^4 \gamma_i}{\gamma_i + \gamma_s}\right)^2 N_s + \frac{4|X_p|^4}{(\gamma_i + \gamma_s)^2}(\gamma_i |X_s|^2 - \gamma_s |X_i|^2)^2 N_p} \tag{5.8}$$

In the formula, we omit the N_i term from (5.7) since $N_i \ll N_s, N_p$. Above threshold the number of polaritons in the pump is almost independent of the excitation power and is given by the threshold value since with increasing power, the additional polaritons are transferred efficiently to the signal and idler [32]. Therefore, the variation of $\tau_{\rm coh}$ with occupation is mainly governed by N_s. N_p is estimated to be $\sim 6{,}000 \pm 3{,}000$ at threshold pump powers of ~3–7 mW. We measure the power of the total emission in the signal beam to be ~200 μW for $P = 30$ mW. This enables us to deduce the average polariton number in a single signal mode (Fig. 5.9) to be $\sim 1{,}000$–$2{,}000$ at $P = 30$ mW. χ^3 is deduced from the energy shift of the signal of ~0.25 meV at $P = 30$ mW. Equation (5.8) was then used to calculate the coherence time $\tau_{\rm coh}$ at high-density N_s. For $N_s = 2{,}000$ at $P = 30$ mW and $N_p = 3{,}000$, close to the values deduced from experiment, we obtain the maximum value of $\tau_{\rm coh}$ of ~450 ps (corresponding to linewidth of ~2 μeV), which is in a good agreement with the experiment.

The increased coherence time of the second-order correlation function (~250 ps), much longer than polariton lifetime, cannot be explained by a semi-classical model. This requires development of a quantum theory of OPO taking into account pump and decay of polariton states like for a non-resonantly pumped condensate in Sect. 5.2.3 [16], which is beyond the scope of the current manuscript.

5.2.5 Conclusion

In conclusion, we report the observation of long decay times for both the first- and the second-order correlation functions of polariton condensates, which arise under the conditions of non-resonant or resonant excitation. In the case of non-resonantly pumped condensates, the decays of both coherence functions, as well as the Gaussian lineshape of $g^{(1)}$, are well explained by a consistent model which takes into account pumping and decay of the coherent mode along with the effect of interactions. More generally, the polariton condensate exhibits properties expected in equilibrium BEC, even though it is subject to gain and loss with its environment. For OPO macroscopically occupied modes, the coherence times of individual signal and idler modes have been observed to be similar over the whole excitation range of powers even though the idler is subject to strong additional scattering. This shows that the temporal coherence is a property of the interacting OPO system, rather than of signal and idler separately. Semi-classical qualitative model is presented to show how the interactions between OPO modes limit its coherence. Overall, $g^{(1)}$ and $g^{(2)}$ decay times and its behaviour with power in non-resonantly and resonantly (OPO) pumped condensates are very similar.

Acknowledgements The work was supported by the EU ITN Clermont 2 and Clermont 4 projects and EPSRC grants GR/S09838/01, GR/S76076/01. D. Krizhanovskii is an EPSRC Advanced Fellow (grant EP/E051448/1).

References

1. J. Kasprzak, M. Richard, S. Kundermann, A. Baas, P. Jeambrun, J.M.J. Keeling, F.M. Marchetti, M.H. Szymanska, R. Andre, J.L. Staehli, V. Savona, P.B. Littlewood, B. Deveaud, Le Si Dang, Nature **443**, 409–414 (2006)
2. M. Richard, J. Kasprzak, R. Andr, R. Romestain, G. Le Si Dang Malpuech, A. Kavokin, Phys. Rev. B **72**, 201301 (2005)
3. R. Balili, V. Hartwell, D. Snoke, L. Pfei_er, K. West, Science **316**, 1007–1010 (2007)
4. D.N. Krizhanovskii, A.P. Love, D. Sanvitto, D.M. Whittaker, M.S. Skolnick, J.S. Roberts, Phys. Rev. B **75**, 233307 (2007)
5. S. Christopoulos, G. Baldassarri von Hogersthal, A.J. Grundy, P.G. Lagoudakis, A.V. Kavokin, J.J. Baumberg, G. Christmann, R. Butte, E. Feltin, J.-F. Carlin, N. Grandjean, Phys. Rev. Lett. **98**, 126405 (2007).
6. G. Christmann, R. Butte, E. Feltin, J.F. Carlin, N. Grandjean, Appl. Phys. Lett. **93**, 051102 (2008)
7. H. Deng et al., Phys. Rev. Lett. **97**, 146402 (2006)
8. R.M. Stevenson, V.N. Astratov, M.S. Skolnick, D.M. Whittaker, M. Emam-Ismail, A.I. Tartakovskii, P.G. Savvidis, J.J. Baumberg, J.S. Roberts, Phys. Rev. Lett. **85**, 3680 (2000)
9. M. Wouters, I. Carussoto, Phys. Rev. Lett. **99**, 140402 (2007)
10. D.N. Krizhanovskii, D.M. Whittaker, R.A. Bradley, K. Guda, D. Sarkar, D. Sanvitto, L. Vina, E. Cerda, P. Santos, K. Biermann, R. Hey, M.S. Skolnick, Phys. Rev. Lett. **104**, 126402 (2010)
11. M.H. Szymanska, J. Keeling, P.B. Littlewood, Phys. Rev. Lett. **96**, 230602 (2006)
12. J. Keeling, N.G. Berlo, Phys. Rev. Lett. **100**, 250401 (2008)
13. D. Sarchi, V. Savona, Phys. Rev. B **75**, 115326 (2007)
14. D. Sanvitto, D.N. Krizhanovskii, D.M. Whittaker, S. Ceccarelli, M.S. Skolnick, J.S. Roberts, Phys. Rev. B **73**, 241308 (2006)
15. D.N. Krizhanovskii, D. Sanvitto, A.P. Love, M.S. Skolnick, D.M. Whittaker, J.S. Roberts, Phys. Rev. Lett. **97**, 097402 (2006)
16. A.P.D. Love, D.N. Krizhanovskii, D.M. Whittaker, R. Bouchekioua, D. Sanvitto, S. Al Rizeiqi, R. Bradley, M.S. Skolnick, P.R. Eastham, R. André, Le Si Dang, Phys. Rev. Lett. **101**, 067404 (2008)
17. K.G. Lagoudakis, M. Wouters, M. Richard, A. Baas, I. Carusotto, R. Andre, B. Le Si Dang Deveaud-Pledran, Nat. Phys. **4**, 706 (2008)
18. D.N. Krizhanovskii, K.G. Lagoudakis, M. Wouters, B. Pietka, R.A. Bradley, K. Guda, D.M. Whittaker, M.S. Skolnick, B. Deveaud-Plédran, M. Richard, R. André, Le Si Dang, Phys. Rev. B **80**, 045317 (2009)
19. F. Tassone, C. Piermarocchi, V. Savona, A. Quattropani, P. Schwendimann, Phys. Rev. B **56**, 7554 (1997)

20. A.I. Tartakovskii, M. Emam-Ismail, R.M. Stevenson, M.S. Skolnick, V.N. Astratov, D.M. Whittaker, J.J. Baumberg, J.S. Roberts, Phys. Rev. B **62**, R2283 (2000)
21. M. Wouters, I. Carusotto, C. Ciuti, Phys. Rev. B **77**, 115340 (2008)
22. P.R. Eastham, Phys. Rev. B **78**, 035319 (2008)
23. R. Loudon, *The Quantum Theory of Light* (Oxford University Press, Oxford, 2000)
24. L.K. Thomsen, H.M. Wiseman, Phys. Rev. A **65**, 063607 and reference there in (2002)
25. M. Kohl, T.W. Hansch, T. Esslinger, Phys. Rev. Lett. **87**, 160404, (2001)
26. A. Ottl, S. Ritter, T. Esslinger, Phys. Rev. Lett. **95**, 090404, (2005)
27. D. Bajoni, P. Senellart, E. Wertz, I. Sagnes, A. Miard, A. Lemaître, J. Bloch, Phys. Rev. Lett. **100**, 047401 (2008)
28. M. Wouters, I. Carusotto, Phys. Rev. A **76**, 043807 (2007)
29. E.A. Cerda-Méndez, D.N. Krizhanovskii, M. Wouters, R. Bradley, K. Biermann, K. Guda, R. Hey, P.V. Santos, D. Sarkar, M.S. Skolnick, Phys. Rev. Lett. **105**, 116402 (2010)
30. M. Gurioli et al., Phys. Rev. Lett. **94**, 183901 (2005)
31. D. Porras, C. Tejedor, Phys. Rev. B **67**, 161310 (R) (2003)
32. D.M. Whittaker, Phys. Rev. B **71**, 115301 (2005)

Chapter 6
Vortices in Polariton OPO Superfluids

Francesca M. Marchetti and Marzena H. Szymańska

Abstract This chapter reviews the occurrence of quantised vortices in polariton fluids, primarily when polaritons are driven in the optical parametric oscillator (OPO) regime. We first review the OPO physics, together with both its analytical and numerical modelling, the latter being necessary for the description of finite size systems. Pattern formation is typical in systems driven away from equilibrium. Similarly, we find that uniform OPO solutions can be unstable to the spontaneous formation of quantised vortices. However, metastable vortices can only be injected externally into an otherwise stable symmetric state, and their persistence is due to the OPO superfluid properties. We discuss how the currents characterising an OPO play a crucial role in the occurrence and dynamics of both metastable and spontaneous vortices.

6.1 Introduction

Quantised vortices are topological defects occurring in macroscopically coherent systems, and as such have been broadly studied in several areas of physics. Their existence was first predicted in superfluids [1, 2], and later in coherent waves [3]. Nowadays, quantised vortices have been the subject of extensive research across several areas of physics and have been observed in type-II superconductors, ^{4}He, ultracold atomic gases, non-linear optical media (for a review, see, e.g., [4]) and very

F.M. Marchetti (✉)
Departamento de Física Teórica de la Materia Condensada, Universidad Autónoma de Madrid, Madrid 28049, Spain
e-mail: francesca.marchetti@uam.es

M.H. Szymańska
Department of Physics, University of Warwick, Coventry, CV4 7AL, UK
e-mail: M.H.Szymanska@warwick.ac.uk

D. Sanvitto and V. Timofeev (eds.), *Exciton Polaritons in Microcavities*,
Springer Series in Solid-State Sciences 172, DOI 10.1007/978-3-642-24186-4_6,

recently microcavity polaritons [5–12], the coherent strong mixing of a quantum well exciton with a cavity photon.

This chapter reviews the occurrence of quantised vortices in polariton fluids, primarily when polaritons are driven in the optical parametric oscillator (OPO) regime. The interest in this area of research is manifold. To start with, the search for condensation in solid-state excitonic systems has been arduous and lasted more than 2 decades: Unambiguous evidence for condensation has been reported for microcavity polaritons for the first time in 2006 [13]. These results have been followed by a wealth of experimental and theoretical advances on aspects related to macroscopic coherence, condensation, superfluidity, quantum hydrodynamics, and pattern formation, just to mention few (for a review, see [14]). Two different schemes of injecting polaritons and spontaneously generating a macroscopically coherent state can be employed: (1) non-resonant pumping, and (2) parametric drive in the OPO regime. What both condensates have in common is the phenomenon of spontaneous phase symmetry breaking (and the consequent appearance of a Goldstone mode) and the non-equilibrium ingredient. However, the way polaritons are pumped has strong effects on the type of condensed regime that can be reached. In both regimes (1i) and (2), the quest for superfluid behaviour has been and is being widely investigated. As it has been recently discussed in [15–17], one of the aspects that makes condensed polariton systems novel compared to known superfluids at thermal equilibrium, is that now all the paradigmatic definitions of a superfluid, such as the appearance of quantised vortices, the Landau criterion, the existence of metastable persistent flow, and the occurrence of solitary waves, have to be singularly examined and might in general be fundamentally different from the equilibrium case. Several of these popular topics are examined in Chapters 1, 3, 5, 7, 9, 10 and 12 of this book.

Resonantly pumped polaritons in the OPO regime [18, 19] have been recently shown to exhibit a new form of non-equilibrium superfluidity [8, 20]. Polaritons continuously injected into the *pump* state, undergo coherent stimulated scattering into the *signal* and *idler* states. Superfluidity has been tested through as frictionless flow [20] of a travelling signal triggered by an additional pulsed probe laser (the triggered OPO [TOPO] regime). In addition, the study of quantised vortices imprinted using pulsed Laguerre-Gauss laser fields has attracted noticeable interest both experimentally [8] and theoretically [21–24], providing a diagnostic for superfluid properties of such a non-equilibrium system. In particular, vorticity has been shown to persist not only in absence of the rotating drive but also longer than the gain induced by the probe, and therefore to be transferred to the OPO signal, demonstrating metastability of quantised vortices and persistence of currents [8,22].

The chapter is arranged as follows: After a very short introduction to microcavity polaritons in Sect. 6.2, we describe the optical parametric oscillator regime in Sect. 6.3, stressing the analogies and differences with an equilibrium weakly interacting Bose–Einstein condensate (Sect. 6.3.2.1) and the numerical modelling that is necessary to use for finite size pumps (Sect. 6.3.3). In Sect. 6.3.5, the occurrence of spontaneous stable vortices in OPO is described for clean cavities, while the case of disordered cavities is studied at the end of Sect. 6.3.3. Next, we describe in general terms the role of adding a pulsed Gaussian probe to the OPO regime (the so-called TOPO regime) in Sect. 6.4, while metastable vortices triggered by a Laguerre-Gauss probe are discussed in Sect. 6.5. Here, in Sect. 6.5.2, we also describe the onset and dynamics of vortex–antivortex pairs. Stability of multiply quantised vortices is analysed in Sect. 6.6, and finally, we mention the occurrence of vortices in polariton fluids in other regimes than OPO in Sect. 6.7.

6.2 A Very Short Introduction to Microcavity Polaritons

Before focusing on the main topic of this review, we give here a very short introduction to microcavity polaritons in order to fix the notation for later on. A more complete introduction can be found in several review articles [14, 25–29] and books [30–33] on microcavity polaritons.

Microcavity polaritons are the normal modes resulting from the strong coupling between quantum well (QW) excitons and cavity photons. In semiconductor microcavities, the mirrors employed to confine the light are distributed Bragg reflectors, i.e., alternating quarter wavelength thick layers of dielectrics with different refractive indices. Between the Bragg reflectors, the cavity light forms a standing wave pattern of confined radiation, which can be described by an approximatively quadratic dispersion, $\omega_{\mathrm{C}}(k) = \omega_{\mathrm{C}}^0 + k^2/(2m_{\mathrm{C}})$ (from here onwards, we fix $\hbar = 1$). Excitons are the hydrogenic bound states of a conductance band electron and a valence band hole; therefore, their mass is much larger than the cavity photon mass (typically $m_{\mathrm{C}} \sim 10^{-5} m_{\mathrm{e}}$, where m_{e} is the free electron mass). For this reason, the exciton dispersion can be neglected, $\omega_{\mathrm{X}}(k) = \omega_{\mathrm{X}}^0$. In microcavities, one or multiple QWs are grown in between the mirrors so that excitons are at the antinodes of the confined light, giving rise to strong coupling. In addition, cavity mirrors are built with a wedge, so as to change the detuning between the normal incidence energy of the cavity field and the exciton one, $\delta \equiv \omega_{\mathrm{C}}^0 - \omega_{\mathrm{X}}^0$. Typical parameter values for a GaAs-based microcavity are listed in Table 6.1.

The polariton normal modes can be found by solving the coupled Schrödinger equations for exciton and photon fields, $\psi_{\mathrm{X,C}} = \psi_{\mathrm{X,C}}(\mathbf{r}, t)$

$$\mathrm{i}\partial_t \begin{pmatrix} \psi_{\mathrm{X}} \\ \psi_{\mathrm{C}} \end{pmatrix} = \hat{H}_0 \begin{pmatrix} \psi_{\mathrm{X}} \\ \psi_{\mathrm{C}} \end{pmatrix} \qquad \hat{H}_0 = \begin{pmatrix} \omega_{\mathrm{X}}^0 - \mathrm{i}\kappa_{\mathrm{X}} & \Omega_{\mathrm{R}}/2 \\ \Omega_{\mathrm{R}}/2 & \omega_{\mathrm{C}}(-\mathrm{i}\nabla) - \mathrm{i}\kappa_{\mathrm{C}} \end{pmatrix}, \tag{6.1}$$

where Ω_{R} is the Rabi splitting and $\kappa_{\mathrm{X,C}}$ are the decay rates of exciton and photon. For an ideal cavity, $\kappa_{\mathrm{X,C}} = 0$, the eigenstates of this equations in momentum space, $\psi_{\mathrm{X,C}}(\mathbf{r}, t) = \mathrm{e}^{\mathrm{i}\omega t} \sum_{\mathbf{k}} \mathrm{e}^{\mathrm{i}\mathbf{k}\cdot\mathbf{r}} \psi_{X,C,\mathbf{k}}$, are the lower (LP) and upper polaritons (UP)

$$\begin{pmatrix} \psi_{X,\mathbf{k}} \\ \psi_{C,\mathbf{k}} \end{pmatrix} = \begin{pmatrix} \cos\theta_k & -\sin\theta_k \\ \sin\theta_k & \cos\theta_k \end{pmatrix} \begin{pmatrix} \psi_{\mathrm{LP},\mathbf{k}} \\ \psi_{\mathrm{UP},\mathbf{k}} \end{pmatrix}, \tag{6.2}$$

$$\cos^2\theta_k, \sin^2\theta_k = \frac{1}{2}\left(1 \pm \frac{\omega_C(k) - \omega_X^0}{\sqrt{(\omega_C(k) - \omega_X^0)^2 + \Omega_R^2}}\right), \tag{6.3}$$

with an energy dispersion given by (see Fig. 6.1):

$$\omega_{LP,UP}(k) = \frac{1}{2}\left[\omega_C(k) + \omega_X^0\right] \mp \frac{1}{2}\sqrt{\left[\omega_C(k) - \omega_X^0\right]^2 + \Omega_R^2}\,. \tag{6.4}$$

At zero detuning ($\delta = 0$) and normal incidence ($k = 0$), polaritons are exactly half-light and half-matter quasiparticles ($\cos^2\theta_0 = 0.5 = \sin^2\theta_0$). The value of the momentum k of polaritons inside the cavity is related to the emission angle φ (with respect to normal incidence) of photons outside the cavity by $ck = \omega_{LP}(\mathbf{k})\sin\varphi$. Thanks to this property, microcavity polaritons can be directly excited by a laser field and detected via reflection, transmission, or photoluminescence measurements. In Fig. 6.1, the energy dispersion of the lower and upper polariton are plotted as a function of both wavevector, k [μm^{-1}], or the emission angle, φ [degree], for typical values of microcavity parameters.

6.2.1 Exciton–Exciton and Exciton–Photon Interaction

A fundamental property of polaritons is their non-linear behaviour inherited from the exciton–exciton interaction and the saturation of the exciton–photon coupling. In this review, we treat excitons as bosonic particles; therefore, the effective exciton–

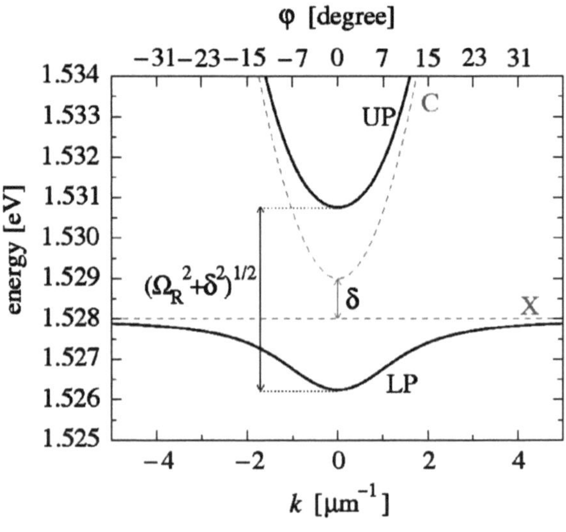

Fig. 6.1 Lower (*LP*) and upper polariton (*UP*) energy dispersions (*solid*) together with the dispersions of the photon (C) and exciton (X) fields (*dashed*) as a function of either the wavevector k [μm^{-1}] or the emission angle φ [degree] for $m_C = 2.3 \times 10^{-5} m_e^0$, $\Omega_R = 4.4$ meV, $\omega_X^0 = 1.5280$ eV, and a detuning $\delta = 1$ meV

Table 6.1 Characteristic parameters of a GaAs-based microcavity, divided between the parameters of the quantum well (*left*) and those describing the microcavity (*right*). Here, $\ell_C = \sqrt{1/(m_C \Omega_R)}$ is a characteristic length for the cavity photons. The photon decay rate κ_C refers to a cavity mirror with typically 25 bottom pairs and 15 lower pairs (see, e.g. [8])

QW	Cavity
$\epsilon \simeq 13$	$\omega_C^0 \simeq \omega_X^0 \simeq 1.53\,\text{eV}$
$m_e = 0.063 m_e^0$	$\delta \in [-10, 10]\,\text{meV}$
$m_h = 0.3 m_e^0$	$m_C = 2.3 \times 10^{-5} m_e^0$
$a_X \simeq 70\ \mathring{A}$	$\ell_C = 0.868\ \mu\text{m}$
$\mathscr{R}y_X \simeq 17\,\text{meV}$	$\Omega_R \simeq 4.4\,\text{meV}$
$\kappa_X \sim \mu\text{eV}$	$\kappa_C = 0.1\,\text{meV}$

exciton interaction can be written as:

$$\mathscr{H}_{XX} = \frac{1}{2A} \sum_{\mathbf{k},\mathbf{k}',\mathbf{q}} V_q \psi^*_{X,\mathbf{k}+\mathbf{q}} \psi^*_{X,\mathbf{k}'-\mathbf{q}} \psi_{X,\mathbf{k}} \psi_{X,\mathbf{k}'},$$

where the effective interaction potential V_q can be determined starting from the microscopic electron–hole Hamiltonian [34, 35]. The typical wavevector involved in the physics described by this review are much smaller than the inverse exciton Bohr radius, $q \ll a_X^{-1}$, where $a_X = \epsilon/(2\mu e^2)$ is the two-dimensional exciton Bohr radius, ϵ the dielectric constant, and $\mu^{-1} = m_e^{-1} + m_h^{-1}$ the electron–hole reduced mass. In this limit, it can be shown [34] that the momentum dependence of V_q can be neglected, thus approximating it with a contact interaction, $V_q \simeq g_X = 6e^2 a_X/\epsilon = 6\,\mathscr{R}y_X\, a_X^2$, where $\mathscr{R}y_X = e^2/(\epsilon a_X) = 1/(2\mu a_X^2)$ is the exciton Rydberg. Typically, for GaAs quantum wells (see Table 6.1), $\epsilon = 13$, $a_X \simeq 7$ nm, and $\mathscr{R}y_X \simeq 17\,\text{meV}$; therefore, $g_X \simeq 0.005\,\text{meV}(\mu\text{m})^2$. We will see, however, that the exact value of the coupling constant g_X has no relevance for the mean field dynamics we are going to describe, i.e., g_X can be rescaled to 1.

The composite nature of excitons, as a bound state of an electron and a hole, is also visible in the saturability of the exciton–photon coupling, resulting in an anharmonic interaction term which adds to the usual harmonic one:

$$\mathscr{H}_{XC} = \frac{\Omega_R}{2} \int d\mathbf{r} \left[\psi_X^*(\mathbf{r})\psi_C(\mathbf{r}) + \psi_C^*(\mathbf{r})\psi_X(\mathbf{r})\right] \left[1 - \frac{|\psi_X(\mathbf{r})|^2}{n_{sat}}\right], \tag{6.5}$$

where $n_{sat} = 7/(16\pi a_X^2)$ is the exciton saturation density [35]. In GaAs, $n_{sat} \simeq 2842\,(\mu\text{m})^{-2}$ and for a Rabi splitting of $\Omega_R = 4.4\,\text{meV}$, the ratio between saturation and exciton–exciton interaction strength:

$$\frac{\Omega_R}{2 g_X n_{sat}} \simeq 0.1,$$

allows us to neglect the anharmonic term in (6.5) for the kind of physics we want to describe in this review.

Therefore, the mean field evolution of the coupled cavity photon–exciton dynamics is described by the following non-linear Schrödinger equation or Gross–Pitaevskii Equation (GPE):

$$i\partial_t \begin{pmatrix} \psi_X \\ \psi_C \end{pmatrix} = \left[\hat{H}_0 + \begin{pmatrix} g_X|\psi_X|^2 & 0 \\ 0 & V_C(\mathbf{r}) \end{pmatrix} \right] \begin{pmatrix} \psi_X \\ \psi_C \end{pmatrix} . \tag{6.6}$$

Here, we have also added an external potential $V_C(\mathbf{r})$ acting on the photon component, which later on will be used to describe the effect of photonic disorder present in the cavity mirrors. Note that (6.6) is a classical field description, which assumes the macroscopic occupation of a finite number of states, each described by a complex classical function ψ.

6.3 Optical Parametric Oscillator Regime

An accurate control of the polariton dynamics can be achieved by directly injecting polaritons at a given wavevector and frequency with a properly tuned external laser—the resonant excitation scheme. Within this scheme, two regimes can be singled out: (1) the regime where only the polariton state generated by the pump is a stable configuration of the system (we refer to this as the *pump-only* state) and (2) the regime where the polaritons continuously injected into the *pump* state undergo coherent stimulated scattering into the *signal* state (close to the normal direction) and the *idler* state (on the other side of the pump). Parametric scattering from pump to signal and idler can be self-induced by the continuous-wave (cw) laser above a pump strength threshold, in which case, one refers to the OPO regime. However, below the threshold for OPO, a second weak probe beam shined close to either the (expected) signal or idler states, can be used to "seed" the parametric scattering processes and amplify the probe; in this case, one refers to the optical parametric amplification (OPA) regime.

We introduce the concept of polariton parametric scattering and review the main experimental results on optical parametric amplification in the next section. In Sect. 6.3.2, we use a simplified theoretical model in terms of plane waves, for both the pump-only resonant state and the OPO state, summarising the main properties of both regimes and drawing an analogy with equilibrium weakly interacting Bose–Einstein Condensates (BECs). Finally, in Sect. 6.3.3, we explain the necessity for carrying out a numerical analysis of the OPO. Much experimental work has been carried out on polaritons in the OPO regime [18, 19, 36–45] (for a review on the experiments, see [46]). We will discuss the experimental achievements along with the theoretical description.

6.3.1 Polariton Parametric Scattering and Optical Parametric Amplification

In the parametric scattering process, two polaritons from a pump mode, with wavevector and frequency $\{\mathbf{k}_\mathrm{p}, \omega_\mathrm{p}\}$, scatter into a lower energy signal mode $\{\mathbf{k}_\mathrm{s}, \omega_\mathrm{s}\}$ and a higher energy idler mode $\{\mathbf{k}_\mathrm{i}, \omega_\mathrm{i}\}$. This scattering process has to conserve energy and momentum, therefore requiring that

$$\begin{aligned} 2\mathbf{k}_\mathrm{p} &= \mathbf{k}_\mathrm{s} + \mathbf{k}_\mathrm{i} \\ 2\omega_\mathrm{p} &= \omega_\mathrm{s} + \omega_\mathrm{i}. \end{aligned} \tag{6.7}$$

This condition cannot be satisfied by any particle dispersion: For example, parametric scattering is forbidden for particles with a quadratic dispersion. In order to check whether parametric scattering is allowed for polaritons, one has to verify if the condition:

$$2\omega_\mathrm{LP}(k_\mathrm{p}) = \omega_\mathrm{LP}(k_\mathrm{s}) + \omega_\mathrm{LP}(|2\mathbf{k}_\mathrm{p} - \mathbf{k}_\mathrm{s}|) \tag{6.8}$$

can be satisfied. If $\mathbf{k}_\mathrm{s} = 0$, then the momenta of pump and idler are uniquely selected (see left panel of Fig. 6.2). In this case, the value of the pumping angle is also referred to as the "magic angle," and is located close to the inflection point of the LP dispersion. However, for a generic signal wavevector, $\mathbf{k}_\mathrm{s} = \mathbf{k} = (k_x, k_y)$ and a fixed pump angle $\mathbf{k}_\mathrm{p}$ (assumed to be oriented along the x-direction, $(k_\mathrm{p}, 0)$, in the right panel of Fig. 6.2), the final states allowed in the parametric scattering process describe a figure-of-eight in momentum space [28, 47, 48].

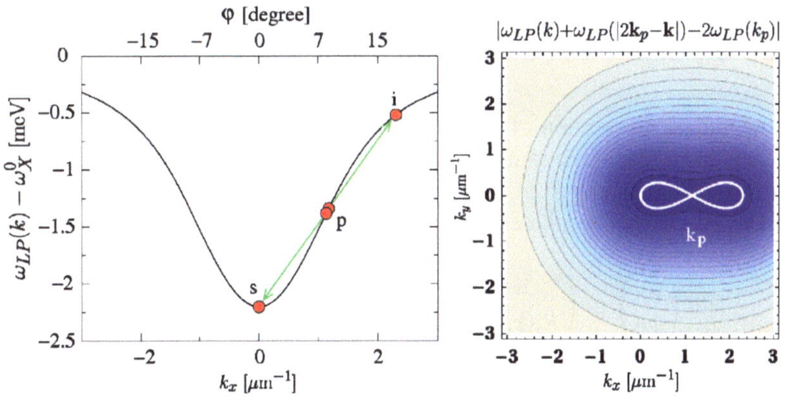

Fig. 6.2 Illustration of the basic idea of parametric scattering. *Left*: 2 LPs scatter from the pump state $\{\mathbf{k}_\mathrm{p}, \omega_\mathrm{p}\}$ towards the signal $\{\mathbf{k}_\mathrm{s}, \omega_\mathrm{s}\}$ (*here at zero momentum*) and the idler state $\{\mathbf{k}_\mathrm{i} = 2\mathbf{k}_\mathrm{p} - \mathbf{k}_\mathrm{s}, \omega_\mathrm{i} = 2\omega_\mathrm{p} - \omega_\mathrm{s}\}$ (*at higher momentum*), conserving momentum and energy. *Right*: Following Refs. [28, 47], we plot $|\omega_{LP}(k) + \omega_{LP}(|2\mathbf{k}_p - \mathbf{k}|) - 2\omega_{LP}(k_p)|$ as a function of $\mathbf{k} = (k_x, k_y)$, where the pump is oriented along the x-direction, $\mathbf{k}_p = (k_p, 0)$. The white line is the zero value of the contour. The parameters used in both panels are the same as the ones of Fig. 6.1

In the case of optical parametric amplification experiments, parametric scattering is stimulated by a weak additional probe field. OPA was first observed in an InGaAs/GaAs/AlGaAs microcavity [49], where a substantial signal gain of up to 70 was measured. Much experimental work has followed this first result [50–59]. Pump-probe parametric amplification of polaritons with an extraordinary gain up to 5000 and at temperatures up to 120 K has been reached in GaAlAs-based microcavities and up to 220 K in CdTe-based microcavities [55]. In three-beam pulsed experiments [51], polaritons scatter from two equal and opposite angles, $\mathbf{k}_p$ and $-\mathbf{k}_p$, into the LP and UP states at $\mathbf{k} = 0$—note that at zero detuning, $\delta = 0$, $2\omega_X^0 = \omega_{LP}(0) + \omega_{UP}(0)$. Interestingly, parametric amplification has been also obtained for ultracold atom pairs confined in a moving one-dimensional optical lattice [60]. The role of the periodic optical lattice is to deform the atom dispersion from the quadratic one, allowing parametric scattering to happen.

The stimulated scattering regime can be reached also in the OPO configuration, i.e., without an additional probe beam. Now, stimulated scattering is self-initiated at pump powers above a threshold intensity, where the final state population is close to one. We will see that in this case, there is no special significance of the "magic angle," rather, a broad range of pumping angles larger than a critical value, ($\theta_p \gtrsim 10°$ for the parameters in [61]) allow OPO with a signal emission close to normal incidence, $\theta_s \simeq 0°$. In addition, for finite size pumping (see later Sect. 6.3.3), the pump, signal, and idler momenta are smeared in a broad interval, while their frequency still satisfy the matching conditions (6.7) exactly. In the next three sections, we will focus mainly on the theoretical description of polariton resonant excitation with a cw laser field, describing the properties of first the pump-only state and then the OPO state.

6.3.2 Bistability and OPO in the Plane-Wave Approximation

The theoretical description of polaritons in the resonant excitation regime can be formulated in terms of the same classical two-field non-linear Schrödinger equation previously introduced in (6.6), where now an external driving field $F_p(\mathbf{r},t)$ is added in order to describe the coherent injection of photons into the cavity:

$$i\partial_t \begin{pmatrix} \psi_X \\ \psi_C \end{pmatrix} = \begin{pmatrix} 0 \\ F_p(\mathbf{r},t) \end{pmatrix} + \left[\hat{H}_0 + \begin{pmatrix} g_X|\psi_X|^2 & 0 \\ 0 & V_C(\mathbf{r}) \end{pmatrix} \right] \begin{pmatrix} \psi_X \\ \psi_C \end{pmatrix}. \tag{6.9}$$

A continuous-wave (cw) pumping laser can be written as:

$$F_p(\mathbf{r},t) = \mathscr{F}_{f_p,\sigma_p}(r) e^{i(\mathbf{k}_p\cdot\mathbf{r} - \omega_p t)}, \tag{6.10}$$

where $\mathscr{F}_{f_p,\sigma_p}(r)$ can either describe a homogeneous pump with strength f_p, $\mathscr{F}_{f_p,\sigma_p}(r) = f_p$, or, as we will assume later in Sect. 6.3.3, a Gaussian or a top-hat spatial profile with strength f_p and full width at half maximum (FWHM) σ_p.

For a homogeneous pump, $\mathscr{F}_{f_p,\sigma_p}(r) = f_p$, and for a clean system, $V_C(\mathbf{r}) = 0$, the conditions under which a stable OPO switches on can be found by making use of an analytical treatment [28, 61–63]. In fact, in this limit, each mode can be approximated as a plane wave. To simplify the analytical expressions, it is useful to rotate (6.9) into the LP and UP basis, as described by (6.2), and to neglect the contribution from the UP states, assuming that the LP and UP branches are not mixed together by the non-linear terms. In this case, working in momentum space, $\psi_{LP}(\mathbf{r},t) = \sum_{\mathbf{k}} e^{i\mathbf{k}\cdot\mathbf{r}} \psi_{LP,\mathbf{k}}(t)$, (6.9) can be written as:

$$i\partial_t \psi_{LP,\mathbf{k}} = [\omega_{LP}(k) - i\kappa(k)]\,\psi_{LP,\mathbf{k}} + \sum_{\mathbf{k}_1,\mathbf{k}_2} g_{\mathbf{k},\mathbf{k}_1,\mathbf{k}_2} \psi^*_{LP,\mathbf{k}_1+\mathbf{k}_2-\mathbf{k}} \psi_{LP,\mathbf{k}_1} \psi_{LP,\mathbf{k}_2}$$
$$+ \sin\theta_k\, f_p e^{-i\omega_p t} \delta_{\mathbf{k},\mathbf{k}_p}, \tag{6.11}$$

where $\kappa(k) = \kappa_X \cos^2\theta_k + \kappa_C \sin^2\theta_k$ is the effective LP decay rate and the interaction strength now reads as $g_{\mathbf{k},\mathbf{k}_1,\mathbf{k}_2} = g_X \cos\theta_k \cos\theta_{|\mathbf{k}_1+\mathbf{k}_2-\mathbf{k}|} \cos\theta_{k_1} \cos\theta_{k_2}$.

If we consider the solutions of Eq. (6.11) where only the pump mode, $\mathbf{k} = \mathbf{k}_p$, is populated, we can find an exact solution in the form:

$$\psi_{LP}(\mathbf{r},t) = p e^{i(\mathbf{k}_p\cdot\mathbf{r} - \omega_p t)}$$
$$\psi_{LP,\mathbf{k}}(t) = p\delta_{\mathbf{k},\mathbf{k}_p} e^{-i\omega_p t}, \tag{6.12}$$

where the complex amplitude p is given by:

$$\left[\omega_{LP}(k_p) - \omega_p - i\kappa(k_p) + g_X \cos^4\theta_{k_p} |p|^2\right] p + \sin\theta_{k_p} f_p = 0\,. \tag{6.13}$$

For practical purposes, one can substitute $g_X \cos^4\theta_{k_p} \mapsto 1$ by redefining the pump strength $f_p' = \sqrt{g_X}\cos^2\theta_{k_p} \sin\theta_{k_p} f_p$ and rescaling the field strength p by $|p'| = \sqrt{g_X}\cos^2\theta_{k_p}|p|$. Note that the $\chi^{(3)}$-non-linear interaction term, $|p'|^2 p'$, renormalises the effective detuning of the pump mode from the LP dispersion:

$$\Delta_p \equiv \omega_p - \omega_{LP}(k_p) - |p'|^2, \tag{6.14}$$

which now includes the blue-shift of the LP dispersion due to interactions.

The general solution of the cubic equation (6.13) is well known, and exhibits a qualitatively different behaviour depending whether the pump frequency is blue- or red-detuned with respect to the LP dispersion. In particular, if $\omega_p - \omega_{LP}(k_p) \leq \sqrt{3}\kappa(k_p)$, the system is in the optical limiter regime, where the population $|p'|^2$ grows monotonically as a function of the pump intensity f_p'. If instead $\omega_p - \omega_{LP}(k_p) > \sqrt{3}\kappa(k_p)$, the system displays bistable behaviour, with a characteristic S-shape of $|p'|^2$ as a function of f_p', the second turning point coinciding with the point where the effective detuning Δ_p (6.14) changes sign (see Fig. 6.3). Because the branch with negative slope is unstable, the polariton density in the pump-only

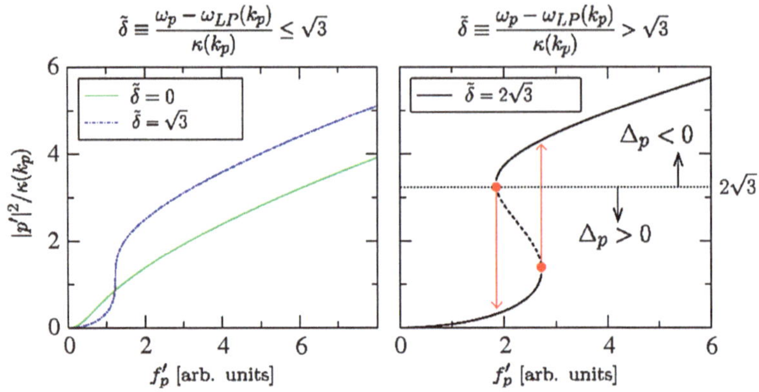

Fig. 6.3 Dimensionless LP population $|p'|^2/\kappa(k_p) = g_X \cos^4\theta_{k_p}|p|^2/\kappa(k_p)$ as a function of the dimensionless pump intensity $f'_p = \sqrt{g_X}\cos^2\theta_{k_p}\sin\theta_{k_p} f_p$ for different values of the parameter $\tilde{\delta} = [\omega_p - \omega_{LP}(k_p)]/\kappa(k_p)$. When $\tilde{\delta} \leq \sqrt{3}$ (*left panel*), the system is in the optical limiter regime, while for sufficiently blue-detuned pump frequencies, $\tilde{\delta} > \sqrt{3}$, a bistable behaviour is obtained (*right*). The sign of the interaction renormalised detuning, Δ_p (6.14), is also given

mode follows a hysteretic behaviour: Increasing the pump intensity, eventually the pump-only mode jumps abruptly into the upper branch, while if the intensity is then decreased, the polariton population decreases and jumps back down to the lower branch for smaller values of the pump intensity. Optical bistability in microcavity polaritons has been observed in [64, 65], with evidence of a hysteresis cycle of the polariton emission as a function of the pump intensity. Multistability of two different polariton states, generated by either populating two different spin states [66–68] or by injecting two states with two different pumps [69], has been also recently proposed and, in the spin case, observed. Part of the interest in this field is to realise all-optical switches [70] and memories.

The dynamical stability of the pump-only solution (6.12) can be established by allowing other states than the pump (i.e., the signal and idler states) to be perturbatively populated via parametric scattering processes:

$$\psi_{\mathrm{LP},\mathbf{k}}(t) = p\delta_{\mathbf{k},\mathbf{k}_p} e^{-i\omega_p t} + s\delta_{\mathbf{k},\mathbf{k}_p-\mathbf{q}} e^{-i(\omega_p-\omega)t} + i^*\delta_{\mathbf{k},\mathbf{k}_p+\mathbf{q}} e^{-i(\omega_p+\omega^*)t}\,, \tag{6.15}$$

where, $\{\mathbf{k}_{s,i} = \mathbf{k}_p \mp \mathbf{q}, \omega_{s,i} = \omega_p \mp \omega\}$ and by assessing whether the time evolution of these states grows exponentially in time or not. Expanding to the first order in s and i, one obtains an eigenproblem for the amplitudes s and i [61–63]:

$$\begin{pmatrix} \omega - \Delta_s - i\kappa(k_s) & g_X c_s c_i c_p^2 p^2 \\ g_X c_s c_i c_p^2 p^{*2} & -\omega - \Delta_i + i\kappa(k_i) \end{pmatrix} \begin{pmatrix} s \\ i \end{pmatrix} = 0\,, \tag{6.16}$$

where $\Delta_{s,i} = \omega_p - \omega_{LP}(k_{s,i}) - 2g_X c_{s,i}^2 c_p^2 |p|^2$ and $c_{p,s,i} = \cos\theta_{k_p,k_s,k_i}$. The complex eigenvalues ω can be obtained imposing that the determinant of the matrix in (6.16) is zero. The dynamical stability is ensured if $\Im(\omega) > 0$. Therefore, the threshold for instability of the pump-only solution (6.12) can be found imposing the condition $\Im(\omega) = 0$. By fixing the pump wavevector and energy $(\mathbf{k}_p, \omega_P)$ and the signal wavevector $\mathbf{k}_s$ (as well as the exciton and photon lifetimes, $\kappa_{X,C}$), the condition above provides a criterion for establishing the boundaries of the instability region, i.e., the lowest and highest values of the LP population $|p|^2$ for which the pump-only solution is not stable. As shown in [61–63], one can classify the instability as a single mode instability when $\mathbf{q} = 0$ and therefore $\mathbf{k}_p = \mathbf{k}_s = \mathbf{k}_i$—the Kerr instability. In particular, the branch with negative slope of the bistable curve (dashed line in Fig. 6.3) is always single mode unstable [61, 63]. If instead $\mathbf{q} \neq 0$, the instability is parametric like. Now, the total extent of the instability region corresponding to different values of $\mathbf{k}_s$ is significantly larger than just the branch with negative slope. In addition, the OPO state does not require a bistable behaviour and can turn on also in the optical limiter case. In particular, it is possible to plot a "phase diagram" [61] of pump energy ω_P as function of pump wavevector $\mathbf{k}_p$, showing the regions where a pump-only solution is always stable, where the OPO switches on, and where instead only a Kerr-type instability is possible. In this way, in [61], it was shown that there is no particular significance to the "magic angle" for the pump. Rather, OPO conditions can be found for all angles larger than a critical value, $\theta_p > \theta_c$ ($\sim 10°$ for the parameters of [61]), as also confirmed experimentally [39, 40]. In addition, the energy renormalisation of the polariton dispersion due to interactions moves the emission angles for the signal always close to $\theta_s \sim 0$ [40,61]. This is also confirmed by the numerical simulations we have carried out and illustrated later on in Sect. 6.3.3.

The method described above implies negligible populations of the signal s and the idler i, and therefore, it allows to find the conditions for the OPO *threshold*. In order to find the OPO states, one cannot linearise in s and i but instead must include the contributions of finite signal and idler populations to the dispersion renormalization [61]. In this way, in the region unstable to parametric scattering determined with the method described above, one can describe first the increase (switch-on) and later the decrease (switch-off) of the signal and idler populations as a function of the pump power. It is interesting to note that by doing that, i.e. by substituting (6.15) into (6.11), "satellite states" oscillating with energies $\omega_{s_2} = 2\omega_s - \omega_p = \omega_p - 2\omega$ and $\omega_{i_2} = 2\omega_i - \omega_p = \omega_p + 2\omega$ automatically appear. In fact, above the OPO threshold when signal and idler populations are not negligible, parametric scattering from the signal (idler) state into the pump and second-signal (second-idler) satellite state take place, i.e., $2s \mapsto p + s_2$ $(2i \mapsto p + i_2)$, and therefore, $2\omega_s = \omega_p + \omega_{s_2}$ $(2\omega_i = \omega_p + \omega_{i_2})$. This is clearly seen in the "exact" OPO solution obtained numerically (see, e.g., Fig. 6.4), as well as it has been observed experimentally (see, e.g., [37]). One has to note, however, that the population of the "satellite states" by multiple scattering processes is always negligible w.r.t. the one of pump, signal, and idler (see right panel of Fig. 6.4).

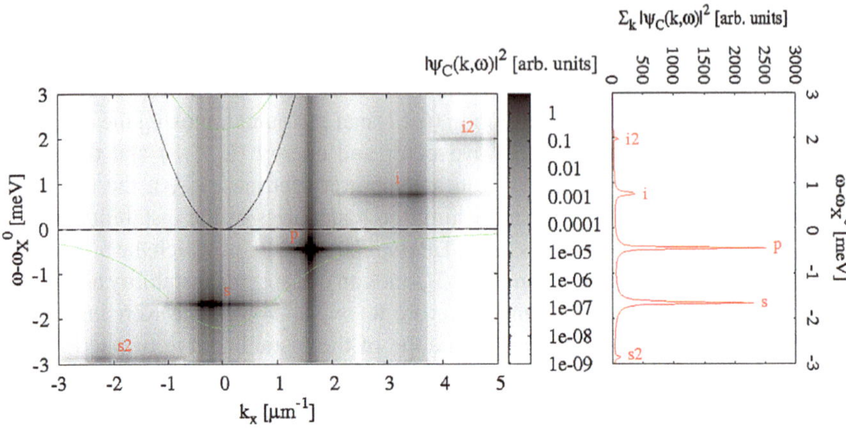

Fig. 6.4 *Left panel*: OPO spectrum $|\psi_{C,X}(\mathbf{k},\omega)|^2$ for a top-hat pump of FWHM $\sigma_p = 70\ \mu m$ and intensity $f_p = 1.25 f_p^{th}$ above the threshold pump power for OPO, $f_p^{(th)}$. For this particular run, we resonantly pump at $k_p = 1.6\ \mu m^{-1}$ in the x-direction, $\mathbf{k}_p = (k_p, 0)$, and at $\omega_p - \omega_X^0 = -0.44$ meV. Polaritons at the pump state undergo coherent stimulated scattering into the signal and idler states, which are blue-shifted with respect to the bare lower polariton (*LP*) dispersion (*green dotted line*) because of interactions. Cavity photon (C) and exciton (X) dispersions are plotted as gray dotted lines. As discussed in the text, above threshold, we observe the population of the satellite states in addition to the one of signal and idler. *Right panel*: momentum integrated spectrum, $\sum_{\mathbf{k}} |\psi_{C,X}(\mathbf{k},\omega)|^2$, as a function of the rescaled energy $\omega - \omega_X^0$. Pump, signal, idler, and satellite states are all equally spaced in energy by roughly 1.19 meV

We will introduce the numerical modelling used to describe the problem for a finite size pump later in Sect. 6.3.3. Before doing that, in the next section, we concentrate on the analogies and differences between an OPO state and an equilibrium weakly interacting BEC.

6.3.2.1 Spontaneous $U(1)$ Phase Symmetry Breaking and Goldstone Mode

The OPO state looks at first sight very different from an equilibrium weakly interacting BEC. In particular, the OPO is an intrinsically non-equilibrium state characterised by the (macroscopic) occupation of three polariton states only, one directly populated by the external pump and the signal and idler states populated by parametric scattering. Contrast this with the thermodynamic phase transition in a BEC, where the macroscopic occupation of the ground state occurs when, for a thermal distribution of bosons, either the temperature is lowered below a critical value or the density is increased. The OPO state does, however, share with a BEC the fundamental property of spontaneous symmetry breaking of the phase symmetry [61, 71]. In fact, the external laser fixes the phase of the pump state ϕ_p and parametric scattering processes constrain the sum of the signal and the idler phase only, $2\phi_p = \phi_s + \phi_i$, but leaves the system to arbitrarily choose the phase difference $\phi_s - \phi_i$. In other words, one can easily show that the system of three equations obtained by imposing the OPO solution (6.15) into the mean field (6.11) is invariant under the simultaneous phase rotation of both signal and idler states:

$$s \mapsto s e^{i\phi}$$
$$i \mapsto i e^{-i\phi}. \tag{6.17}$$

This $U(1)$ phase rotation symmetry gets spontaneously broken in the OPO regime, where the signal and idler spontaneously select their phase, though not independently. Note that in this respect the OPO regime differs very much from the OPA regime, where both signal and idler phases are fixed by the external probe, and therefore the $U(1)$ phase rotation symmetry is explicitly broken by the probe and no phase freedom is left in the system.

Goldstone's theorem states that the spontaneous symmetry breaking of the $U(1)$ phase symmetry in OPO is accompanied by the appearance of a gapless soft mode, i.e., a mode $\omega(\mathbf{k})$ whose both frequency $\Re[\omega(\mathbf{k})]$ and decay rate $\Im[\omega(\mathbf{k})]$ tend to zero in the long wavelength $\mathbf{k} \to 0$ limit. The dispersion for the Goldstone mode in OPO has been derived in [71], where also an experimental setup to probe its dispersion has been proposed. In addition, the appearance of spontaneous coherence in OPO has been shown via quantum Monte Carlo simulations [72] through the divergence of the coherence length when the pump intensity approaches the threshold. In contrast, in the OPA regime, where the phase rotation symmetry is explicitly broken by the probe, there is no Goldstone mode, and a gap opens in the imaginary part of the elementary excitation dispersion, $\Im[\omega(\mathbf{k})]$.

We would like to stress here that even though an equilibrium weakly interacting BEC and an OPO state share the fundamental property of spontaneous symmetry breaking of the phase symmetry, some care needs to be applied in pushing this analogy further. In particular, the existence of a free phase alone is not sufficient to ensure the paradigmatic properties of a superfluid, such as the Landau criterion, the stability of quantised vortices, and the persistency of metastable flow. For example, let us consider here the case of the Landau criterion: In an equilibrium weakly interacting BEC, the existence of the soft Goldstone mode (the Bogoliubov mode), with its characteristic linear dispersion for $\mathbf{k} \to 0$, $\omega(\mathbf{k}) \simeq c_s k$ implies the existence of a critical velocity, $v_c \equiv \min_{\mathbf{k}} \omega(\mathbf{k})/k = c_s$ (the speed of sound), below which a perturbative defect dragged through the fluid cannot dissipate energy (superfluid regime). In the non-equilibrium OPO regime instead, similarly to what happens for incoherently pumped polaritons condensates [21, 73, 74], the unusual form of the excitation spectrum—diffusive at small momenta—poses fundamental questions on the fulfilling of the Landau criterion and the possibility of dissipationless superflow.

Similarly, properties such as the appearance and stability of quantised vortices and the persistency of metastable flow need to be independently assessed in polariton fluids in the three different pumping schemes available—(1) non-resonant pumping, (2) parametric drive in the optical parametric oscillator regime, (3) coherent drive in the pump-only configuration. In fact, in the case of an equilibrium condensate, the ground state is flow-less, i.e. a vortex solution is unstable in non rotating condensates.[1] In contrast, in a polariton fluid, its intrinsic non-equilibrium nature implies the presence of a flow even when a steady state regime is reached. In this sense, not always the presence of vortices can be ascribed to the superfluid property of the system. We will discuss these aspects more in depth later in Sect. 6.5.

6.3.3 Numerical Modelling

We have seen in Sect. 6.3.2 that, for homogeneous pumps, $\mathscr{F}_{f_p,\sigma_p} = f_p$, the conditions under which a stable OPO switches on can be found analytically by assuming that pump, signal, and idler states can be described by plane wave fields (6.15) and therefore are characterised by single wavevectors $\mathbf{k}_{p,s,i}$ and by uniform currents, the intensity and direction of which are given by $\mathbf{k}_{p,s,i}$.[2] However, for pumping lasers with a finite excitation spot, $\mathscr{F}_{f_p,\sigma_p}(r)$, such as the ones employed in experiments, one can only resort to a numerical analysis [78] of the coupled equations (6.9). A finite size pump implies that, in the OPO regime, pump, signal, and idler states are broaden in momentum; as a consequence, these states are going to be characterised by non-trivial configurations of the currents (see Fig. 6.5). We will see later on that these currents play a crucial role in the occurrence and dynamics of both metastable and spontaneous vortices in OPO.

In particular, we numerically solve (6.9) on a 2D grid of typically $N \times N = 2^8 \times 2^8$ points and a separation of 0.47 μm (i.e., in a box $L \times L = 140 \times 140$ μm) by using a 5th-order adaptive-step Runge–Kutta algorithm. We have checked that our results are converged with respect to both the resolution in space L/N and the one in momentum π/L. Note also that of course the extension of the momentum box $k_{\max} = \pi N/L$ has to be big enough to contain the idler state. In the specific case of Figs. 6.4, 6.5, and 6.6, we have chosen a smoothed top-hat profile $\mathscr{F}_{f_p,\sigma_p}(r)$ with FWHM $\sigma_p = 70$ μm and (maximum) strength f_p (later for Fig. 6.8, we have chosen instead a FWHM $\sigma_p = 35$ μm). Considering the case of zero detuning, $\delta = 0$, we pump at $k_p = 1.6$ μm^{-1} in the x-direction, $\mathbf{k}_p = (k_p, 0)$, and at $\omega_p - \omega_X^0 = -0.44$ meV, i.e. roughly 0.5 meV above the bare LP dispersion, and gradually increase the pump strength until the OPO switches on. We find that broader LP linewidths imply a wider range in pump strength of stable OPO, and for this reason, we fix $\kappa_X = \kappa_C = 0.26$ meV in these particular runs. We define f_p^{th}

[1] In rotating condensates, a vortex can be created if the angular velocity is higher than a critical value [75, 76]. When the rotation is halted, then the vortex will spiral out of the condensate [77].

[2] Given a complex field or wavefunction, $|\psi(\mathbf{r},t)|e^{i\phi(\mathbf{r},t)}$, describing either a quantum particle of mass m or a macroscopic number of particles condensed in the same quantum state, the current is defined as [76]:

$$\mathbf{j}(\mathbf{r},t) = \frac{\hbar}{m}|\psi(\mathbf{r},t)|^2\nabla\phi(\mathbf{r},t) = |\psi(\mathbf{r},t)|^2\mathbf{v}_s(\mathbf{r},t), \tag{6.18}$$

where $\mathbf{v}_s(\mathbf{r},t)$ is the flow velocity. In the following, with a slight abuse of notation, we will refer to the current as the gradient of the phase only, $\nabla\phi(\mathbf{r},t)$.

as the pump strength threshold for OPO emission—here and in the following, we only select OPO solutions which reach a dynamically stable steady state (dynamical stability is studied in Sect. 6.3.5.1). In the case of Figs. 6.4 and 6.5, the pump strength is fixed just above threshold, $f_p = 1.25 f_p^{th}$.

The numerical analysis provides the time evolution of both photon and exciton fields either in space, $\psi_{C,X}(\mathbf{r}, t)$, or in momentum, $\psi_{C,X}(\mathbf{k}, t)$. The OPO implies the simultaneous presence of (at least) three states emitting at different momenta, and therefore, at a fixed time t, the full emission $\psi_{C,X}(\mathbf{r}, t)$ is characterised by interference fringes. Because, like for the pump, the dominant wavevectors for signal and idler are in the x-direction, the fringes are vertical, i.e. predominantly oriented along the y-axis (see first panel of Fig. 6.5). We plot the photon component only, which is what can be measured experimentally. Note, however, that in cw experiments, emission is always integrated in time, which clearly washes away the interference fringes. The OPO phase information can instead be recovered by obtaining interference fringes with a reference beam in a Michelson configuration. In addition to the full emission, either in space or momentum, one can also evaluate the spectrum resolved in momentum $\psi_{C,X}(\mathbf{k}, \omega)$ by taking the Fourier transform in time of $\psi_{C,X}(\mathbf{k}, t)$ (in Fig. 6.4, a grid in time of 2^9 points spaced by 0.3 ps has been used). As shown in Fig. 6.4, for the chosen parameters, a signal at $\omega_s - \omega_X^0 = -1.66$ meV and an idler at $\omega_s - \omega_X^0 = 0.75$ meV appear with a sharp δ-like emission in energy, which satisfies exactly the energy matching condition (6.7), $2\omega_p = \omega_s + \omega_i$, as clearly shown by the momentum-integrated spectrum on the right panel of Fig. 6.4. In contrast, the momentum distribution is broad (because of the pump being finite size) and peaked respectively at $k_s \simeq -0.2\ \mu m^{-1}$ and $k_i \simeq 3.5\ \mu m^{-1}$, which only roughly satisfies the momentum matching condition, $2\mathbf{k}_p = \mathbf{k}_s + \mathbf{k}_i$. Note that the idler intensity is always weaker than the signal one because of the small photonic component at the idler. Further, note that, in addition to signal and idler, the spectrum also shows the appearance of satellite states ($s_2, s_3, \ldots$ and $i_2, i_3, \ldots$) all equally spaced of around 1.19 meV one from the other. As discussed at the end of Sect. 6.3.2, their presence is a consequence of the

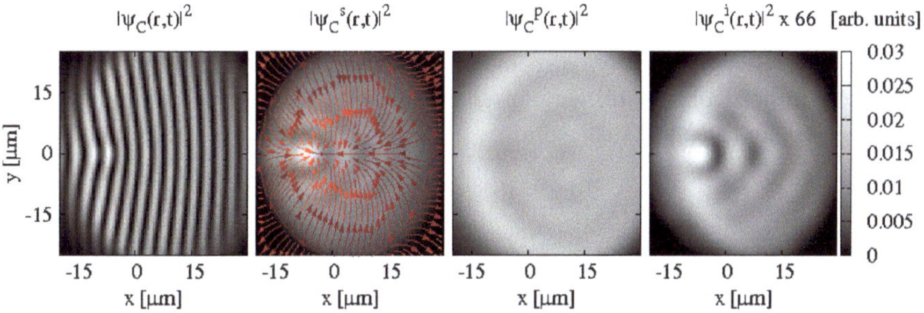

Fig. 6.5 Full emission (*first panel, with superimposed currents*) and filtered emission of signal (*second*), pump (*third*), and idler (*fourth*) states for the same parameters of Fig. 6.4. We plot the rescaled currents of the signal in the second panel by subtracting the dominant uniform flow, i.e., by plotting $\nabla\phi^s_{C,X} - \mathbf{k}_s$

secondary parametric scattering processes $2s \mapsto s_2+p$, $2i \mapsto i_2+p$, $2s_2 \mapsto s_2+s_3$, and so on, which trigger on automatically as soon as signal and idler have finite populations. The occupation of the satellite states gets gradually suppressed the further we move higher in energy above the idler and lower in energy below the signal—which is why they are usually neglected in the plane wave approximation, as discussed in the end of Sect. 6.3.2. Note also that the satellite states just described do not imply the presence of phase symmetries additional to the $U(1)$ one described in Sect. 6.3.2.1. These satellite states therefore differ from the states which one could obtain as a result of secondary instabilities, e.g. $2s \mapsto s_2' + s_2''$ with $s_2'' \neq p$ and $2i \mapsto i_2' + i_2''$ with $i_2'' \neq p$, and successive spontaneous symmetry breaking mechanism [79].

In order to analyse the OPO properties, similarly to what is done in experiments, it is also useful to filter the full emission in order to select only the emission coming from the signal, pump or idler. This can be equivalently done either filtering in momentum space in a cone around the momenta $\mathbf{k}_{\mathrm{p,s,i}}$ or filtering in energy, bringing to the same results. We indicate the filtered spatial profiles of pump, signal, and idler by $|\psi_{\mathrm{C,X}}^{\mathrm{p,s,i}}(\mathbf{r},t)|\mathrm{e}^{\mathrm{i}\phi_{\mathrm{C,X}}^{\mathrm{p,s,i}}(\mathbf{r},t)}$. The associated currents, $\nabla\phi_{\mathrm{C,X}}^{\mathrm{p,s,i}}$, are a superposition of a dominant uniform flow $\mathbf{k}_{\mathrm{p,s,i}}$ (which is subtracted from the images of the second panel of Fig. 6.5) and more complex currents (caused by the system being finite size), which move particles from gain to loss dominated regions. Note that because we select only steady state OPO solutions, the profiles of pump, signal, and idler, $|\psi_{\mathrm{C,X}}^{\mathrm{p,s,i}}(\mathbf{r},t)|$, are time independent. In addition, note that the fact that the pump is shined on the microcavity with a finite angle with respect to the normal incidence, implies that, for rotationally symmetric pump profiles, the symmetry inversion $\mathbf{r} \mapsto -\mathbf{r}$ is broken in the direction of the pump wavevector $\mathbf{k}_{\mathrm{p}}$. For example, if the pump is shined on the x-direction, $\mathbf{k}_{\mathrm{p}} = (k_{\mathrm{p}}, 0)$, as in the case of Fig. 6.5, only the symmetry $y \mapsto -y$ is left intact. Clearly, this symmetry, while allowing vortex–antivortex pairs, does not in principle permit OPO solutions carrying single vortices, which can spontaneously appear in presence of a symmetry breaking perturbation, such as disorder (next paragraph) or a noise pulse (see Sect. 6.3.5.1).

The typical changes of the signal space profile as the pump power is increased above threshold, together with the pump, signal, and idler intensities, are shown in Fig. 6.6. For these runs we fix the parameters, such as ω_{p}, $\mathbf{k}_{\mathrm{p}}$, and the pumping spot size σ_{p}, as in Fig. 6.4, but we also include a static photonic disorder potential—see (6.6). In particular, here, we consider a disorder potential with zero average, $\langle V_{\mathrm{C}}(\mathbf{r})\rangle = 0$, and a spatial distribution:

$$\langle V_{\mathrm{C}}(\mathbf{r}) V_{\mathrm{C}}(\mathbf{r}')\rangle = \sigma_d^2 \mathrm{e}^{-|\mathbf{r}-\mathbf{r}'|^2/2\ell_d^2}, \tag{6.19}$$

with a correlation length $\ell_d \simeq 20\ \mu$m and strength $\sigma_d \simeq 0.1$ meV. Below threshold, the system is in a pump-only state. By increasing the pump power f_{p}, above threshold, the OPO signal first switches on only in a small (compared with the pump spot FWHM $\sigma_{\mathrm{p}} = 70\ \mu$m) region (see inset 1). At $f_{\mathrm{p}} = 1.2 f_{\mathrm{p}}^{\mathrm{th}}$ (inset 2), the signal becomes large and quite homogeneous, though, already at $f_{\mathrm{p}} = 2.3 f_{\mathrm{p}}^{\mathrm{th}}$, the OPO

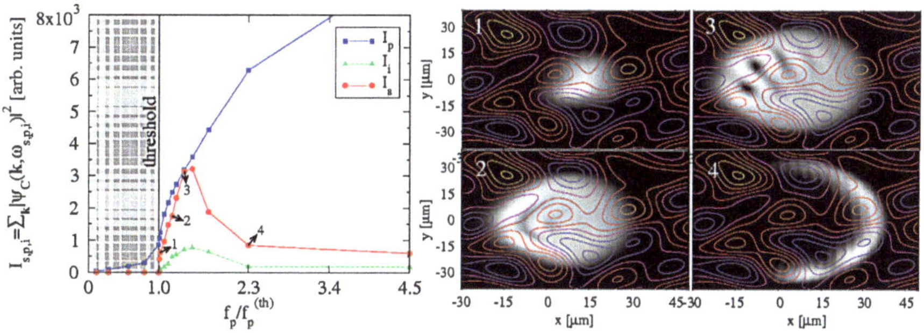

Fig. 6.6 Evolution of the signal, idler, and pump state intensities as a function of the pump intensity f_p/f_p^{th} (*left*). Space profiles of the filtered signal at different values of the pump intensity (*right*). The parameters are the same as in Fig. 6.4, with the addition of a photonic disorder potential $V_C(\mathbf{r})$, with a correlation length $\ell_d \simeq 20\ \mu$m, and strength $\sigma_d \simeq 0.1$ meV (*contour-level lines in the panels on the right*)

signal starts switching off in the middle (inset 4), and then, it slowly switches off everywhere. A similar behaviour has been found in the numerical simulations of [78] (though there a smaller pump beam of FWHM $\sigma_p \sim 20\ \mu$m has been used), as well as observed experimentally in [42].

The qualitative behaviour of the signal (as well as the idler) profiles, in particular their switching on and then off, as a function of the pump power that we have just described for a disordered sample is very similar to the case of an OPO in a homogeneous sample [3] (i.e., with no photonic disorder, $V_C(\mathbf{r}) = 0$). One of the main differences is that for homogeneous samples, the profiles are $y \mapsto -y$ symmetric, while this symmetry is explicitly broken by the photonic disorder. In addition, the fundamental difference between the homogeneous and the disordered case is that the presence of photonic disorder promotes stable vortex solutions in large pump spot OPOs at intermediate pumping strengths, $f_p \simeq 1.4 f_p^{th}$—such as the one shown in panel 3 of Fig. 6.6 which carries two vortices. Single or multiple vortex solutions are generally not allowed in the homogeneous case because of the $y \mapsto -y$ symmetry, which instead only allows pairs of vortex–antivortex solutions which are $y \mapsto -y$ symmetric. In large pump spots, such as the one of Fig. 6.6, vortex–antivortex solutions in the clean case tend to be dynamically unstable, i.e. easily destabilised by a weak noise pulse, while, as analysed later in Sect. 6.3.5, spontaneous vortex solutions in homogeneous cavities can be stabilised by a small pump spot (see Fig. 6.8) confining the vortex inside. Note finally that spontaneous vortices in disordered cavities with a large pump spots are not pinned into minima of the disorder potential, rather, as analysed in the Sect. 6.5.2.2, are the OPO steady state currents in the signal to play an essential role in the stabilisation of vortices.

[3] Note also that we find that the value of the pump threshold for OPO is not altered by the presence of a weak photonic disorder.

6.3.4 Vortex Phase and Profile

Before moving on to describe the occurrence of stable vortices in OPO and, later, the onset and dynamics of metastable vortices, let us briefly remind the definition of a quantised vortex in an *irrotational* fluid. In general, a quantised vortex with charge m is described by a wavefunction:

$$\psi(\mathbf{r}) = \psi_0(r)e^{im\varphi(\mathbf{r})}, \tag{6.20}$$

the phase of which, $m\varphi(\mathbf{r})$, linearly winds around the vortex core from 0 to $2\pi m$ (with m integer)—i.e. in cylindrical coordinates centered at the vortex core, φ is the azimuthal angle. This implies that the vortex carries a quantised angular momentum, $\hbar m$. In addition, the phase has a branch cut and therefore is not defined at the vortex core, implying the vortex wavefunction has to be zero at the vortex core. An example of an $m = -1$ vortex, with $\psi_0(r) = re^{-r^2/(2\sigma_v^2)}$, has been plotted on the left panel of Fig. 6.7. Here, the phase winds clockwise around the core, from 0 to 2π, and therefore the vortex current:

$$\nabla\varphi(\mathbf{r}) = \frac{\hat{\varphi}}{r}, \tag{6.21}$$

is constant at fixed distances from the vortex core, r, while decreases like the inverse distance (right panel of Fig. 6.7). Contrast this with the case of a *rotational* vortex in a classical fluid which rotates as a solid body with an angular velocity Ω: Now, the fluid tangential velocity is zero at the vortex core and increases linearly with the distance, i.e., $\mathbf{v}_\varphi = \Omega r\hat{\varphi}$. Quantised vortices can be detected in interference fringes (middle panel of Fig. 6.7) as fork-like dislocations, the difference in arms giving the charge $|m|$ of the vortex.

6.3.5 Stable Vortices in a Small-Sized OPO

As explained later in Sect. 6.5, spontaneous stable vortices differ from metastable vortices (described in Sect. 6.5.1): Metastable vortices can only be injected exter-

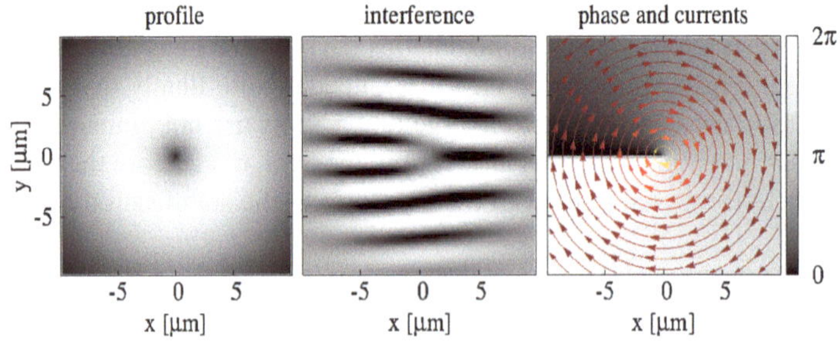

Fig. 6.7 Typical profile (*left*), phase and currents (6.21) (*right*), and interference fringes (*middle*) of an $m = -1$ vortex (6.20)

nally, for example, by an additional Laguerre-Gauss beam probe, into an otherwise stable symmetric state, and their persistence is due to the OPO superfluid properties [8, 22]. The metastable vortex is a possible but not unique stable configuration of the system. In contrast, as for non-resonantly pumped polaritons [5, 80], the appearance of spontaneous vortices is not a consequence of the polariton condensate being superfluid, but rather of the presence of currents related to the non-equilibrium nature of these condensates. This strongly differs from the case of equilibrium superfluids, the ground state of which is flow-less. Later, in Sect. 6.7 we will briefly discuss how, for polaritons non-resonantly injected into a microcavity, the presence of a confining potential can generate currents favourable to the spontaneous formation of vortices [5, 81] and vortex lattices [80].

For resonant excitation, currents arise in the OPO regime due to the simultaneous presence of pump, signal, and idler emitting at different momenta, as well as due to the fact the system is finite size (see Fig. 6.5). We have seen in Fig. 6.6 that, similarly to non-resonantly pumped polaritons, the presence of a disorder potential can lead to the spontaneous appearance of vortices. However, it is remarkable that, even in the absence of disorder or trapping potentials, the OPO system can undergo spontaneous breaking of the $y \mapsto -y$ symmetry and becomes unstable towards the formation of a quantised vortex state with charge $m = \pm 1$ if the size of the OPO is small enough [22]. This is the subject of this section. Further, as discussed in some detail later in Sect. 6.3.5.2, like for equilibrium superfluids, both stable and metastable vortices are characterised by a healing length which is determined by the parameters of the OPO system alone. Spontaneous stable vortex solutions are robust to noise (Sect. 6.3.5.1) and to any other external perturbation and thus should be experimentally observable. However, while spontaneous vortex solutions in OPO have been observed for a toroidal pump spot,[4] so far, they have not been observed in OPO with a "simply connected" pump profile, for example, either a Gaussian or a top-hat.

6.3.5.1 Dynamical Stability

As mentioned in Sect. 6.3.3, if the pump is shined on the x-direction, $\mathbf{k}_p = (k_p, 0)$, only the symmetry $y \mapsto -y$ is left intact in the system. Clearly, this symmetry, allows for OPO solutions where the signal (and therefore also the idler) have vortex–antivortex pairs, with the vortex core position at (x_c, y_c) and the antivortex core position at $(x_c, -y_c)$. However, both single and multiple vortex solutions explicitly break the $y \mapsto -y$ symmetry and cannot be accessed by the dynamics—note that two vortices located at opposite sides with respect to the x-axis break the $y \mapsto -y$ symmetry because of their currents.

In order to check the dynamical stability of the OPO states, one has to add small fluctuations to the steady state mean field solution: The existence of modes with

[4] D. Sarkar (University of Sheffield), private communication.

positive imaginary part in the excitation spectrum indicate dynamical instability towards the growth of different modes. The dynamical stability analysis for OPO described within the plane-wave approximation of Sect. 6.3.2 has been discussed in [61, 71]. Equivalently, dynamical stability can be numerically checked by introducing a weak noise. In particular, we add white noise as a quick (δ-like in time) pulse at a certain time t_0 to both modulus and phase of excitonic and photonic wavefunctions in momentum space, $|\psi_{X,C}(\mathbf{k},t)|e^{i\phi_{X,C}(\mathbf{k},t)}$:

$$|\psi_{X,C}(\mathbf{k},t_0)| \mapsto |\psi_{X,C}(\mathbf{k},t_0)| + \delta|\psi_{X,C}(\mathbf{k})|$$
$$\phi_{X,C}(\mathbf{k},t_0) \mapsto \phi_{X,C}(\mathbf{k},t_0) + \delta\phi_{X,C}(\mathbf{k}) \ .$$

Both $\delta|\psi_{X,C}(\mathbf{k})|$ and $\delta\phi_{X,C}(\mathbf{k})$ are white noise functions, with an amplitude 2π for the phase $\delta\phi_{X,C}(\mathbf{k})$, while the amplitude of the noise in the modulus $\delta|\psi_{X,C}(\mathbf{k})|$ is specified in units of the maximum of the pump intensity in momentum space.

Following this procedure, we have been able to single out symmetric OPO states, as shown in Fig. 6.8a, which are unstable towards the spontaneous formation of stable vortex solutions. After the $y \mapsto -y$ symmetry is broken by the noise pulse, we have observed a vortex with quantised charge $m = \pm 1$ ($m = \mp 1$) entering and stabilising into the OPO signal (idler)—note that parametric scattering constrains the phases of pump, signal, and idler by $2\phi_p = \phi_s + \phi_i$ (see Sect. 6.3.2.1); therefore, an $m = -1$ vortex in the signal at a given position implies an $m = 1$ antivortex in the idler at the same position and vice versa. In the case of Fig. 6.8 and the right panel of Fig. 6.9, the noise strength is 0.01 and 432 ps after the noise pulse, a vortex with $m = -1$ ($m = +1$) enters the signal (idler) and stabilises. The strength of the noise has no relevance on the final steady state, and, notably, it can be infinitesimally weak. Different noise strengths do only affect the *transient* time the system needs to accommodate the vortex and to reach the new steady configuration. We have in addition examined whether the vortex steady state is dynamically stable by applying an additional noise pulse. For weak noise, with a strength up to 0.1 (in units of the maximum of the pump intensity in momentum space), the vortex is stable and can only drift around a little before settling again into the same state. For strong noise, with strength 1 and above, the vortex gets washed away, but after a transient period, the very same state enters and stabilises again into the signal, with the possibility of flipping vorticity.[5] Different noise strengths do not affect the final steady state, but only the transient time.

As discussed later in Sect. 6.5, one can alternatively break the $y \mapsto -y$ symmetry by a pulsed vortex probe (6.24) and assess whether the stable steady state is in any

[5] When generated by a noise pulse, both stable and metastable vortices have equal probability to have either charge ± 1. Similarly, when vortices are triggered via a Laguerre-Gauss probe, their vorticity can flip during the transient period. In particular, flipping can follow the appearance of two antivortices at the edge of the signal, one recombining with the triggered vortex. Note that the vorticity flipping conserves the total orbital angular momentum, in the sense that when for the signal m flips, say, from $+1$ to -1, for the idler the opposite happens, i.e. m flips from -1 to $+1$.

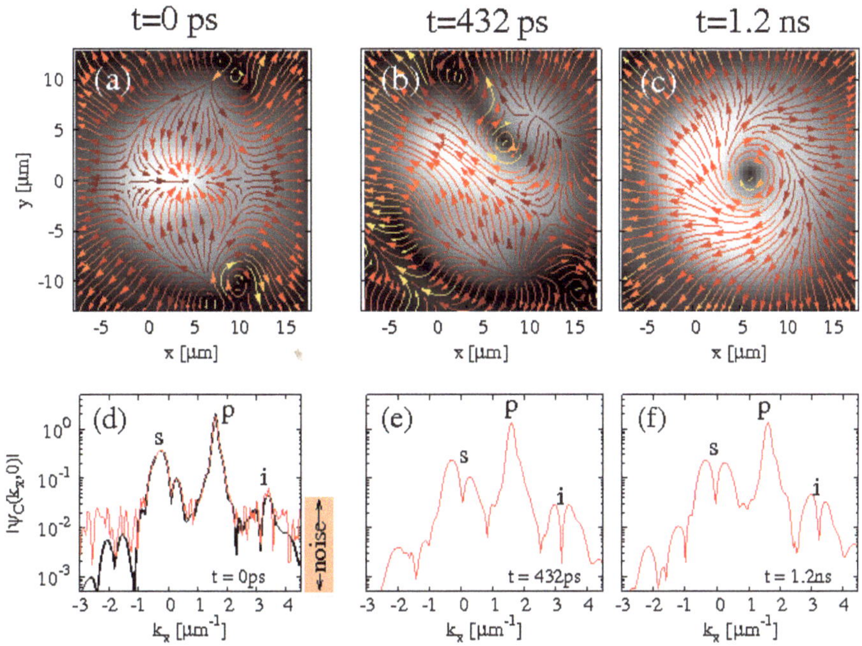

Fig. 6.8 Appearance of a spontaneous stable vortex solution in a homogeneous small-sized OPO. Filtered signal profile $|\psi_C^s(\mathbf{r},t)|$ with superimposed currents $\nabla\phi_C^s(\mathbf{r},t)$ (*upper panels* (**a**–**c**)) and full momentum emission $|\psi_C(k_x,0,t)|$ (*lower panels* (**d**, **f**), in arb. units) at three different times: $t = 0$ (**a**, **d**), $t = 432$ ps (**b**, **e**), and 1.2 ns (**c**, **f**). At $t = 0$, a pulsed weak random noise of strength 0.01 (see text) is added to the OPO steady state (in (**d**) both OPO momentum profiles without and with the added noise are shown for comparison) and at $t = 432$ ps a vortex, with $m = -1$, enters the signal and settles into a steady state. Note that, because of phase matching conditions, the presence of an $m = -1$ vortex in a signal implies the presence of an $m = 1$ antivortex in the idler. A vortex (antivortex) in the signal (idler) space emission appears also as a dip in momentum space at the signal (idler) momentum (**e**, **f**). Parameters used: smoothed top-hat pump with FWHM $\sigma_p = 35\mu$m, pump strength $f_p = 1.12 f_p^{(th)}$, $k_p = 1.6\,\mu$m^{-1} in the x-direction, $\omega_p - \omega_X^0 = -0.44$ meV, zero detuning $\delta = 0$, and $\kappa_X = \kappa_C = 0.22$ meV. Adapted from [22]

way dependent on the external perturbation. The homogeneous OPO states which are unstable towards the spontaneous formation of stable vortices following a white noise pulse, exhibit the same instability following a vortex Laguerre-Gauss (LG) probe pulse (see the left panel of Fig. 6.9). The steady state vortex is independent on both the probe intensity f_{pb} and size σ_{pb}; however, the weaker the probe the longer the vortex takes to stabilise, between 30 and 400 ps for our system parameters. As shown in Fig. 6.9, the stable vortex following the LG probe is exactly the same as the one triggered by a weak white noise, indicating that the probe acts only as a symmetry breaking perturbation.

Summarising, one can find OPO conditions where the $y \mapsto -y$ symmetric solution is dynamically unstable, and any symmetry breaking perturbation allows the signal and idler to relax into a stable steady state carrying a vortex with

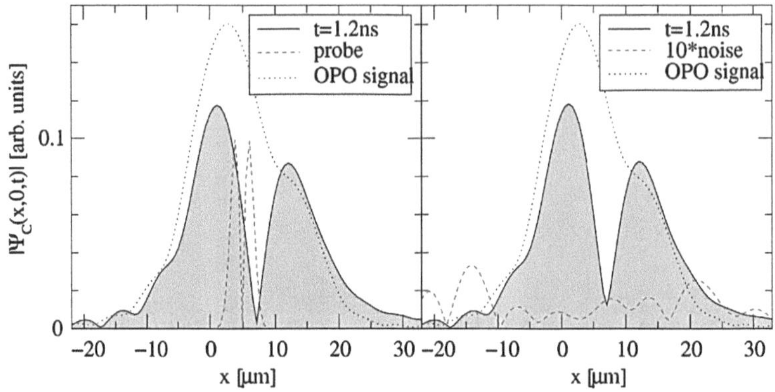

Fig. 6.9 Steady state filtered signal profile (*dotted line*) $\psi_C^s(x,0,t)$ for $y \simeq 0$ before the arrival of either a Laguerre-Gauss vortex probe (6.24) with $\sigma_{pb} \simeq 1\mu$m (*left panel*, *red dashed line*) or a noise pulse of strength 0.01 (*right panel*, *red dashed line*)—same OPO conditions as Fig. 6.8. After the arrival of any perturbation breaking the $y \mapsto -y$ symmetry, the same vortex with charge $m = \pm 1$ (*solid shaded curve*) stabilises into the signal. From [22]

charge ± 1. For homogeneous cavities, i.e., in absence of any disorder or confining potential, we found that this requires either a small Gaussian or small top-hat like pump spot which can confine the vortex inside or a doughnut-shape pump spot. Instability of the uniform state to spontaneous pattern (e.g. vortex) formation is a typical feature of systems driven away from equilibrium [79]. Similarly, we find conditions for which the uniform OPO solution is unstable to spontaneous formation of a quantised vortex. In alternative, a disorder potential breaks the symmetry explicitly and allows the pinning of stable vortex solutions in OPO, (like in the case cosindered in Fig. 6.6) which is less surprising.

6.3.5.2 Healing Length

In contrast to their classical counterpart, quantised vortices with the same angular momentum $|m|$ are all identical, with a size (or healing length) determined by the system non-linear properties [76]. In the case of a superfluid in equilibrium with a typical interaction energy gn (n is the average density) and mass m, the healing length, $\xi = 1/\sqrt{2mgn}$, is the typical distance over which the condensate wave function recovers its "bulk" value around a perturbation. In particular, for an $|m| = 1$ vortex (6.20), ξ is the typical size of the vortex.

Similarly, in OPO, one case show that, like in equilibrium superfluids, both stable (see Sect. 6.3.5) and metastable (see Sect. 6.5) vortices are characterised by a healing length which is determined by the parameters of the OPO system alone. In particular, shape and size of the metastable vortices described in Sect. 6.5 are independent on the external probe. In the case of vortices in OPO, an approximate analytical expression for the vortex healing length can be derived for homogeneous pumping [9, 22], assuming that only signal and idler can carry angular momentum with opposite sign, $\pm m$, $\psi^{\rm s,i}(\mathbf{r}) = \sqrt{n_{\rm s,i}} e^{i\mathbf{k}_{\rm s,i}\cdot\mathbf{r}} e^{\pm im\varphi} \psi^{\rm s,i}(r)$, while the pump remains in a plane-wave state, $\psi^{p}(\mathbf{r}) = \sqrt{n_p} e^{i\mathbf{k}_{\rm p}\cdot\mathbf{r}}$, as also supported by our numerical analysis. For pump powers close to the OPO threshold, it can be shown [9, 22] that signal and idler steady state spatial profiles are locked together and satisfy the following complex GP equation:

$$\left[-\frac{1}{2m_{\rm C}}\left(\frac{d^2}{dr^2} + \frac{1}{r}\frac{d}{dr} - \frac{m^2}{r^2}\right) + \alpha\left(\psi^{s2} - 1\right)\right]\psi^{\rm s} = 0,$$

where $|\alpha| \simeq g_{\rm X}\sqrt{n_{\rm s} n_{\rm i}}$.[6] This equation describes a vortex profile [76] with a healing length given by:

$$\xi = \left(2m_{\rm C} g_{\rm X}\sqrt{n_{\rm s} n_{\rm i}}\right)^{-1/2} . \tag{6.22}$$

This expression is similar to the one of an equilibrium superfluid, with the condensate density replaced by the geometric average of signal and idler densities. Further above threshold, one can show that signal and idler profiles are no longer locked together, and that they start to develop different radii. In both the simulations of Figs. 6.9 and 6.6, we find $\xi \simeq 4\mu$m compatible with the estimate (6.22).

In [9], vortices in OPO have been created in a controlled manner by adding a weak continuous probe in resonance with the signal. Even if the phase freedom of the OPO system is explicitly broken in this configuration by the vortex cw probe, because the ratio of the probe to signal power density is low, the size of the vortex has been demonstrated to be determined by the OPO non-linear properties only rather than by the imprinting probe. In particular, a systematic study of the decrease of the vortex core radius with increasing pump power above threshold has allowed to confirm the behaviour described by the (6.22).

6.4 Triggered Optical Parametric Oscillator Regime

Before moving on into the description of metastable vortices in OPO, i.e. vortices which are transferred by a pulsed vortex probe to the OPO signal and idler, and their relation to superfluidity (Sect. 6.5), we summarise here first the effect of an additional pulsed probe on OPO in general terms. As described previously, in the OPO regime, polaritons are continuously injected into the pump state, and undergo coherent stimulated scattering into the signal and idler states. The OPO is a steady state regime, where the filtered profiles of signal, idler, and pump, $|\psi^{\rm p,s,i}_{\rm C,X}(\mathbf{r},t)|$, are time independent. This also is reflected in the typical flat dispersion around pump,

[6]This is in reality an oversymplified version of the full equation satisfied by $\psi^{s,i}(\mathbf{r})$, because one can show that there are small terms breaking the rotational symmetry, implying the vortex is not a pure angular momentum state, though it still has a definite winding number [D. Whittaker, private communication]. This, however, does not affect the expression of the healing length.

signal, and idler which can be observed in the OPO spectra (see Fig. 6.4). The *group velocity* of pump, signal, and idler, defined as the derivative of the energy dispersion at $\mathbf{k}_{\mathrm{p,s,i}}$, is therefore zero. This, however, does not mean that there is no flow of polaritons, which instead is described by the *phase velocity* or current, $\nabla\phi_{\mathrm{C,X}}^{\mathrm{p,s,i}}$ (see footnote 2 on page 186), with a dominant uniform flow given approximatively by $\mathbf{k}_{\mathrm{p,s,i}}$.

In resonantly pumped polaritons, in order to initiate a travelling wave packet characterised by a finite group velocity, one needs to use an additional pulsed laser beam on top of the cw one. The description of the system is therefore still in terms of the (6.9), with a total pump term given by the sum of the cw laser (6.10) and a probe beam $F_{\mathrm{pb}}(\mathbf{r},t)$:

$$F(\mathbf{r},t) = F_{\mathrm{p}}(\mathbf{r},t) + F_{\mathrm{pb}}(\mathbf{r},t) \ . \tag{6.23}$$

For the moment we will consider the generic case of a pulsed probe with a Gaussian space profile, shined at a momentum and energy $\{\mathbf{k}_{\mathrm{pb}},\omega_{\mathrm{pb}}\}$ [7]:

$$F_{\mathrm{pb}}(\mathbf{r},t) = f_{\mathrm{pb}} e^{-|\mathbf{r}-\mathbf{r}_{\mathrm{pb}}|^2/(2\sigma_{\mathrm{pb}}^2)} e^{i(\mathbf{k}_{\mathrm{pb}}\cdot\mathbf{r}-\omega_{\mathrm{pb}}t)} e^{-(t-t_{\mathrm{pb}})^2/(2\sigma_t^2)} \ . \tag{6.24}$$

A pulse duration of 3 ps (defined as the FWHM in time of $F_{\mathrm{pb}}(\mathbf{r},t)$) corresponds to $\sigma_t = 1.3$ ps. The idea, first introduced by [20], is that the pulsed probe triggers parametric scattering [8] between the probe state at momentum and energy $\{\mathbf{k}_{\mathrm{pb}},\omega_{\mathrm{pb}}\}$ and a *conjugate* state at $\{\mathbf{k}_{\mathrm{c}} = 2\mathbf{k}_{\mathrm{p}} - \mathbf{k}_{\mathrm{pb}}, \omega_{\mathrm{c}} = 2\omega_{\mathrm{p}} - \omega_{\mathrm{pb}}\}$—because one can either have $\mathbf{k}_{\mathrm{pb}} > \mathbf{k}_{\mathrm{p}}$ or $\mathbf{k}_{\mathrm{pb}} < \mathbf{k}_{\mathrm{p}}$, we use the state labels "probe" and "conjugate," rather than "signal" and "idler"; by doing so, one also does not confuse the states generated by the OPO (signal and idler) with the additional ones generated by the probe (probe and conjugate). Both probe and conjugate states are travelling decaying states which can evolve freely from the laser probe constraints once the pulse switches off. Such states are referred to as TOPO states. Note that a TOPO can be triggered in two regimes: either (1) in a regime where the cw laser drives the system above threshold for OPO, in which case the probe and conjugate states are the extra population states on top of the steady state OPO signal and idler states, or (2) when no OPO is present, i.e. for the cw pump strength below threshold. For simplicity, the numerical analysis discussed below in Sect. 6.4.1 is conducted in the regime (2), but we have checked that the qualitative results also hold in the regime (1)—where, now, the steady state OPO population needs to be subtracted so that one studies the properties of the population triggered by the probe only.

---

[7] Note that, differently from the cw laser beam, the energy distribution spectrum of which is essentially δ-like, a pulsed beam has an intrinsic width in energy, proportional to the inverse pulse duration, σ_t^{-1}.

[8] If the cw pump drives the system into the OPO regime, then the parametric scattering triggered by the pulsed probe will emerge in addition to the one related to OPO. However, as discussed later, the TOPO regime can be reached also in absence (below threshold) of the OPO.

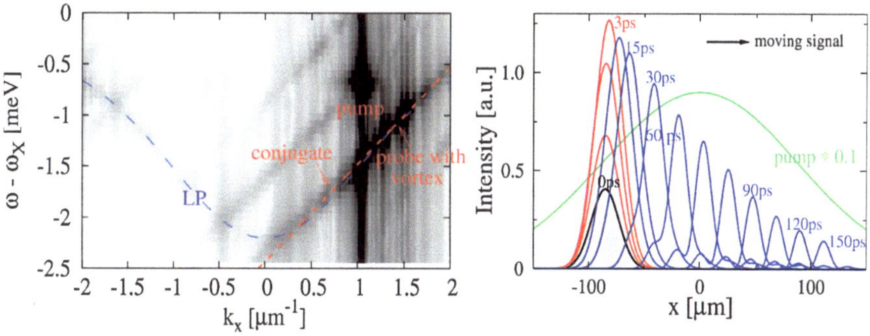

Fig. 6.10 Spectrum (*left*) and spatial profiles of pump and filtered signal $|\psi_C^s(x,0,t)|$ (*right*) for the TOPO regime. A short, $\sigma_t = 1$ ps, $m = 2$ Laguerre-Gauss (6.25) (*left*) or Gaussian $m = 0$ (6.24) (*right*) probe shined at $\mathbf{k}_{pb} = (1.4, 0)\ \mu\text{m}^{-1}$ triggers the propagating probe and conjugate states, which lock to the same group velocity (for these simulations we fix $\kappa_X = 0$ and $\kappa_C = 0.02$ meV). Adapted from [23]

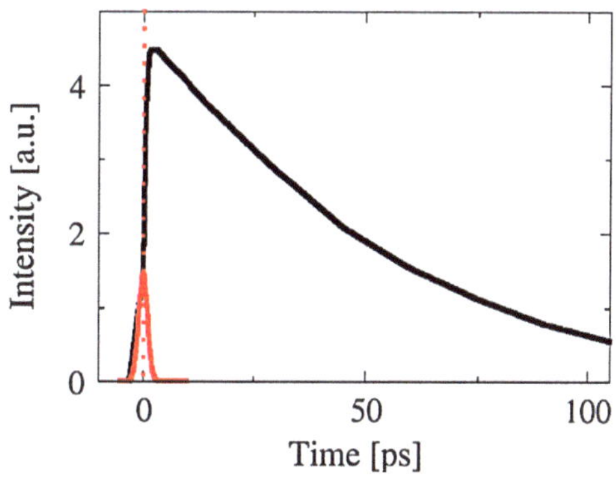

Fig. 6.11 TOPO signal intensity (*black line*) and the intensity of an external Gaussian probe (*red line*) as a function of time. Parameters are the same ones of the right panel in Fig. 6.10. The TOPO signal gets initially strongly amplified, then decays slowly, and finally exponentially

6.4.1 *Theoretical Description of the TOPO*

In order to analyse the dynamical evolution of a TOPO wave packet, we study numerically the time-dependent solutions of the equations (6.9), with a total pump given by (6.23) and (6.24). The probe triggers parametric scattering between a probe state and a conjugate state. In the majority of cases, as discussed in [23], the parametric scattering is too weak to induce any significant amplification, and an exponential decay of both probe and conjugate populations is observed immediately after the probe $F_{pb}(\mathbf{r}, t)$ switches off. Here, the spectrum shows a strong emission from the pump state and a weak emission from the LP states mainly at momenta $\mathbf{k}_{pb}$ and $\mathbf{k}_c$.

However, we have found conditions under which both signal and conjugate states get initially strongly amplified by the parametric scattering from the pump, then decay slowly and, only at later times, decay exponentially (see Figs. 6.10

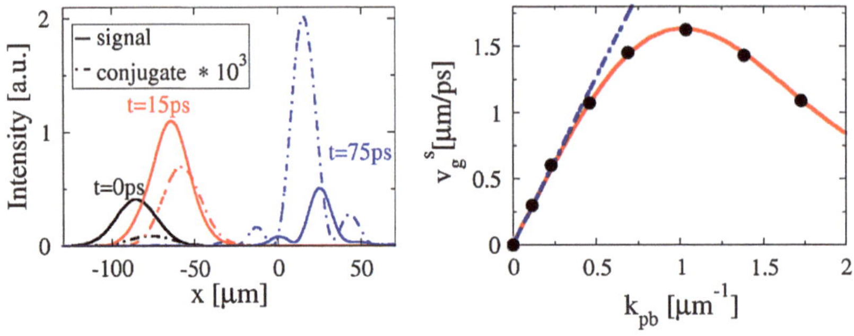

Fig. 6.12 *Left panel*: probe (*solid lines*) and conjugate (*dashed lines*) density profiles at different times after the arrival of the probe (same parameters as in Fig. 6.10). *Right panel*: Group velocity, v_g^s, of the propagating probe state as a function of the probe momentum k_{pb}. The black dots are determined from simulations, whereas the solid (*red*) line is the derivative of the LP dispersion evaluated at k_{pb}, $v_{k_{\text{pb}}}^{\text{LP}}$. The blue dashed line is a guide for eye to indicate where the LP dispersion deviates from the quadratic. Adapted from [23]

and 6.11)—we refer to this as the "proper" TOPO regime. A similar behaviour has been also observed in experiments.[9] Now, the spectrum is observed to be linear, $\omega = \mathbf{v}_g \cdot \mathbf{k}$ (see Fig. 6.10). A linear spectrum can be explained by the fact that, in order to have efficient parametric scattering, probe and conjugate state must have a large spatial overlap and therefore similar group velocities. Thus, signal and conjugate group velocities need to lock, which is only possible if the dispersion becomes linear—a similar result has been also found in 1D simulations (see Chap. 1) as well as in experiments [20]. The group velocity is defined as the derivative of the energy dispersion with respect to the momentum. However, we can also measure it from the probe and the conjugate density variations in time, i.e., as $\mathbf{v}_g^{\text{pb,c}} = d\mathbf{r}_m^{\text{pb,c}}/dt$, where $\mathbf{r}_m^{\text{pb,c}}$ is the maximum of either the probe or conjugate spatial profile, which we use as a reference position. By analysing the change in time of the spatial profile of the TOPO probe state $|\psi_{\text{C,X}}^{\text{pb}}(\mathbf{r},t)|$, it is possible to show [23] that its group velocity v_{g}^{pb} is given exactly by the derivative of the lower polariton (LP) dispersion evaluated at $\mathbf{k}_{\text{pb}}$, i.e., for zero detuning and low densities by $v_{k_{pb}}^{LP} \equiv k_{pb}/(2m_C) - k_{pb}^3/(2m_C\sqrt{k_{pb}^4 + 4m_C^2\Omega_R^2})$ (see Fig. 6.12). This behaviour is consistent with the form of the spectrum shown in Fig. 6.10.[10] Further, we have been able to determine [23] that the TOPO linear dispersion is tangential to the LP branch at k_{pb}; thus, its slope is given in this case also by $v_{k_{\text{pb}}}^{\text{LP}}$.

[9] See, for example, Fig. 3 of [82], where the intensity maximum of the extra population is reached within 4 ps after the maximum of the pulsed probe, is followed by a slow decay.

[10] In the regime where the probe generates only a weak parametric scattering, aside the strong emission from the pump state, the dispersion is simply that of the LP and thus is not surprising that the signal propagates with a group velocity given by $v_{k_{\text{pb}}}^{\text{LP}}$. Remember that here, the cw pump is below the threshold for OPO.

From the PL spectrum, we can also deduce the nature of the wave packet propagation. For systems characterised by a linear dispersion, like in the TOPO regime, one expects a soliton-like behaviour, where probe and conjugate states propagate without changing neither their shape nor intensity. For quadratic dispersion, a Gaussian wave packet moves at a constant velocity $\mathbf{v}^{LP}_{k_{pb}} = \mathbf{k}_{pb}/(2m_C)$, and it preserves its overall shape in time, but its width grows $\left(\text{FWHM} = (\sigma^2_{pb} + (t/(2m_C\sigma_{pb}))^2)^{1/2}\right)$ [76]. Note, however, that, due to the finite polariton lifetime, the total density decays exponentially, with a rate given by $(\kappa_C + \kappa_X)/2$ at zero detuning. Finally, for non-quadratic dispersion, propagation becomes complex: The wave packet gets distorted and there are beatings in the spatial profiles. In general, due to the dynamical nature of the TOPO state, the system evolves between these different scenarios. In particular, only in the strong amplification regime the spectrum is linear, while it evolves back to the LP one at longer times. However, for a one-dimensional version of the equations (6.9), and for uniform, infinitely extended in space, pumping spots, non-decaying, soliton solutions have been recently found [83]. This has been also generalised to a two-dimensional infinite systems [84], where, for some narrow range of pumping strengths, a soliton-like behaviour has been predicted for $k_{\mathrm{pb}} = 0$ and $k_{\mathrm{pb}} = k_{\mathrm{p}}$. However, to date, a non-decaying wave packet propagation has not been found in experiments.

Finally, we would like to note that, the typical behaviour of probe and conjugate states (left panel of Fig. 6.12) is analogous to the one discussed in four-wave-mixing experiments [85, 86]: When the probe arrives, and shortly after that, the conjugate propagates faster then the probe, before getting locked to it with a small spatial shift of their maximum intensities. At later times, when the density drops and the parametric process becomes inefficient, the two wave packets start unlocking—the conjugate slows down with respect to the signal if $k_{\mathrm{c}} < k_{\mathrm{pb}}$ as in Fig. 6.12, or it moves faster when $k_{\mathrm{c}} > k_{\mathrm{pb}}$.

6.4.2 *Experiments*

The TOPO regime has been recently studied in experiments in [20,45] (for a review, see [87]). As previously described, the additional pulsed probe has been used to create a travelling, long-living, coherent polaritons signal, continuously fed by the OPO. A large increase of the signal lifetime has been observed for a pump intensity approaching and exceeding the OPO threshold [45]. This observation can be explained in terms of a critical slowing down of the dynamics following appearance of a soft Goldstone mode in the spectrum close to threshold. It is also consistent with the nature of wave packet propagation in systems with linear dispersion. This has been used to interpret subsequent experiments, where the linearisation of the dispersion leads to the suppression of weak scattering and therefore to a polariton motion without any dissipation [20, 87] (see Chap. 1). Due to the finite size of the excitation spot, the travelling TOPO signal lives only as long as it reaches the edge of the excitation spot. However, as discussed in detail in [87], in order to asses the sustainability in time of the TOPO process, both pump and probe beams can be chosen so that the probe state forms at $k_{pb} \sim 0$. In such a case, the polaritons in the probe state are not travelling, and it is thus possible to measure the lifetime of the TOPO wave packet, which is of the order of a nanosecond. The decay of the TOPO population in time, as well as the finite lifetime of TOPO pulses, indicate that the soliton behaviour predicted in [84] is not the explanation of the current experiments. However, the linearisation of the system's dispersion due to the parametric process, as well as the appearance of the Goldstone mode provide a sufficient explanation of dissipationless propagation in free space, as well as frictionless flow against an obstacle, during the part of the dynamics when parametric processes are strong and the spectrum linear.

6.5 Triggered Metastable Vortices as a Diagnostic of the OPO Superfluid Properties

OPO condensates, as well as polariton condensates pumped incoherently, share with weakly interacting Bose–Einstein condensates at equilibrium phenomena such as the spontaneous breaking of the phase symmetry and the appearance of a Goldstone mode (see Sect. 6.3.2.1). However, being intrinsically non-equilibrium, all polaritonic systems need continuous pumping to balance the fast decay and maintain a steady state regime. In strong contrast with equilibrium superfluids, the ground state of which is flow-less, pump and decay lead to currents that carry polaritons from gain- to loss-dominated regions. This can lead to the spontaneous formation of vortices: The presence of currents in polariton condensates can lead to the spontaneous appearance of vortices without invoking any superfluid properties. This is true for incoherently pumped polaritons in presence of a confining potential [5, 80, 81], as well as for polaritons in the OPO regime, with the difference that here, even in the absence of disorder or a trapping potential, the system becomes unstable towards the formation of a quantised vortex state with charge $m = \pm 1$ (see Sect. 6.3.5). In addition, the hydrodynamic nucleation of quantised vortices can appear as a consequence of the collisions of a moving polariton fluid with an obstacle, as it will be briefly discussed in Sect. 6.7. Therefore, in general, for polaritonic systems, one has to apply some care when using the appearance of vortices as a diagnostic for the superfluid properties of such a non-equilibrium system.

In the case of equilibrium superfluids, the rotation of a condensate is accompanied, above a critical angular velocity [75, 76], by the creation of quantised vortices. Here, vortices are stable as far as the system is kept rotating and become unstable when the imposed rotation is halted [77]. However, persistent flow can be observed when a BEC is confined into a toroidal trap and the quantised rotation is initiated by a pulsed Laguerre-Gauss beam [88–90]. The toroidal trap is essential to allow the vortex stability because of the energy cost of the vortex core to move through the high density region from the centre of the torus where the density is zero. The very same idea of questioning the persistency of flow in a BEC via a pulsed Laguerre-Gauss beam as a diagnostic for superfluidity, can be applied to polaritons.[11] As recently proposed for non-resonantly pumped polariton condensates in [21], this definition of superfluidity as metastable flow is equally meaningful in non-equilibrium systems as in equilibrium ones. However, as we will see, the important difference is that, in the OPO regime, flow persistency is possible even in a simply connected geometry, i.e., without the need of a toroidal trap which pins the vortex.

A pulsed Laguerre-Gauss (LG) probe beam carrying a vortex of charge m can be described by:

$$F_{\rm pb}(\mathbf{r},t) = f_{\rm pb}|\mathbf{r}-\mathbf{r}_{\rm pb}|^{|m|} e^{\mathrm{i}m\varphi} e^{-|\mathbf{r}-\mathbf{r}_{\rm pb}|^2/(2\sigma_{\rm pb}^2)} e^{\mathrm{i}(\mathbf{k}_{\rm pb}\cdot\mathbf{r}-\omega_{\rm pb}t)} e^{-(t-t_{\rm pb})^2/(2\sigma_t^2)}, \quad (6.25)$$

where $\{\mathbf{k}_{\rm pb}, \omega_{\rm pb}\}$ can be tuned resonantly with either the OPO signal or idler. As discussed in the next section, by using a pulsed LG beam (6.25), vorticity has been shown to persist not only in absence of the rotating drive, but also longer than the gain induced by the probe, and therefore to be transferred to the OPO signal, demonstrating metastability of quantised vortices and persistence of currents in OPO. Experiments and theory will be discussed in the next Sect. 6.5.1.

6.5.1 Theory and Experiments

In the case of metastable vortex solutions, the symmetric vortex-less OPO steady state is dynamically stable but, because of its superfluid properties, can support persistent metastable currents injected externally. From a theoretical point of view, metastable solutions can be equally induced by either a vortex probe pulse (6.25) or a noise pulse (of the same kind of the one considered in Sec. 6.3.5.1). However, differently from the case of stable vortices, metastable solutions require a threshold in the perturbation breaking the system $y \mapsto -y$ symmetry. For the simulations of [8], as we were interested mainly into the transfer of angular momentum from the probe to the OPO signal and idler, we have been considering conditions where the parametric scattering induced by the probe is too weak to induce any significant long-lasting amplification, and the gain introduced by the probe on top of the OPO disappears quite quickly. We have found conditions where the vortex is transferred

[11] Note, however, that even if in the atomic and polaritonic cases the same Laguerre-Gauss laser field is used, the mechanism of spinning the BEC atoms is different from the one which rotates polaritons.

from the probe to the signal [12] (and antivortex in the idler) immediately [13] when the probe is shined. The transfer is followed by a transient time during which the imprinted vortex drifts around inside the signal and in certain cases settles into a metastable solution. Similarly to what happens to stable vortex solutions, we have found that the spatial position of the metastable steady state vortices is close to the position where the OPO signal has the currents pointing inwards (see second panel of Fig. 6.5). The influence of currents on the formation of vortices is discussed further in Sect. 6.5.2. Such metastable solutions do not always exist: If the probe is positioned well inside a wide OPO signal, as the creation of a vortex is accompanied by the creation of an antivortex (see Sect. 6.5.2), often, the vortex–antivortex pair quickly recombines; in other cases, during the transient period, the excited vortex can spiral out of the signal. Finally note that, as discussed in Sect. 6.3.5.2, the shape and the size of metastable vortices are independent on the external probe but are only determined by the parameters of the OPO.

In the experiment of [8] also shown in Fig. 6.13, a vortex is excited by a probe smaller than that of the signal to allow free motion of the vortex within the condensate. Vortices are detected, and their evolution in time followed by a streak camera, in interference images, generated by making interfere the OPO signal with a constant phase reference beam in a Michelson interferometer (second row of Fig. 6.13). As single shot measurements would give a too low signal to noise ratio, every picture is the result of an average over many pulsed experiments taken always for the same OPO conditions. The probe triggers a TOPO response, creating a strong gain and an extra decaying population on top of the OPO signal (TOPO). In experiments, different regimes have been investigated. In particular, it has been possible to establish that, only under high pump power and at specific points in the sample, the vorticity was transferred from the TOPO to the OPO signal, generating a metastable vortex solution. This not only demonstrates that the OPO polariton condensate can show unperturbed rotation but also that a vortex can be another metastable solution of the final steady state, demonstrating therefore the superfluid behaviour in the non-equilibrium polariton OPO system. After the vortex is imprinted into the OPO signal, it has been possible to observe the vortex core slowly drifting, changing in shape and moving with different velocities. Note that a minimum probe power is required for the vortex to be imprinted to the OPO. However, once the vortex is imprinted, the probe power does not change significantly the depth of the vortex in the steady state.

---

[12] We checked that $m = \pm 1$ ($m = \mp 1$) vortex solutions can appear only into the OPO signal (idler). A vortex probe pulse of any charge m injected resonantly to the pump momentum, and energy gets immediately transferred to an $m = \pm 1$ ($m = \mp 1$) vortex in the signal (idler), leaving the pump vortex-less.

[13] Later, in Sect. 6.6.1, in connection to the stability of multiply quantised vortices, we also describe vortices in the TOPO regime, where we follow the vortex dynamics not of the OPO like here, but of the extra population only.

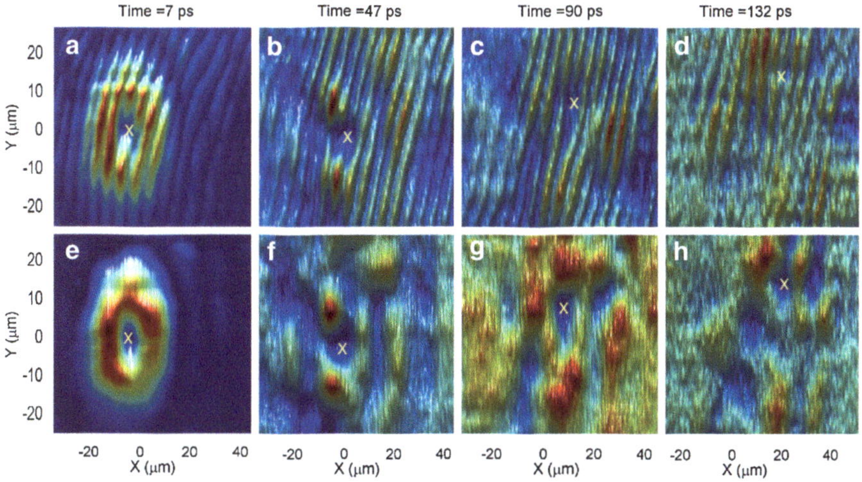

Fig. 6.13 Time evolution of the polariton signal following the arrival of a LG pulsed beam carrying a vortex of $m = 1$. The first row shows the interference images obtained by overlapping the vortex with a small expanded region of the same image far from the vortex core, where the phase is constant, while the second row shows the space profiles of the signal. The sequence demonstrates that the vortex remains steady as a persisting metastable state for times much longer than the extra population created by the probe pulse and eventually gets imprinted in the steady state of the OPO signal. This is revealed by the strong contrast of the fork in the interference images for as long as the core remains within the condensate area. From [8]

6.5.2 *Onset and Dynamics of Vortex–Antivortex Pairs*

There is an aspect that we have been neglecting in the discussion of the previous section on the occurrence of metastable vortex solutions in OPO triggered by an external LG probe. If the extension of the probe carrying a vortex with charge $m = +1$ is smaller than the size of the vortex-free OPO signal, continuity of the polariton wavefunction requires that necessarily an antivortex with charge $m = -1$ has to form at the edge of the probe (see Fig. 6.14). Indeed, "unintended" antivortices have been shown to appear in the signal at the edge of the imprinting vortex probe, and we have explained in [91], both theoretically and via experiments, the origin of the deterministic behaviour of the antivortex onset and dynamics, i.e. where antivortices are more likely to appear in terms of the currents of the imprinting probe and the ones of the underlying OPO.

6.5.2.1 Random Phase Between Pump and Probe

As mentioned earlier in Sect. 6.5.1, single shot measurements would give a too low signal to noise ratio; therefore, an average is performed over many pulsed

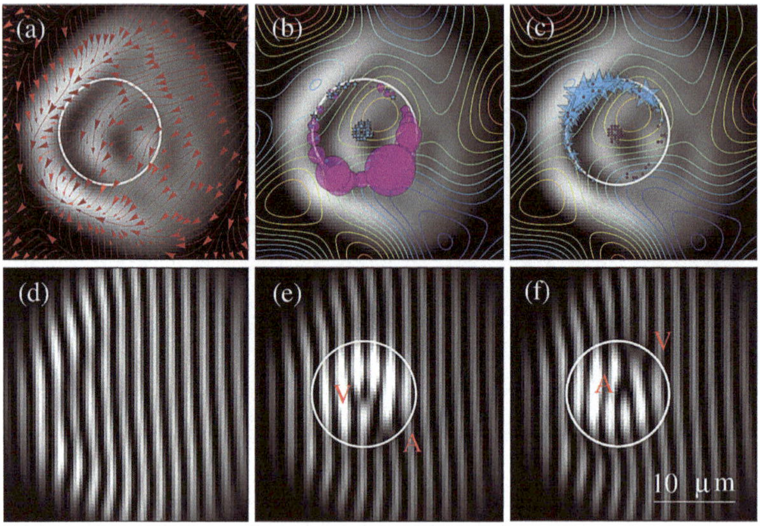

Fig. 6.14 Profile and currents of the steady state OPO signal before the arrival of the probe (**a**) and associated interference fringes (**d**)—parameters for OPO are exactly the same as the ones of the inset 2 in Fig. 6.6 ($f_{\rm p} = 1.2 f_{\rm p}^{\rm th}$). Location of antivortices (*dots (b)*) and vortices (*stars (c)*) at the arrival of a vortex (*stars (b)*) or an antivortex (*dots (c)*) probe, for 1,000 realisations of the random relative phase between pump and probe, $\Phi_{\rm rdm}$. The size of dots in (**b**) (*stars in (c)*) is proportional to the number of times the antivortices (vortices) appear in that location. Panel (**e**) (**f**) shows single shot interference fringes relative to the plot in (**b**) (**c**). Contour-level lines in (**b**) and (**c**) represent the photonic disorder $V_{\rm C}(\mathbf{r})$. The white circle represents the edge of the probe. From [91]

experiments taken always for the same OPO conditions. What differs at each probe arrival is the random relative phase $\Phi_{\rm rdm}$ between pump and probe:

$$F(\mathbf{r},t) = F_{\rm p}(\mathbf{r},t) + F_{\rm pb}(\mathbf{r},t)e^{i\Phi_{\rm rdm}}, \qquad (6.26)$$

with $\Phi_{\rm rdm}$ uniformly distributed between 0 and 2π. We simulate the dynamics of the vortex-free signal OPO (same conditions of Fig. 6.6 at $f_{\rm p} = 1.2 f_{\rm p}^{\rm th}$) following the arrival of a vortex probe (6.24) for 1000 realisations of $\Phi_{\rm rdm}$ and then average the complex wavefunctions over such realisations at fixed time and space, $\langle|\psi_{\rm C}^{s}(\mathbf{r},t)|e^{i\phi_{\rm C}^{s}(\mathbf{r},t)}\rangle_{\Phi_{\rm rdm}}$.

The steady state currents of the OPO signal before the arrival of the probe have a dominant component pointing leftwards and an equilibrium position where all currents point inwards (bottom left part of the panel (a) in Fig. 6.14). In single-shot simulations of Fig. 6.14d,f (one realisation of the phase $\Phi_{\rm rdm}$), we find that if the probe is positioned well inside the OPO signal, then the imprinting of a vortex $m = +1$ (antivortex $m = -1$) forces the system to generate, at the same time, an antivortex $m = -1$ (vortex $m = +1$) at the edge of the probe. This is a consequence of the continuity of the polariton wavefunctions: If the signal OPO phase is homogeneous and vortex-free before the arrival of the probe, then imposing a topological defect, i.e. a branch cut, on the signal phase at the probe core, requires the branch cut to terminate where the phase is not imposed by the probe any longer and has to continuously connect to the freely chosen OPO signal phase, i.e. at the edge of the probe. As repeatedly mentioned in this review, OPO parametric scattering processes constrain the sum of signal and idler phases to the phase of the laser pump by $2\phi_p = \phi_s + \phi_i$. Thus, at the same positions where the V–AV pair appears in the signal, an AV–V pair appears in the idler so that locally the phase constraint described above is satisfied. This agrees with the experiments in [9], though there only a single V (AV) in the signal (idler) could be detected, because the signal size was comparable to the probe one.

Different relative phases Φ_{rdm} cause the antivortex (vortex) to appear in different locations around the vortex (antivortex) probe. However, over 1000 realisations of the random phase uniformly distributed between 0 and 2π, we observe that the antivortices (vortices) are more likely to appear on positions where the current of the steady state OPO signal before the probe arrival and the probe current are opposite. For example, for the $m = +1$ ($m = -1$) probe of Fig. 6.14c (Fig. 6.14e), the current constantly winds anticlockwise (clockwise); therefore, comparing with the signal current of Fig. 6.14a, the two are antiparallel in the bottom right (top left) region on the probe edge, region where is very likely that an antivortex (vortex) is formed. Note also that the onset of antivortices (vortices) privileges regions where the steady OPO signal has a minimal intensity. This agrees remarkably well with what it has been recently measured experimentally in [91].

6.5.2.2 Multi-Shot Averaged Dynamics

Crucially, via numerical simulations, we elucidate the reason why an experimental average over many shots allows detecting a vortex by direct visualisation in density and phase profiles. Recently, it has been suggested by stochastic simulations [21] that vortices in non-resonantly pumped polariton condensates undergo a random motion which will hinder their direct detection, unless they are close to be pinned by the stationary disorder potential and thus follow a deterministic trajectory [7]. In the case considered here of a superfluid generated by OPO, we can instead explain a deterministic dynamics of the V–AV pair in terms of the OPO steady state currents, which determine a unique trajectory for the pair, allowing their observation in multi-shot measurements.

By averaging the 1000 images obtained at the probe arrival, for example, in Fig. 6.14c, neither the imprinted vortex nor the antivortex can be detected: Both phase singularities are washed away by averaging the differently positioned branch cuts. However, the steady state signal currents push the V and AV, initially positioned in different locations, towards the same equilibrium position where all currents point inwards. Thus, exactly at the time where the probe is shined, on average, there is no V–AV pair; after $\sim$10 ps, both V and AV appear and last $\sim$75 ps (see [91]), till they eventually annihilate.

It is interesting to note that it has been experimentally shown [91] that the onset of vortices in polariton superfluids does not require a LG imprinting beam, but instead, vortex–antivortex pairs can be also generated when counter-propagating currents are imposed, similarly to what happens in normal (classical) fluids. In [91] a Gaussian pulsed beam has been shined either at rest with respect to the OPO signal, $\mathbf{k}_{\mathrm{pb}} = \mathbf{k}_{\mathrm{s}} \simeq 0$, or moving $\mathbf{k}_{\mathrm{pb}} \neq \mathbf{k}_{\mathrm{s}}$. While no vortex–antivortex pair appears in the first case, in the second, a vortex–antivortex pair appears on opposite sides of the probe edge.

6.6 Stability of Multiply Quantised Vortices

The energy of a vortex is proportional to its quantum of circulation m (see equation (6.20)) squared [76], m^2. Thus, ignoring interactions, a doubly charged $m = 2$ vortex has higher energy than two single $m = 1$ vortices. However, including interactions between vortices, the energy of an $m = 2$ vortex turns to be the same as the energy of two $m = 1$ interacting vortices close together. The behaviour of doubly quantised vortices has been the subject of intensive research in the context of ultra cold atomic gases. In particular, it has been established that the nature of the splitting is the dynamical instability. Nevertheless, $m = 2$ vortices have been predicted to be stable for specific ranges of density and interaction strength [92, 93], though, so far, they have not been observed experimentally [94]. As for single vortices, multiply quantised vortices can be, however, stabilised in multiply connected geometry. Indeed, stable pinned $m = 2$ persistent vortices have been recently observed [89] by using a toroidal pinning potential generated by an external optical plug and demonstrated to split soon after the plug was removed. In this case, the presence of a plug beam at the vortex centre can pin both $m = 1$ and $m = 2$ vortex states and stabilise them against respectively spiralling out of the condensate for $m = 1$ and splitting for $m = 2$. In other words, the external trap mechanically prevents the persistent flow to undergo any movement.

In contrast to equilibrium superfluids, such as atomic gases, both stable and unstable $m = 2$ vortices have been experimentally realised in polariton OPO superfluids [8]. In this section, we provide a theoretical explanation of the stability and splitting of doubly charged polariton vortices. As done previously, vortices in OPO are generated by an external pulsed probe (6.25). As such, we classify the response of the system to an $m = 2$ LG probe, depending whether the probe generates a TOPO state (and the vortex is only carried by the extra population but is not transferred to the OPO signal), as described in Sect. 6.6.1, or instead is transferred to the OPO signal (Sect. 6.6.2).

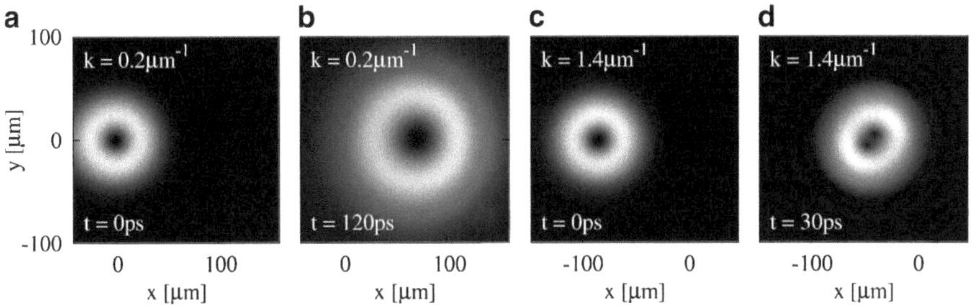

Fig. 6.15 Intensity of the TOPO probe profile $|\psi_C^{pb}(\mathbf{r}, t)|$ after the arrival (at $t = 0$) of an $m = 2$ vortex pulsed probe (6.25) with $\sigma_{pb} \simeq 87\mu$m. The case of a small k_{pb} is shown in panels (**a**) and (**b**), while a large k_{pb} in panels (**c**) and (**d**). While in the case (**a, b**), the $m = 2$ vortex does not split within its lifetime; in (**c, d**), the vortex splits soon after the probe arrives. The intensity scale in (**b**) is 200 times smaller then in (**a**)—the signal expands and the density drops two orders of magnitude (see text). Adapted from [23]

6.6.1 TOPO Regime

We first consider the TOPO regime (see Sect. 6.4), i.e., when the vortex propagates inside the triggered probe and conjugate wave packets, and it get not transferred to the OPO signal and idler. It has been found [8, 23] that, in the TOPO regime, $m = 2$ vortices are stable within their lifetime when triggered at small momenta k_{pb} (see Fig. 6.15 panels (a) and (b)), while they split into two $m = 1$ vortices for large values of k_{pb} (see Fig. 6.15 panels (c) and (d)). This conclusion was reached both by experimental observations [8] and theoretical analysis [8, 23]. The numerical analysis shows that the crossover from non-splitting to splitting happens for the probe momenta where the LP dispersion deviates from the quadratic one (see Fig. 6.12). The two different cases are shown in Fig. 6.15: For $k_{pb} = 0.2\,\mu\text{m}^{-1}$, at short times, the probe propagates without changing shape and with little change in intensity (not shown in Fig. 6.15), consistent with the linear dispersion of spectrum characterising this regime. However, at longer times, the density of the triggered probe and conjugate states drops more then two orders of magnitude, the dispersion changes to the quadratic one and the wave packet expands (panel (b)). A uniform expansion of the wave packet leads to the decrease of the probe and the conjugate polariton densities and thus to an increase of the vortex core, but it does not cause the vortex to split. In contrast, for $k_{pb} = 1.4\,\mu\text{m}^{-1}$, where the LP dispersion is not quadratic, the $m = 2$ vortex state splits into two $m = 1$ vortices shortly after the arrival of the probe (panel (d)). This behaviour can be understood by analysing the evolution of the system's excitation spectrum in time: The dispersion of the time-dependent TOPO evolves from LP (before the probe arrival) to linear (at early times after the probe arrival, when the stimulated scattering is strong), and back to the LP at later times. For large k_{pb}, the LP dispersion deviates strongly from quadratic (see Fig. 6.12 right panel). Wave packets propagating with non-quadratic dispersion do not keep their shapes (as discussed in Sect. 6.4.1), and the simulations show that the distortion can be very pronounced in particular at later times of the evolution. The distortion during the early times of the propagation leads to the mechanical splitting of an $m = 2$ vortex, analogous to the structural instability discussed in [95]. Additionally, as discussed in [8], for small k_{pb}, within the quadratic part of the dispersion, the group velocity of the wave packet carrying the vortex equals the velocity of the net supercurrent (given by k_{pb}) associated with phase variations. This is not, however, the case for larger k_{pb}, beyond the quadratic part of the dispersion. In this case, the propagating vortex feels a net current in its moving reference frame, which may provide additional mechanism for splitting.

6.6.2 OPO Regime

In contrast to the TOPO regime described above, it has been shown both experimentally and theoretically [8, 23] that $m = 2$ vortices that do get imprinted into the steady state OPO signal are never stable and split into two $m = 1$ vortices almost immediately, even before the probe reaches its maximum intensity (see Fig. 6.16). By analysing the system's dispersion in different regimes, as well as the dynamics of currents visible in the simulations, we have been able to identify several causes for the splitting: Before the arrival of the probe, the steady state OPO dispersion is flat around the pump, signal, and idler. However, the triggering probe favours the signal and conjugate to lock and propagate with the same velocity $v^{\text{LP}}_{k_{\text{pb}}}$. This behaviour corresponds to a linear dispersion. Further, once the vortex gets imprinted into the stationary OPO signal and idler, the system's dispersion changes back to be flat. The evolution of the dispersion between flat, linear, and again flat leads to a complicated dynamics of both signal and idler (the *transient period* described in [22] and previously in Sect. 6.3.5.1), causing the structural instability and splitting of the $m = 2$ vortex during the transient time. Another reason for the structural instability

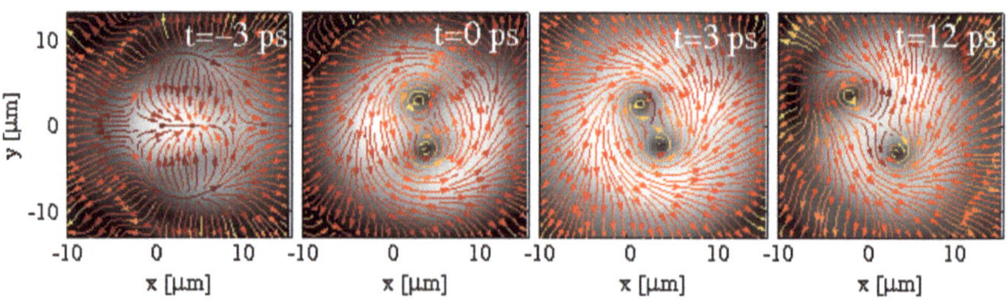

Fig. 6.16 Filtered signal profile and currents above threshold for OPO ($f_{\text{p}} = 1.12 f_{\text{p}}^{(\text{th})}$) for a top-hat pump with FWHM $\sigma_{\text{p}} = 35\ \mu\text{m}$ at $t = -3$ ps (*first panel*) before the arrival of an $m = 2$ probe. The doubly quantised vortex gets transferred from the probe to the OPO signal and splits into two $m = 1$ vortices even before the probe reaches its maximum intensity at $t = 0$ ps (*second panel*). In this simulation, the vortices coexist for sometime (roughly 15 ps), then one gets expelled from the signal (*fourth panel*). Adapted from [23]

and splitting are the non-uniform currents (see Fig. 6.16) present in the OPO signal caused by the interplay between spatial inhomogeneity, pump and decay, which the OPO vortex experiences in its reference frame.

6.7 Vortices in Other Polariton Fluids

We do not pretend to give an exhaustive review of the broad field which studies vortices in polariton fluids in this chapter, where we have mostly focussed on the occurrence of vortices in polariton OPO superfluids. However, we would like at least to briefly mention what happens for polariton fluids other than OPO.

6.7.1 Spontaneous Vortices in Trapped Incoherently Pumped Polaritons

For incoherently pumped polaritons, the presence of a harmonic trapping potential, can make the non-rotating solution unstable to the spontaneous formation of a vortex lattice [80] (This work has been generalised to include the effects of polarisation in [96]). In experiments, spontaneous vortices in incoherently pumped systems, have been observed in [5, 97] and their existence explained in terms of pinning by the disorder present in the CdTe sample. Adding the polarisation degrees of freedom, can give rise to the appearance of half vortices [6, 98]. Polariton vortices have been also observed in cavity mesas [81]. (see Chap. 3).

Vortex–antivortex pairs have been observed in the non-resonantly pumped experiments of [10], where the mechanism of V–AV generation is explained in terms of density fluctuations originating from the cw multi-mode pumping laser, while for a single-mode laser, no V–AV pairs have been observed. In this sense, their motivation and interpretation is in terms of the BKT transition. The pair dynamics in the condensate has been studied in [99]. (see Chap. 4)

Finally, as mentioned previously in Sects. 6.5 and 6.5.2.2, generation and detection of metastable vortices have been also recently discussed for polariton condensates generated by toroidal non-resonantly pumping in [21], where vortices have been seeded with an external LG probe. Interestingly, very recently in [100], it has been observed an all-optical spontaneous pattern formation in a polariton condensate non-resonantly pumped with a ring geometry.

6.7.2 Resonantly Pumped-Only Polaritons

In [11, 12], the hydrodynamic nucleation of V–AV pairs is studied by making the polariton fluid collide with a large defect. In particular, polaritons are resonantly (coherently) injected with a pulsed laser beam, creating a population in the pump state only. The focus and interest of these studies are the possibility of exploring quantum turbulence, the appearance of dissipation and drag above a critical velocity because of the nucleation of vortices in the wake of the obstacle.

Acknowledgements We would like to acknowledge fruitful collaborations with A. Berceanu, E. Cancellieri, D. Sanvitto, C. Tejedor, and D. M. Whittaker on some of the topics discussed in this review, as well as the collaboration with the experimental group at UAM in Madrid (C. Antón, M. Baudisch, G. Tosi, L. Viña). We are particularly grateful to I. Carusotto and J. Keeling for stimulating discussions and for the critical reading of this manuscript. We also would like to thank C. Creatore, B. Deveaud-Plédran, M. Maragkou, and Y. Yamamoto for useful suggestions and points of discussion. F.M.M. acknowledges the financial support from the programs Ramón y Cajal and POLATOM (ESF). This work has been also supported by the Spanish MEC (MAT2008-01555, QOIT-CSD2006-00019), CAM (S-2009/ESP-1503), and FP7 ITN "Clermont4" (235114).

References

1. L. Onsager, Nuovo Cimento **6**, 249 (1949)
2. R. Feynman, *Application of Quantum Mechanics to Liquid Helium*, *Progress in Low Temperature Physics*, vol. 1 (Elsevier, North Holland, Amsterdam, 1955)
3. J.F. Nye, M.V. Berry, Proc. R. Soc. A **336**, 165 (1974)
4. K. Staliunas, V.J. Sanchez-Morcillo, *Transverse Patterns in Nonlinear Optical Resonators* (Springer, Berlin, 2003)
5. K.G. Lagoudakis, M. Wouters, M. Richard, A. Baas, I. Carusotto, R. André, L.S. Dang, B. Deveaud-Plédran, Nature Phys. **4**, 706 (2008)
6. K.G. Lagoudakis, T. Ostatnický, A.V. Kavokin, Y.G. Rubo, R. André, B. Deveaud-Plédran, Science **326**, 974 (2009)
7. K.G. Lagoudakis, F. Manni, B. Pietka, M. Wouters, T.C.H. Liew, V. Savona, A.V. Kavokin, R. André, B. Deveaud-Plédran, Phys. Rev. Lett. **106**, 115301 (2011)
8. D. Sanvitto, F. Marchetti, M. Szymańska, G. Tosi, M. Baudisch, F. Laussy, D. Krizhanovskii, M. Skolnick, L. Marrucci, A. Lemaître, J. Bloch, C. Tejedor, L.V. na, Nature Phys. **6**, 527 (2010)
9. D.N. Krizhanovskii, D.M. Whittaker, R.A. Bradley, K. Guda, D. Sarkar, D. Sanvitto, L. Viña, E. Cerda, P. Santos, K. Biermann, R. Hey, M.S. Skolnick, Phys. Rev. Lett. **104**, 126402 (2010)
10. G. Roumpos, M.D. Fraser, A. Löffler, S. Höfling, A. Forchel, Y. Yamamoto, Nature Phys. **7**, 129 (2011)
11. G. Nardin, G. Grosso, Y. Leger, B. Pietka, F. Morier-Genoud, B. Deveaud-Plédran, Nature Phys. 635–641 (2011). DOI doi:10.1038/nphys1959
12. D. Sanvitto, S. Pigeon, A. Amo, D. Ballarini, M.D. Giorgi, I. Carusotto, R. Hivet, F. Pisanello, V.G. Sala, P.S. Soares-Guimaraes, R. Houdre, E. Giacobino, C. Ciuti, A. Bramati, G. Gigli, Nature Photonics **5**, 610–614 (2011)
13. J. Kasprzak, M. Richard, S. Kundermann, A. Baas, P. Jeanbrun, J.M.J. Keeling, F.M. Marchetti, M.H. Szymańska, R. André, J.L. Staehli, V. Savona, P.B. Littlewood, B. Devaud, L.S. Dang, Nature **443**, 409–414 (2006)
14. H. Deng, H. Haug, Y. Yamamoto, Rev. Mod. Phys. **82**, 1489 (2010)
15. I. Carusotto, presented at ICSCE4, Cambridge, UK (available at http://tcm.phy.cam.ac.uk/BIG/icsce4/talks/carusotto.pdf) (2008)
16. J. Keeling, N.G. Berloff, Nature **457**, 273 (2009)
17. J. Keeling, N. Berloff, Contemp. Phys. **52**, 131 (2011)

18. R.M. Stevenson, V.N. Astratov, M.S. Skolnick, D.M. Whittaker, M. Emam-Ismail, A.I. Tartakovskii, P.G. Savvidis, J.J. Baumberg, J.S. Roberts, Phys. Rev. Lett. **85**, 3680 (2000)
19. J.J. Baumberg, P.G. Savvidis, R.M. Stevenson, A.I. Tartakovskii, M.S. Skolnick, D.M. Whittaker, J.S. Roberts, Phys. Rev. B **62**, R16247 (2000)
20. A. Amo, D. Sanvitto, F.P. Laussy, D. Ballarini, E. del Valle, M.D. Martin, A. Lemaitre, J. Bloch, D.N. Krizhanovskii, M.S. Skolnick, C. Tejedor, L.V. na, Nature **457**, 291 (2009)
21. M. Wouters, V. Savona, Phys. Rev. B **81**, 054508 (2010)
22. F.M. Marchetti, M.H. Szymańska, C. Tejedor, D.M. Whittaker, Phys. Rev. Lett. **105**, 063902 (2010)
23. M.H. Szymańska, F.M. Marchetti, D. Sanvitto, Phys. Rev. Lett. **105**, 236402 (2010)
24. A.V. Gorbach, R. Hartley, D.V. Skryabin, Phys. Rev. Lett. **104**, 213903 (2010)
25. M.S. Skolnick, T.A. Fisher, W.D. M., Semicond. Sci. Technol. **13**, 645 (1998)
26. V. Savona, C. Piermarocchi, A. Quattropani, P. Schwendimann, F. Tassone, Phase Transitions **68**, 169 (1999)
27. G. Khitrova, H.M. Gibbs, F. Jahnke, M. Kira, S.W. Koch, Rev. Mod. Phys. **71**, 1591 (1999)
28. C. Ciuti, P. Schwendimann, A. Quattropani, Semicond. Sci. Technol. **18**, S279 (2003)
29. J. Keeling, F.M. Marchetti, M.H. Szymańska, P.B. Littlewood, Semicond. Sci. Technol. **22**, R1 (2006)
30. Y. Yamamoto, A. İmamoğlu, *Mesoscopic Quantum Optics* (Wiley, New York, 1999)
31. Y. Yamamoto, F. Tassone, H. Cao, *Semiconductor Cavity Quantum Electrodynamics*, *Springer Tracts in Modern Physics*, vol. 167 (Springer, Berlin, 2000)
32. A. Kavokin, G. Malpuech, *Cavity Polaritons*, *Thin Films and Nanostructures*, vol. 32 (Elsevier, NY, 2003)
33. A.V. Kavokin, J.J. Baumberg, G. Malpuech, F.P. Laussy, *Microcavities* (Oxford University Press, Oxford, 2007)
34. C. Ciuti, V. Savona, C. Piermarocchi, A. Quattropani, P. Schwendimann, Phys. Rev. B **58**, 7926 (1998)
35. G. Rochat, C. Ciuti, V. Savona, C. Permarocchi, A. Quattropani, P. Schwendimann, Phys. Rev. B **61**, 13856 (2000)
36. R. Houdré, C. Weisbuch, R.P. Stanley, U. Oesterle, M. Ilegems, Phys. Rev. Lett. **85**, 2793 (2000)
37. A.I. Tartakovskii, D.N. Krizhanovskii, D.A. Kurysh, V.D. Kulakovskii, M.S. Skolnick, J.S. Roberts, Phys. Rev. B **65**, 081308 (2002)
38. D.N. Krizhanovskii, A.I. Tartakovskii, V.D. Kulakovskii, M.S. Skolnick, J.S. Roberts, Phys. Rev. B **66**, 165329 (2002)
39. R. Butté, M.S. Skolnick, D.M. Whittaker, D. Bajoni, J.S. Roberts, Phys. Rev. B **68**, 115325 (2003)
40. N.A. Gippius, S.G. Tikhodeev, V.D. Kulakovskii, D.N. Krizhanovskii, A.I. Tartakovskii, EPL Europhys. Lett. **67**, 997 (2004)
41. A. Baas, J.P. Karr, M. Romanelli, A. Bramati, E. Giacobino, Phys. Rev. Lett. **96**, 176401 (2006)
42. D. Sanvitto, D.N. Krizhanovskii, D.M. Whittaker, S. Ceccarelli, M.S. Skolnick, J.S. Roberts, Phys. Rev. B **73**, 241308 (2006)
43. D.N. Krizhanovskii, D. Sanvitto, A.P.D. Love, M.S. Skolnick, D.M. Whittaker, J.S. Roberts, Phys. Rev. Lett. **97**, 097402 (2006)
44. D.N. Krizhanovskii, S.S. Gavrilov, A.P.D. Love, D. Sanvitto, N.A. Gippius, S.G. Tikhodeev, V.D. Kulakovskii, D.M. Whittaker, M.S. Skolnick, J.S. Roberts, Phys. Rev. B **77**, 115336 (2008)
45. D. Ballarini, D. Sanvitto, A. Amo, L. Viña, M. Wouters, I. Carusotto, A. Lemaitre, J. Bloch, Phys. Rev. Lett. **102**, 056402 (2009)
46. M.S. Skolnick, D.M. Whittaker, R. Butt, A.I. Tartakovskii, Semicond. Sci. Tech. **18**, S301 (2003)
47. C. Ciuti, P. Schwendimann, A. Quattropani, Phys. Rev. B **63**, 041303 (2001)
48. W. Langbein, Phys. Rev. B **70**, 205301 (2004)

49. P.G. Savvidis, J.J. Baumberg, R.M. Stevenson, M.S. Skolnick, D.M. Whittaker, J.S. Roberts, Phys. Rev. Lett. **84**, 1547 (2000)
50. P.G. Savvidis, J.J. Baumberg, R.M. Stevenson, M.S. Skolnick, D.M. Whittaker, J.S. Roberts, Phys. Rev. B **62**, R13278 (2000)
51. R. Huang, F. Tassone, Y. Yamamoto, Phys. Rev. B **61**, R7854 (2000)
52. G. Dasbach, T. Baars, M. Bayer, A. Larionov, A. Forchel, Phys. Rev. B **62**, 13076 (2000)
53. J. Erland, V. Mizeikis, W. Langbein, J. Jensen, N. Mortensen, J. Hvam, Phys. Status Solidi B **221**, 115 (2000)
54. G. Messin, J.P. Karr, A. Baas, G. Khitrova, R. Houdré, R.P. Stanley, U. Oesterle, E. Giacobino, Phys. Rev. Lett. **87**, 127403 (2001)
55. M. Saba, C. Ciuti, J. Bloch, V. Thierry-Mieg, R. André, L.S. Dang, S. Kundermann, A. Mura, G. Bongiovanni, J.L. Staehli, B. Deveaud, Nature **414**, 731 (2001)
56. P.G. Savvidis, C. Ciuti, J.J. Baumberg, D.M. Whittaker, M.S. Skolnick, J.S. Roberts, Phys. Rev. B **64**, 075311 (2001)
57. S. Kundermann, M. Saba, C. Ciuti, T. Guillet, U. Oesterle, J.L. Staehli, B. Deveaud, Phys. Rev. Lett. **91**, 107402 (2003)
58. A. Huynh, J. Tignon, O. Larsson, P. Roussignol, C. Delalande, R. André, R. Romestain, L.S. Dang, Phys. Rev. Lett. **90**, 106401 (2003)
59. C. Diederichs, J. Tignon, G. Dasbach, C. Ciuti, A. Lamaitre, J. Bloch, P. Roussignol, C. Delalande, Nature **440**, 904 (2006)
60. G.K. Campbell, J. Mun, M. Boyd, E.W. Streed, W. Ketterle, D.E. Pritchard, Phys. Rev. Lett. **96**, 020406 (2006)
61. D.M. Whittaker, Phys. Rev. B **71**, 115301 (2005)
62. C. Ciuti, I. Carusotto, Phys. Status Solidi B **242**, 2224 (2005)
63. M. Wouters, I. Carusotto, Phys. Rev. B **75**, 075332 (2007)
64. A. Baas, J.P. Karr, M. Romanelli, A. Bramati, E. Giacobino, Phys. Rev. B **70**, 161307 (2004)
65. L. Cavigli, M. Gurioli, Phys. Rev. B **71**, 035317 (2005)
66. N.A. Gippius, I.A. Shelykh, D.D. Solnyshkov, S.S. Gavrilov, Y.G. Rubo, A.V. Kavokin, S.G. Tikhodeev, G. Malpuech, Phys. Rev. Lett. **98**, 236401 (2007)
67. C. Adrados, A. Amo, T.C.H. Liew, R. Hivet, R. Houdré, E. Giacobino, A.V. Kavokin, A. Bramati, Phys. Rev. Lett. **105**, 216403 (2010)
68. T.K. Paraïso, M. Wouters, Y. Léger, F. Morier-Genoud, B. Deveaud-Plédran, Nat. Mater. **9**, 655 (2010)
69. E. Cancellieri, F.M. Marchetti, M.H. Szymańska, C. Tejedor, Phys. Rev. B **83**, 214507 (2011)
70. A. Amo, S. Pigeon, C. Adrados, R. Houdré, E. Giacobino, C. Ciuti, A. Bramati, Phys. Rev. B **82**, 081301 (2010)
71. M. Wouters, I. Carusotto, Phys. Rev. A **76**, 043807 (2007)
72. I. Carusotto, C. Ciuti, Phys. Rev. B **72**, 125335 (2005)
73. M.H. Szymańska, J. Keeling, P.B. Littlewood, Phys. Rev. Lett. **96**, 230602 (2006)
74. M. Wouters, I. Carusotto, Phys. Rev. Lett. **99**, 140402 (2007)
75. G.B. Hess, W.M. Fairbank, Phys. Rev. Lett. **19**, 216 (1967)
76. L.P. Pitaevskii, S. Stringari, *Bose-Einstein Condensation*, (Clarendon Press, Oxford, 2003)
77. D.S. Rokhsar, Phys. Rev. Lett. **79**, 2164 (1997)
78. D.M. Whittaker, Phys. Stat. Sol. C **2**, 733 (2005)
79. M.C. Cross, P.C. Hohenberg, Rev. Mod. Phys. **65**, 851 (1993)
80. J. Keeling, N.G. Berloff, Phys. Rev. Lett. **100**, 250401 (2008)
81. G. Nardin, K.G. Lagoudakis, B. Pietka, F.M.C. Morier-Genoud, Y. Léger, B. Deveaud-Plédran, Phys. Rev. B **82**, 073303 (2010)
82. G. Tosi, M. Baudisch, D. Sanvitto, A. Lemaitre, J. Bloch, E. Karimi, B. Piccirillo, L. Marrucci, J. Phys. Conf. Series **210**, 012023 (2010)
83. O.A. Egorov, D.V. Skryabin, A.V. Yulin, F. Lederer, Phys. Rev. Lett. **102**(15), 153904 (2009)
84. O.A. Egorov, A.V. Gorbach, F. Lederer, D.V. Skryabin, Phys. Rev. Lett. **105**, 073903 (2010)
85. V. Boyer, C.F. McCormick, E. Arimondo, P.D. Lett, Phys. Rev. Lett. **99**, 143601 (2007)

86. A.M. Marino, V. Boyer, R.C. Pooser, P.D. Lett, K. Lemons, K.M. Jones, Phys. Rev. Lett. **101**, 093602 (2008)
87. A. Amo, D. Sanvitto, L.V. na, Semicond. Sci. Technol. **25**, 043001 (2010)
88. M.F. Andersen, C. Ryu, P. Cladé, V. Natarajan, A. Vaziri, K. Helmerson, W.D. Phillips, Phys. Rev. Lett. **97**, 170406 (2006)
89. C. Ryu, M.F. Andersen, P. Cladé, V. Natarajan, K. Helmerson, W.D. Phillips, Phys. Rev. Lett. **99**, 260401 (2007)
90. A. Ramanathan, K.C. Wright, S.R. Muniz, M. Zelan, W.T. Hill, C.J. Lobb, K. Helmerson, W.D. Phillips, G.K. Campbell, Phys. Rev. Lett. **106**, 130401 (2011)
91. G. Tosi, F.M. Marchetti, D. Sanvitto, C. Antón, M.H. Szymańska, A. Berceanu, C. Tejedor, L. Marrucci, A. Lemaître, J. Bloch, L. Viña, Phys. Rev. Lett. **107**, 036401 (2011)
92. H. Pu, C.K. Law, J.H. Eberly, N.P. Bigelow, Phys. Rev. A **59**, 1533 (1999)
93. M. Möttönen, T. Mizushima, T. Isoshima, M.M. Salomaa, K. Machida, Phys. Rev. A **68**, 023611 (2003)
94. Y. Shin, M. Saba, M. Vengalattore, T.A. Pasquini, C. Sanner, A.E. Leanhardt, M. Prentiss, D.E. Pritchard, W. Ketterle, Phys. Rev. Lett. **93**, 160406 (2004)
95. J.J. García-Ripoll, G. Molina-Terriza, V.M. Pérez-García, L. Torner, Phys. Rev. Lett. **87**, 140403 (2001)
96. M.O. Borgh, J. Keeling, N.G. Berloff, Phys. Rev. B **81**, 235302 (2010)
97. K.G. Lagoudakis, F. Manni, B. Pietka, M. Wouters, T.C.H. Liew, V. Savona, A.V. Kavokin, R. André, B. Deveaud-Plédran, Phys. Rev. Lett. **106**, 115301 (2011)
98. Y.G. Rubo, Phys. Rev. Lett. **99**, 106401 (2007)
99. M.D. Fraser, G. Roumpos, Y. Yamamoto, New J. Phys. **11**, 113048 (2009)
100. F. Manni, K.G. Lagoudakis, T.C.H. Liew, R. André, B. Deveaud-Plédran, Phys. Rev. Lett. **107**, 106401 (2011)

Chapter 7
Superfluidity and Hydrodynamic Topological Excitations of Microcavity Polaritons

A. Amo and A. Bramati

Abstract Bosonic condensates subject to interactions may give rise to the phenomenon of superfluidity. This is the case of polaritons in semiconductor microcavities, in which superfluidity is manifested in a number of effects like the frictionless flow, persistence of currents, or the quantisation of the angular momentum of the fluid. The dissipative nature of polaritons, determined by their finite lifetime, results in specific properties diverting notably from equilibrium systems like Bose–Einstein condensates of alkali atoms. In the first part of this chapter we describe the superfluidity of polaritons attending to the propagation characteristics of these gases in the presence of a potential barrier. We concentrate our analysis on the body of available experimental results, which can be well understood in terms of the shape of the spectra of excitations. We devote the second part of the chapter to the conditions for the break up of superfluidity via the nucleation of hydrodynamic topological excitations, i.e. quantised vortex pairs and solitons. We discuss how the out-of-equilibrium nature of polariton condensates favours the nucleation of these excitations, opening the way to novel turbulent regimes.

7.1 Introduction

Superfluidity, the ability to flow without friction [1], is one of the most striking manifestations of quantum phenomena at the macroscopic scale. It was first observed in 1938 as an abrupt drop of the viscosity below some critical temperature

A. Amo (✉)
CNRS-Laboratoire de Photonique et Nanostructures, Marcoussis, France
e-mail: alberto.amo@lpn.cnrs.fr

A. Bramati
Laboratoire Kastler Brossel, Université Pierre et Marie Curie-Paris 6, École Normale Supérieure et CNRS, Paris, France

D. Sanvitto and V. Timofeev (eds.), *Exciton Polaritons in Microcavities*, Springer Series in Solid-State Sciences 172, DOI 10.1007/978-3-642-24186-4_7,

(the lambda point, 2.17 K) when circulating ^{4}He through very thin capillaries. In fact, superfluidity manifests in a number of effects. Some examples are the so-called Hess–Fairbanks effect and the quantization of circulation, observed when exploring the rotational properties of a superfluid. For instance, if liquid ^{4}He above the lambda temperature is introduced in a toroidal container set in rotation, within a few minutes the whole liquid will achieve a steady angular velocity equal to that of the container. If we now cool down the liquid below the lambda point, the angular velocity of the liquid, the superfluid, can only get discrete quantized values (including staying at rest if the initial angular velocity is below some critical quantity). These effects can be understood from the fact that the superfluid is described by a single macroscopic wavefunction with the property of being irrotational, and they have been experimentally observed not only in superfluid ^{4}He [2–4] but also in Bose–Einstein condensates of alkali atoms [5–7].

However, the most popular image of superfluidity is that of the flow without friction, which was first evidenced in the above-mentioned reports of 1938. This phenomenon can be intuitively described in terms of the Landau criterion, which will be treated in detail in the next section. The Landau model considers the linearized spectrum of excitations characteristic of a Bose–Einstein condensate subject to particle–particle repulsive interactions. The slope of the spectrum of excitations is known as the speed of sound of the system, c_s, as it sets the group velocity at which density sound-like perturbations propagate in the fluid.

A most striking feature of superfluidity is the absence of density excitations when the fluid hits a localised static obstacle at flow speeds v_f below some critical velocity v_c. For small potential barriers, this critical speed is given by c_s. On the other hand, for supersonic flows ($v_f > c_s$), the presence of a small obstacle induces dissipation (drag) via the emission of sound waves. Superfluidity in the framework of the Landau criterion has been experimentally observed in ^{4}He, both in bulk and in droplets by measuring the kinetic energy loss of ions and atoms traversing the fluid at controlled velocities [8, 9]. In atomic condensates, a critical speed [10, 11] and the emission of sound-like shockwaves in supersonic flows have been reported [12]. Additionally, if the superflow is made to rotate in a toroidal container, the quenching of excitations results in a metastable state holding a persistent current for an, in principle, infinitely long time [13, 14].

The situation with respect to the fundamental excitations of a superfluid is even more interesting when the barrier is strong and its size is larger than the healing length of the fluid. The healing length is the characteristic distance at which the interparticle interaction energy is of the same order as the kinetic energy related to the spatial variations of the condensate density. In other words, it is the minimum distance over which the density of the condensate can vary, limited by the repulsive interactions. The presence of a large barrier in the flow path results in flow velocities around the barrier which can be as high as twice the flow speed far away from the barrier [15]. The velocity and density gradients induced by the obstacle can give rise to the nucleation of topological excitations, such as vortices [15] and solitons [16], which carry energy away and induce drag [17]. These are the fundamental quantum hydrodynamic excitations and are not accounted for by the Landau criterion. Their nucleation results in critical velocities smaller than c_s, a feature already observed both in He and atomic condensates [10, 11, 18, 19].

In the panorama of quantum gases, polaritons in semiconductor microcavities have provided novel theoretical and experimental insights on the physics of superfluidity and quantum hydrodynamics. Since the first reports of polariton condensation in 2006 [20], coherent propagation and quantum fluid effects have been thoroughly studied. The continuous escape of photons from the cavity allows for the direct visualisation of the polariton wave function with a high spatial and temporal resolution using standard optical imaging techniques. Additionally, under resonant excitation, polariton condensates of controlled frequency, momentum and density can be created. These assets have played a very important role in the experimental development of quantum fluid experiments in polaritons. Some of the observations include the formation of quantised vortices [21, 22], long-range order [20, 23, 24], Goldstone mode physics [25, 26], persistent currents [27–29], superfluidity according to the Landau criterion in the presence of small defects [30–32], supersonic Cerenkov wavefronts [32] and the hydrodynamic nucleation of vortices [33,34] and solitons in the wake of a potential barrier [35,36]. Additionally, the relatively high optical losses set polariton condensates out of equilibrium, giving rise to richer possibilities in the spectrum of excitations as compared to equilibrium systems [25,30,37–40] and resulting in new quantum fluid effects [29,41–44]. Some of these specific properties will be treated in this chapter.

7.2 Superfluidity According to the Landau Criterion

A very intuitive picture of the flow without friction characteristic of a superfluid was that introduced by Landau in what is now known as the Landau criterion for superfluidity. Let us consider a uniform fluid in its ground state at rest being traversed by an impenetrable obstacle, a point-like potential barrier, at a speed v. In this reference frame, the fluid has energy E_0. Let us apply a Galilean boost to the reference frame in which the obstacle is at rest ($\boldsymbol{r} \to \boldsymbol{r} - \boldsymbol{v}t$). In this case, the energy of the system will be [45]:

$$E\left(\boldsymbol{v}\right) = E_0 + \frac{mv^2}{2} \tag{7.1}$$

If we now add to the fluid an excitation with energy $\varepsilon_{\mathrm{ex}}$ and momentum $\boldsymbol{p_{ex}}$, the energy of the system will now be:

$$E'(\boldsymbol{v}) = E_0 + \varepsilon_{\mathrm{ex}} + \frac{(\boldsymbol{p_{ex}})^2}{2m} = E_0 + \frac{mv^2}{2} + \varepsilon_{\mathrm{ex}} - \boldsymbol{p_{ex}} \cdot \boldsymbol{v} \tag{7.2}$$

If we subtract the energy with and without excitation, we see that an excitation in the fluid can be created only if:

$$\varepsilon_{\mathrm{ex}} = \boldsymbol{p_{ex}} \cdot v = p_{\mathrm{ex}} v \cos\theta, \tag{7.3}$$

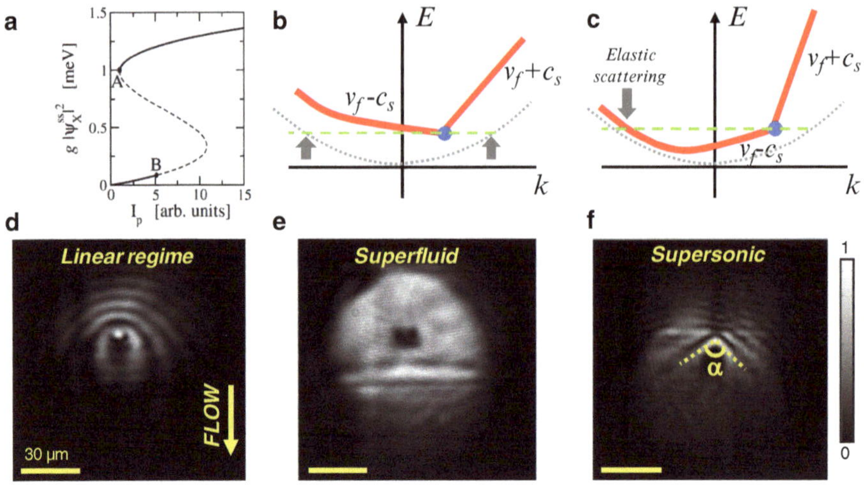

Fig. 7.1 (**a**) Calculated polariton density as a function of excitation power for the case of quasi-resonant excitation of the lower polariton branch with $k = 0.3\,\mu\text{m}^{-1}$, showing a bi-stable behaviour. Reprinted with permission from [30], copyright (2004) by the American Physical Society. (**b–d**) *Solid lines*: schematic representation of the spectrum of excitation of a subsonic (**b**) and supersonic (**c**) polariton quantum fluid. The *dotted lines* show the single particle parabolic dispersion of a polariton gas at very low density. (**d–f**) Real-space emission of a resonantly excited polariton fluid in the presence of an obstacle. At low density (**d**), polaritons elastically scatter from the obstacle, resulting in the parabolic waves observed upstream. At high density and low flow velocity (**e**), $v_f < c_s$, the flow shows a superfluid character. This corresponds to the situation plotted in (**b**). At high-flow velocities (**f**, **c**), $v_f > c_s$, the supersonic regime is attained, with the formation of linear sound wavefronts. From [32]

where θ is the angle between the fluid velocity and the momentum of the excitation. In a classical fluid of non-interacting particles, the spectrum of excitations corresponds to the single particle one, it has a parabolic form $\varepsilon_{ex} = p_{ex}^2/2m$, and we can always find a value of $\boldsymbol{p_{ex}}$ satisfying (7.3). This is illustrated in short dashed lines in Fig. 7.1b. As the obstacle is static in its reference frame, it cannot transfer energy to the fluid. However, elastic scattering of fluid particles to the states marked with an arrow is possible, resulting in a depletion of the original fluid state (black dot).

In general, condition (7.3) is satisfied for velocities greater than a critical value, v_c, given by $v_c = \min(\varepsilon_{ex}/p_{ex})$. For a fluid with a parabolic spectrum of excitations, $v_c = 0$, but in a quantum fluid subject to repulsive interparticle interactions, the spectrum of excitations becomes linear, and $v_c \neq 0$. As illustrated by the solid lines in Fig. 7.1b, the critical velocity corresponds to the slope of the spectrum of excitations (the so-called speed of sound: $\varepsilon_{ex} = c_s \cdot p_{ex}$), while at large p_{ex}, the single particle parabolic dispersion is recovered. Below v_c, the fluid is thus able to flow without friction in the presence of an obstacle, as elastic scattering is simply not allowed due to the absence of final states.

The superfluid regime corresponds to subsonic flow speeds $v_f < c_s$. The situation of supersonic flow $v_f > c_s$ is represented in Fig. 7.1c. In this case, elastic scattering is allowed as final states are available. According to (7.3), excitations are possible with momenta forming an angle $\cos(\varepsilon_{ex}/p_{ex}v)$ with respect to the flow direction. In real space, this is manifested in the form of a conic shockwave, with its apex located at the obstacle. In two dimensions, this reduces to two linear wavefronts with an angle of aperture given by $\cos(\alpha/2) = c_s/v_f$, as shown in Fig. 7.1f.

In atomic condensates, which are in thermal equilibrium, a linear spectrum of excitations has indeed been observed [46]. Its slope sets the value of the speed of sound of the fluid and it is given by:

$$c_s = \sqrt{\hbar g|\psi|^2/m}, \tag{7.4}$$

where $|\psi|^2$ and m are the particle density and mass, respectively, and g is the inter-particle interaction constant. While the linearization of the spectrum of excitation has also been experimentally observed for polariton condensates at $k = 0$ [47, 48], the presence of pump and decay has been predicted to result in richer configurations [37, 38, 40]. We can distinguish two different situations corresponding to the cases of resonant and non-resonant excitation of the polariton gas. The main difference between the two cases is that under resonant excitation, the phase of the condensate is imposed by the pump laser via the direct coupling to the photon part of the polariton wave function. As we will see, this sets important differences with the case of non-resonant pumping, in which the phase of the condensate is spontaneously established as a consequence of the break-up of the U(1) symmetry of the system.

7.2.1 Resonant Excitation of the Lower Polariton Branch

By choosing the angle of incidence and the frequency of the excitation laser, a polariton gas with a well-defined momentum and density can be directly injected by resonant excitation of the lower polariton branch. For microcavities of low polariton lifetime (up to a few picoseconds), the polariton gas is basically restricted to the excitation spot with a phase set by the laser. The system can be well described by the Gross–Pitaevskii equation including pump and decay:

$$i\hbar\frac{\partial}{\partial t}\psi(\boldsymbol{r},t) = \left[-\frac{\hbar^2\nabla}{2m} + V(\boldsymbol{r}) + \hbar g|\psi|^2 + i(\gamma_p - F)\right]\psi(\boldsymbol{r},t) \tag{7.5}$$

$\psi(\boldsymbol{r},t)$ is the polariton wave function, γ_p is the polariton lifetime, and F is the pump field. $V(\boldsymbol{r})$ is an additional potential energy with which we can model the polariton landscape. For instance, it can take the form of a localised barrier. In the case of resonant excitation in a finite size Gaussian spot of width σ_0 centred at $\boldsymbol{r_0}$, F takes the form [32]:

$$F(\boldsymbol{r},t) = F_0 \exp\left[i\hbar(\boldsymbol{k_0}\cdot\boldsymbol{r} - \omega_0 t)\right]\exp\left[-(\boldsymbol{r}-\boldsymbol{r_0})^2/2\sigma_0^2\right], \tag{7.6}$$

where F_0 is the amplitude, $\boldsymbol{k_0}$ is the wave vector and ω_0 is the frequency of the laser. Calculations based on the solution of (7.5) [30] show that if the excitation laser is slightly blue-detuned from the lower polariton branch, a strong bi-stable behaviour in the polariton density is observed [see Fig. 7.1a]. This bi-stability is accompanied by a dramatic change in the spectrum of excitations: at low power, interparticle interactions are negligible, and excitations present the single particle parabolic shape [dashed lines in Fig. 7.1b]; above a given power threshold, interparticle interactions results both in a blueshift of the spectrum of excitations—rendering the excitation on resonance, and in its linearization [solid lines in Fig. 7.1b]. The slope defined by the linear part of these spectra defines the speed of sound of the fluid [30]. We obtain then a situation that can be directly addressed in terms of the Landau criterion.

Figure 7.1d–f displays the experimental realisation of the above described situation. A polariton gas is quasi-resonantly injected in a spot of about 100 μm in diameter with a potential barrier located in its centre. Such an obstacle arises from a photonic defect formed during the microcavity growth due to the relaxation of the strain accumulated in the multilayer Bragg mirrors. The details of the experiment can be found in [32]. For a low injected in-plane momentum ($k = 0.34\,\mu\mathrm{m}^{-1}$) and low excitation density [Fig. 7.1d], the spectrum of excitations is well described by the parabolic dispersion shown in dashed lines [Fig. 7.1b]. Polaritons are strongly scattered by the obstacle and give rise to upstream parabolic wavefronts. They arise from the interference between the injected polaritons (in a quasi-plane wave) and those scattered by the defect (in circular waves) [37].

At high excitation density, polariton–polariton interactions result in the blueshift and linearization of the spectrum of excitations as sketched by the solid lines in Fig. 7.1b. Now the flow around the defect evidences its superfluid character via the absence of density modulations in real space [Fig. 7.1e] and by the collapse of the Raleigh ring in momentum space [32]. In contrast, if the pump is prepared with a large momentum, the high-density regime results in a supersonic situation $v_f > c_s$, and Cerenkov-like shockwaves with its characteristic linear shape are formed around the defect as depicted in Fig. 7.1f.

The theory of polariton superfluidity in terms of the Landau criterion under resonant excitation has been thoroughly developed in [37] in the linear Bogoliubov approximation, considering an infinite system (plane-wave approximation). It is worth noting that in these conditions, the specific superfluid situation with a well-defined speed of sound, as described in the upper panels of Fig. 7.1, is only attained at a particular density for each given choice of frequency and momentum, as long as this momentum is low enough. This point is located close to the low-density edge of the upper branch of the bistability curve [point A in Fig. 7.1a]. Nevertheless, at higher densities, the renormalisation of the lower polariton branch opens a full gap in the spectrum of excitations. This preserves the frictionless flow as no states are available within the gap for the elastic scattering of polaritons.

Full numerical calculations with a finite size spot show that actually, the pumped dissipative nature of the polariton system results in a residual drag force present even at subsonic flow speeds [49]. Nonetheless, the threshold-like behaviour with

$v_c^* = c_s$ described by the Landau criterion is still present, resulting in a strong inhibition of the drag force for speeds lower than c_s [29].

So far, we have described polariton fluids created by the quasi-resonant excitation of the lower polariton branch in terms of the Gross–Pitaevskii equation with pump and decay (7.5). This description is quite convenient in order to compare the polariton physics with that of ultracold atomic condensates in which concepts such as speed of sound, fluid velocity or spectrum of excitations are commonly used. However, let us note that an analogous description could also be performed in terms of non-linear optics. The non-equilibrium Gross–Pitaevskii equation is formally equivalent to the monochromatic response of a $\chi^{(3)}$ (or Kerr) non-linear medium. In this direction, there have been theoretical proposals showing that superfluidity of photons could be observed in a situation analogous to that shown in Fig. 7.1 if the medium presents a strong non-linearity [50–52].

7.2.2 Non-resonant Excitation

Polariton condensates can also be created under non-resonant excitation (see also the Chaps. 3, 4 and 6 of this book). In this case, condensation is accompanied by the spontaneous break-up of a U(1) symmetry, setting up the phase of the condensate. Contrary to the resonant excitation situation, this phase is not imposed by the pump beam, which photocreates polaritons at much higher energy than the condensate. Under continuous excitation, the steady state of the system arises from the interplay between pumping, carrier relaxation and decay. Again, the dissipative nature of polaritons gives rise to important specificities in the shape of the spectrum of excitations with respect to the equilibrium case. For instance, for a condensate spontaneously formed at rest at the bottom of the lower polariton branch, the spectrum of excitations has been predicted to present a flat horizontal shape around $k = 0$ [see Fig. 7.2a], [38, 40]. According to the Landau criterion we have just described, this would mean that the critical velocity would be equal to zero. However, as thoroughly discussed in [29], one should also take into account the non-trivial shape of the imaginary part of the spectrum of excitations. As illustrated in Fig. 7.2b, the imaginary part may have negative values which approaches zero when $k \to 0$. For the case of the condensate at rest, this gives rise to a Goldstone mode at $k = 0$ with a diffusive character and, more importantly, the density excitations get damped out of $k = 0$ [40].

The situation is analogous when the considered condensate is in motion with a finite momentum [29]. Similarly to the $k = 0$ case, it has been calculated that a flat region in the real part of the spectrum of excitations appears around the condensate's momentum. Drag is now possible at any fluid velocity due to the availability of final states in the scattering process. However, the damping of the density excitations results in a phenomenology similar to that described by the Landau model for an equilibrium system: above a critical velocity, a potential barrier present in the flow path gives rise to drag and to density modulations propagating upstream from the

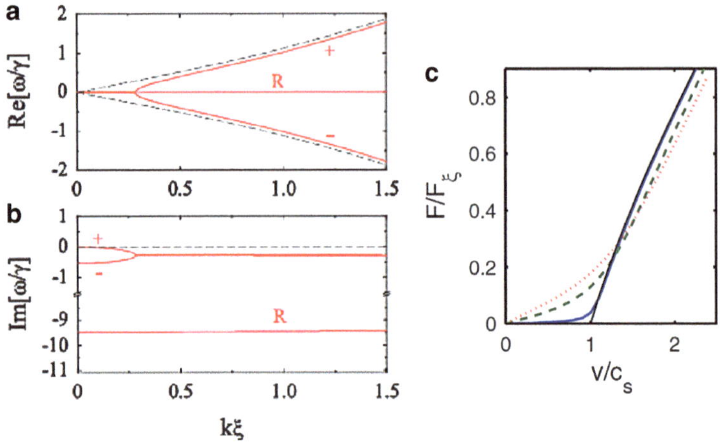

Fig. 7.2 Calculated real (**a**) and imaginary (**b**) part of the spectrum of excitations of a polariton condensate at $k = 0$ created under non-resonant excitation. The *dashed lines* show the result for a fluid in equilibrium (i.e. infinite lifetime). Reprinted with permission from [40]. Copyright (2007) by the American Physical Society. (**c**) Calculated drag force on an obstacle by a polariton fluid created non-resonantly, as a function of the flow speed. The different curves show the results for decreasing polariton lifetime, from the *thin solid line* representing the case of infinite lifetime, to the *dotted lines* with low polariton lifetime. Reprinted with permission from [29]. Copyright (2008) by the American Physical Society

obstacle [analogous to the case shown in Fig. 7.1f]; below this critical velocity, while drag is still present, its value is greatly reduced, and the density perturbation created by the barrier is restricted to a small region around it. For small obstacles, the critical velocity separating the two regimes corresponds to the sound speed calculated from the polariton density [(7.4); see Fig. 7.2c]. As proposed by Wouters and Carusotto [29], though drag is always present, the existence of a "critical" velocity can be considered as a generalised Landau criterion.

Experimentally, condensates with finite momentum can be obtained with an out-of-resonance excitation in several manners. One of them is via accelerating ramps generated by the strongly interacting reservoir [24]. The excitation laser can create a dense exciton reservoir which induces a local blueshift of the polariton energy. The reservoir is basically restricted to the excitation spot, from where the polariton condensate accelerates to the adjacent spatial regions. Polariton gases can also be set in motion by triggering condensation with an optical seed with a pre-selected momentum. One of the most remarkable examples is the triggered optical parametric oscillation [31, 53]. In this configuration, a light pulse of a given energy and momentum triggers the parametric oscillation between the "signal" condensate, a continuously pumped "pump" state, and an "idler" state. The polariton condensate created by the pulse seed continues living after the pulse is gone and can propagate for large distances, as long as it is fed by the pump polaritons via stimulated scattering. Persistent currents, one of the landmarks of superfluidity, have been demonstrated in this way [27], while the first experiments on the frictionless flow of polaritons were actually performed in this configuration [31].

7.3 Hydrodynamic Effects

A complementary approach to the physics of boson condensates can be followed from the point of view of fluid dynamics. The link between the equilibrium Gross–Pitaevskii equation and Bernoulli's equations describing standard fluids has been well established [54]. The usual approach is to use the Madelung transformation [17]:

$$\psi(\boldsymbol{r},t) = \sqrt{n(\boldsymbol{r},t)}\exp(\mathrm{i}\phi(\boldsymbol{r},t)), \tag{7.7}$$

where n is the particle density and the fluid velocity is given by the phase gradient $v_f \equiv (\hbar/m)\nabla\phi$. For polariton fluids, the derivation of Bernoulli's equation has been treated in [41, 42]. Let us consider a steady state situation such that neither the density nor the velocity changes in time. In this case, $\mathrm{i}\hbar\partial_t\psi(r,t) = \mu\psi(r,t)$, where μ is the chemical potential determined by the interplay between pump (F) and losses (γ_p). By inserting (7.7) into (7.5), we obtain the following equations for the imaginary and real parts:

$$\frac{\hbar^2}{2m}\nabla(n\nabla\phi) = (\gamma_\mathrm{p} - F)n, \tag{7.8}$$

$$\mu = \frac{\hbar^2}{2m}\left(|\nabla\phi|^2 - \frac{\nabla^2\sqrt{n}}{\sqrt{n}}\right) + V + \hbar g n. \tag{7.9}$$

The first one is a continuity equation, while the second one corresponds to Bernoulli's equation. Even in the absence of a pump source ($F = 0$), we can readily notice that currents are expected to flow from different density regions. In this way, the presence of an impenetrable obstacle in the flow path of a fluid will give rise to transverse currents and density gradients which will in its turn induce new currents [see Fig. 7.3a]. When the density variations induced by these currents are of the order of the healing length of the fluid, the nucleation of topological excitations is possible. In a two-dimensional fluid, two basic types of topological excitations can be created: quantised vortices and solitons.

Vortices are characterised by rotating currents around an empty dark core. They present a continuous change of the phase of the condensate wave function from 0 to $2s\pi$ after a complete turn around the core [21] [Fig. 7.3b], whose size is given by the healing length. s is the winding number, and in principle, it can have any integer value, though vortices with $s > 1$ tend to decay into s vortices of $s = 1$, as they have a greater stability. The phase pattern associated to the vortex allows for their generation by the phase imprinting of the polariton condensate

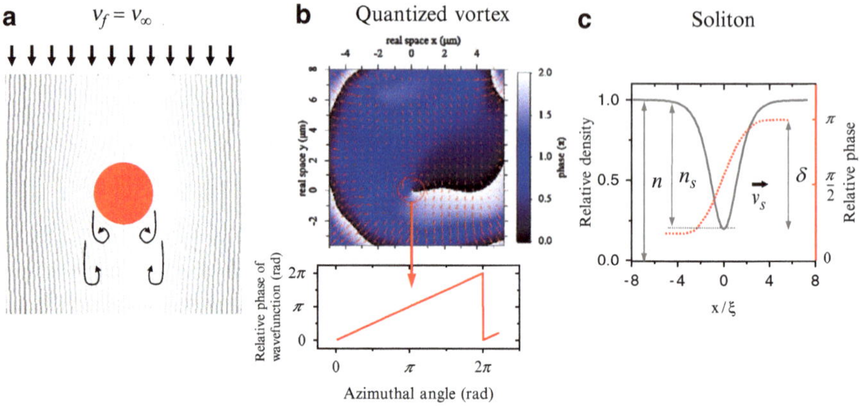

Fig. 7.3 (**a**) Schematic representation of the currents formed in the fluid passing around a large obstacle, giving rise to turbulence in its wake. (**b**) Calculated phase pattern showing the presence of a vortex as a point-like phase dislocation. From [21]. (**c**) Density and phase profile of a 1D soliton obtained from (7.10), with a relative depth of n_s/n and phase jump δ

with an external laser [27, 28, 55, 56]. In this way, the creation of vortices with very long lifetimes and up to $s = 2$ has been demonstrated [27]. In this section, we will concentrate on the spontaneous nucleation of vortices via the hydrodynamic mechanism introduced above.

The second type of basic topological excitation is the soliton, solitary density waves whose shape and width is preserved while they travel on the fluid, thanks to the compensation of dispersion with the non-linearities arisen from interparticle interactions. In 2D and for interparticle repulsive interactions, as it is the case of polaritons ($g > 0$), only dark solitons are sustained if the fluid posses a positive effective mass. Solitonic solutions to the condensate wave function were first introduced in 1D in the absence of decay by Tsuzuki [57]. The density profile of the soliton has the form:

$$|\psi(x,t)| = n\left[\frac{v_s^2}{{c_s}^2} + \left(1 - \frac{v_s^2}{c_s^2}\right)\tanh^2\left(\frac{x - v_s t}{\sqrt{2}\xi}\sqrt{1 - \frac{v_s^2}{c_s^2}}\right)\right] \tag{7.10}$$

where $\xi = (\hbar/2mng)^{1/2}$ is the healing length and n is the density far from the soliton, where the fluid is not perturbed. Equation (7.10) shows that solitons move on top of the condensate with a velocity v_s (always smaller than c_s), and they are accompanied by a phase jump δ in the condensate wave function across the soliton of up to π [58] [see Fig. 7.3c]. Their depth, phase jump and propagation speed follow the relation [59]:

$$\cos\left(\frac{\delta}{2}\right) = \left(1 - \frac{n_s}{n}\right)^{1/2} = \frac{v_s}{c_s}, \tag{7.11}$$

where n_s is the depth of the density notch [see Fig. 7.3c]. The soliton width can be calculated from (7.10), and it is given by the healing length multiplied by the factor $1 - v_s/c_s$. Very dark solitons (n_s/n close to 1) present a large phase jump (close to π) and stay almost at rest, while shallow solitons travel at velocities close to the sound speed. Analogously to the case of vortices, phase imprinting with an external laser has also been used to generate solitons in quasi-one-dimensional Bose–Einstein condensates of atoms [60, 61].

Equations (7.10) and (7.11) have been derived analytically for 1D gases in the absence of losses, but they can be directly transferred to the 2D case by adding a transverse dimension perpendicular to the soliton propagation direction. Numerically, soliton solutions have been found in 2D both in atoms [16] and polaritons [35]. In the 2D case however, the solutions are only stable for hydrodynamic solitons generated in the wake of a potential barrier present in the path of a flowing condensate, as we will see below. In the following, we will concentrate on this type of solitons, which have been experimentally observed in polariton superfluids [36] and comment on the differences between the equilibrium (atomic condensates) and out-of-equilibrium cases (polaritons).

7.4 Free Flow Around a Large Obstacle

As we can see from (7.8), (7.9) and (7.11), hydrodynamic features can only appear in the fluid if the phase of the condensate is free to evolve. The generation of transverse currents when the fluid encounters a potential barrier results in its turn in transverse phase gradients which can eventually accommodate solitons and vortices. These excitations themselves present particular phase structures, as shown in Fig. 7.3. Thus, the hydrodynamic regime cannot be attained in the conditions of resonant pumping described above, where the phase of the polariton fluid is imposed by the excitation laser. For this reason, in the conditions of Fig. 7.1 where the resonant laser spot is centred on the defect, while a clear distinction between superfluid (subsonic) and supersonic regimes can be made, hydrodynamic excitations cannot be observed.

Still, from an experimental point of view, hydrodynamic effects would be advantageously accessed in the case of resonant excitation, as both the density and momentum of the fluid can be independently controlled. In order for the fluid's phase to be free to evolve, the pump field should be spatially separated from the part of the fluid where hydrodynamic features are to be observed. This was the proposal of [35], in which the fluid's phase is free to evolve around an obstacle positioned downstream from the injection area [see Fig. 7.3a]. Several regimes where predicted, ranging from superfluidity to vortex streets and soliton nucleation, and experimentally observed in [36], while the hydrodynamic nucleation of single vortex dipoles was evidenced in [33] and [34].

In order to provide an intuitive approach to some of the different accessible hydrodynamic regimes, let us consider a polariton flow resonantly injected in the microcavity with velocity v_∞, in a region upstream but not far from an obstacle (a defect in the microcavity). This is the situation depicted in Fig. 7.4 [36], in which

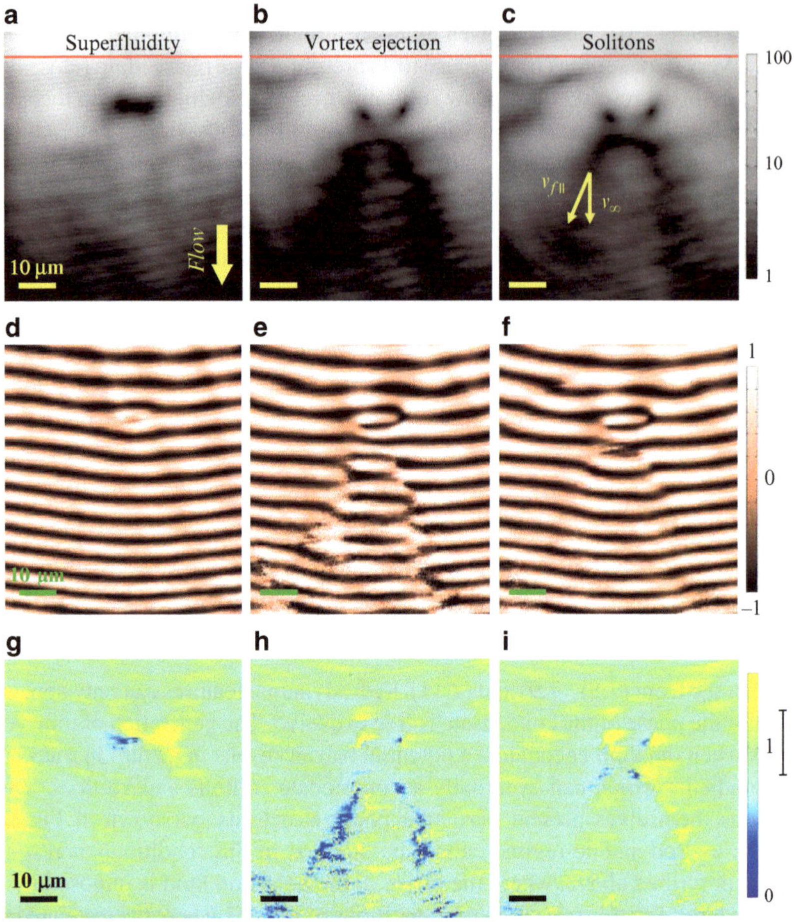

Fig. 7.4 (**a–c**) Real-space-emitted intensity from a polariton fluid resonantly injected above the *red line*, showing the regimes of superfluidity (**a**), ejection of vortices (**b**) and soliton formation (**c**). The polariton density decreases from (**a**) to (**c**). (**d–e**) Interferograms obtained from the interference of the emission in (**a–c**) with a reference beam of homogeneous phase. (**g–i**) Time-averaged visibility of the interference patterns shown in (**d–e**). The continuous passage of vortices is evidenced in (**h**) as a reduced value of the visibility along the vortex streets. A mean error bar in the measurement of the visibility is shown next to the colour scale. From [36]

polaritons are created above the red line in panels (a), (b) and (c). We will assume that the obstacle is larger than the healing length, which is the minimum spatial size of the topological excitations of the fluid. If the polariton lifetime is long enough, the injected gas will flow from the excited area and encounter the obstacle. In order for the overall mass flow to be conserved, the fluid speed around the obstacle is expected to be higher than far from it. In the considered case of an impenetrable barrier, the flow velocity at its equator can be as high as twice v_∞ [15]. This implies that the

Landau criterion for superfluidity needs to be considered locally: even for $v_\infty < c_s$, there might be supersonic regions around the obstacle. This is very different to the case described in Fig. 7.1, where the phase set by the laser imposes a constant speed around the obstacle. Here, the induced currents around the barrier can nucleate vortices at flow speeds as low as $v_\infty \approx 0.4c_s$, as calculated for atomic condensates [15, 17].

At very low velocities ($v_\infty < c_s$), the gas is in the superfluid regime and flows around the obstacle without scattering, as depicted in Fig. 7.4a. A real-space interferometric study of the emission [21] reveals the absence of phase dislocations in the condensate wave function [Fig. 7.4d] and an extended high degree of first order coherence [Fig. 7.4g]. At higher flow velocities or, alternatively, lower polariton densities (lower ratio v_∞/c_s), two dark channels appear in the wake of the barrier, accompanied by extended phase dislocations [Fig. 7.4e]. This regime corresponds to the nucleation of pairs of vortices and antivortices in the wake of the obstacle, being dragged away by the flow along two vortex streets. The continuous passage of individual vortices along these channels results in the low degree of average coherence shown in Fig. 7.4h in these time-integrated measurements. This turbulent situation is attained at $v_\infty \approx 0.4c_s$ ($\bar{c}_s$ being the average sound speed of the fluid around the barrier), a value that agrees well with the predictions for the onset of the emission of vortex trains in equilibrium systems [15], and coincides with the value of v_∞ at which superfluidity is locally broken around the obstacle [33, 34].

At even higher velocities (or lower densities), the rate of vortex emission is so high that the vortex cores start to overlap, giving rise to the formation of a pair of oblique dark solitons. Once the soliton is formed, the repulsive interparticle interactions stabilise its shape as it propagates. This is the case depicted in Fig. 7.4c where the phase jump characteristic of solitons can be well identified in the interference images [Fig. 7.4f]. In this case, we recover a steady state laminar flow with a high degree of coherence [Fig. 7.4i]. If either the fluid velocity or the obstacle size is further increased, the nucleation of higher order soliton multiplets can take place [16], as recently demonstrated in [36].

It is interesting to discuss the stability of the dark f soliton solutions supported by (7.5) in a 2D degenerate gas. While vortex solutions are stable, solitons in a fluid at rest are instable. This is due to the fact that the soliton energy E_s decreases when the propagation velocity v_s increases [62]: $E_s \propto (c_s^2 - v_s^2)^{3/2}$. Simultaneously, as extracted from (7.11), its depth decreases. For instance, if we create a line soliton on a condensate at rest by phase engineering of the wave function, any perturbation in a section of the soliton will tend to accelerate it and make it shallower until it disappears. A careful analysis of the dynamics of the system has shown that the soliton actually decays into pairs of vortices in a phenomenon called the snake instability [63–65].

As proposed by Kamchatnov and Pitaevskii, the snake instability can be circumvented in a 2D gas if the fluid is set in motion in the presence of an obstacle. This is treated in detail in [66, 67] for the case of atomic condensates in equilibrium. The potential barrier does not only create the conditions for the soliton nucleation, as discussed in the previous paragraphs, but also plays an important role in its stability. Once the oblique dark soliton is nucleated in the wake of the barrier, it starts to propagate downstream, thanks to the non-zero component of the fluid velocity $v_{f||}$ along the soliton line [67] [see Fig. 7.4c]. If $v_{f||}$ is larger than the velocity for the propagation of the snake instability along the soliton, the soliton line can indefinitely increase its length starting from the obstacle. In other words, any local perturbation at a given point of the soliton is drifted away by the flow. Thus, the flow of polaritons results, at a given point in space, in the effective damping of perturbations to the solitons, the absolute snake instability being transformed into a convective instability [66].

Recent numerical simulations have shown that in the case of atomic condensates, the minimum flow velocity for dark solitons to be stable is [67] $v_f \approx 1.43c_s$. The case of polaritons is of particular interest in terms of the soliton stability. Polaritons are intrinsically out of equilibrium due to their limited lifetime. This results in an additional damping which should lower the critical velocity for the nucleation of stable dark solitons. First, theoretical and experimental results point into this direction, as it has been shown that oblique dark solitons are stable at subsonic speeds [35, 36].

7.4.1 Non-equilibrium Hydrodynamics

We would finally like to briefly mention other hydrodynamic effects that have been predicted to arise spontaneously from the intrinsic lossy nature of polaritons. Some of the first proposals were introduced by Keeling and Berloff in 2D systems [41, 42], and they have been adapted to 1D by Cuevas and co-workers [43]. The hydrodynamic equations (7.8) and (7.9) show that polariton currents can be generated between pumped and non-pumped regions. This is more easily seen in the case of non-resonant pumping, in which the feeding of the polariton condensate from the optically pumped reservoir can be modelled by an effective non-linear loss term. In (7.8), this can be expressed as [40, 41] $F \rightarrow F \cdot n$. Loss and gain in the condensate are then dependent on its local density and give rise to polariton currents which in its turn affect the local density. Therefore, we see that by using a non-homogeneous pump spot and/or in the presence of an external potential, strong currents can spontaneously appear in the condensate. A very interesting situation is found when the size of the pump spot is different that of a given confining potential, for instance a 2D harmonic trap. In this case, it has been predicted that lattices of vortices are the stable solutions, spontaneously breaking the radial symmetry of the system [41], as depicted in Fig. 7.5a. The non-conservative open-dissipative character of polariton condensates allows for the appearance of a net angular momentum. Analogously, in a 1D harmonic trap, the supercurrents between high- and low-density regions give rise to several regimes, including strong turbulence and the formation of standing solitons [43], as shown in Fig. 7.5b.

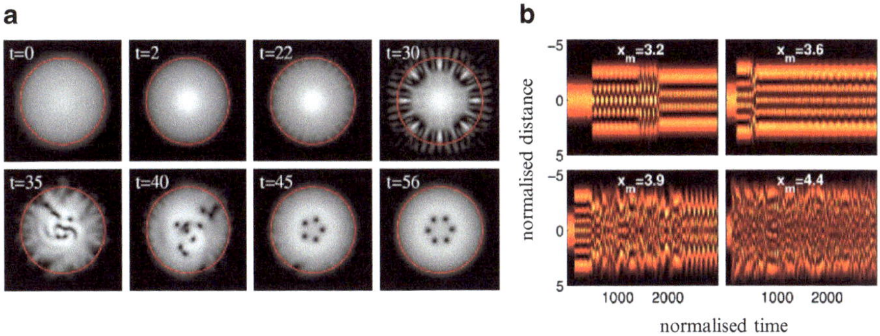

Fig. 7.5 (**a**) Simulations showing the real-space emission at different times from a non-resonantly excited microcavity with a Gaussian spatial potential whose width is marked by the red lines. The spontaneous formation of a lattice of vortices can be observed. From [41], copyright (2008) by the American Physical Society. (**b**) Simulated real-space emission from a 1D polariton microwire with a longitudinal harmonic potential, for different relative sizes of the excitation spot x_m. Depending on the ratio between the size of the ground state of the harmonic potential and the excitation spot, spontaneous soliton formation (*upper panels*) or turbulence (*lower panels*) is expected. From [43]

These examples show new interesting phenomena arising from the non-equilibrium nature of the polariton system. Indeed, the open-dissipative character of polariton condensates favours the spontaneous nucleation of topological excitations, whose stability is stronger than in the archetypical equilibrium case of ultracold atomic condensates. This is probably one of the reasons why oblique dark solitons have been first observed in polariton condensates, though their existence had been originally predicted for atomic BEC. This fact can therefore be a very important advantage in the study of turbulence in quantum fluids which, in the case of polaritons, has been predicted to have a very rich phenomenology [42, 68]. Current hydrodynamic studies take advantage of the possibility to optically control the polariton landscape [69] to create different obstacle geometries. For instance, Gaussian-type potential profiles (instead of abrupt ones as those considered in Figs. 7.1 and 7.4) have shown the nucleation of vortex pairs upstream of the barrier (instead of downstream) [34]. Interesting effects are also expected when taking into account the polariton spin degree of freedom. Examples are the so-called half vortices, in which there is a simultaneous rotation of the phase and the pseudospin of the polariton fluid around the vortex core [70] and which were recently observed [22]. Half-solitons [71] and black hole models [44] in spinor polariton condensates are among some of the exciting phenomena to be studied in the near future, with no analogue in atomic condensates.

We would like to thank N. G. Berloff, E.Cancellieri, A. M. Kamchatnov, G. Malpuech and N. Pavloff for stimulating discussions during the writing of this material.

References

1. A.J. Leggett, Superfluidity. Rev. Mod. Phys. **71**, S318 (1999)
2. W.F. Vinen, The detection of single quanta of circulation in liquid helium II. Proc. R. Soc. Lond. A **260**, 218 (1961)
3. G.B. Hess, W.M. Fairbank, Measurements of angular momentum in superfluid helium. Phys. Rev. Lett. **19**, 216 (1967)
4. R.E. Packard, T.M. Sanders, Observations on single vortex lines in rotating superfluid helium. Phys. Rev. A **6**, 799 (1972)
5. F. Chevy, K.W. Madison, J. Dalibard, Measurement of the angular momentum of a rotating Bose–Einstein condensate. Phys. Rev. Lett. **85**, 2223 (2000)
6. K.W. Madison, F. Chevy, W. Wohlleben et al., Vortex formation in a stirred Bose–Einstein condensate. Phys. Rev. Lett. **84**, 806 (2000)
7. J.R. Abo-Shaeer, C. Raman, J.M. Vogels et al., Observation of vortex lattices in Bose–Einstein condensates. Science **292**, 476 (2001)
8. D.R. Allum, P.V.E. McClintock, A. Phillips et al., The breakdown of superfluidity in liquid ^{4}He: an experimental test of Landau's theory. Phil. Trans. R. Soc. Lond. A **284**, 179 (1977)
9. M. Hartmann, F. Mielke, J.P. Toennies et al., Direct spectroscopic observation of elementary excitations in superfluid He droplets. Phys. Rev. Lett. **76**, 4560 (1996)
10. C. Raman, M. Köhl, R. Onofrio et al., Evidence for a critical velocity in a Bose–Einstein condensed gas. Phys. Rev. Lett. **83**, 2502 (1999)
11. R. Onofrio, C. Raman, J.M. Vogels et al., Observation of superfluid flow in a Bose–Einstein condensed gas. Phys. Rev. Lett. **85**, 2228 (2000)
12. I. Carusotto, S.X. Hu, L.A. Collins et al., Bogoliubov–Cerenkov radiation in a Bose–Einstein condensate flowing against an obstacle. Phys. Rev. Lett. **97**, 260403 (2006)
13. S.C. Whitmore, W. Zimmermann, Observation of quantized circulation in superfluid helium. Phys. Rev. **166**, 181 (1968)
14. C. Ryu, M.F. Andersen, P. Clade et al., Observation of persistent flow of a Bose–Einstein condensate in a toroidal trap. Phys. Rev. Lett. **99**, 260401 (2007)
15. T. Frisch, Y. Pomeau, S. Rica, Transition to dissipation in a model of superflow. Phys. Rev. Lett. **69**, 1644 (1992)
16. G.A. El, A. Gammal, A.M. Kamchatnov, Oblique dark solitons in supersonic flow of a Bose–Einstein condensate. Phys. Rev. Lett. **97**, 180405 (2006)
17. T. Winiecki, B. Jackson, J.F. McCann et al., Vortex shedding and drag in dilute Bose–Einstein condensates. J. Phys. B: At. Mol. Opt. Phys. **33**, 4069 (2000)
18. S. Inouye, S. Gupta, T. Rosenband et al., Observation of vortex phase singularities in Bose–Einstein condensates. Phys. Rev. Lett. **87**, 080402 (2001)
19. T.W. Neely, E.C. Samson, A.S. Bradley et al., Observation of vortex dipoles in an oblate Bose–Einstein condensate. Phys. Rev. Lett. **104**, 160401 (2010)
20. J. Kasprzak, M. Richard, S. Kundermann et al., Bose–Einstein condensation of exciton polaritons. Nature **443**, 409 (2006)
21. K.G. Lagoudakis, M. Wouters, M. Richard et al., Quantized vortices in an exciton–polariton condensate. Nat. Phys. **4**, 706 (2008)
22. K.G. Lagoudakis, T. Ostatnicky, A.V. Kavokin et al., Observation of half-quantum vortices in an exciton–polariton condensate. Science **326**, 974 (2009)
23. C.W. Lai, N.Y. Kim, S. Utsunomiya et al., Coherent zero-state and π-state in an exciton–polariton condensate array. Nature **450**, 529 (2007)
24. E. Wertz, L. Ferrier, D.D. Solnyshkov et al., Spontaneous formation and optical manipulation of extended polariton condensates. Nat. Phys. **6**, 860 (2010)
25. M. Wouters, I. Carusotto, Goldstone mode of optical parametric oscillators in planar semiconductor microcavities in the strong-coupling regime. Phys. Rev. A **76**, 043807 (2007)
26. D. Ballarini, D. Sanvitto, A. Amo et al., Observation of long-lived polariton states in semiconductor microcavities across the parametric threshold. Phys. Rev. Lett. **102**, 056402 (2009)

27. D. Sanvitto, F.M. Marchetti, M.H. Szymanska et al., Persistent currents and quantized vortices in a polariton superfluid. Nat. Phys. **6**, 527 (2010)
28. F.M. Marchetti, M.H. Szymanacuteska, C. Tejedor et al., Spontaneous and triggered vortices in polariton optical-parametric-oscillator superfluids. Phys. Rev. Lett. **105**, 063902 (2010)
29. M. Wouters, I. Carusotto, Superfluidity and critical velocities in nonequilibrium Bose–Einstein condensates. Phys. Rev. Lett. **105**, 020602 (2010)
30. I. Carusotto, C. Ciuti, Probing microcavity polariton superfluidity through resonant rayleigh scattering. Phys. Rev. Lett. **93**, 166401 (2004)
31. A. Amo, D. Sanvitto, F.P. Laussy et al., Collective fluid dynamics of a polariton condensate in a semiconductor microcavity. Nature **457**, 291 (2009)
32. A. Amo, J. Lefrère, S. Pigeon et al., Superfluidity of polaritons in semiconductor microcavities. Nat. Phys. **5**, 805 (2009)
33. G. Nardin, G. Grosso, Y. Leger et al., Hydrodynamic nucleation of quantized vortex pairs in a polariton quantum fluid. Nat. Phys. **7**, 635 (2011)
34. D. Sanvitto, S. Pigeon, A. Amo et al., All-optical control of the quantum flow of a polariton superfluid. Nat. Phot. **5**, 610 (2011)
35. S. Pigeon, I. Carusotto, C. Ciuti, Hydrodynamic nucleation of vortices and solitons in a resonantly excited polariton superfluid. Phys. Rev. B: Condens. Matter Mater. Phys. **83**, 144513 (2011)
36. A. Amo, S. Pigeon, D. Sanvitto et al., Polariton superfluids reveal quantum hydrodynamic solitons. Science **332**, 1167 (2011)
37. C. Ciuti, I. Carusotto, Quantum fluid effects and parametric instabilities in microcavities. Phys. Status Solid B **242**, 2224 (2005)
38. M.H. Szymanska, J. Keeling, P.B. Littlewood, Nonequilibrium quantum condensation in an incoherently pumped dissipative system. Phys. Rev. Lett. **96**, 230602 (2006)
39. J. Keeling, F.M. Marchetti, M.H. Szymanska et al., Collective coherence in planar semiconductor microcavities. Semicond. Sci. Technol. **22**, R1 (2007)
40. M. Wouters, I. Carusotto, Excitations in a nonequilibrium Bose–Einstein condensate of exciton polaritons. Phys. Rev. Lett. **99**, 140402 (2007)
41. J. Keeling, N.G. Berloff, Spontaneous rotating vortex lattices in a pumped decaying condensate. Phys. Rev. Lett. **100**, 250401 (2008)
42. N.G. Berloff, Turbulence in exciton–polariton condensates (2010), arXiv:1010.5225
43. J. Cuevas, A.S. Rodrigues, R. Carretero-Gonzalez et al., Nonlinear excitations, stability inversions and dissipative dynamics in quasi-one-dimensional polariton condensates Phys. Rev. B: Condens Matter Mater. Phys. **83**, 245140 (2011).
44. D.D. Solnyshkov, H. Flayac, G. Malpuech, Black holes and wormholes in spinor polariton condensates (2011), arXiv:1104.3013v1
45. C.J. Pethick, H. Smith, *Bose–Einstein Condensation in Dilute Gases* (Cambridge University Press, Cambridge, 2002)
46. J. Steinhauer, R. Ozeri, N. Katz et al., Excitation spectrum of a Bose–Einstein condensate. Phys. Rev. Lett. **88**, 120407 (2002)
47. S. Utsunomiya, L. Tian, G. Roumpos et al., Observation of Bogoliubov excitations in exciton–polariton condensates. Nat. Phys. **4**, 700 (2008)
48. V. Kohnle, Y. Léger, M. Wouters et al., From single particle to superfuid excitations in a dissipative polariton gas (2011), arXiv:1103.1488v1
49. E. Cancellieri, F.M. Marchetti, M.H. Szymanska et al., Superflow of resonantly driven polaritons against a defect. Phys. Rev. B: Condens. Matter Mater. Phys. **82**, 224512 (2010)
50. R.Y. Chiao, J. Boyce, Bogoliubov dispersion relation and the possibility of superfluidity for weakly interacting photons in a two-dimensional photon fluid. Phys. Rev. A **60**, 4114 (1999)
51. E.L. Bolda, R.Y. Chiao, W.H. Zurek, Dissipative optical flow in a nonlinear Fabry–Pérot cavity. Phys. Rev. Lett. **86**, 416 (2001)
52. P. Leboeuf, S. Moulieras, Superfluid motion of light. Phys. Rev. Lett. **105**, 163904 (2010)

53. D. Sanvitto, D.M. Whittaker, M.S. Skolnick et al., Continuous wave pump–probe experiment on a planar microcavity. Phys. Status Solid A **202**, 353 (2005)
54. P. Nozieres, D. Pines, *Theory of Quantum Liquids* (Westview Press, Boulder, CO, USA, 1999)
55. D.N. Krizhanovskii, D.M. Whittaker, R.A. Bradley et al., Effect of interactions on vortices in a nonequilibrium polariton condensate. Phys. Rev. Lett. **104**, 126402 (2010)
56. M. Wouters, V. Savona, Superfluidity of a nonequilibrium Bose–Einstein condensate of polaritons. Phys. Rev. B: Condens Matter Mater. Phys. **81**, 054508 (2010)
57. T. Tsuzuki, Nonlinear waves in the Pitaevskii–Gross equation. J. Low Temp. Phys. **4**, 441 (1971)
58. F. Dalfovo, S. Giorgini, L.P. Pitaevskii et al., Theory of Bose–Einstein condensation in trapped gases. Rev. Mod. Phys. **71**, 463 LP (1999)
59. A.D. Jackson, G.M. Kavoulakis, C.J. Pethick, Solitary waves in clouds of Bose–Einstein condensed atoms. Phys. Rev. A **58**, 2417 (1998)
60. S. Burger, K. Bongs, S. Dettmer et al., Dark solitons in Bose–Einstein condensates. Phys. Rev. Lett. **83**, 5198 (1999)
61. J. Denschlag, J.E. Simsarian, D.L. Feder et al., Generating solitons by phase engineering of a Bose–Einstein condensate. Science **287**, 97 (2000)
62. Y.S. Kivshar, B. Luther-Davies, Dark optical solitons: physics and applications. Phys. Rep. **298**, 81 (1998)
63. D.E. Pelinovsky, Y.A. Stepanyants, Y.S. Kivshar, Self-focusing of plane dark solitons in nonlinear defocusing media. Phys. Rev. E: Stat. Phys. Plasmas Fluids Relat. Interdiscip. Topics **51**, 5016 LP (1995)
64. E.A. Kuznetsov, S.K. Turitsyn, Instability and collapse of solitons in media with a defocusing nonlinearity. Sov. Phys. JETP **67**, 1583 (1988)
65. Z. Dutton, M. Budde, C. Slowe et al., Observation of quantum shock waves created with ultra-compressed slow light pulses in a Bose–Einstein condensate. Science **293**, 663 (2001)
66. A.M. Kamchatnov, L.P. Pitaevskii, Stabilization of solitons generated by a supersonic flow of Bose–Einstein condensate past an obstacle. Phys. Rev. Lett. **100**, 160402 (2008)
67. A.M. Kamchatnov, S.V. Korneev, Condition for convective instability of dark solitons Phys. Lett. A **375**, 2577 (2011)
68. J. Keeling, N.G. Berloff, Controllable half-vortex lattices in an incoherently pumped polariton condensate (2011), arXiv:1102.5302v1
69. A. Amo, S. Pigeon, C. Adrados et al., Light engineering of the polariton landscape in semiconductor microcavities. Phys. Rev. B: Condens Matter Mater. Phys. **82**, 081301 (2010)
70. Y.G. Rubo, Half vortices in exciton polariton condensates. Phys. Rev. Lett. **99**, 106401 (2007)
71. H. Flayac, D.D. Solnyshkov, G. Malpuech, Oblique half-solitons and their generation in exciton–polariton condensates. Phys. Rev. B: Condens Matter Mater. Phys. **83**, 193305 (2011)

Chapter 8
Spin Effects in Exciton–Polariton Condensates

Alexey Kavokin

Abstract Exciton–polaritons in microcavities form an unusual gas of weakly interacting bosons. It has no direct analogy in cold atomic gases, superfluids or superconductors due to its two-component spin structure: in typical planar microcavities the polaritons have two allowed spin projections to the structure axis. This is why the order parameter of a polariton condensate is a complex spinor. The magnitude and, possibly, sign of polariton–polariton interaction constant depends on the spin state of polaritons. The energy of an exciton–polariton condensate is also spin-dependent. These specific features make polariton condensates a unique laboratory for studies of spin effects in interacting Bose gases. Several new spin-dependent effects in polariton condensates have been theoretically predicted and experimentally observed during the recent decade. This review chapter addresses some of these effects: polarisation multistability, spin switching, spin rings and spin Meissner effect. In the last section we address the perspective of observation of spin superfluidity in microcavities.

8.1 Introduction

Powering the optical fibre internet with its huge global reach, photonics has changed our lives. Optical fibres snake across continents and oceans carrying terabits per second of data in a vast information network that brings untold human connectivity. Capacity demand continues to grow at a startling rate, doubling every two years, while the internet is estimated as burning 4% of world energy usage. The optical

A. Kavokin (✉)
Laboratoire Charles Coulomb, CC074, Universite de Montpellier II, Place Eugene de Bataillon, Montpellier, Cedex 34095, France

Physics and Astronomy School, University of Southampton, Highfield, Southampton, SO171BJ, UK
e-mail: A.Kavokin@soton.ac.uk

D. Sanvitto and V. Timofeev (eds.), *Exciton Polaritons in Microcavities*, Springer Series in Solid-State Sciences 172, DOI 10.1007/978-3-642-24186-4_8,

Internet is reaching its capacity limits. The solution to these consequences of unbridled demand is *spin/polarisation encoding* of information to overcome the existing bottlenecks and employing next-generation optical components.

Encoding of information with a vector polarisation of mixed light-matter quasiparticles, exciton–polaritons, is one of the goals of polaritonics. Polaritonics proposes a research programme on the frontier between semiconductor optics and quantum transport, involving the mesoscopic and statistical physics. The transport of mixed light-matter quasiparticles which carry an integer spin and have bosonic statistics appears to be a valuable alternative to traditional spintronics, and it may allow for dramatic improvement of capacity and efficiency of optical communication lines. This short review addresses the recent progresses in the field of polaritonics with an emphasis on the spin interaction effects.

Strong light matter coupling in planar semiconductor microcavities has been reported for the first time in 1992 [1]. This observation has resulted in the creation of a new research field which has continuously grown till today. The early 1990s has been mainly devoted to the observation and theoretical description of the linear optical properties of cavity exciton–polaritons [2–4]. In 1996, Imamoglu [5, 6] was the first to propose the use of the bosonic character of cavity polaritons to build up an exciton–polariton condensate that would emit coherent laser light spontaneously. Two years later, Le Si Dang et al. [7] reported the non-linear emission of a II–VI microcavity under non-resonant excitation and in 2000 Savvidis, Baumberg et al. [8] demonstrated, under resonant excitation, the bosonic behaviour of cavity polaritons. These breakthroughs have engendered an intense experimental and theoretical activity which has recently culminated by the report of the Bose–Einstein condensation of polaritons, in different systems [9–11], and up to room temperature [12].

An important property of the cavity polaritons is their spin, which is directly connected with the polarisation of the light absorbed or emitted by a microcavity. One of the first theoretical works focused on this problem has been published in 1999 [13] and was aiming to describe polariton polarisation states in the linear regime. In 2002, the first experimental measurements of polariton spin dynamics in the non-linear regime have been published [14, 15]. They were showing striking effects clearly linked with the interplay between polariton spin relaxation and bosonic stimulated scattering but which, at that time, were not clearly explained. These first chapters have been followed by an intense theoretical and experimental activity. The spin dynamics of cavity polaritons has been found to play a key role in the Bose–Einstein condensation of cavity polaritons and associated phenomena. A large variety of new fundamental effects has been proposed and observed. These effects can be practically used in future spin-dependent optoelectronic or "spin-optronic" devices. Recently, the experimental realisation of polariton polarisation multistability [16] and optical spin switches [17] has been reported. This opens way to realisation of polariton transistors and optical integrated circuits.

Spin-optronics is based on a concept of bosonic spin transport. It studies electrically neutral short-living spin carriers: exciton–polaritons. In contrast, traditional

Table 8.1 Essential characteristics of spintronic and spin-optronic devices

	Spintronics	Spin-optronics
Spin carrier	Electrically charged fermion	Electrically neutral boson
Control tools	External electric and magnetic fields	Polarisation, frequency incidence angle of light
Output	Electronic spin polarisation	Polarised emitted light
Size	Nanometers	Microns
Operation temperature	Cryogenic temperatures	Possibly room temperature

spintronics relies on the fermionic electrically charged spin carriers (electrons and holes). This fundamental difference dictates essentially different characteristics of *spintronic* and *spin-optronic* devices summarised in Table 8.1.

One can see that spin-optronics brings advantages of high operation temperature (possibly, room temperature) and full optical control. It also allows for the fast switching (theoretically, the switching times can be as short as several picoseconds) and low switching power. Among disadvantages of spin-optronic devices is their relatively large size (of the order of microns compared to the expected nanometre size of future spintronic devices). It is unlikely that one day spin-optronics can fully replace traditional electronics, while it is most likely that in 5–15 years spin-optronic devices will bring revolutionary changes to the market of optical logic elements, optical circuits, switches, etc.

Here, we review some effects in spin physics of polariton condensates. In particular, attention will be focussed on the spin Meissner effect, spin switching and spin superfluidity.

8.2 Spin Meissner Effect

Spin Meissner effect consists in full suppression of Zeeman splitting of the lowest energy eigenstate in a microcavity due to the Bose–Einstein condensation of exciton–polaritons in this state. The effect predicted by Rubo et al. in 2006 [18] implies thermal equilibrium in the system. Assuming zero temperature, one can represent the free energy density in the system as

$$F = -\mu n - g\mu_B B S_z + \frac{1}{4}(\alpha_1 + \alpha_2)n^2 + (\alpha_1 - \alpha_2)S_z^2, \tag{8.1}$$

where μ is the chemical potential, $n = n_\uparrow + n_\downarrow$ is the sum of densities of spin-up and spin-down polaritons $n_\uparrow$and $n_\downarrow$, respectively, and $S_z = \frac{1}{2}(n_\uparrow - n_\downarrow)$, g is the exciton g-factor, μ_B is Bohr magneton, α_1 and α_2 are interaction constants of polaritons with parallel and antiparallel spins, respectively.

Qualitatively, the spin Meissner effect can be understood in the following way. In the absence of external magnetic field, an exciton–polariton condensate in a microcavity is expected to be linearly polarised. The linear polarisation ($S_z = 0$) minimises the free energy of the condensate if the polaritons with parallel spins repel each other stronger than the polaritons with antiparallel spins ($\alpha_1 > \alpha_2$). The condensate emits light at the energy given by its chemical potential. From the condition

$$\frac{\partial F}{\partial n} = 0 \tag{8.2}$$

we easily obtain

$$\mu = \frac{\alpha_1 + \alpha_2}{2} n. \tag{8.3}$$

Once an external magnetic field is applied, the condensate tends to orient all polariton spins along the field due to the Zeeman effect. However, at low magnetic fields, it proves impossible as such spin orientation would result in the increase of the free energy of the system due to strong repulsion of polaritons with parallel spins. Therefore, up to some critical field

$$B_c = \frac{(\alpha_1 - \alpha_2)n}{\mu_B g}. \tag{8.4}$$

the condensate remains partially spin-polarised and emits an elliptically polarised light. The circular polarisation degree of emission is given by

$$\rho = \frac{2S_z}{n} = \frac{\mu_B g B}{(\alpha_1 - \alpha_2)n}. \tag{8.5}$$

This implied that spin-up and spin-down components of the condensate remain fully degenerate, and the order parameter remains a two-dimensional complex vector.

The chemical potential and emission energy are still given by (8.3) in this regime. Above the critical field, the condensate is fully spin-polarised, it emits the circularly polarised light, and its order parameter is a scalar. The chemical potential linearly decreases with increase of the magnetic field as

$$\mu = \frac{\alpha_1 + \alpha_2}{2} n - \frac{1}{2} g \mu_B (B - B_c). \tag{8.6}$$

Zeeman splitting of the polariton condensate is exactly zero at $B < B_c$. Above the critical field, the degeneracy of two spin components is lifted and Zeeman splitting

$$\Delta = g\mu_B(B - B_c) \tag{8.7}$$

can be observed. This effect has been experimentally observed by Larionov et al. [19] in 2010 (Fig. 8.1, right panel).

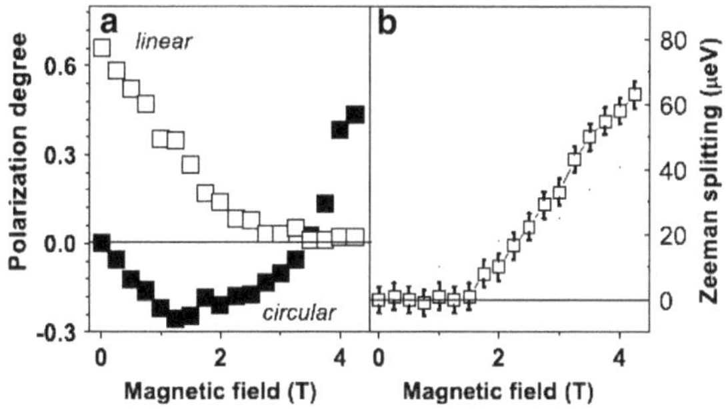

Fig. 8.1 Experimental manifestation of the spin Meissner effect in [19]. The microcavity has been pumped non-resonantly, and formation of the exciton–polariton BEC has been observed above some excitation threshold. Full suppression of Zeeman splitting has been observed up to the critical magnetic field dependent on the polariton population ($B_c = 1.8$ T in the right panel of this figure). The behaviour of polarisation of the condensate as a function of magnetic field (*left panel*) deviated from predictions of (8.5) probably due to the non-equilibrium character of the polariton condensate at low magnetic fields

8.3 Spin Switching

Spin switching is based on polarisation multistability in microcavities, an effect which has been theoretically predicted by Gippius et al. [20] in 2007 and experimentally observed by Paraiso et al. [16] in 2010. It is based on a high optical non-linearity of microcavities which opens way to coexistence of up to four stable polarisation states of a resonantly driven polariton condensate. The idea of spin switching is very simple. Suppose that the energy of cw excitation laser is blue detuned from the lowest polariton state in a microcavity by the energy δ. The laser light is elliptically polarised so that it creates non-equal concentrations of spin-up and spin-down polaritons $n_\uparrow \neq n_\downarrow$ in general. In this case, the blueshifts of energies of spin-up and spin-down polaritons due to their interactions are

$$\delta_\uparrow = \alpha_1 n_\uparrow + \alpha_2 n_\downarrow, \quad \delta_\downarrow = \alpha_1 n_\downarrow + \alpha_2 n_\uparrow, \tag{8.8}$$

respectively. If $\delta_\uparrow, \delta_\downarrow < \delta$, the excitation of both types of polaritons is non-resonant and absorption of light is relatively low. We will call this state (0,0), indicating that the densities of spin-up and spin-down polaritons are low. Now, suppose that we send a short pulse of right-circularly polarised light which induces a rapid increase of $n_\uparrow$, so that the condition $\delta_\uparrow = \delta$ is fulfilled. This means that the eigenstate of spin-up polaritons is brought into resonance with the *cw* laser. The efficiency of absorption strongly increases, and the system passes to (1,0) state, where the density of spin-up polaritons is high, while the density of spin-down polaritons remains low. The system would emit right circular polarised light in this regime. The trigger pulse of the opposite circular polarisation would bring the system to (0,1) state for the same reasons. It would emit left circular polarised light in this case. Finally, a

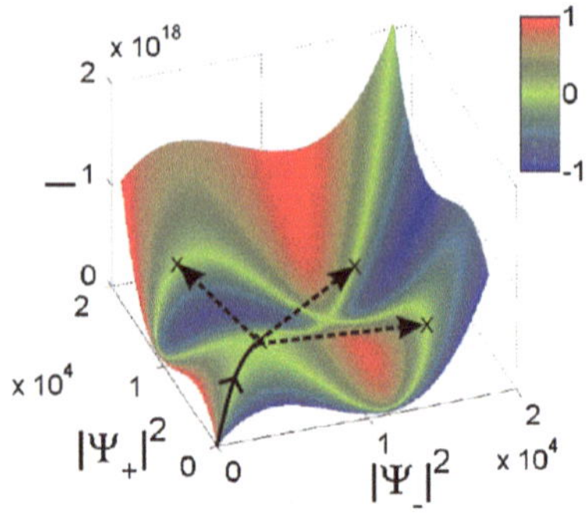

Fig. 8.2 Polarisation multistability (from [20]). $|\Psi_+| = n_\uparrow$, $|\Psi_-| = n_\downarrow$, crosses indicate (0,0), (1,0), (0,1) and (1,1) states

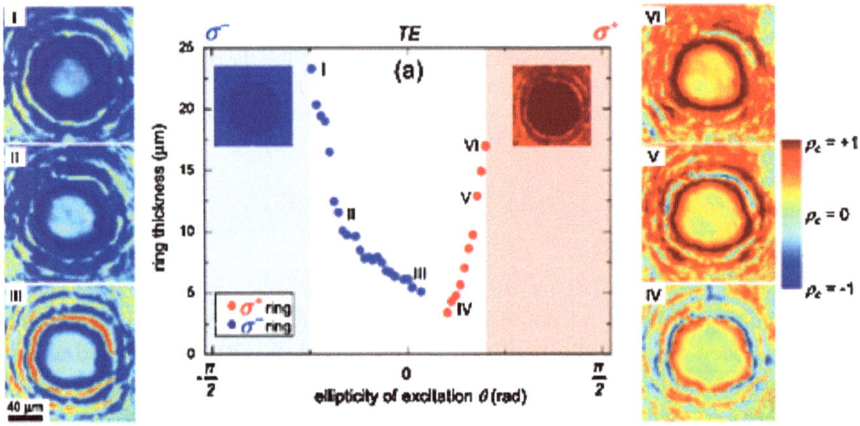

Fig. 8.3 Polariton spin rings predicted by Shelykh et al. [21] have been recently observed by Sarkar et al. [22] and Adrados et al. [23]. They manifest themselves as circular polarisation patterns in near-field photoluminescence spectra of microcavities

trigger pulse of linear polarisation may bring the system to (1,1) state, in which case it would emit linearly polarised light. Remarkably, once brought to a given state, the system remains in this state until the next trigger pulse arrives or until the *cw* pump is switched off (Fig. 8.2).

In the case of spatially inhomogeneous pumping, the polarisation switching effect may lead to formation of peculiar polarisation patterns or "spin rings" (Fig. 8.3). Spin rings are excited by elliptically polarised laser beams which have a Gaussian intensity profile in the plane of the sample. If the intensity of pump and its detuning from the polariton branch δ are chosen correctly, (1,1) state is realised in the centre of the excitation spot. It is surrounded by a ring of (1,0) or (0,1) condensates which emit right- or left-circularly polarised light (depending on the sign of the circular polarisation degree of pump). Outside the ring, the system is in a (0,0) state. While the polarisation multistability, spin switching and spin rings have been experimentally observed, the problem of switching the system back from, say, (1,0) to (0,0) state or from (0,1) to (0,0) state seems to require a supplementary investigation. In theory, the back switching can be achieved by sending a trigger pulse which would cause the temporary decrease of concentration of polaritons in a given spin component.

Liew et al. [24] proposed using a trigger pulse coming with the opposite phase with respect to the *cw* pump. This method requires a careful phase tuning of the trigger pulse, which is hard to realise experimentally. Another, seemingly simple, way to realise back switching would be by sending a trigger pulse of the opposite circular polarisation with respect to one which needs to be switched off. Equation (8.8) shows that introduction of supplementary, say, spin-up polaritons would induce reduction of the blueshift for the polaritons of the opposite spin if the interaction constant $\alpha_2 < 0$. In this case, the polariton branch in spin-down polarisation would be brought out of resonance with the excitation pulse, and the system would pass to (0,0) state. Recently, this kind of effect has been observed by the group of EPFL [25]; however, its interpretation is still uncertain. The uncertainty arises because the sign of α_2 is a subject of controversy and in any case it may strongly vary as a function of exciton–photon detuning in microcavities [26].

8.4 Spin Superfluidity

Having a light mass and significant coherence length (up to 0.1–0.2 mm), the polaritons ballistically propagate over hundreds of microns in the plane of microcavities [27]. The optical spin Hall effect (OSHE) proposed theoretically in 2005 [28] and experimentally observed in 2007 [29] demonstrated a possibility of creation of the coherent polariton spin currents due to conversion of linear to circular polarisation because of TE-TM splitting of exciton–polariton modes (Fig. 8.4).

These observations encourage development of Spin-optronics, a new supradisciplinary research area on the interface between spintronics and solid-state optics. The exciton–polaritons are electrically neutral spin carriers. As such, they cannot be directly manipulated by external electric or magnetic fields. On the other hand, they are efficiently controlled by light, having their polarisation and spin directly dependent on the polarisation, frequency, and the incidence angle of light illuminating the cavity. Another important difference between polaritons and conventional spin carriers (electrons and holes) comes from the finite lifetime of the polaritons, which limits their mean free path in the cavities sometimes stronger than the disorder scattering. Furthermore, contrary to electrons and holes the polaritons obey bosonic statistics, they may form Bose–Einstein condensates and superfluids. This opens an intriguing perspective of creation of the superfluid spin currents, which would be impossible to realise in fermionic systems. On the theoretical side, the transport properties of exciton–polaritons have not yet been studied in detail. Glazov and Golub have proposed a model which describes the resonant backscattering of spin-polarised polaritons [30]. Shelykh et al. have introduced the topological (Berry) phase of exaction–polaritons and proposed the Aharonov–Bohm

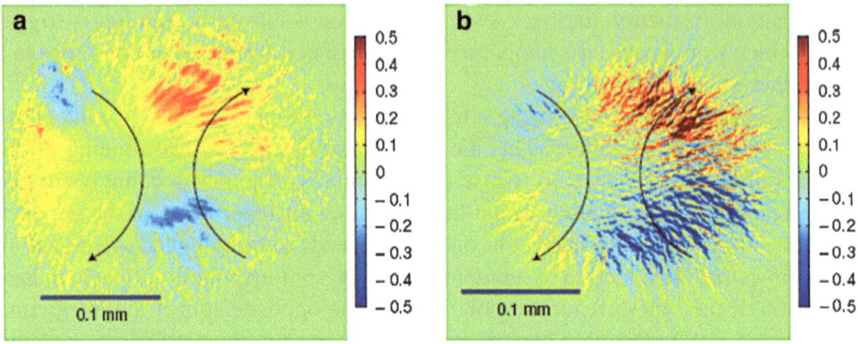

Fig. 8.4 Polariton spin currents induced in plane of a microcavity by a linearly polarised light: experiment (**a**) and theory (**b**). The colour scale shows the circular polarisation degree of the propagating polaritons (from [29]). The effect is based on polarisation conserving resonant Rayleigh scattering which populates the "elastic circle" (i.e. an equi-energy circle in the reciprocal space) with linearly polarised polariton. The TE-TM splitting of exciton–polariton modes induces conversion of linear to circular polarisation, while the sign and the magnitude of circular polarisation degree is governed by orientation of the wave vectors of propagating polaritons

ring interferometer based on exciton–polaritons [31]. These first works show an importance of creation of the theory of polariton spin transport.

Here we attempt to formulate a rough concept of spin superfluidity in a few lines.

If the propagating polaritons form a coherent state (condensate) characterised by a complex order parameter

$$\Psi = \begin{bmatrix} \sqrt{n_\uparrow} e^{i\varphi_+} \\ \sqrt{n_\downarrow} e^{i\varphi_-} \end{bmatrix}, \tag{8.9}$$

Let us define the spin current density j as the concentration of spins which cross a unity surface in a unit of time in a given direction. This definition allows us to introduce the polariton spin current density as

$$\mathbf{j} = \frac{\hbar}{m^*} \begin{pmatrix} n_\uparrow \nabla\varphi_+ \\ n_\downarrow \nabla\varphi_- \end{pmatrix} = \hat{\sigma} \begin{pmatrix} \Delta\mu_+ \\ \Delta\mu_- \end{pmatrix}. \tag{8.10}$$

where m^* is the polariton effective mass (if the current is conducted by polaritons situated in the non-parabolic part of the dispersion, a more general definition involving the flux operator may be given). $\hat{\sigma}$ is the spin conductivity tensor, μ_+ and μ_- are the chemical potentials for spin-up and spin-down polaritons.

The components of the spin conductivity tensor may be measured if two spin-polarised polariton condensates are linked by a sufficiently long one-dimensional channel. The condensates may be excited by cw lasers blue detuned from the lowest energy polariton states by the energies δ_1 and δ_2, respectively. Assume that both condensates are in "switched-on" states characterised by identical circular polarisation, so that their energies coincide with the energies of the excitation lasers. In this case, the diagonal elements of the spin conductivity tensor may be defined as

$$\sigma_{++} = \sigma_{--} = \left. \frac{j}{\delta_2 - \delta_1} \right|_{\delta_2 \to \delta_1}. \tag{8.11}$$

In the superfluid regime, σ_{++} would tend to infinity due to the stationary Josephson effect between two condensates. The magnitude of spin current would be dependent on the relative phases of two pumps as well as on the concentration and polarisation of polaritons in the channel. In the simplest case of two circularly polarised condensates of phases φ_1 and φ_2, the current in the channel of length L would be given by

$$j = \frac{\hbar n(\phi_2 - \phi_1)}{m^* L}, \tag{8.12}$$

where n is the polariton density in the channel. Note that this consideration neglects radiative losses in the channel.

Non-diagonal elements of the spin conductivity tensor could be experimentally measured if two pumps exciting the polariton condensates on both sides of the channel have opposite circular polarisations. The value and behaviour of these elements would be strongly dependent on the strength of interaction between the polaritons of opposite spins given by the constant α_2 and on the external fields mixing polariton spin states in the channel. Typically, an external effective magnetic field able to induce conversion of spin-up to spin-down polaritons and vice versa may come from the splitting of linearly polarised polariton states due to the strain or optical anisotropy of the system.

In order to detect superfluid spin currents, spin selective near-field interferometry experiments seem to be the most suitable. Figure 8.5 shows the experimental set-up used by Lagoudakis et al. [32] in order realise such kind of experiment. The polarised interferometry allows for restoring spatial dependencies of the phases φ_+ and φ_-. Once the spatial variation of phases is known, the spin current can be obtained using (8.10).

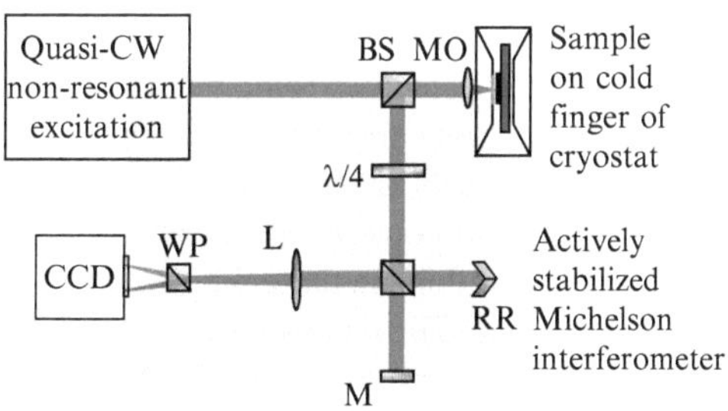

Fig. 8.5 Experimental set-up of the polarisation resolved interferometry (supplementary information to [32]). Two polarisation components σ^+ and σ^- get separated by means of the Wollaston prism, and they are sent on the two opposite sides of the same CCD. *BS*, beam splitter; *M*, mirror; *MO*, microscope objective; $\lambda/4$, quarter wave plate; *RR*, retroreflector; *L*, lens; *WP*, Wollaston prism and *CCD*, charge-coupled device

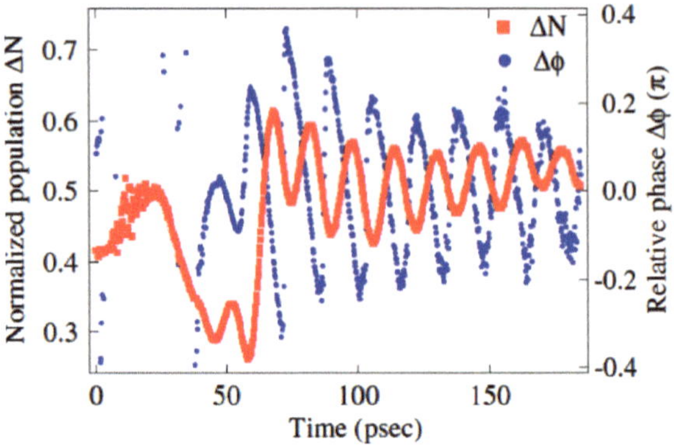

Fig. 8.6 Relative population and relative phase of two Josephson coupled condensates (from [33]). Observed oscillations evidence the non-stationary Josephson effect

In the experiment [32], the condensate of exciton–polaritons has been created by stationary non-resonant pumping. In this configuration, the Josephson spin currents cannot be measured. On the other hand, this experiment presented the first experimental evidence of half-quantum vortices (HQV), which are elementary excitations of spinor condensates. The signature of HQV is a vortex in one of the spin components coexisting with no vortex in the other spin component (for more details see the chapter "Vortices in spontaneous Bose–Einstein condensates of exciton polaritons" in this volume). In the later work [33], the oscillating Josephson currents between two localised polariton condensates have been recorded by time-resolved interferometry (Fig. 8.6). These measurements pave way to observation of the spin superfluidity, which would be manifested by appearance of dissipationless spin currents. For a definitive prove of spin superfluidity, one should study the spin conductivity in one-dimensional channels as a function of temperature, in our opinion.

8.5 Conclusions

Bosonic spin transport is a young and extremely promising area of solid-state physics. The theories of mesoscopic transport of charge carriers and quantum transport in one-dimensional channels are among the most interesting chapters of modern physics. Substitution of fermions by bosons and of a scalar electric charge by a spin vector cannot be formally done in these theories. Basically, all mesoscopic and quantum transport effects need to be reconsidered if we speak about electrically neutral bosonic spin carriers like exciton–polaritons. Moreover, due to the finite polariton life time, the most basic in conventional transport theory requirements of charge and flux conservation cannot be fulfilled in the polariton systems. This is why the area of "spin-optronics" essentially remains *terra incognita*. Experimentally, it is much easier to study the steady state of polariton condensates than their motion. Nevertheless, recent years are manifested by numbers of publications on dissipationless propagation of exciton–polaritons measured by time- and spatially resolved spectroscopies [34, 35]. Moreover, one-dimensional channels or "polariton wires" have been successfully realised and the first studies revealed the spontaneous macroscopic coherence of exciton–polaritons [36]. The rapid progress in structure growth and experimental methods allows us to hope that very soon the polariton spin superfluidity will be experimentally observed and new interesting spin transport effects will be discovered.

Acknowledgements This work has been supported by the EU ITN project "CLERMONT4" and the IRSES project "POLAPHEN". The author is deeply grateful to Yura Rubo, Tim Liew, Ivan Shelykh, Kirill Kavokin, Masha Vladimirova, Alberto Bramati, Alberto Amo, Daniele Sanvitto, Nikolay Gippius, Dima Krizhanovskii, Maurice Skolnick, Mike Kaliteevski, Konstantinos and Pavlos Lagoudakis, Luis Vina, Jeremy Baumberg and Jacqueline Bloch for many years of fruitful collaboration in the area of spin-related effects in microcavities.

References

1. C. Weisbuch, M. Nishioka, A. Ishikawa, Y. Arakawa, Phys. Rev. Lett. **69**, 3314 (1992)
2. R. Houdré, C. Weisbuch, R.P. Stanley, U. Oesterle, P. Pellandini, M. Ilegems, Phys. Rev. Lett. **73**, 2043 (1994)
3. V. Savona, L.C. Andreani, P. Schwendimann, A. Quattropani, Solid State Commun. **93**, 733 (1995)
4. A.V. Kavokin, M. Kaliteevski, Solid State Commun. **95**, 859 (1995)
5. A. Imamoglu, J.R. Ram, Phys. Lett. A **214**, 193 (1996)
6. A. Imamoglu, J.R. Ram, S. Pau, Y. Yamamoto, Phys. Rev. A **53**, 4250 (1996)
7. L.S. Dang, D. Heger, R. Andre, F. Boeuf, R. Romestain, Phys. Rev. Lett. **81**, 3920 (1998)
8. P.G. Savvidis, J.J. Baumberg, R.M. Stevenson, M.S. Skolnick, D.M. Whittaker, J.S. Roberts, Phys. Rev. Lett. **84**, 1547 (2000)
9. J. Kasprzak et al., Nature **443**, 409 (2006)
10. R. Balili et al., Science **316**, 1007 (2007)
11. C.W. Lai et al., Nature **450**, 529 (2007)
12. S. Christopoulos et al., Phys. Rev. Lett. **98**, 126405 (2007)
13. G. Panzarini et al., Phys. Rev. B **59**, 5082 (1999)
14. P.G. Lagoudakis, P.G. Savvidis, J.J. Baumberg, D.M. Whittaker, P.R. Eastham, M.S. Skolnick, J.S. Roberts, Phys. Rev. B: Condens Matter Mater. Phys. **65**, 161310 (2002)
15. M.D. Martín, G. Aichmayr, L. Viña, R. André, Phys. Rev. Lett. **89**, 077402 (2002)
16. T.K. Paraïso, M. Wouters, Y. Léger, F. Morier-Genoud, B. Deveaud-Plédran, Nat. Mater. **9**, 655 (2010)
17. A. Amo, T.C.H. Liew, C. Adrados, R. Houdré, E. Giacobino, A.V. Kavokin, A. Bramati, Nat. Photon. **4**, 361 (2010)
18. Y.G. Rubo, A.V. Kavokin, I.A. Shelykh, Phys. Lett. A **358**, 227 (2006)
19. A.V. Larionov, V.D. Kulakovskii, S. Hofling, C. Schneider, L. Worschech, A. Forchel, Phys. Rev. Lett. **105**, 256401 (2010)
20. N.A. Gippius, I.A. Shelykh, D.D. Solnyshkov, S.S. Gavrilov, Y.G. Rubo, A.V. Kavokin, S.G. Tikhodeev, G. Malpuech, Phys. Rev. Lett. **98**, 236401 (2007)
21. I.A. Shelykh, T.C.H. Liew, A.V. Kavokin, Phys. Rev. Lett. **100**, 116401 (2008)
22. D. Sarkar, S.S. Gavrilov, M. Sich, J.H. Quilter, R.A. Bradley, N.A. Gippius, K. Guda V.D. Kulakovskii, M.S. Skolnick, D.N. Krizhanovskii, Phys. Rev. Lett. **105**, 216402 (2010)
23. C. Adrados, A. Amo, T.C.H. Liew, R. Hivet, R. Houdré, E. Giacobino, A.V. Kavokin, A. Bramati, Phys. Rev. Lett. **105**, 216403 (2010)

24. T.C.H. Liew, A.V. Kavokin, I.A. Shelykh, Phys. Rev. Lett. **101**, 016402 (2008)
25. R. Cerna, T.K. Paraiso, Y. Leger, M. Wouters, F. Morier-Genoud, M.T. Portella-Oberli, B. Deveaud-Pledran, privat communication.
26. M. Vladimirova, S. Cronenberger, D. Scalbert, K.V. Kavokin, A. Miard, A. Lemaitre, J. Bloch, D. Solnyshkov, G. Malpuech, A.V. Kavokin, Phys. Rev. B **82**, 075301 (2010)
27. T. Freixanet, B. Sermage, A. Tiberj, R. Planel, Phys. Rev. B **61**, 7233 (2000)
28. A. Kavokin, M. Glazov, G. Malpuech, Phys. Rev. Lett. **95**, 136601 (2005)
29. C. Leyder et al., Nat. Phys. **3**, 628 (2007)
30. M.M. Glazov, L.E. Golub, Phys. Rev. B **77**, 165341 (2008)
31. I.A. Shelykh et al., Phys. Rev. Lett. **102**, 046407 (2009)
32. K. Lagoudakis, T. Ostatnicky, A.V. Kavokin, Y.G. Rubo, R. Andre, B. Deveaud-Pledran, Science **326**, 974 (2009)
33. K.G. Lagoudakis, B. Pietka, M. Wouters, R. André, B. Deveaud-Plédran, Phys. Rev. Lett. **105**, 120403 (2010)
34. A. Amo et al., Nature **457**, 291 (2009)
35. A. Amo et al., Nat. Phys. **5**, 805 (2009)
36. E. Wertz et al., Nat. Phys. **6**, 860 (2010)

Chapter 9
Disorder Effects on Exciton–Polariton Condensates

G. Malpuech and D. Solnyshkov

Abstract The impact of a random disorder potential on the dynamical properties of Bose–Einstein condensates is a very wide research field. In microcavities, these studies are even more crucial than in the condensates of cold atoms, since random disorder is naturally present in the semiconductor structures. In this chapter, we consider a stable condensate, defined by a chemical potential, propagating in a random disorder potential, like a liquid flowing through a capillary. We analyze the interplay between the kinetic energy, the localization energy, and the interaction between particles in 1D and 2D polariton condensates. The finite lifetime of polaritons is taken into account as well. In the first part, we remind the results of [Malpuech et al., Phys. Rev. Lett. 98, 206402 (2007)] where we considered the case of a static condensate. In that case, the condensate forms either a glassy insulating phase at low polariton density (strong localization), or a superfluid phase above the percolation threshold. We also show the calculation of the first-order spatial coherence of the condensate versus the condensate density. In the second part, we consider the case of a propagating noninteracting condensate which is always localized because of Anderson localization. The localization length is calculated in the Born approximation. The impact of the finite polariton lifetime is taken into account as well. In the last section, we consider the case of a propagating interacting condensate where the three regimes of strong localization, Anderson localization, and superfluid behavior are accessible. The localization length is calculated versus the system parameters. The localization length is strongly modified with respect to the noninteracting case. It is infinite in the superfluid regime, whereas it is strongly reduced if the fluid flows with a supersonic velocity (Čerenkov regime).

G. Malpuech (✉) · D. Solnyshkov
LASMEA, Nanostructures and Nanophotonics group, Clermont Université,
Université Blaise Pascal, CNRS, France
e-mail: malpuech@lasmea.univ-bpclermont.fr

D. Sanvitto and V. Timofeev (eds.), *Exciton Polaritons in Microcavities*,
Springer Series in Solid-State Sciences 172, DOI 10.1007/978-3-642-24186-4_9,

9.1 Introduction

In a normal fluid, the viscosity arises because of the elastic scattering of the particles which compose it. This includes both the scattering on the external potential, for example, the walls of the capillary, and the scattering of the particles on each other, if their velocities are different. In contrast to that, for a condensate of weakly interacting bosons (a Bose–Einstein condensate – BEC), which will be the main object studied in this chapter, single independent particles are replaced by collective sonic-like excitations [1–3]. As a result, such condensate propagating with a velocity smaller than the speed of sound cannot dissipate its kinetic energy by scattering on a disorder potential or on the noncondensed particles. This collective behavior results in a vanishing mechanical viscosity called superfluidity.

However, sometimes the potential fluctuations can be large enough to destroy the superfluid behavior of a Bose condensate by provoking its complete localization. The question of the interplay between kinetic energy, localization energy, and the interaction between particles has been widely studied in solid-state physics since the seminal work of Anderson [4] which described the localization of electrons in a disordered media. Some works have addressed these questions for a gas of bosons in the eighties [5], and this activity took an enormous theoretical and experimental expansion since the observation of the BEC of cold atoms [6, 7]. Particularly interesting to study is the simple case of a 1D weakly interacting Bose gas moving in a disorder potential. Two different model frameworks are typically considered: discrete and continuous. The discrete lattice models usually employ the Bose–Hubbard Hamiltonians by which Mott insulator, Bose Glass, superfluid, or Anderson localized phases are described [8]. The continuous models are usually employed for the description of a relatively weak and smooth potential, where one cannot apply the tight-binding approximation. The theoretical modeling can be performed in this case with the Gross-Pitaevskii equation [7, 8, and refs. therein].

Exciton–polaritons are the quasiparticles formed of cavity photons strongly coupled with quantum well excitons, which are expected to behave as weakly interacting bosons, at least at relatively low densities. Despite their short lifetime, they can thermalize to a quasithermal (Bose) distribution [9–13] which can, in principle, allow the polariton gas to undergo a Berezinskii–Kosterlitz–Thouless phase transition toward a superfluid state [14–16]. In CdTe or GaN cavities, this superfluid behavior of the condensed phase was not observed because of the presence of a strong in-plane disorder which tends to localize the condensate, leading to the formation of a glassy phase [17]. In cleaner GaAs-based samples, the generation of a superfluid is, in principle, simpler and the observation of a renormalized linear dispersion above the condensation threshold has been reported [18].

Another possible way for generating a polariton superfluid besides the BKT (equilibrium) phase transition is to use the resonant excitation configuration as proposed in 2004 [19] and in 2008 for spinor polaritons [20]. The idea is to pump a polariton state with a laser, which should be slightly blue-detuned from the bare polariton dispersion. If the blue shift induced by the inter-particle interactions in the macrooccupied pumped state exactly compensates the detuning, the laser and the polariton mode become resonant, and the dispersion of elementary excitations is similar to the equilibrium case, and the pump state can propagate as a superfluid. This configuration has been recently used [21, 22] to generate a high-density flux of moving polaritons and to study their elastic scattering on a large in-plane defect. A substantial decrease of the flux dissipation by elastic scattering has been observed, but the expected singular character of the superfluid formation under resonant pumping has not been evidenced. However, this type of experiment is really opening a new research field. It reveals the enormous potential of the polariton system to study quantum hydrodynamic effects when a moving quantum fluid hits a large defect (typically larger than the fluid healing length). As predicted [23, 24], this configuration has allowed the observation of oblique solitons [25] (two-dimensional stable solitons [26]), whereas the accounting of the spin degree of freedom allowed to predict the formation of oblique half-solitons [27]. Another very promising configuration is given by the fabrication of high-quality 1D GaAs microwires [28]. In these samples, the radiative particle lifetime can reach 30 ps, which is one order of magnitude longer than in other material systems. Under nonresonant excitation, the 1D character allows the formation of a high-density nonequilibrium condensate moving along the wire, spatially independent from the pumping region. Because of the long lifetime, the propagation for large distances can take place without a substantial decay of the particle density. This is, therefore, a quite ideal configuration, where the motion of a condensate in a random continuous potential can be studied versus the velocity and the density of particles.

In this chapter, we do not consider the specific case of a condensate hitting a single potential barrier and we do not study the formation of topological defects, such as solitons. Parametric instabilities are also not taken into account. We consider a stable condensate, defined by a chemical potential, propagating in a random disorder potential, like a liquid flowing through a capillary. This chapter is organized as follows: in the first section, we give an overview and a critical discussion of the literature devoted to the disorder effects on polaritons. In the second section, we recall the main expected properties of a static Bose–Einstein condensate placed in a disorder potential, analyzing the interplay between localization and interaction effects. In the last section, we finally consider a propagating condensate, first in the linear noninteracting limit, mainly discussing Anderson localization of polaritons, and then taking into account the interactions. We then discuss the critical condition required for the occurrence of superfluidity. We finally address the question of the interplay between the kinetic energy, the localization energy, the interaction between particles, and the lifetime. To summarize the different possibilities for our interesting system, we plot a phase diagram.

9.2 Historical Overview

The role of the structural disorder on the linear optical properties of microcavities was first evidenced by resonant Rayleigh scattering experiments. An exciton–polariton eigenstate is normally characterized by a wave vector $\vec{k}$ which defines the polariton energy through the dispersion relation $E(k)$. Because the polariton is a mixed state of two particles having different masses, its dispersion is not parabolic. If the in-plane translational invariance is broken by the presence of disorder (acting on one or both polariton components), the polariton wave $\vec{k}$ is not anymore a good eigenstate. Such wave, for example, resonantly created by a laser, scatters toward the "elastic ring" of isoenergetic polariton states as shown experimentally in [29]. This perturbative description is correct if the disorder amplitude is small with respect to the kinetic energy. If the kinetic energy is small compared with the disorder amplitude, the particles become strongly localized which provokes a strong change of the particle dispersion, as discussed, for example, by Agranovich and Litinskaia in 2001 [30]. In the hypothetic case where the decay processes such as lifetime or phonon scattering are negligible, this process leads to an isodistribution of particles on the elastic ring. If the coherence is sufficient, this process should lead to a weak localization of the polariton waves called Anderson localization. So far, this process has not been yet neither observed nor described in polaritonic systems. It is typically dominated by the short radiative life time of particles which limits the spatial extension of a polariton wave much more than localization effects. The other consequence of the effect of disorder is the inhomogeneous broadening of the polariton line. As a result, the sum of the widths of the lower and upper polariton modes is not constant versus the exciton–photon detuning but shows a minimum [31]. This result was interpreted in the 1990s as a "motional narrowing" effect which led to some controversy [32]. Another important aspect relies on the type of material used to grow the structure which strongly affects the amplitude of the disorder potential. GaAs-based samples show the best structural qualities with the inhomogeneous broadening of the polariton line which can be as low as 0.1 meV. InGaAs QWs are a bit less good, with values of the order of 0.5 meV. In CdTe-based structures, the inhomogeneous broadening value is typically a few millielectron volts. It is typically 10 times larger in GaN-based samples and again about 5–10 times larger in organic-based structures. Disorder affects both the excitonic and photonic parts of the polariton modes, but the typical correlation lengths for both are different.

After the study of linear properties of the microcavities, the nonlinear optical response of microcavities has been explored under resonant and nonresonant excitation. Under nonresonant excitation, the goal of experimentalists was to achieve polariton lasing, first suggested by Imamoglu in 1996 [33]. A nonresonant laser pulse creates high-energy electron–hole pairs which bind to form an incoherent exciton reservoir which in turn forms an exciton–polariton condensate in the ground state. The polariton condensation is possible because of exciton–exciton and exciton–phonon interaction. Because of the finite polariton lifetime and limited efficiency of the relaxation processes, the polariton condensation is, in principle, an out-of-equilibrium process. However, different regimes can be distinguished depending on the type of materials used, on the exciton–photon detuning, on the polariton lifetime, and on the size of the pumping spot. A thermodynamic regime can be defined [34], corresponding to the achievement of a quasithermal distribution function. In such a case, important features, such as the critical condensation density, or the polarization of the polariton condensate, can be extracted from thermodynamic calculations, which often have the advantage of being analytical. On the other hand, another regime, called kinetic, does exist as well, where the condensate features are fully governed by the dynamics of the system. This feature and the existence of the two regimes in a given structure with a possible transition between them have been demonstrated in all types of semiconductor microcavities: CdTe [34], GaAs [35], and GaN [12]. Technically, the first clear evidence of the feasibility of the Imamoglu's proposal has been published by Le Si Dang in 1998 [36]. However, the work, which is now mostly cited and recognized as being the one where polariton condensation was observed, is the Nature paper of 2006 by Kasprzak et al. [11]. The sample and the experiment performed were the very same as in 1998, but three new measurements were added. First, the distribution function was measured and found to be close to an equilibrium distribution function. Second, the spatial coherence was found to pass from 0% to 30% at distances of about 5–10 μm. Third, the condensate was found to be linearly polarized above threshold, which is another confirmation of the condensation taking place, because the polarization is the order parameter of such phase transition in a spinor system [37]. With these new data, the observation of polariton condensation close to thermal equilibrium ("polariton BEC") was claimed. Since that time, there is a strong tendency to state that polariton condensation is a nonequilibrium process and that the achievement of equilibrium (which was one of the important results of [11]) is unimportant or impossible. If this is indeed true, it would be probably fair to slightly rewrite the history. The buildup of linear polarization, pinned along crystallographic axis in the polariton laser regime, was demonstrated before the [11], by the group of Luis Vina [38]. Also, a "nonequilibrium" condensation had been reported earlier, in two papers of 2005 by M. Richard et al. In [39], condensation took place in finite-k states because of the use of a small pumping spot, as it was understood later. The coherence between different k states was evidenced directly. In [40], condensation was taking place in the ground state but was stated as nonequilibrium, because of the use of a pulsed pumping laser. The condensate was found to be spatially localized in several different spots linked with the presence of an in-plane disorder potential. However, the angular width of the emission remained narrow, well below the diffraction limit imposed by a single emitting spot. This evidenced that these different spots were emitting in phase together, demonstrating the onset of spatial coherence, one year before it was made by interferometric technique in the Nature of 2006. The most convincing evidence of the buildup of a spatially coherent condensate was given later, in the Nature Physics of 2010 by Wertz et al. [20], where a coherence degree larger than 80% was found for distances over 200 μm (50 times the de Broglie wavelength). In this last work however and similarly in the PRL of

2005 [39], the condensates generated are completely out of equilibrium because of the use of a small pumping spot which limits the overlap between the thermal exciton reservoir and the polariton condensate itself.

The conclusion one can draw from this brief historical overview is that if the achievement of quasiequilibrium is absolutely uninteresting, as suggested in many recent works, then possibly other works than [11] could be cited as the first evidence of the polariton condensation achievement, depending on the importance given to the achievement of spatial coherence. One could interestingly notice that spontaneous symmetry breaking is often referred to as the "smoking gun" of Bose condensation. From that point of view, the clearest evidence of polariton condensation could be the J.J. Baumberg's PRL of 2008 [41], where the buildup of a condensate polarization above threshold was observed, with a polarization direction varying randomly from one experiment to another, and not pinned to any crystallographic axis.

Let us now go back to the main topic of this chapter, which is the effect of disorder on polariton condensates. From this point of view, the experiments performed on CdTe-based samples are really of strong interest because of the relatively large disorder present in these samples. In [39] already, the condensate was found to be strongly inhomogeneous in real space, peaked around the in-plane fluctuations of the potential. The formation of vortices pinned to these defects was already suggested, which, we remind, was later demonstrated clearly in the Science [42]. In the Nature paper [34], evidences of localization were even stronger, with the appearance of flat dispersion around $k = 0$.

9.3 Static Condensate in a Disorder Potential

Let us first consider a static gas of bosons at 0 K in a random potential as discussed in [16]. A realization of this potential is shown on Fig. 9.1. The ground state of this system can be found by solving the Gross–Pitaevskii equation by minimization of the free energy for a fixed number of particles. In a noninteracting case, at thermal equilibrium, all particles are strongly localized in the deepest site, so-called Lifshits state. In practice, thermal activation or the out-of-equilibrium system character allows the occupation of many localized states having possibly different energies (Lifshits tail). Such case is shown on Fig. 9.2a, b, demonstrating the spatial distribution of particles and their dispersion, which is quasiparabolic near the ground state, characterized by substantial inhomogeneous broadening. If one considers a weakly interacting Bose gas, it fills all localized states having their energy below the chemical potential values μ. The condensate forms independent lakes separated by potential barriers (Fig. 9.2c) and forms a glassy phase.

This glassy phase of a Bose condensate can be called a Bose glass, but this term is usually applied to the glassy phase formed due to excessive interparticle interactions, which inhibit the hopping between different sites in lattice models, which is never the case for polaritons for reasonable parameters. Our glassy phase,

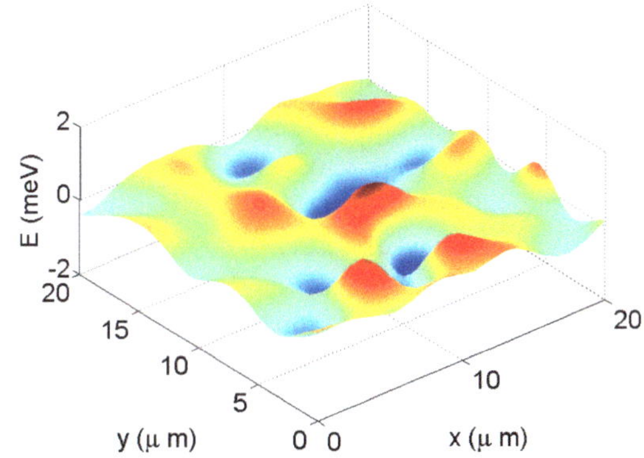

Fig. 9.1 A typical disorder potential characterized by rms fluctuation V0 = 0.5 meV and correlation length lc = 2 μm

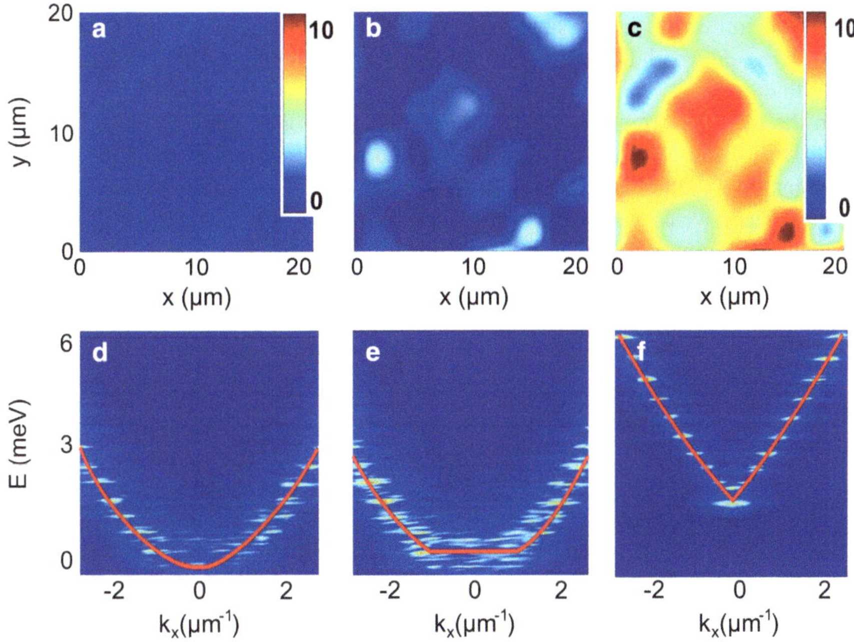

Fig. 9.2 Spatial images (*top panel*) and quasiparticle spectra (*bottom panel*) for a realistic disorder potential. The figures shown correspond to densities 0, 6×10^{10} and 2×10^{12} cm. The *red lines* are only guides to the eye, showing parabolic, flat, and linear-type dispersions. Colormap of the panel (**b**) is the same as (**c**)

therefore, has to be called Anderson glass, and the distinction between the two glasses has become clear since [8], where the transitions between all four possible phases (Anderson glass, superfluid, Bose glass, Mott insulator) were studied.

Even if the system is not in thermodynamic equilibrium and several independent condensates are formed in the potential minima, depending on the particle density, the neighboring sites having nonconnected noninteracting ground states can start to overlap to finally synchronize, as shown in [43]. However, in 2D, the randomness of the potential forbids the creation of a conduction band for the particles until the classical percolation threshold is reached. The spectrum of elementary excitations of the condensate is shown on Fig. 9.2d. It shows the flat dispersion observed experimentally. The superfluid fraction, calculated using the twisted boundary condition technique, is close to zero. This picture corresponds to the one experimentally observed. If one increases the density further, the classical percolation threshold is reached and a fraction of the fluid becomes superfluid as demonstrated by the linear dispersion of the elementary excitations shown on Fig. 9.2f.

A phase diagram from [17] summarizing these results is shown in Fig. 9.3. It demonstrates that exciton–polaritons in a disordered microcavity can present several different phases, and the superfluidity can be recovered only above certain density threshold, linked with the rms amplitude of the disorder. Since the polariton density is limited by the possible loss of the strong coupling, only relatively good material systems with low disorder, such as GaAs, can be expected to show signs of superfluidity.

In the Fig. 9.4, we show a calculation of the first-order coherence for a finite-size 1D system, assuming a thermal uncondensed fraction of particles with a homogeneous spatial distribution (high kinetic energy of uncondensed particles provides their weak sensitivity to the disorder). The coherence degree fluctuates, showing maximum values at the potential minima, because the density of the condensate is higher in these minima. This is again in good agreement with the last work of the Deveaud's group on the very same CdTe sample [44]. The density and coherence fluctuations of the condensate associated with the presence of the

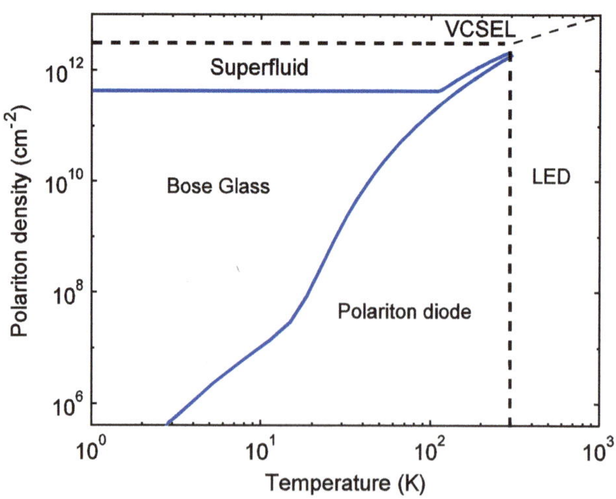

Fig. 9.3 Phase diagram for exciton–polaritons in a disordered microcavity in the thermodynamic limit. The Bose glass phase can also be called Anderson glass. Superfluidity is only possible above certain density, depending on the disorder rms fluctuations

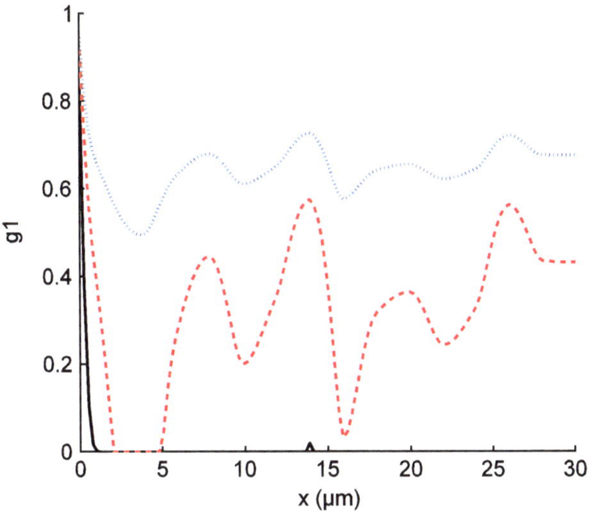

Fig. 9.4 Coherence degree for three different condensate densities. In the linear case (almost zero density) the coherence length is given approximately by the de Broglie wavelength. The condensed part is assumed to be 100% coherent, but its density varies from point to point due to the presence of the disorder potential

disorder potential is measured and related to the formation of a Bose glass phase (which, we remind, should rather be called Anderson glass). For some puzzling reasons, the authors insist that this effect is the result of the strong nonequilibrium nature of their system, whereas it is obviously not the case.

Similar works, considering 1D condensates in disordered systems, have then been published by the group of Vincenzo Savona [8, 45]. They reached the same conclusion that the insulating to superfluid phase transition occurs at the percolation threshold. However, one should notice that in an infinite 1D disorder potential with a Gaussian distribution, there is no percolation threshold. This transition therefore only exists either in a finite system or in a system with a well-defined higher bound for the potential fluctuations.

One should point out that, in 2007, an alternative explanation to the absence of superfluidity in polariton condensates was proposed both by the Cambridge group of P. Littlewood [46] and the Trento group of I. Carusotto [47] who attributed the formation of a diffusive (flat) dispersion to the nonequilibrium nature of the polariton condensate. It is indeed well known that in the *resonant* pumping case, the external laser driving the system strongly affects the dispersion of the excitations of the driven mode which can be diffusive, parabolic, or even linear for some precise value of the pumping intensity [16, 17]. In [47], a simple reservoir-condensate model was proposed which gave essentially the same result as the resonant pumping scheme, being very similar to the latter from the mathematical point of view. Depending on the decay time constant of the reservoir, the dispersion of the elementary excitations of the condensate can be parabolic, linear, and, when using an unphysically short decay time for the reservoir (five times shorter than for the condensate), diffusive. This result is mathematically interesting, but it is clearly completely unrelated to the observed experimental result. The latter statement is also supported by several experimental studies. In [15], a GaAs structure less affected by disorder than CdTe sample has been studied. The dispersion of elementary excitations was claimed to pass from parabolic to linear above threshold. The data could be judged not so convincing, but it is however clear that the dispersion was not diffusive in that case. Finally, tons of experimental works demonstrating the strong disorder effect in CdTe cavities came. One can mention the observation of the condensation in a single potential trap of a few microns by Sanvitto et al. [48], or the observation of a Josephson-like dynamics between two localized states where the condensation is taking place [49]. Also, observation of vortices [50] and half vortices [42] was made possible because of the pinning of the flux by the disorder potential.

9.4 Localization and Superfluidity of a Moving Condensate

In this second section, we discuss a bit more subtle problem of a moving fluid, interacting or not, flowing against a defect or through a disorder potential. This has been a very active field of research since the seminal work of Anderson [4]. It is by far beyond our capacities to give a complete review on this topic, and we are going to concentrate on a few specific results.

9.4.1 Anderson Localization

Brownian motion of classical particles in a random potential leads to diffusive dynamics. A particle freely propagates for a typical distance (mean free path, a distance between two impurities) and is scattered in a random direction. The width of the particle distribution around the initial position increases as $\sqrt{Dt}$, where D is the diffusion coefficient and t the time. The probability distribution therefore infinitely broadens in time. In his initial work, Anderson considered a 3D lattice model. The energies of the sites are random. They are connected by some matrix element of hopping between the sites. Anderson shown that in such a model, the wave function of a single quantum particle does not always spread with time but sometimes remains exponentially localized on the initial site. The transparent explanation given by Anderson is that the electron cannot hop to the neighboring site if the energy mismatch exceeds the hopping matrix element. In other words, the diffusion coefficient is zero. In 3D, the existence of a mobility edge has been demonstrated. However, in 1D and 2D, any weak random potential is expected to localize a quantum particle, whatever its kinetic energy, unless some extra terms are added to the Hamiltonian, such as the spin–orbit coupling or the interactions. Localization has been evidenced for light waves[51–55], microwaves [56,57], sound waves [58], and matter waves [59, 60]. Anderson localization is strongly associated with the concept of phase coherence as we are going to explain below, which is not evident from the initial interpretation given by Anderson.

Let us consider a particle moving in a random 1D potential. This particle hits a defect and is either transmitted or reflected with certain probabilities. In a classical scheme, the reflected fraction will follow a classical random walk, but at a later time, it is expected to collide with the same barrier again and participate in the transmission. For a quantum particle, this random walk will lead to destructive interferences, and the reflected part does not give any contribution to the transmission. The wave is therefore attenuated at each barrier, which leads to an exponential decay of the wave function. The effect is really subtle because the role of interference is to cancel all multi-reflection channels for the transmission, which play a crucial role in the classical diffusive motion.

If one considers 1D systems, a standard approach used [61, 62] to calculate the localization length is the so-called phase formalism [63], where the Schrödinger equation is solved in a Born approximation using a localized wave function ansatz. Here, for the sake of demonstration, we choose to use a slightly different approach.

We consider a propagating plane wave with a wave vector k_O. The bare dispersion of the particles $E(k_0)$ is assumed to be quasiparabolic. The wave is propagating in a continuous disorder potential $V(\vec{r})$ with a Gaussian random distribution of energies with a root mean square fluctuation V_0 and a correlation length L_c. The action of this potential on the propagating particles is treated as a perturbation. It does not provoke any change in the dispersion of the particles but simply induces some elastic (Rayleigh) scattering processes which are assumed to provoke a decay of the number of the propagating particles (scattering to the other k states). This assumption is based on the fact that, in the Anderson picture, the backscattered part of the wave will never interfere again constructively with the propagating wave.

From the Fermi golden rule, the decay rate of the propagating particles reads

$$\frac{1}{\tau} = \frac{2\pi}{\hbar} \sum_f |M|^2 \delta\left(E_i - E_f\right) \tag{9.1}$$

where $M = \langle \psi_k | V_{k-k'} | \psi_{k'} \rangle$, $|k| = |k'|$, and V_k is the Fourier transform of the potential. The final result essentially depends on the density of final states, which depends itself on the dimensionality of the problem:

$$\frac{1}{\tau_d} = |x|^2 \frac{2^{d/2} L_c^d}{\hbar} V_0^2 D_d \left(E = E(k_0)\right) \tag{9.2}$$

where d is the dimensionality, k_0 is the wave vector, $|x|^2$ the exciton fraction. $D^d(E = E(k_0))$ is the density of states at the energy of the propagating wave.

From (9.2), one can extract the inverse of the propagation length as

$$\gamma_d(k_0) = \frac{1}{v\tau_d} = \frac{m}{\hbar k_0 \tau_d} = |x|^2 \frac{2^{d/2} m L_c^d}{\hbar^2 k_0} V_0^2 D_d(E = E(k_0)) \quad (9.3)$$

In 1D and 2D, and for parabolic dispersions, this formula reads:

$$\gamma_{1D} = |x|^2 \frac{\sqrt{2} m^2 L_c}{\hbar^4 k_0^2} V_0^2 \quad (9.4)$$

$$\gamma_{2D} = |x|^2 \frac{2 m^2 L_c^2}{\hbar^4 k_0} V_0^2 \quad (9.5)$$

The formula (9.4) is similar to formula (9.13) of [14], derived from the phase formalism, except that our numerical prefactor $\sqrt{2}$ is replacing $\frac{\pi}{2}$.

In a system with a finite lifetime, a supplementary decay source should be considered and the total decay rate can be written as

$$\gamma_d(k_0) = |x|^2 \frac{2^{d/2} m L_c^d}{\hbar^2 k_0} V_0^2 D_d(E = E(k_0)) + \frac{m\left(1 - |x|^2\right)}{\hbar k_0 \tau_R} \quad (9.6)$$

where τ_R is the photon lifetime in the cavity. The scattering of excitons by phonons is neglected, but it could be easily introduced, that would bring additional temperature dependence into the formula. Figure 9.5 shows, in blue and red for 1D and 2D respectively, the localization lengths of polaritons for a rms potential

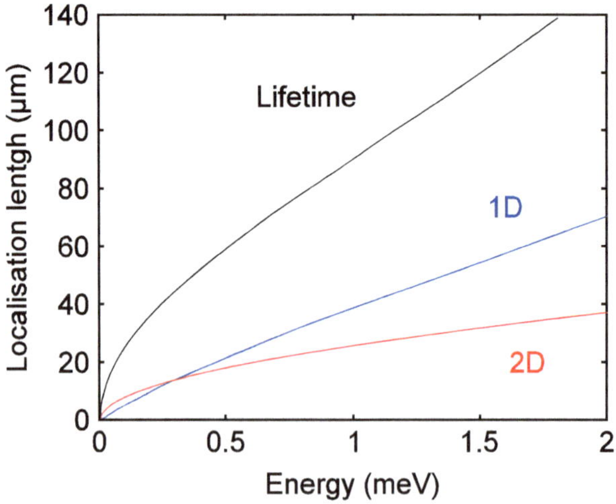

Fig. 9.5 Localization length γ_d^{-1} of the propagating polariton condensate (from formula (9.6)). The *black curve* shows the localization length in a disorder-free structure (lifetime only)

fluctuation of 0.25 meV and a correlation length of 1 micron. The Rabi splitting is taken 20 meV and the exciton–photon detuning is zero. The cavity photon lifetime is taken 30 ps. The black line shows the localization length induced by the life time. For these parameters, Anderson localization of photons can be evidenced both in 2D and 1D, despite the considerable effect of the lifetime. The 1D and 2D cases do not appear extremely different. Particles are more localized in 1D than in 2D for small energies. The situation is opposite for higher energies. The exact position of the crossing of the two curves depends on the coherence length of the disorder.

9.4.2 *Superfluidity*

A superfluid is a fluid moving without mechanical viscosity. As first understood by Landau [1], this extraordinary property can arise in a fluid if the dispersion of the elementary excitations of the fluid is linear. In that case, the fluid can flow without friction as long as its velocity remains below the speed of sound c_s given by the slope of this linear dispersion. In 1947, Bogoliubov [2, 3] established the link between superfluidity and condensation by showing that the dispersion of the excitations of weakly interacting bosons follows, at low k, the linear shape proposed by Landau. Physically, the suppression of viscosity is associated with the suppression of elastic scattering processes provoked by the roughness of the surface along which the fluid in propagating.

Below, we will demonstrate the Bogoliubov linearization procedure for a condensate of weakly interacting bosons propagating with a certain wave vector. Let us consider the Gross–Pitaevskii equation:

$$i\hbar \frac{\partial \psi(r,t)}{\partial t} = T\psi(r,t) + \alpha \left|\psi(r,t)\right|^2 \psi(r,t) \tag{9.7}$$

where T is the kinetic energy operator, namely, $-(\hbar^2/2m)\Delta$ for a parabolic dispersion.

A plane wave $\psi_0 e^{i(k_0 r - \mu t)}$ is a solution of (9.7) with

$$\mu = \omega_0(k_0) + \alpha \left|\psi_0\right|^2 \tag{9.8}$$

where $\omega_0(k_0)$ is the kinetic energy.

We then consider a small perturbation of the condensate wave function:

$$\psi\,(r,t) = \psi_0 + \delta\psi.$$

The non linear term of (9.1) reads

$$\left|\psi(r,t)\right|^2 \psi(r,t) \approx \left|\psi_0\right|^2 \psi_0 + 2\left|\psi_0\right|^2 \delta\psi + 2\psi_0^2 \delta\psi*$$

which means that the non-linear term couples $\delta\psi$ and $\delta\psi^*$.

Therefore, the solutions of (9.7) can be written as

$$\psi = \psi_0 e^{i(k_0 r - \mu t)} \left(1 + A e^{i(kr - \omega t)} + B^* e^{-i((2k_0 - k)r - \omega t)}\right) \tag{9.9}$$

Inserting (9.9) into (9.7), neglecting high-order terms yields the linearized system of equations below:

$$\begin{aligned} A(\mu + \omega) &= \left(\omega_0(k) + 2\alpha|\psi_0|^2\right) A + \alpha\psi_0^2 B \\ B(\mu - \omega) &= \left(\omega_0(2k_0 - k) + 2\alpha|\psi_0|^2\right) B + \alpha\psi_0^{*2} A \end{aligned} \tag{9.10}$$

The solutions of the system (9.10) are given by

$$\omega_\pm = \frac{-\Delta_1 - \Delta_2}{2} \pm \sqrt{\left(\frac{\Delta_1 - \Delta_2}{2}\right)^2 + \alpha|\psi_0|^2(\Delta_1 - \Delta_2)} \tag{9.11}$$

where

$$\begin{aligned} \Delta_1 &= \omega_0(2k_0 - k) - \omega_0(k_0) \\ \Delta_2 &= \omega_0(k_0) - \omega_0(k) \end{aligned}$$

For a parabolic dispersion for $k \sim k_0$

$$\omega_\pm = \left(\frac{\hbar k_0}{m} \pm \sqrt{\frac{\alpha|\psi_0|^2}{m}}\right) \hbar(k - k_0) \tag{9.12}$$

$$|A_\pm|^2 \equiv \frac{\left(\omega_\pm - \omega_0(k_0) + \omega_0(2k_0 - k) + \alpha|\psi_0|^2\right)^2}{\left(\left(\omega_\pm - \omega_0(k_0) + \omega_0(2k_0 - k) + \alpha|\psi_0|^2\right)^2 + \left(\alpha|\psi_0|^2\right)^2\right)}$$

$$|B_\pm|^2 \equiv \frac{\left(\omega_\pm + \omega_0(k_0) - \omega_0(k) - \alpha|\psi_0|^2\right)^2}{\left(\left(\omega_\pm + \omega_0(k_0) - \omega_0(k) - \alpha|\psi_0|^2\right)^2 + \left(\alpha|\psi_0|^2\right)^2\right)} \tag{9.13}$$

Figure 9.6 shows the dispersions of elementary excitations of a polariton condensate propagating with a wave vector $10^6\,\mathrm{m}^{-1}$. Panel (a) illustrates the Čerenkov regime, where the elastic scattering toward an excited state with the same energy as the condensate is possible because the renormalization of the dispersion due to the interactions is not strong enough. Panel (b) corresponds to the superfluid case, where the dispersion is so strongly renormalized that the elastic scattering toward any excited state is no more possible, and therefore, due to the Landau criterion, the viscosity of the propagating polariton fluid drops to zero.

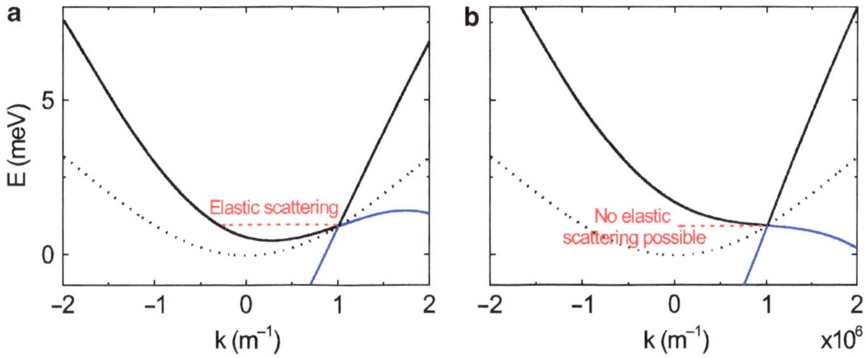

Fig. 9.6 Bare polariton dispersion (*black dotted line*) and the dispersions of elementary excitations (*black* and *blue/light gray solid lines*) in the Čerenkov (**a**) and superfluid (**b**) regimes. Elastic scattering towards excited states is indicated by a *red arrow*

The solution (9.12) tells us that the excitations of the condensate are created by a kind of parametric process symmetric with respect to the energy and the wave vector of the condensate. However, when the final state is far from the initial dispersion (far means that the energy difference becomes comparable or larger than μ), the amplitude of the corresponding state vanishes, which means that this state does not exist and that the excitation is just a plane wave (free particle), essentially unperturbed by the presence of the condensate. The condensate itself is expected to be stable against parametric processes if $\omega_{\pm}$ are real values, which is always the case for a parabolic dispersion in 1D.

OPO-like processes are very well known in the polariton community because of the nonparabolicity of the polariton branch which makes them resonant. They have therefore been strongly studied since the year 2000 and the pioneering works of Savvidis [64], Stevenson [65], and the strong theoretical contribution of Ciuti [66–68]. After the claiming by the Yamamoto's group of the observation of this linear Bogoliubov dispersion [15], some doubts came from the fact that they did not observe the negative part of the dispersion. Recently, the Lausanne group attacked this question, by trying to observe this "ghost branch" by heterodyne detection [69]. One should notice that off-resonant emission coming from "ghost branches" has already been evidenced [66, 67] under resonant excitation. In fact, even if these new branches (eigenvalues) come from OPO-like processes, such processes are not balanced. Figure 9.7 below shows the expected emission spectra for the same parameters as for Fig. 9.6. The dispersion is indeed perturbed, but the "ghost branch" is not easily visible.

The second comment is that superfluidity is fully linked with interparticle interactions. A noninteracting gas moving with any velocity in a weak random potential gets exponentially localized because of Anderson localization. If interactions are considered, the same gas can show a superfluid behavior which means that it does not even "feel" the presence of the disorder during the propagation (Fig. 9.8).

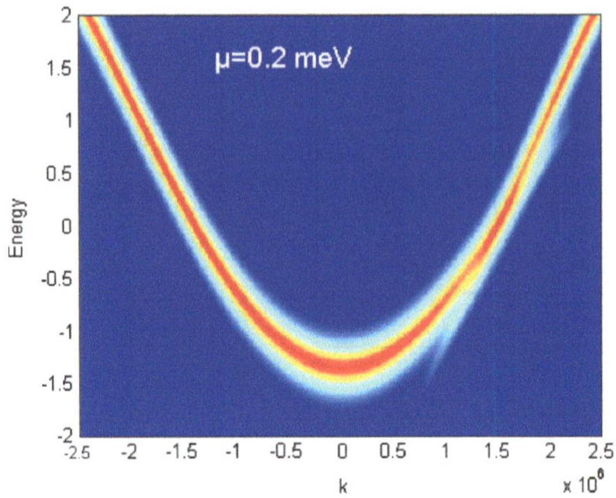

Fig. 9.7 Emission from the elementary excitations of a polariton condensate, calculated using formulas (9.11) and (9.13). The condensate has an in-plane wave vector $1.5 \times 10^6\,\mathrm{m}^{-1}$. The interaction energy $\alpha|\psi_0|^2$ is 0.2 meV. All states are assumed to emit light proportionally to the coefficients calculated with the formula (9.13), including a linewidth of 0.1 meV. The "ghost branch" can be guessed to the right of the main dispersion

Fig. 9.8 Phase diagram of condensed polaritons in the presence of disorder. $V_0 = 1\,\mathrm{meV}$, $m = 5 \times 10^{-5}\,\mathrm{m}_0$. Everything below the *red line* is localized. Above the *red* and *blue*: delocalized or Čerenkov. Above the *red* and below the *blue*: superfluid

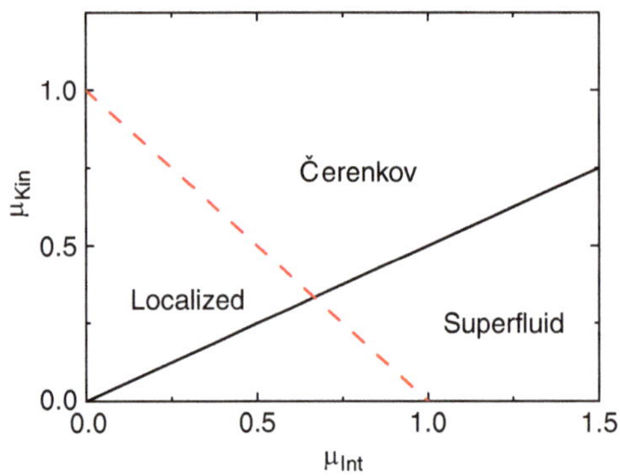

9.4.3 *Phase Diagram*

In what follows, we consider a 1D condensate moving in a random disorder potential. This condensate is characterized by a fixed energy, which means that we neglect the possible occurrence of parametric instabilities, and the possible formation of topological defects, such as solitons. This is a strong approximation. In [70, 71] for example, such situation is considered numerically and the condensate is found "unstable" in a wide range of parameters, which means that the propagating flux cannot be characterized by a single frequency. Therefore, in what follows, we assume that the only possible dissipative process for the propagating condensate is backscattering processes, which could also be called "emission of Čerenkov waves.”. The qualitative picture we propose is summarized on the phase diagram shown on Fig. 9.3. As said in the previous section, at negligible interaction energy, bosons in a 1D system are known to be always localized in a random potential, whatever their kinetic energy. Two different localization regimes can however be distinguished. If the kinetic energy ω_0 is much smaller than the amplitude of the random potential V_0, the condensate is classically localized in some independent localized states, referred to as Liftshits states in [38]. The localization length is typically given by the correlation length L_c of the disorder potential. If $\omega_0 > V_0$, the bosons are still localized but the physical mechanism is more subtle, relying on Anderson localization, as discussed in the previous sections. A propagating plane wave is backscattered by the disorder with some finite scattering time proportional to the density of states at the propagating energy $\tau_{1D} = \frac{\hbar^3 k_0}{\sqrt{2m L_c V_0^2}}$, where k_0 is the wave vector of the wave, m – the mass and L_c – the correlation length of the disorder potential. The localization length is given by $\frac{\hbar k_0}{m}\tau_{1D}$ and is no more directly proportional to the coherence length of the disorder potential. The smooth crossover between the two regimes takes place approximately when $\omega_0 = V_0$.

This picture is dramatically modified by the interactions. Their first role is to bleach the disorder potential, whose effective height is given by $V_0 - \alpha|\psi|^2$. This reduces the kinetic energy at which the transition between strong and Anderson localization takes place to $\omega_0 = V_0 - \alpha|\psi_0|^2$. Second, the interactions provide a finite superfluid velocity $v_s = \sqrt{\frac{\alpha|\psi_0|^2}{m}}$. A condensate propagating with a velocity smaller than v_s shows no viscosity because of the suppression of the elastic backscattering—the mechanism, which is also responsible for the occurrence of the Anderson localization. However, if the system is strongly localized, it cannot be called superfluid because there is no motion of the fluid in this case. Therefore, superfluidity can take place only if the kinetic energy of the condensate is large enough for the particles not to be strongly localized but small enough to have the condensate velocity below the speed of sound in the media, which is expressed by the condition $V_0 - \alpha|\psi_0|^2 < \omega_0 < \alpha|\psi_0|^2/2$. If $\omega_0 > \alpha|\psi_0|^2/2$, a superfluid enters the so-called Čerenkov regime. We show that in this regime, the backscattered amplitude is enhanced with respect to the noninteracting case, providing a reduced localization length.

An important remark regards the difference of this model with the one describing disordered lattices. In the lattices, the coupling between regularly spaced localized states can result in the formation of a conduction band allowing superfluidity. In the case of a continuous random potential, the delocalization threshold corresponds to the percolation of the wave function, and the formation of a finite conduction band below this threshold is not expected.

Quantitatively, one can calculate the localization length using the formula (9.6) in which the renormalized density of states calculated from the dispersion (9.11) is used. This approach is perturbative. The hypothesis made is that the interparticle interaction is the main process leading to a well-defined renormalized dispersion for the elementary excitations of the condensate, which is stable against the parametric processes. Second, the disorder is also a perturbation, which can only provoke scattering processes within this dispersion. In fact, there is a wide range of parameters [42], where both mechanisms should be taken into account simultaneously, leading to the instability of the condensate, something which we neglect here.

For a parabolic dispersion (a valid approximation for polaritons with relatively small wave vectors), one can obtain a simple analytical expression for the density of excited states at the energy of the propagating condensate:

$$D_{1D}\left(E=E(k_0)\right)=\frac{1}{\frac{\hbar^2 k_0}{m}-\frac{\mu}{k_0}} \quad \text{if} \quad \left(\frac{\hbar^2 k_0^2}{m}\right)>\mu \tag{9.14}$$

$$D_{1D}\left(E=E(k_0)\right)=0 \quad \text{if} \quad \left(\frac{\hbar^2 k_0^2}{m}\right)<\mu \tag{9.15}$$

Since the density of available states for the subsonic case is zero, γ_{1D} drops to zero as well, and the condensate can propagate infinitely far without any scattering. For supersonic waves, γ_{1D} is given by

$$\gamma_{1D}\left(k_0,\mu\right)=|x|^2\frac{\sqrt{2}m^2 L_c V_0^2}{\hbar^4 k_0^2-m\hbar^2\mu}\Theta\left(\hbar^4 k_0^2-m\hbar^2\mu\right) \tag{9.16}$$

where Θ is the Heaviside function.

The denominator describes the enhancement of the scattering by disorder in the Čerenkov regime. One should note that γ_{1D} goes to infinity close to the transition to a superfluid state which corresponds to a fully localized system. The hypothesis we made to derive the formula is that the disorder is acting as a weak perturbation on the propagating wave. This hypothesis breaks down close to the transition point between the superfluid and Čerenkov regimes.

In a system with a finite lifetime, a supplementary decay source should be considered and the total decay rate can be written as

$$\gamma_{1D}\left(k_0,\mu\right)=|x|^2\frac{\sqrt{2}m^2 L_c V_0^2}{\hbar^4 k_0^2-m\hbar^2\mu}+\left(1-|x|^2\right)\frac{m}{\hbar k_0 \tau_R} \tag{9.17}$$

Figure 9.9 shows a plot of γ_{1D}^{-1} which is the propagation length for interaction energy of 1 meV. In the low-energy range, the kinetic energy is at least twice smaller than the interaction energy and polaritons are superfluid. Their propagation is only limited by their lifetime. When the speed of sound is reached (kinetic energy is one half of the interaction energy), the density of states available for elastic scattering processes passes from zero to infinity and the condensate becomes very strongly localized. First, we can expect that this abrupt transition can be smoothed in a real system because of finite linewidth, finite-size effects, and so on. Second, it is not clear at all if our approximation (stability of the condensate against parametric processes) remains valid in this case. The dashed line shows a plot of the decay in the noninteracting case (same as for Fig. 9.5) for comparison. The extra localization induced by the renormalization of the dispersion in the Čerenkov regime is clearly

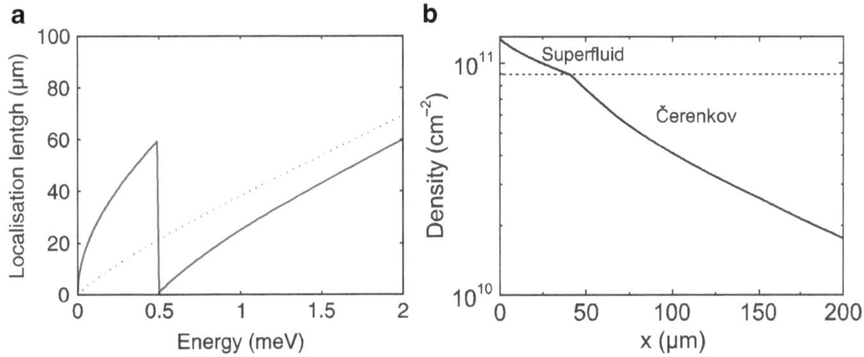

Fig. 9.9 (**a**) Propagation length of a 1D polariton condensate as a function of its energy in the presence of disorder $L_c = 1\,\mu\text{m}$, $V_0 = 0.25\,\text{meV}$. Solid line: $\alpha|\psi_0|^2 = 1\,\text{meV}$, *dashed line*, no interaction (same case as for Fig. 9.5). (**b**) Density profile of a 1D polariton condensate propagating in the presence of disorder. The transition from superfluid to Čerenkov regime is visible as a cusp

evidenced. It is interesting to notice that for infinite lifetime particles, our approach gives at the transition point a localization length passing from infinity to zero.

We show the results of a hydrodynamic simulation of the propagation of a polariton condensate with a wave vector in a disordered system, taking into account the finite particle lifetime and the backscattering induced by the disorder, both included in (9.17). The condensate is injected (presumably by nonresonant pumping) at the point $x = 0\,\mu\text{m}$ and propagates to the right. The potential $U(x)$ felt by the condensate at the point x is composed of the disorder potential and the interaction energy linked with the local density. Figure 9.9b shows the steady-state situation, obtained when the interaction energy significantly exceeds the kinetic one at the injection point, and the condensate is therefore initially superfluid. Since the density is decreasing with the coordinate even in the superfluid regime because of the finite lifetime, the condensate propagates down a potential slope (created by the gradient of the density) and is therefore accelerated, until it reaches the (local) speed of sound. At this point, the condensate becomes supersonic and backscattering starts to play a major role, strongly reducing the localization length. When the potential energy is completely transferred into kinetic energy, the condensate starts to move faster and therefore the backscattering is reduced again. The localization length increases almost back to the value induced by the finite lifetime.

The transition between the superfluid and supersonic phases is a horizon at which strong instabilities may develop, but their study will be a subject of a separate work.

9.5 Conclusion and Perspectives

The main result of this chapter is to calculate the localization length of an interacting polariton condensate, moving in a weak random potential. To our knowledge, this type of calculation was never published before for this type of system. We consider the transition from a superfluid to an Anderson localized condensate. The localization length is found to be reduced in the Čerenkov regime. Until 2009, polariton lifetime was typically below 10 ps, and the propagation length of particles was typically limited by this factor. Another case realized is the one of large disorder were the condensate is classically localized. The fabrication of new structures, of high quality, and showing polariton lifetime of 30 ps [27] and even more [Bloch, private communication] should allow to study in details localization and superfluidity phenomena. Another interesting direction relates to the realization of periodically modulated samples [72,73]. In such type of structures, a wide variety of new phenomena (such as Bloch oscillations [74] for instance) can take place.

References

1. L.D. Landau, The theory of superfluidity of helium II, J. Phys. USSR **5**, 71 (1941)
2. N.N. Bogoliubov, On the theory of superfluidity, J. Phys. USSR, **11**, 23 (1947)
3. N.N. Bogoliubov, *Lectures on Quantum Statistics*, Vol 1 *Quantum Statistics*, (Gordon and Breach Science Publisher, New York, London, Paris, 1970)
4. P.W. Anderson, Phys. Rev. **109**, 1492, (1958)
5. M.P.A. Fisher, P.B. Weichman, G. Grinstein, D.S. Fisher, Phys. Rev. B **40**, 546 (1989)
6. M.N. Anderson et al., Science **269**,198 (1995)
7. K.B. Davis et al., Phys. Rev. Lett. **75**, 3669 (1995)
8. R.T. Scalettar, G.G. Batrouni, G.T. Zimanyi, Phys. Rev. Lett. **66**, 3144 (1991)
9. G. Malpuech, A. Kavokin, A. Di Carlo, J.J. Baumberg, Phys. Rev. B **65**, 153310 (2002)
10. D. Porras, C. Ciuti, J.J. Baumberg, C. Tejedor, Phys. Rev. B 66, 085304 (2002)
11. J. Kasprzak et al., Nature **443**, 409 (2006)
12. J. Kazprzak, D.D. Solnyshkov, R. André, L.S. Dang, G. Malpuech, Phys. Rev. Lett. **101**, 146404 (2008)
13. J. Levrat, R. Butté, E. Feltin, J.-F. Carlin, N. Grandjean, D. Solnyshkov, G. Malpuech, Phys. Rev. B **81**, 125305 (2010)
14. G. Malpuech, Y.G. Rubo, F.P. Laussy, P. Bigenwald, A. Kavokin, Semicond. Sci. Technol. 18, Special issue on microcavities, edited by J.J. Baumberg and L. Vina, S 395 (2003)
15. J. Keeling, Phys. Rev. B **74**, 155325 (2006)
16. G. Malpuech, D. Solnyshkov, http://arxiv.org/abs/0911.0807 (2009).
17. G. Malpuech, D. Solnyshkov, H. Ouerdane, M. Glazov, I. Shelykh, Phys. Rev. Lett. **98**, 206402 (2007)
18. S. Utsunomiya et al., Nat. Phys. **4**, 700 (2008)
19. I. Carusotto, C. Ciuti, Phys. Rev. Lett. **93**, 166401 (2004)
20. D.D. Solnyshkov, I.A. Shelykh, N.A. Gippius, A.V. Kavokin, G. Malpuech, Phys. Rev. B **77**, 045314 (2008)
21. A. Amo et al., Nature **457**, 291 (2009)
22. A. Amo et al., Nat. Phys. **5**, 805 (2010)
23. S. Pigeon et al., Phys. Rev. B, **83**, 144513 (2011)
24. G.A. El et al., Phys. Rev. Lett. **97**, 180405 (2006)
25. A. Amo et al., Science **332**, 1167 (2011)
26. A.M. Kamchatnov et al., Phys. Rev. Lett. **100**, 160402 (2008)
27. H. Flayac, D. Solnyshkov, G. Malpuech, Phys. Rev. B **83**, 193305 (2011)
28. E. Wertz et al., Nat. Phys. **6**, 860 (2010)
29. R. Houdré, C. Weisbuch, R.P. Stanley, U. Oesterle, M. Ilegems, Phys. Rev. B **61**, R13333 (2000)

30. M. Litinskaia, G.C. La Rocca, V.M. Agranovich, Phys. Rev. B **64**, 165316 (2001)
31. D.M. Whittaker, P. Kinsler, T.A. Fisher, M.S. Skolnick, A. Armitage, A.M. Afshar, M.D. Sturge, J.S. Roberts, Phys. Rev. Lett. **77**, 4792 (1996)
32. A. Kavokin, G. Malpuech, *Cavity Polaritons*, chapter 3 (Elsevier, Amsterdam, 2003).
33. A. Imamoglu, J.R. Ram, Phys. Lett. A **214**, 193 (1996)
34. L. Fontanesi, M. Wouters, V. Savona, Phys. Rev. A 83, 033626 (2011).
35. E. Wertz et al., Appl. Phys. Lett. **95**, 051108 (2009)
36. L.S. Dang et al., Phys. Rev. Lett. **81** 3920 (1998)
37. I.A. Shelykh et al., Phys. Rev. Lett. **97**, 066402 (2006)
38. L. Klopotowski et al., Solid State Commun. **139** 511 (2006)
39. M. Richard et al., Phys. Rev. Lett. **94** 187401 (2005)
40. M. Richard et al., Phys. Rev. B **72**, 201301 (2005)
41. J.J. Baumberg et al., Phys. Rev. Lett. **101**, 136409 (2008)
42. K.G. Lagoudakis et al., Science **326**, 974 (2009)
43. M. Wouters, Phys. Rev. B **77**, 121302 (2008)
44. F. Manni et al., Phys. Rev. Lett. **106**, 176401 (2011)
45. L. Fontanesi, M. Wouters, V. Savona, Phys. Rev. A **81**, 053603 (2010)
46. F.M. Marchetti, J. Keeling, M.H. Szymaska, P.B. Littlewood, Phys. Rev. Lett. **96**, 066405 (2006)
47. M. Wouters, I. Carusotto, Phys. Rev. Lett. **99**, 140402 (2007)
48. D. Sanvitto, A. Amo, L. Viña, R. André, D. Solnyshkov, G. Malpuech,Phys. Rev.B **80**, 045301 (2009)
49. K.G. Lagoudakis, B. Pietka, M. Wouters, R. André, B. Deveaud-Plédran, Phys. Rev. Lett. **105**, 120403 (2010)
50. K.G. Lagoudakis et al., Nat. Phys. **4**, 706 (2008)
51. D.S. Wiersma, P. Bartolini, A. Lagendijk, R. Righini, Nature **390**, 671–673 (1997)
52. F. Scheffold, R. Lenke, R. Tweer, G. Maret, Nature **398**, 206–270 (1999)
53. M. Storzer, P. Gross, C.M. Aegerter, G. Maret, Phys. Rev. Lett. **96**, 063904 (2006)
54. T. Schwartz, G. Bartal, S. Fishman, M. Segev, Nature **446**, 52–55 (2007)
55. Y. Lahini et al., Phys. Rev. Lett. **100**, 013906 (2008)
56. R. Dalichaouch, J.P. Armstrong, S. Schultz, P.M. Platzman, S.L. McCall, Nature **354**, 53–55 (1991)
57. A.A. Chabanov, M. Stoytchev, A.Z. Genack Nature **404**, 850–853 (2000)
58. R.L. Weaver, WaveMotion **12**, 129–142 (1990)
59. J. Billy et al., Nature **453**, 891 (2008)
60. G. Roati et al., Nature **453**, 895 (2008)
61. L. Sanchez Palencia et al., Phys. Rev. Lett. **98**, 210401 (2007)
62. G. Madugno, Rep. Prog. Phys. **73** (2010) 102401 and ref. Therein
63. I.M. Lifshits et al., *Introduction to the Theory of Disordered Systems* (Wiley, New York, 1988)
64. P.G. Savvidis et al., Phys. Rev. Lett. **84**, 1547 (2000)
65. R.M. Stevenson et al., Phys. Rev. Lett. **85**, 3680 (2000)
66. C. Ciuti et al., Phys. Rev. B **62** R4825 (2000)
67. C. Ciuti et al., Phys. Rev. B **63**, 041303(R) (2001)
68. P.G. Savvidis et al., Phys. Rev. B **63** 041303(R) (2001)
69. V. Kohnle et al., Phys. Rev. Lett. **106**, 255302 (2011)
70. T. Paul, M. Albert, P. Schlagheck, P. Leboeuf, N. Pavloff, Phys. Rev. A **80**, 033615 (2009)
71. M. Albert, T. Paul, N. Pavloff, P. Leboeuf, Phys. Rev. A **82**, 011602 (2010)
72. C.W. Lai et al., Nature **450**, 529 (2007)
73. E.A. Cerda-Mendez, D.N. Krizhanovskii, M. Wouters, R. Bradley, K. Biermann, K. Guda, R. Hey, P.V. Santos, D. Sarkar, M.S. Skolnick, Phys. Rev. Lett. **105**, 116402 (2010)
74. H. Flayac, D.D. Solnyskov, G. Malpuech, Phys. Rev. B. **83**, 045412 (2011)

Chapter 10
Truncated Wigner Approximation for Nonequilibrium Polariton Quantum Fluids

Michiel Wouters and Vincenzo Savona

Abstract In this chapter we review the stochastic approach that we recently developed to model the kinetics of polariton Bose–Einstein condensation, based on a truncated Wigner approximation. The approach consists in neglecting the third-order term appearing in the master equations for the Wigner distribution of the quantum field. The resulting Fokker–Planck equation can be modeled by numerically solving the corresponding stochastic Langevin equation, coupled to a phenomenological diffusion equation for the excitonic reservoir that provides the gain-loss mechanism. This approach is particularly well suited for polaritons, in which the neglected term is often negligible compared to the intrinsic loss rates of the polariton field. We apply our model to typical experimental situations and discuss the results, with particular focus on the dynamics of phase fluctuations and the possibility to observe a Berezinski-Kosterlitz-Thouless crossover in the polariton superfluid.

10.1 Introduction

One of the most striking manifestations of a quantum degenerate Bose gas is certainly that of polaritons in planar semiconductor microcavities. Polaritons are composite quasiparticles, arising from the normal-mode coupling between an exciton in a semiconductor quantum well and a photon in the resonant mode of

M. Wouters (✉)
TQC, Universiteit Antwerpen, Universiteitsplein 1, 2610 Antwerpen, Belgium
e-mail: michiel.wouters@ua.ac.be

V. Savona
Institute of Theoretical Physics, Ecole Polytechnique Fédérale de Lausanne EPFL, CH-1015 Lausanne, Switzerland
e-mail: vincenzo.savona@epfl.ch

D. Sanvitto and V. Timofeev (eds.), *Exciton Polaritons in Microcavities*, Springer Series in Solid-State Sciences 172, DOI 10.1007/978-3-642-24186-4_10,

the surrounding cavity. The excitonic part in the normal mode is responsible for the polariton two-body interactions, arising from the Coulomb interaction and the Pauli exclusion. The resulting Bose gas is a surprisingly interesting system for studying fundamental properties and for possible applications in the domain of quantum photonics. In particular, following the first experimental proofs of polariton Bose–Einstein condensation [10, 35] and polariton parametric oscillation [76], several phenomena expected from a quantum fluid were observed, including superfluidity [5, 6], quantized vortices [40, 41, 44, 56, 67], spontaneous Josephson oscillations [43], and solitons [4]. Thanks to the unique spectroscopical detection tools available for polaritons, the polariton condensate is also opening the way to the first direct observation of phenomena predicted for quantum fluids and never observed before, such as the half-quantum vortices predicted for a spinor Bose gas [42]. Several reviews have recently appeared, focusing on various aspects of polaritons as a Bose quantum fluid [15, 37, 74, 75].

Many are the features that make this system so appealing. First is the fact that quantum degeneracy in a polariton Bose gas arises at relatively high temperatures, in the range of tens to hundreds of Kelvins. This is mostly due to the extremely light effective mass that characterizes the polariton quasiparticle close to the bottom of its energy-momentum dispersion curve, four orders of magnitude smaller than the free electron mass. In addition, polaritons constitute an ideally two-dimensional system, as motion along the third direction is frozen due to the strong confinement of the electromagnetic cavity mode, thus offering a considerable advantage with respect to ultracold atomic clouds for the study of low-dimensional systems. Additional sample processing has recently provided close-to-ideal one-dimensional [84] and fully quantum-confined structures [9, 11, 13, 20, 34]. Finally, the finite polariton lifetime offers an invaluable investigation tool: measuring the light field resulting from the polariton radiative decay gives direct access to the full spectral distribution of the polariton gas. In this way, quantities such as the one- and two-body density matrices can be easily measured by means of optical spectroscopy.

The finite lifetime of polariton quasiparticles has, however, an important consequence: all experiments are carried out in a regime that significantly deviates from thermal equilibrium. In a typical experiment under steady-state non-resonant excitation, the stationary polariton field results from the balance between the gain provided by relaxation from high-energy laser-driven states and the losses due to both relaxation to lower energy states and the radiative process. The peculiar shape of the polariton energy-momentum dispersion contributes to this non-universal feature: beyond a characteristic momentum, the much heavier effective exciton mass is recovered, and a large density of exciton-like states thus exists. In these states, a large population builds up due to the very inefficient relaxation mechanisms. This *relaxation bottleneck* is an omnipresent feature of experiments under nonresonant excitation [78, 80]. More precisely, the typical experimental situation is that in which the polariton occupation follows a thermal distribution only within the strong coupling region close to zero in-plane momentum, characterized by an effective temperature higher than that of the semiconductor lattice [35], while a large occupation is present outside this region due to the polariton bottleneck. Most of the bottleneck region cannot be probed optically, but evidence of the population buildup is provided by the spectral blueshift of the polariton condensate, and by its dependence on the excitation power [21], showing that the polariton condensate has a major contribution to the blueshift only well above the condensation threshold. A further deviation from thermal equilibrium is provided by the inhomogeneous spatial profile that characterizes all experiments, due both to the presence of disorder [26, 45, 70] and to the finite size of the excitation area. Combined with the gain and loss processes, a spatial inhomogeneous profile will usually give rise to a flow of the polariton fluid, that can result in the formation of quantized vortices [44] or in the propagation far out of the excitation area [84].

Thermal equilibrium can only partially be approached by increasing the polariton radiative lifetime (through optimal design of the cavity mirrors) or by tuning the laser excitation directly to the bottleneck region (quasi-resonant excitation) [16]. This latter attempt in particular was rather effective, as suggested by the power-dependent blueshift whose slope decreases above threshold [82] as opposed to the increased slope observed by Ferrier et al. [21]. In spite of this encouraging trend, the short polariton lifetime still prevents a full thermalization that would be required to use polaritons for studying the universal properties and the critical phenomena of a Bose gas at thermal equilibrium.

The polariton quantum fluid is, however, a promising system for the investigation of a rich variety of non-equilibrium phenomena, including vortex dynamics [38, 40–42, 44, 51, 56, 60, 67], Josephson oscillations [43], multistability [1, 7, 8, 25, 59, 69], solitons [4, 60], and more exotic physics like for example Hawking radiation [66]. Thanks to the possibility of engineering nanostructures [11, 13, 34], the polariton quantum fluid has also been considered as a ground for various kinds of device operation [3, 30, 36, 46, 48–50, 55, 58, 72].

In a two-dimensional quantum degenerate Bose gas, fluctuations play a dominant role and determine the nature of the collective state [64]. The Mermin–Wagner theorem forbids the occurrence of a true condensate with off-diagonal long-range order at any finite temperature. The quantum degenerate state is called quasi-condensate and is characterized by a power-law decay of the one-body density matrix in space, induced by the low-energy phase fluctuations. These same fluctuations will eventually drive the Berezinski-Kosterlitz-Thouless phase transition to a non-degenerate gas as the temperature is increased [62]. Also in the absence of thermal equilibrium, one expects fluctuations to play an important role in two-dimensions. Hence, a theoretical formalism for nonequilibrium polariton condensation should correctly account for fluctuations of the condensate field. Most current experiments rely on the measurement of the time-integrated polariton density profile, namely a mean-field feature scarcely affected by the nature of fluctuations. Many experiments, however, include the measurement of the one-body density matrix in space [17, 35, 84, 85], through the interference of the field emitted from different positions on the sample. The decay of this quantity is solely determined by the size and nature of fluctuations of the polariton condensate, and showed a strong dependence on the experimental conditions, although a recent experimental finding brought evidence of a regular power-law decay reminding of the thermal-equilibrium situation [17]. Direct evidence of fluctuations was observed in the spontaneous transient motion of quantized vortices [41], while it was shown that they can affect the initial conditions of certain dynamical evolution patterns, even when only mean-field features are measured, as was the case in the study of polariton Josephson oscillations [43].

A correct account of fluctuations in a low-dimensional Bose gas is a challenging task already at thermal equilibrium, where many-body perturbation theory generally introduces an unphysical infrared divergence [62] and ad-hoc methods [2, 54] or a first principle Monte Carlo approach [32, 61, 65] must be adopted. When deviations from thermal equilibrium are present, several approaches to a kinetic theory of the Bose gas have been proposed. For ultracold atomic clouds, where kinetic features are often small deviations from the equilibrium gas, kinetic theories have been based on the symmetry-breaking assumption, as in the works by Nikuni–Zaremba–Griffin [57] or Gardiner–Zoller [24]. These methods mostly describe the kinetics of fluctuations in presence of a stationary condensate. A generalization is possible, if the condensate dynamics is restricted to a few modes, as in a recent model of the dynamics of Josephson oscillations [81], but this method cannot be applied to polaritons, where the modal structure of the condensate is often determined self-consistently by the gas kinetics. Studies of polariton kinetics in microcavities have been carried out very early, with the main objective to characterize the energy relaxation mechanisms mediated by polariton–phonon, polariton–polariton [12, 19, 53, 79], and polariton–electron [53] scattering. To this purpose, the method of election was the solution of the semiclassical Boltzmann equations with microscopically computed scattering amplitudes. The Boltzmann kinetics can also predict, to some extent, the occurrence of a condensate, by way of the bosonic final-state stimulation in the scattering process. If on one side this approach can describe several aspects of polariton condensation, it fails to account for collective excitations and is thus not suited for describing low-energy fluctuations. Possible extensions of the Boltzmann approach include the Boltzmann-master equations [18], for describing the density fluctuations, and a Langevin–Gross–Pitaevski approach to phase fluctuations [28]. A common feature to all these approaches, however, is the fact that only single particle excitations of the condensate are modeled, characterized by a parabolic energy-momentum dispersion, while the occurrence of collective Bogoliubov excitations is neglected. Attempts to merge the Boltzmann approach with collective excitations have been proposed [68], but they rely on the assumption of a spatially homogeneous system and do not account for the collective excitation spectrum in a self-consistent fashion. Indeed, in presence of gain and losses, the spectrum of collective excitations is considerably different from the standard Bogoliubov spectrum, showing a soft Goldstone mode that extends to finite momenta [77, 87], significantly affecting even basic properties such as superfluidity [88].

In the light of these considerations, a radically different approach to condensation kinetics—the truncated Wigner approximation (TWA)—turns out to be very effective for polaritons. The TWA was initially developed to model the condensation kinetics of ultracold atomic gases [14, 31, 33, 73]. It is based on the mapping of the many-body problem onto a set of third-order partial differential equations for the Wigner distribution of the quantum field. Unfortunately, the third-order term makes these equations intractable in most interesting situations. If this term is neglected, the remaining terms constitute a set of Fokker–Planck equations for the field, that can be solved directly or by a numerical Monte Carlo solution of the corresponding quantum Langevin equations. The effectiveness of this approximation can be intuitively understood if one considers that the Fokker–Planck equation is the stochastic description of the dynamics of a classical field, as, for example, in classical laser theory [71]. If one restricts the analysis of a quantum degenerate gas to a coarse-grained grid in real space, then the grid step can always be chosen large enough to ensure that each grid point is occupied by a number of particles $N(\mathbf{r}) \gg 1$, namely the classical limit of a Bose gas. The price payed for this assumption is that the model can only describe situations with sufficiently long-rage spatial variations of the density. Translated in reciprocal space, this implies an upper cutoff of the momentum distribution, which is likely to be a very effective approximation if the gas is close to a quantum degenerate thermal distribution. We have recently generalized the TWA to the case of polaritons [90], where it turns out that losses extend the limits of its validity [89, 90], resulting in a particularly well suited theoretical tool for modeling dynamical aspects of polariton condensates. In particular, fluctuations are correctly modeled by TWA, and the approach could be applied to model the slow decay of the one-body density matrix in space, or the random path followed by a vortex in a polariton superfluid [91].

When applying TWA to polaritons, however, it should be kept in mind that a dominant role in the kinetics is played by energy relaxation from the highly occupied bottleneck region, where polaritons are exciton-like with vanishing photon component. In presence of a polariton condensate, this relaxation is what provides the gain mechanism that counterbalances the losses due to the finite polariton lifetime. It can occur via exciton-phonon scattering or, most effectively, via an exciton–exciton scattering process in which one exciton scatters to a polariton state at lower momentum while the second is excited to a higher energy-momentum state [63]. As the TWA is mostly suited to a situation where high-energy states are scarcely occupied, the bottleneck effect must be described by assuming a distinct exciton field, that is, modeled by a diffusion equation for the density, coupled to the equation for the polaritons, similarly to well-established mean-field approaches [38, 87]. In this picture, the only missing mechanism is the residual energy relaxation between modes of the polariton field. To account for this effect, the TWA can be completed by a term that models the intra-condensate relaxation. We have recently shown [86] that this intra-condensate relaxation is the dominant effect in presence of strongly inhomogeneous density profile, as in the downhill relaxation of polaritons in a smooth trap [10, 75], or in the recently observed multimode condensation in one-dimensional polariton wires [84].

In this work, we review the theoretical formalism based on the TWA, including the intra-condensate relaxation mechanism, and illustrate its range of validity by modeling some typical experimental situations.

10.2 Derivation of the Stochastic Equations of Motion

As we have mentioned in the introduction, the TWA is only applicable to the polariton close to the bottom of the dispersion. We therefore divide the phase space in two parts: (1) the low energy region, where the quasi-particles have a significant photonic component, that will be called the lower polariton region and (2) the higher energy excitonic part, where the photonic content can be neglected, that will be called the exciton "reservoir" states. The dynamics of the polaritonic region is the one of interest, but under nonresonant excitation, the excitonic states are of great importance as well because they replenish the polaritons that are lost from the microcavity. Although polariton and exciton states are instances of the same quantum field, within our approximation, they will be modeled as two different fields, whose field operators are denoted as $\psi(\mathbf{x})$ and $\phi(\mathbf{x})$ for polaritons and excitons respectively.

The full Hamiltonian can then be naturally divided into three parts:

$$H = \int \mathrm{d}\mathbf{x}\,[H_{\mathrm{LP}}(\mathbf{x}) + H_{\mathrm{R}}(\mathbf{x}) + H_{R,\mathrm{LP}}(\mathbf{x})]\,. \tag{10.1}$$

Here, the lower polariton Hamiltonian density is the usual

$$H_{\mathrm{LP}}(\mathbf{x}) = \psi^\dagger(\mathbf{x})\epsilon_{\mathrm{LP}}(-\mathrm{i}\nabla)\psi(\mathbf{x}) + \frac{g}{2}\psi^\dagger(\mathbf{x})\psi^\dagger(\mathbf{x})\psi(\mathbf{x})\psi(\mathbf{x}), \tag{10.2}$$

where g quantifies the strength of the polariton–polariton interactions that are well approximated by a zero-range potential, and $\epsilon_{\mathrm{LP}}(\mathbf{k})$ describes the lower polariton energy-momentum dispersion. The exciton reservoir Hamiltonian is given by

$$H_{\mathrm{R}}(\mathbf{x}) = \phi^\dagger(\mathbf{x})\epsilon_X(-\mathrm{i}\nabla)\phi(\mathbf{x}) + \frac{g}{2}\phi^\dagger(\mathbf{x})\phi^\dagger(\mathbf{x})\phi(\mathbf{x})\phi(\mathbf{x}), \tag{10.3}$$

where $\epsilon_X(\mathbf{k})$ is the exciton dispersion.

Of the interaction terms, we retain only those that contribute to energy-conserving processes:

$$H_{\mathrm{R,LP}}(\mathbf{x}) = H^{\mathrm{loss}}_{\mathrm{R,LP}}(\mathbf{x}) + H^{\mathrm{gain}}_{\mathrm{R,LP}}(\mathbf{x}) + H^{\mathrm{sc}}_{\mathrm{R,LP}}(\mathbf{x}). \tag{10.4}$$

Lower polaritons are created by the term

$$H^{\mathrm{gain}}_{\mathrm{R,LP}}(\mathbf{x}) = g\,\phi^\dagger(\mathbf{x})\phi(\mathbf{x})\phi(\mathbf{x})\psi^\dagger(\mathbf{x}), \tag{10.5}$$

whereas they are destroyed by

$$H^{\mathrm{loss}}_{\mathrm{R,LP}}(\mathbf{x}) = g\,\phi^\dagger(\mathbf{x})\phi^\dagger(\mathbf{x})\phi(\mathbf{x})\psi(\mathbf{x}). \tag{10.6}$$

Mean field shifts, relaxation and heating of the lower polaritons due to scattering processes between one polariton and one exciton particle, are described by the Hamiltonian

$$H^{\text{sc}}_{\text{R,LP}}(\mathbf{x}) = g\, \phi^{\dagger}(\mathbf{x})\phi(\mathbf{x})\psi^{\dagger}(\mathbf{x})\psi(\mathbf{x}). \tag{10.7}$$

In this model, we have artificially extended the phase space by introducing the two-particles representation. In particular, the excitonic phase space is extended down to $k = 0$ and the polaritonic phase space to arbitrarily large momenta. The first extension adds negligibly few states with respect to the overall large density of excitonic states at the energy of the exciton reservoir. The extension of the polariton phase space is on the other hand significant, but its effect will be suppressed by an exponential cutoff that enters the scattering rates as detailed below.

Starting from the Liouville equation of motion for the density matrix

$$\dot{\rho} = -\text{i}[H, \rho]. \tag{10.8}$$

and following the usual steps in the derivation of the quantum optical master Eq. [83], one obtains

$$\begin{aligned}\rho(t) = \rho(t_0) \; & -\text{i}\int_{t_0}^{t} \text{d}t'[H_{\text{R,LP}}(t'), \rho_{\text{LP}}(t')] \\ & - \int_{t_0}^{t} \text{d}t' \int_{t_0}^{t'} \text{d}t''[H_{\text{R,LP}}(t'), [H_{\text{R,LP}}(t''), \rho(t'')]].\end{aligned} \tag{10.9}$$

In order to model the kinetics of the polariton subsystem, the trace of this equation must be taken over the degrees of freedom of the exciton reservoir, resulting in a master equation for the reduced density of states ρ_{LP}. In this equation will appear terms of the form

$$\text{Tr}_R\{H^{\text{gain/loss}}_{\text{R,LP}} H^{\text{gain/loss}}_{\text{R,LP}}\, \rho\}. \tag{10.10}$$

Many of these terms are zero:

$$\text{Tr}_R\{H^{\text{gain}}_{\text{R,LP}} H^{\text{gain}}_{\text{R,LP}} \rho_{\text{LP}}\} = \text{Tr}_R\{H^{\text{loss}}_{\text{R,LP}} H^{\text{loss}}_{\text{R,LP}} \rho_{\text{LP}}\} = 0. \tag{10.11}$$

A nonzero term is, for example, given by

$$R_1 = \int \text{d}t'\; \text{d}t''\text{Tr}_R\{H^{\text{gain}}_{\text{R,LP}}(t') H^{\text{loss}}_{\text{R,LP}}(t'')\; \rho(t'')\}, \tag{10.12}$$

and we will show how it can be computed by introducing a Wigner transform of the reservoir propagator. We introduce relative and center of mass coordinates

$$\mathbf{X} = \frac{\mathbf{x}' + \mathbf{x}''}{2}, \qquad \mathbf{x} = \mathbf{x}' - \mathbf{x}'', \tag{10.13}$$

$$T = \frac{t' + t''}{2}, \qquad t = t' - t''. \tag{10.14}$$

As in [23], we define the Wigner transform of the reservoir propagator as

$$F^{\mathrm{W}}(\mathbf{X}, \mathbf{k}, T, \omega) = \int \mathrm{d}t \ \mathrm{d}\mathbf{x} \ \mathrm{e}^{\mathrm{i}\omega t} \mathrm{e}^{-\mathrm{i}\mathbf{k}\mathbf{x}} \ \mathrm{Tr}_R\{\phi^\dagger(\mathbf{X} + \mathbf{x}/2, T + t/2) \\ \times \phi(\mathbf{X} - \mathbf{x}/2, T - t/2)\rho\}. \tag{10.15}$$

With the inverse transformation, we obtain for R_1 defined in (10.12)

$$R_1 = \frac{1}{\Omega^3} \sum_{\mathbf{k}_{1,2,3}} \int \mathrm{d}\mathbf{X} \, \mathrm{d}\mathbf{x} \, \mathrm{e}^{\mathrm{i}(\Delta \mathbf{k} \cdot \mathbf{x} - \Delta \epsilon \, t)} \ \Pi_f(\mathbf{X}, \mathbf{k}_{1,2,3}) \\ \times \psi^\dagger(\mathbf{X} + \mathbf{x}/2, T + t/2) \psi(\mathbf{X} - \mathbf{x}/2, T - t/2) \rho(T - t/2), \tag{10.16}$$

where Ω is the area of our system, $\Delta\mathbf{k} = \mathbf{k}_2 + \mathbf{k}_3 - \mathbf{k}_1$, $\Delta\epsilon = \epsilon_X(\mathbf{k}_2) + \epsilon_X(\mathbf{k}_3) - \epsilon_X(\mathbf{k}_1)$ and $\Pi_f(\mathbf{X}, \mathbf{k}_{1,2,3})$ is a typical Boltzmann collision rate

$$\Pi_f(\mathbf{X}, \mathbf{k}_{1,2,3}) = f(\mathbf{X}, \mathbf{k}_1, T)[f(\mathbf{X}, \mathbf{k}_2, T) + 1][f(\mathbf{X}, \mathbf{k}_3, T) + 1]. \tag{10.17}$$

We have used the quasi-particle approximation [29]

$$F^{\mathrm{W}}(\mathbf{X}, \mathbf{k}, T, \omega) = (2\pi \mathrm{i})\delta(\omega - \epsilon_X(\mathbf{k})) f(\mathbf{X}, \mathbf{k}, T), \tag{10.18}$$

where f is here the occupation number of the mode with wave vector $\mathbf{k}$ at position $\mathbf{X}$ in the usual sense of the usual position dependent Boltzmann equation.

The time evolution of the LP field operators is approximately given by

$$\psi^\dagger(\mathbf{X} + \mathbf{x}/2, T + t/2) \simeq \frac{1}{\Omega} \sum_{\mathbf{Q}} \mathrm{e}^{\mathrm{i}\mathbf{Q}(\mathbf{X}+\mathbf{x}/2)} \mathrm{e}^{\mathrm{i}\epsilon_{\mathrm{LP}}(\mathbf{Q})t} \psi^\dagger(\mathbf{Q}, T - t/2). \tag{10.19}$$

Substituting this time evolution in (10.12), we obtain for R_1:

$$R_1 = \pi g^2 T \int \mathrm{d}\mathbf{X} \frac{1}{\Omega^4} \sum_{\mathbf{k}_{1,2,3}, \mathbf{Q}} \delta_{\Delta\mathbf{k},0} \delta_{\Delta\epsilon, \epsilon_{\mathrm{LP}}(\mathbf{q})} \\ \times \Pi_f(\mathbf{X}, \mathbf{k}_{1,2,3}) \psi^\dagger_{\mathbf{Q}}(T) \psi_{\mathbf{X}}(T) \rho(T). \tag{10.20}$$

The effect of this term is to remove a particle from the lower polariton region with a momentum dependent probability. Among the other terms that result from (10.9), there is one that leads to the creation of lower polaritons, whereas others are responsible for the relaxation and excitation of polaritons due to the interactions with the excitonic reservoir.

Collecting all terms, we obtain

$$\frac{d}{dt}\rho_{LP}(t) = -i[H_{LP}, \rho_{LP}] + K_{in}(\rho_{LP}) + K_{out}(\rho_{LP}) + K_{relax}(\rho_{LP}). \tag{10.21}$$

Here, the density matrix evolves under the in-scattering as

$$K_{in}(\rho) = \frac{1}{2}\sum_{\mathbf{q}} \int d\mathbf{x}\; R_{in}[n_R(\mathbf{x}), \epsilon_{LP}(\mathbf{q})] \big[e^{i\mathbf{q}\mathbf{x}} \psi^\dagger(\mathbf{x})\rho\psi(\mathbf{q}) \\ - e^{i\mathbf{q}\mathbf{x}}\psi(\mathbf{q})\psi^\dagger(\mathbf{x})\rho + \text{h.c.}\big] \tag{10.22}$$

and under out-scattering as

$$K_{out}(\rho) = \frac{1}{2}\sum_{\mathbf{q}} \int d\mathbf{x}\; R_{out}[n_R(\mathbf{x}), \epsilon_{LP}(\mathbf{q})] \big[e^{i\mathbf{q}\mathbf{x}} \psi(\mathbf{q})\rho\psi^\dagger(\mathbf{x}) \\ - e^{i\mathbf{q}\mathbf{x}}\psi^\dagger(\mathbf{x})\psi(\mathbf{q})\rho + h.c.\big]. \tag{10.23}$$

The rates $R_{in/out}$ are given by the usual semiclassical Boltzmann rates. Neglecting stimulated processes in the reservoir, R_{in} and R_{out} depend on the reservoir density, respectively, as n_R^2 and n_R. We therefore write

$$R_{in}(n_R, \epsilon_{LP}(\mathbf{q})) = n_R^2 R_{in}(\epsilon_{LP}(\mathbf{q})), \tag{10.24}$$

$$R_{out}(n_R, \epsilon_{LP}(\mathbf{q})) = n_R R_{out}(\epsilon_{LP}(\mathbf{q})). \tag{10.25}$$

Actually, another loss mechanism for the lower polariton field is present: leakage of the photon out of the imperfect microcavity mirrors, which gives a finite line width γ to the lower polariton. This loss mechanism has a negligible energy and momentum dependence and can be added to the model by simply adding the constant term γ to R_{out}.

The effect of relaxation is described by

$$K_{relax}(\rho) = \sum_{\mathbf{q}_1\mathbf{q}_2} \int d\mathbf{x}_1 d\mathbf{x}_2 d\mathbf{x}\; e^{-i\mathbf{q}_1\mathbf{x}_1 + i\mathbf{q}_2\mathbf{x}_2} R_{relax}[n_R(\mathbf{x}), \epsilon_{\mathbf{q}_1} - \epsilon_{\mathbf{q}_2}]\big[\psi^\dagger(\mathbf{q}_1) \\ \times\psi(\mathbf{q}_2)\psi^\dagger(\mathbf{x})\psi(\mathbf{x})\rho + \rho\psi^\dagger(\mathbf{x})\psi(\mathbf{x})\psi^\dagger(\mathbf{q}_1)\psi(\mathbf{q}_2) + \psi^\dagger(\mathbf{q}_1) \\ \times\psi(\mathbf{q}_2)\rho\psi^\dagger(\mathbf{x})\psi(\mathbf{x}) + \psi^\dagger(\mathbf{x})\psi(\mathbf{x})\rho\psi^\dagger(\mathbf{q}_1)\psi(\mathbf{q}_2)\big]. \tag{10.26}$$

In the following, we will not include the relaxation between lower polariton states. We refer the reader to [86], where the effects of relaxation on polariton condensates in polariton wires was included in mean field theory.

10.3 Truncated Wigner Approximation

An exact numerical solution of the full master (10.21) is only possible for systems with few particles on a few spatial modes. Large systems can still be treated approximately by the use of quasi-probability distributions, that have been introduced in quantum optics, starting from the work by Glauber and Sudarshan [83]. The general strategy is to use the quasi-probability distribution in order to convert the quantum mechanical master equation into a partial differential equation (PDE) of the Fokker–Planck type. This Fokker–Planck equation can be simulated by the corresponding Langevin equation. In the presence of sufficient dissipation, the Wigner quasi-probability distribution provides a robust scheme for the solution of the master equation: the third order derivatives appearing in the PDE can be neglected with respect to the second order ones that originate from the losses.

The introduction of a spatial grid with nonvanishing elementary area ΔS is essential for the validity of this truncation approximation because the condition for its validity is given by $g/\Delta S \ll \gamma$. In physical terms, the blueshift due to a single polariton on a grid cell should be much smaller than the polariton line width. Under this condition, the interactions are not able to build up significant amounts of entanglement that cause the Wigner function to become negative and that invalidate the truncation of the PDE. In the following, we will thus use a discrete index for the real space dependence of the polariton field.

The Langevin-type stochastic differential equation that sample the Wigner quasi-probability distribution consists of a deterministic F_{det} and a stochastic part $\mathrm{d}W$:

$$\mathrm{d}\psi(\mathbf{x},t) = F_{\text{det}}\,\mathrm{d}t + \mathrm{d}W, \tag{10.27}$$

where the deterministic evolution is

$$F_{\text{det}} = -\mathrm{i}\left[\epsilon_{\text{LP}}(-\mathrm{i}\nabla) + \frac{\mathrm{i}(\mathscr{R}_{\text{in}} - \mathscr{R}_{\text{out}} - \gamma)}{2} + g|\psi(\mathbf{x})|^2\right]\psi(\mathbf{x}). \tag{10.28}$$

The in/out operators act on the polariton field as

$$\mathscr{R}_{\text{in}}\psi(\mathbf{x}) = n_{\text{R}}^2(\mathbf{x})\sum_{\mathbf{qx}'} \mathrm{e}^{\mathrm{i}\mathbf{q}(\mathbf{x}-\mathbf{x}')} R_{\text{in}}[\epsilon_{\text{LP}}(\mathbf{q})]\psi(\mathbf{x}'), \tag{10.29}$$

$$\mathscr{R}_{\text{out}}\psi(\mathbf{x}) = n_{\text{R}}(\mathbf{x})\sum_{\mathbf{qx}'} \mathrm{e}^{\mathrm{i}\mathbf{q}(\mathbf{x}-\mathbf{x}')} R_{\text{out}}[\epsilon_{\text{LP}}(\mathbf{q})]\psi(\mathbf{x}'). \tag{10.30}$$

The stochastic part $\mathrm{d}W(\mathbf{x},t)$ in the Langevin-type equation is a white-noise force field with a spatial correlation function $\langle \mathrm{d}W(\mathbf{x})\mathrm{d}W^*(\mathbf{x}')\rangle$ which is defined as a function of the scattering rates in order to fulfil the fluctuation-dissipation theorem [90]

$$\langle \mathrm{d}W(\mathbf{x})\mathrm{d}W(\mathbf{x}')\rangle = 0, \tag{10.31}$$

$$\langle dW(\mathbf{x})dW^*(\mathbf{x}')\rangle = \frac{dt}{2\Delta V}\left(\langle \mathbf{x}|\mathscr{R}^S_{\text{in}} + \mathscr{R}^S_{\text{out}}|\mathbf{x}'\rangle + \gamma\delta_{\mathbf{x},\mathbf{x}'}\right), \tag{10.32}$$

where $\mathscr{R}^S_{\text{in,out}} = [\mathscr{R}_{\text{in,out}} + (\mathscr{R}_{\text{in,out}})^{\text{T}}]/2$ are the symmetrized kernels.

In our description of the microcavity dynamics, the exciton-like particles are treated as a classical reservoir. This allowed to trace out the excitonic degrees of freedom and to isolate the quantum dynamics of the polaritons from the classical exciton dynamics. This approximation, however, misses the reservoir depletion when a lower polariton is created. Missing this effect makes the model unphysical, because without the reservoir depletion (gain saturation in laser physics terminology) the polariton dynamics does not reach a steady state above the condensation threshold, because it is the suppression of the gain due to the reservoir that stops the exponential increase of the polariton density.

We propose to include the reservoir depletion by coupling its dynamics to the equation of motion for the classical polariton field in the following way

$$\frac{dn_{\text{R}}}{dt} = -\gamma_{\text{R}}[n_{\text{R}} - n^o_{\text{R}}(I_{\text{p}}, \psi)]. \tag{10.33}$$

Similar ideas have been implemented in [47], where the dynamics of a single condensate mode was coupled to a Boltzmann equation for the excited states, and in [52], where the reservoir was modeled by a saturable gain medium, a model widely used in laser physics.

In (10.33), $n^o_{\text{R}}(I_{\text{p}}, \psi)$ is the average steady state value of the reservoir density in the presence of a pump with intensity I_{p} and a lower polariton field ψ. The relaxation time γ_R^{-1} is a measure of the time it takes for the reservoir density to adjust to a new environment (I_{p}, ψ). Spatial diffusion of the reservoir excitons is expected to be a small effect [87] and was therefore neglected. For the steady state value of the reservoir density, we assume that it is simply proportional to the balance of incoming and outgoing particles:

$$n^o_{\text{R}}(P, \psi) = \beta\left(I_{\text{p}} - \frac{d}{dt}\langle\psi^\dagger\psi\rangle|_{\text{res}}\right), \tag{10.34}$$

where $\frac{d}{dt}\langle\psi^\dagger\psi\rangle|_{\text{res}} = 2\text{Re}[\psi^*(\mathscr{R}_{\text{in}} - \mathscr{R}_{\text{out}})\psi]$ is the net scattering rate from the reservoir into the lower polariton branch. It is instructive to substitute (10.34) into (10.33):

$$\frac{dn_{\text{R}}}{dt} = P - \gamma_{\text{R}}n_{\text{R}} - \beta\gamma_{\text{R}}\frac{d}{dt}\langle\psi^\dagger\psi\rangle|_{\text{res}}, \tag{10.35}$$

where $P = \beta I_{\text{p}}$ is the effective pump term for the active reservoir polaritons. The parameter β quantifies the back action of the condensate on the reservoir. This back action is needed to obtain a steady state for the dynamical equations above the threshold, where for $n_{\text{R}} = P/\gamma_{\text{R}}$ the in-scattering rate exceeds the out-scattering rate. In mean field theory, the reservoir density n_{R} is clamped to its threshold value $n_{\text{R},mf}$ that satisfies for homogeneous systems $n^2_{\text{R},mf}R_{\text{in}}(0) - n_{\text{R},mf}R_{\text{out}}(0) = \gamma$.

If we rewrite the motion equations for n_{R} in terms of the renormalized $\tilde{n}_{\mathrm{R}} = n_{\mathrm{R}}/n_{\mathrm{R},mf}$, we have

$$\frac{\mathrm{d}\tilde{n}_{\mathrm{R}}}{\mathrm{d}t} = \tilde{P} - \gamma_{\mathrm{R}}\tilde{n}_{\mathrm{R}} - \alpha\frac{\mathrm{d}}{\mathrm{d}t}\langle\psi^{\dagger}\psi\rangle|_{\mathrm{res}}, \tag{10.36}$$

where $\alpha = \beta\gamma_{\mathrm{R}}/n_{\mathrm{R},mf}$. Also in the presence of fluctuations, the dimensionless reservoir density $\tilde{n}_{\mathrm{R}}$ is close to one above threshold, in order for the gain to compensate for the losses. The factor α plays an important physical role because the back action of the condensate on the reservoir tends to damp the condensate fluctuations. If the condensate density is at some time larger than average, the reservoir will be depleted, $R_{\mathrm{in}} - R_{\mathrm{out}}$ decreases, and the deterministic part in the equations of motion for the condensate will decrease the amplitude of the fluctuation. In principle, the parameter α could be calculated from the Boltzmann equation. We prefer, however, to study the physics in terms of this parameter, because it gives a good insight in the nonequilibrium aspects of the coherence.

In the truncated Wigner approximation, the density of polaritons is related to ψ as $n = |\psi|^2 - 1/(2\Delta V)$, or in words, the classical field ψ contains half a particle per mode of zero point fluctuations. These fluctuations should be taken into account when evaluating the last term in (10.36). For the out-scattering, the zero-point fluctuations do not contribute and should be subtracted, whereas for the in-scattering, the zero-point fluctuations give rise to only half of the spontaneous in-scattering. The remaining part should be added. The equation of motion for the reduced reservoir density then finally reads

$$\frac{\mathrm{d}\tilde{n}_{\mathrm{R}}}{\mathrm{d}t} = \tilde{P} - \gamma_{\mathrm{R}}\tilde{n}_{\mathrm{R}} - \alpha\frac{\mathrm{d}}{\mathrm{d}t}\langle\psi^{\dagger}\psi\rangle|_{\mathrm{res},W} - \frac{\alpha}{2\Delta V}\sum_{\mathbf{k}}\left[R_{\mathrm{in}}(\epsilon_{\mathrm{LP}}(\mathbf{k}) + R_{\mathrm{out}}(\epsilon_{\mathrm{LP}}(\mathbf{k}))\right]. \tag{10.37}$$

10.4 Numerical Results

We present here numerical simulations based on the TWA formalism, that highlight the potential of this kinetic approach. To this purpose, we need to quantify the in- and out-scattering rates linking the dynamics of the polariton and the reservoir fields. The energy dependence of the in- and outscattering rates $R_{\mathrm{in}}(E)$ and $R_{\mathrm{out}}(E)$ can be determined by a microscopic model of the scattering process. However, some general features can be already determined by the requirement that the high-energy tail of the polariton distribution reproduce a thermal distribution at some fixed temperature T_{R}. By fulfilling this latter prescription, we can fix the rates phenomenologically. For reasons of stability of the numerical simulation, we also introduce a cutoff in the rates at finite energy, as otherwise unphysically high scattering rates would arise in the model and make the fluctuations numerically intractable. When including this cutoff, the out-scattering rate is $R_{\text{out}}(E) = \min[2\gamma e^{2E/T_R}, 3.3\,\text{meV}]$. The in-scattering rate is instead chosen as $R_{\text{in}}(E) = [R_{\text{out}}(E) + \gamma]e^{-E/T_R}$.

The stochastic equations of motion can be simulated by Monte Carlo techniques. As we have already mentioned in the introduction, the nonequilibrium condition of the polariton condensates implies that the effect of the reservoir on the correlation functions cannot be characterized by the temperature alone. We will discuss below two other physical quantities that determine the degree of coherence in the polariton condensate: the feedback parameter α and the out-scattering rate R_{out}. The other parameters, we keep fixed for all simulations: $\gamma = 0.5\,\text{meV}$, $m_{\text{LP}}/\hbar = 1\ \mu\text{m}^{-2}\,\text{meV}^{-1}$, $g/\hbar = 0.03\ \mu\text{m}^2$ and $k_B T_R = 2\,\text{meV}$. The simulations were done on a 32- $\times$ 32-point grid with physical dimension of $66 \times 66\mu\text{m}^2$ and periodic boundary conditions. The deterministic evolution of the polariton field 10.28 was calculated with a split-step method, evaluating the kinetic energy in momentum space and the other terms in real space, transforming back and forth using FFT algorithms. The stochastic fields with the spatial correlation functions $\langle dW(\mathbf{x})dW^*(\mathbf{x}')\rangle$ were constructed from unit variance uncorrelated complex Gaussian noise fields $\xi(\mathbf{x})$ by the linear transformation $d\mathbf{W} = \mathbf{D}\boldsymbol{\xi}$, where we used a vector notation for the values of $dW(\mathbf{x})$ on the discrete space grid, and $\mathbf{D}$ is the matrix square root of the noise correlation $\mathbf{D} = \sqrt{\langle d\mathbf{W}d\mathbf{W}^\dagger\rangle}$. The numerical computation of this square root is an important addition to the numerical complexity in computing the time evolution of the lower polariton field.

We first present results obtained by numerically integrating (10.27) and (10.37), assuming a spatially uniform pump term. Figure 10.1 illustrates single Monte Carlo realizations of the classical field $\psi(\mathbf{x})$ for different excitation parameters. They represent snapshots of the polariton density and phase below (panels a and b) and slightly above (panels c and d) the threshold. Even though these images have, strictly speaking, no direct physical meaning, they already illustrate qualitatively the coherence properties of the polariton condensate. The phase profile of panel (d) shows that the phase ordering is only partial. Vortex-antivortex pairs appear to exist at densities well above the stimulation threshold. This is an indication that the physics of the Berezinskii–Kosterlitz–Thouless type [27, 39] could occur in polariton condensates.

Typical experiments are carried out using a circular cross-sectional shape of the excitation laser beam, which induces a nonuniform density profile and consequently a nontrivial spatial kinetics of the polariton condensate. Figure 10.2 shows two examples for a finite excitation spot for pump intensities below (panels a and b) and above the threshold (panels c and d). At low density, both the density and phase fluctuations are large, whereas the phase fluctuations are clearly suppressed in the high-density regime. Panel (d) shows that phase coherence exists all over the extent of the excitation region. The concentric phase profile originates from the repulsive polariton–polariton interaction that causes an outward flow of polaritons [89]. Note that while the density under the finite excitation spot (Fig. 10.2c) is not higher than in the simulation with a homogeneous pumping profile (Fig. 10.1c), the phase is much more ordered in the finite spot case. The reason for this effect is that the high

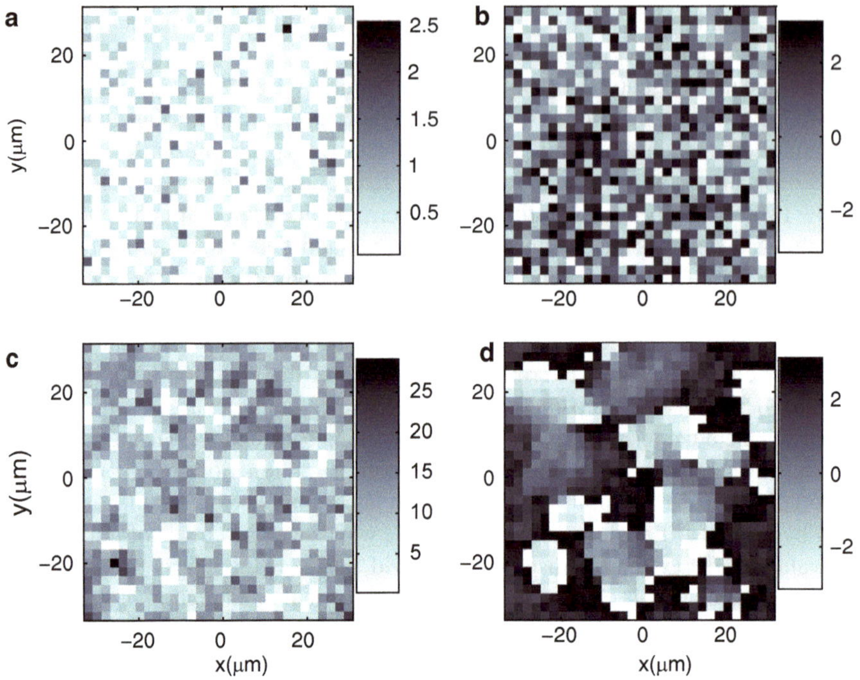

Fig. 10.1 Snapshots of a single Monte Carlo realization of the density (**a**, **c**) and phase (**b**, **d**) for excitation parameters below (**a**, **b**) and above threshold (**c**, **d**) of a polariton condensate excited by a uniform excitation laser with periodic boundary conditions

energy excitations can escape from the system in the case of the finite spot excitation through a kind of evaporative cooling.

Three momentum distributions for increasing pump intensity are shown in Fig. 10.3. As expected, our model shows the build up of a large occupation of the low momentum states for increasing pump intensity. The momentum distributions appear to be rather well fitted by a Bose–Einstein function (full line). It is important to mention here the important role of the reservoir relaxation rates. We have chosen them in such a way that a thermal distribution is obtained even in the absence of collisions between lower polaritons. In simulations with energy independent relaxation rates (not shown) and a large, yet realistic,[1] polariton–polariton interaction strength, we have obtained a constant instead of exponential decay at large momenta.

Note that the temperature extracted from the fits of the tails to a Maxwellian is lower than the reservoir temperature T_R (2 meV for the present simulations),

[1]An experimental upper bound to the blue shift due to polariton–polariton interaction is given by the total blue shift, which is less than 1 meV.

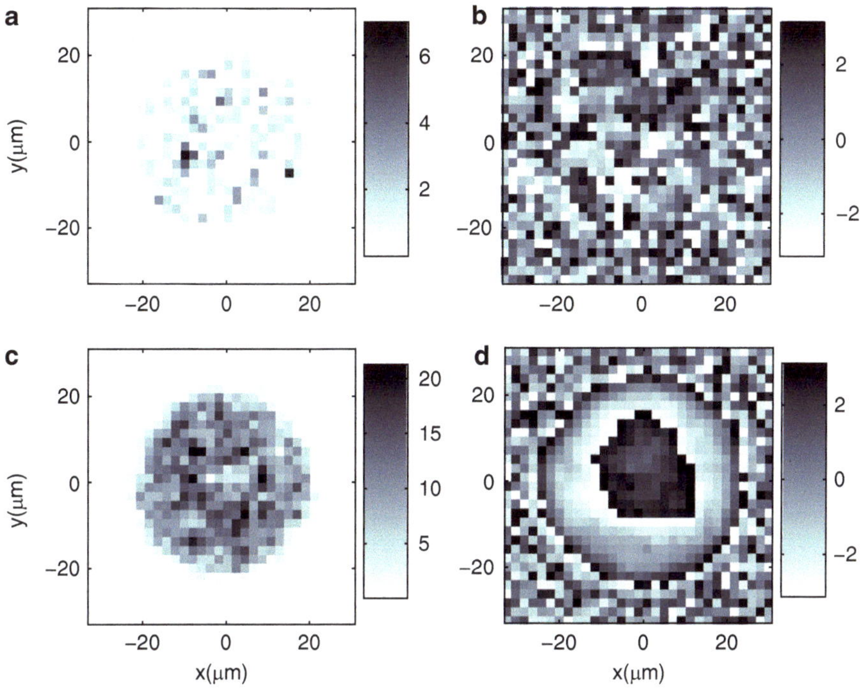

Fig. 10.2 The same as Fig. 10.1, but for a finite spot laser excitation

that enters the rates $R_{\text{in,out}}$ according to (10.25): the nonlinearity modifies the temperature that is expected in the linear regime. We remind the reader that T_{R} does not coincide with the lattice temperature and that $T_{\text{fit}} < T_{\text{R}}$ does not imply that the polariton temperature is lower than the lattice temperature.

The subtle features of long range coherence are much clearer in the Fourier transform of the momentum distribution, that is, the first order spatial coherence function. In Fig. 10.4 two values of α are compared. Below the condensation threshold, the gain saturation parameter α (see (10.37)) has no influence and the fit of the coherence by the one-body density matrix of the noninteracting Bose gas yields the reservoir temperature of 2 meV. For the simulation above the threshold, a higher value of α improves the long range coherence. Both instances of the one-body density matrix are relatively well fitted by the one of a noninteracting Bose gas. Both temperatures are below the reservoir temperature. The lowest effective temperature is obtained for the largest feedback parameter α.

In the simulations of Fig. 10.4, the polariton out-scattering rate was set to zero. In the simulations presented in Fig. 10.5, we have instead included a finite out-scattering rate according to (10.25). The comparison between panels (a) and (b) shows that the out-scattering has a dramatic effect on the coherence function. This should not come as a surprise, because the out-scattering increases the fluctuations (physically shot noise due to the discrete nature of the polariton field). Keeping

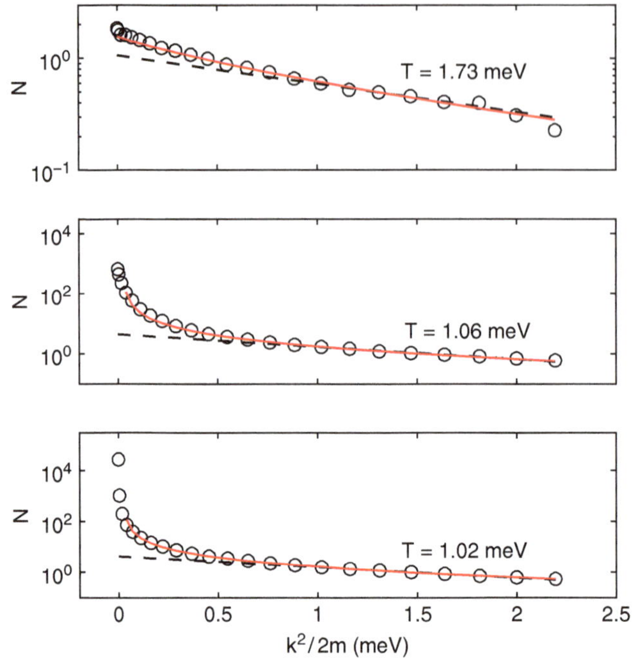

Fig. 10.3 Momentum distribution of the polariton field for various pump intensities. The open circles are the result of the Monte Carlo simulations and the full line is a fit to a BE distribution. Statistical errors of the Monte Carlo simulations are within the symbol size

$\alpha = 0.01\mu\text{m}^{-1}$ as in Fig. 10.4, but including some out-scattering, the coherence in panel (a) is dramatically decreased. As compared to the simulations of Fig. 10.4, the effect of α is much more pronounced.

10.5 Conclusions

In conclusion, we have reviewed the theoretical formalism based on the TWA as a model of the kinetics of a polariton quantum fluid produced under nonresonant incoherent excitation. This approach turns out to be extremely well adapted to the driven-dissipative nature of the polariton quantum-degenerate gas, where the dissipation rate dominates over the characteristic polariton correlation energy (charging energy in one element of the spatial grid). In condensates of ultracold atoms, this condition is always violated – as the dissipation vanishes – and the TWA can only be applied if the more restrictive condition of vanishing high-momentum components of the density distribution is fulfilled. In the polariton case, the restriction on momentum can instead be conveniently handled by effectively separating the physical field in a coherent polariton field coupled to an incoherent

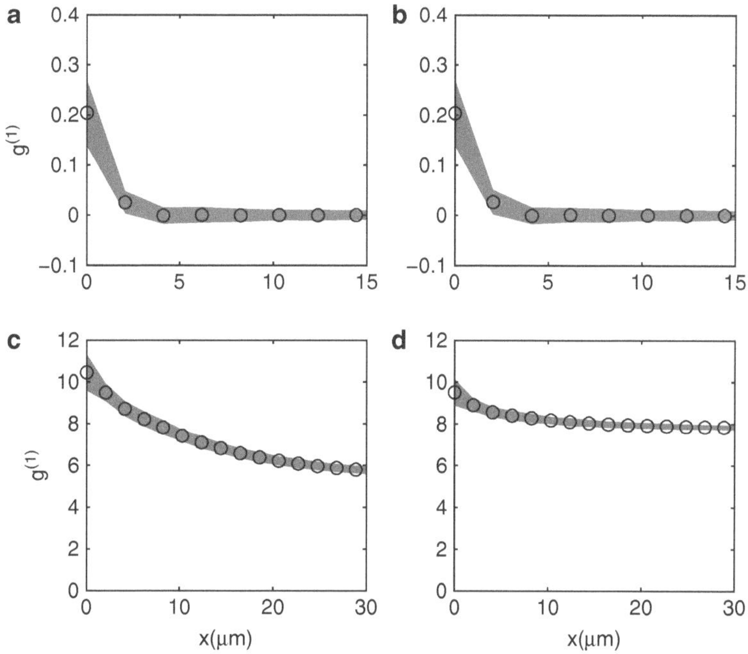

Fig. 10.4 Spatial coherence (*open circles*, the grey band indicates the error on the Monte Carlo data) for two values of the feedback from the condensate on the reservoir: from left to right $\alpha = 0.01\mu\text{m}^2$ (**a**, **c**) and $\alpha = 0.1\mu\text{m}^2$ (**b**, **d**). *Upper panels*: pump below the condensation threshold. *Lower panels*: above threshold

exciton reservoir at higher energy. By employing this prescription, numerical simulations based on the TWA are very successful in predicting the phase transition to a quantum degenerate gas, the phase fluctuations, and the consequent behaviour of the one-body density matrix in space. Fluctuations in particular are included by solving the Langevin equation associated with the Fokker–Planck equation resulting from the TWA, using Monte Carlo simulations. When, however, only mean field effects must be described, as in most recent applications, solving directly the mean-field time-dependent equations is a much less demanding computational task. A possible extension of the TWA formalism for polaritons consists in including an energy relaxation term within the polariton field that can be inferred from the Boltzmann kinetic theory [86]. This energy relaxation process turns out to play an important role in the condensate kinetics, especially in its spatial dynamics, sometimes resulting in a multimode condensate [84]. In very specific situations, the model of the excitonic reservoir must also be refined, for example, by accounting for two coupled reservoirs, one accommodating the external pump term while the other determines the density fluctuations of the polariton condensate. A weakness of the approach consists in the phenomenological way the coupling between the

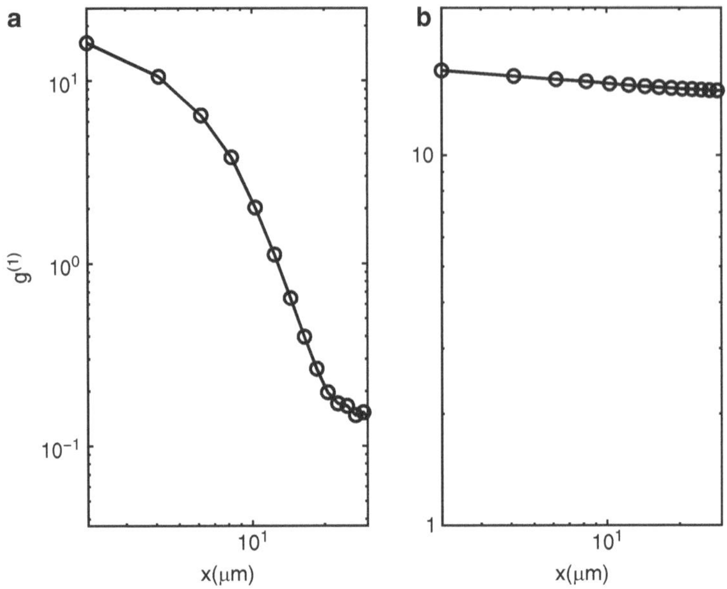

Fig. 10.5 The spatial coherence function as in Fig. 10.4, but including out-scattering. Simulations with $\alpha = 0.01\mu\text{m}^2$ (**a**) and $\alpha = 0.1\mu\text{m}^2$ (**b**) are shown. Both plots correspond to a situation above the condensation threshold

polariton and the exciton fields is introduced. In particular, we have shown that the back action of the polariton field on the reservoir plays an important role in determining the fluctuations and the spatial coherence. It is possible to determine the coupling model from a kinetic description based on microscopically computed scattering coefficients, but this analysis has not yet been carried out.

The only fundamental physical limitation of the TWA is of course its inability to describe most kinds of quantum correlations within the polariton field. Striking examples of quantum correlations in a condensate are for instance those predicted for arrays of closely-packed vortices, for which states analogous to fractional quantum-Hall states can arise [22]. This physical limit of a superfluid represents today both an experimental challenge and a very promising situation for the study of strongly-correlated systems. In this case, a classical field approach is clearly of no use, and a full quantum many-body theory would be required instead. It is unlikely that current polariton systems can reach this regime, given their strongly dissipative nature that fights against the onset of quantum correlations. It is, however, possible that strong correlations might arise in future systems optimized for a much longer polariton lifetime, opening an exciting perspective for the study of these intriguing phenomena.

References

1. C. Adrados, A. Amo, T.C.H. Liew, R. Hivet, R. Houdre, E. Giacobino, A.V. Kavokin, A. Bramati, Spin rings in bistable planar semiconductor microcavities. Phys. Rev. Lett. **105**(21), 116402 (2010)
2. U. Al Khawaja, J.O. Andersen, N.P. Proukakis, H.T.C Stoof, Low dimensional bose gases. Phys. Rev. A, **66**(1), 013615 (2002)
3. A. Amo, T.C.H. Liew, C. Adrados, R. Houdre, E. Giacobino, A.V. Kavokin, A. Bramati, Exciton-polariton spin switches. Nat. Photon. **4**(6), 361–366 (2010)
4. A. Amo, S. Pigeon, D. Sanvitto, V.G. Sala, R. Hivet, I. Carusotto, F. Pisanello, G. Lemenager, R. Houdre, E. Giacobino, C. Ciuti, A. Bramati, Hydrodynamic solitons in polariton superfluids. Science, **332**, 1167 (2011)
5. A. Amo, D. Sanvitto, F.P. Laussy, D. Ballarini, E. del Valle, M.D. Martin, A. Lemaitre, J. Bloch, D.N. Krizhanovskii, M.S. Skolnick, C. Tejedor, L. Vina, Collective fluid dynamics of a polariton condensate in a semiconductor microcavity. Nature, **457**(7227), 291–U3 (2009)
6. A. Amo, J. Lefrere, S. Pigeon, C. Adrados, C. Ciuti, I. Carusotto, R. Houdre, E. Giacobino, A. Bramati, Superfluidity of polaritons in semiconductor microcavities. Nat. Phys. **5**(11), 805–810 (2009)
7. A. Baas, J.P. Karr, H. Eleuch, E. Giacobino, Optical bistability in semiconductor microcavities. Phys. Rev. A **69**(2) 023809 (2004)
8. D. Bajoni, E. Semenova, A. Lemaitre, S. Bouchoule, E. Wertz, P. Senellart, S. Barbay, R. Kuszelewicz, J. Bloch, Optical bistability in a gaas-based polariton diode. Phys. Rev. Lett. **101**(26), 267404 (2008)
9. D. Bajoni, P. Senellart, E. Wertz, I. Sagnes, A. Miard, A. Lemaitre, J. Bloch, Polariton laser using single micropillar gaas-gaalas semiconductor cavities. Phys. Rev. Lett. **100**(4), 047401 (2008)
10. R. Balili, V. Hartwell, D. Snoke, L. Pfeiffer, K. West, Bose–einstein condensation of microcavity polaritons in a trap. Science **316**(5827), 1007–1010 (2007)
11. J. Bloch, R. Planel, V. Thierry-Mieg, J.M. Gerard, D. Barrier, J.Y. Marzin, E. Costard, Strong-coupling regime in pillar semiconductor microcavities. Superlattices Microst. **22**(3), 371–374 (1997)
12. H. Thien Cao, T.D. Doan, D.B. Tran Thoai, H. Haug, Condensation kinetics of cavity polaritons interacting with a thermal phonon bath. Phys. Rev. B **69**(24), 245325 (2004)
13. G. Dasbach, M. Schwab, M. Bayer, A. Forchel, Parametric polariton scattering in microresonators with three-dimensional optical confinement. Phys. Rev. B **64**(20), 201309 (2001)
14. M.J. Davis, S.A. Morgan, K. Burnett, Simulations of bose fields at finite temperature. Phys. Rev. Lett. **87**(16), 160402 (2001)
15. H. Deng, H. Haug, Y. Yamamoto, Exciton-polariton bose–einstein condensation. Rev. Mod. Phys. **82**(2), 1489–1537 (2010)
16. H. Deng, D. Press, S. Gotzinger, G.S. Solomon, R. Hey, K.H. Ploog, Y. Yamamoto, Quantum degenerate exciton-polaritons in thermal equilibrium. Phys. Rev. Lett. **97**(14), 146402 (2006)
17. H. Deng, G.S. Solomon, R. Hey, K.H. Ploog, Y. Yamamoto, Spatial coherence of a polariton condensate. Phys. Rev. Lett. **99**(12), 126403 (2007)
18. T.D. Doan, H. Thien Cao, D.B. Tran Thoai, H. Haug, Coherence of condensed microcavity polaritons calculated within boltzmann-master equations. Phys. Rev. B **78**(20), 205306 (2008)
19. T.D. Doan, H. Thien Cao, D.B. Tran Thoai, H. Haug, Condensation kinetics of microcavity polaritons with scattering by phonons and polaritons. Phys. Rev. B **72**(8), 085301 (2005)
20. L. Ferrier, S. Pigeon, E. Wertz, M. Bamba, P. Senellart, Isabelle Sagnes, Aristide Lemaitre, Cristiano Ciuti, and Jacqueline Bloch. Polariton parametric oscillation in a single micropillar cavity. Appl. Phys. Lett. **97**(3), 031105 (2010)
21. L. Ferrier, E. Wertz, R. Johne, D.D. Solnyshkov, P. Senellart, I. Sagnes, A. Lemaitre, G. Malpuech, J. Bloch, Interactions in confined polariton condensates. Phys. Rev. Lett. **106**(12), 126401 (2011)

22. A.L. Fetter, Rotating trapped bose–einstein condensates. Rev. Mod. Phys. **81**(2), 647–691 (2009)
23. C.W. Gardiner, M.J. Davis, J. Phys. B **36**, 4731 (2003)
24. C.W. Gardiner, P. Zoller, Quantum kinetic theory: A quantum kinetic master equation for condensation of a weakly interacting bose gas without a trapping potential. Phys. Rev. A **55**(4), 2902–2921 (1997)
25. N.A. Gippius, S.G. Tikhodeev, V.D. Kulakovskii, D.N. Krizhanovskii, A.I. Tartakovskii, Nonlinear dynamics of polariton scattering in semiconductor microcavity: Bistability vs. stimulated scattering. Europhys. Lett. **67**(6), 997–1003 (2004)
26. M. Gurioli, F. Bogani, D.S. Wiersma, P. Roussignol, G. Cassabois, G. Khitrova, H. Gibbs, Experimental study of disorder in a semiconductor microcavity. Phys. Revi. B **64**(16), 165309 (2001)
27. Z. Hadzibabic, P. Krueger, M. Cheneau, B. Battelier, J. Dalibard, Berezinskii - kosterlitz - thouless crossover in a trapped atomic gas. Nature **441**, 1118 (2006)
28. H. Haug, H. Thien Cao, D.B. Tran Thoai, Coherence and decoherence of a polariton condensate. Phys. Rev. B **81**(24), 245309 (2010)
29. H. Haug, A.-P. Jauho, *Quantum Kinetics in Transport and Optics of Semiconductors*. (Springer, Berlin, 1997)
30. R. Johne, I.A. Shelykh, D.D. Solnyshkov, G. Malpuech, Polaritonic analogue of datta and das spin transistor. Phys. Rev. B **81**(12), 125327 (2010)
31. D. Kadio, M. Gajda, K. Rzazewski, Phase fluctuations of a bose–einstein condensate in low-dimensional geometry. Phys. Rev. A **72**(1), 013607 (2005)
32. Yu. Kagan, V.A. Kashurnikov, A.V. Krasavin, N.V. Prokof'ev, B.V. Svistunov, Quasicondensation in a two-dimensional interacting bose gas. Phys. Rev. A **61**(4), 043608 (2000)
33. Yu. Kagan, B.V. Svistunov, Evolution of correlation properties and appearance of broken symmetry in the process of bose–einstein condensation. Phys. Rev. Lett. **79**(18), 3331–3334 (1997)
34. R. Idrissi Kaitouni, O. El Daif, A. Baas, M. Richard, T. Paraiso, P. Lugan, T. Guillet, F. Morier-Genoud, J.D. Ganiere, J.L. Staehli, V. Savona, B. Deveaud, Engineering the spatial confinement of exciton polaritons in semiconductors. Phys. Rev. B **74**(15) 155311 (2006)
35. J. Kasprzak, M. Richard, S. Kundermann, A. Baas, P. Jeambrun, J.M.J. Keeling, F.M. Marchetti, M.H. Szymanska, R. Andre, J.L. Staehli, V. Savona, P.B. Littlewood, B. Deveaud, Le Si Dang, Bose–einstein condensation of exciton polaritons. Nature **443**(7110), 409–414 (2006)
36. K.V. Kavokin, M.A. Kaliteevski, R.A. Abram, A.V. Kavokin, S. Sharkova, I.A. Shelykh, Stimulated emission of terahertz radiation by exciton-polariton lasers. Appl. Phys. Lett. **97**(20), 1111 (2010)
37. J. Keeling, F.M. Marchetti, M.H. Szymanska, P.B. Littlewood, Collective coherence in planar semiconductor microcavities. Semicond. Sci. Tech. **22**(5), R1–R26 (2007)
38. J. Keeling, N.G. Berloff, Spontaneous rotating vortex lattices in a pumped decaying condensate. Phys. Rev. Lett. **100**(25), 250401 (2008)
39. J.M. Kosterlitz, D.J. Thouless, Ordering, metastability and phase transitions in two-dimensional systems. J. Phys. C **6**, 1181 (1973)
40. D.N. Krizhanovskii, D.M. Whittaker, R.A. Bradley, K. Guda, D. Sarkar, D. Sanvitto, L. Vina, E. Cerda, P. Santos, K. Biermann, R. Hey, M.S. Skolnick, Effect of interactions on vortices in a nonequilibrium polariton condensate. Phys. Rev. Lett. **104**(12), 126402 (2010)
41. K.G. Lagoudakis, F. Manni, B. Pietka, M. Wouters, T.C.H. Liew, V. Savona, A.V. Kavokin, R. Andre, B. Deveaud-Pledran, Probing the dynamics of spontaneous quantum vortices in polariton superfluids. Phys. Rev. Lett. **106**(11), 115301 (2011)
42. K.G. Lagoudakis, T. Ostatnicky, A.V. Kavokin, Y.G. Rubo, R. Andre, B. Deveaud-Pledran, Observation of half-quantum vortices in an exciton-polariton condensate. Science **326**(5955), 974–976 (2009)
43. K.G. Lagoudakis, B. Pietka, M. Wouters, R. Andre, B. Deveaud-Pledran, Coherent oscillations in an exciton-polariton josephson junction. Phys. Rev. Lett. **105**(12), 120403 (2010)

44. K.G. Lagoudakis, M. Wouters, M. Richard, A. Baas, I. Carusotto, R. Andre, L.E.S.I. Dang, B. Deveaud-Pledran, Quantized vortices in an exciton-polariton condensate. Nat. Phys. **4**(9), 706–710 (2008)
45. W. Langbein, J.M. Hvam, Elastic scattering dynamics of cavity polaritons: Evidence for time-energy uncertainty and polariton localization. Phys. Rev. Lett. **88**(4), 047401 (2002)
46. F.P. Laussy, A.V. Kavokin, I.A. Shelykh, Exciton-polariton mediated superconductivity. Phys. Rev. Lett. **104**(10), 106402 (2010)
47. F.P. Laussy, G. Malpuech, A. Kavokin, P. Bigenwald, Spontaneous coherence buildup in a polariton laser. Phys. Rev. Lett. **93**(1), 016402 (2004)
48. C. Leyder, T.C.H. Liew, A.V. Kavokin, I.A. Shelykh, M. Romanelli, J. Ph. Karr, E. Giacobino, A. Bramati, Interference of coherent polariton beams in microcavities: Polarization-controlled optical gates. Phys. Rev. Lett. **99**(19), 196402 (2007)
49. T.C.H. Liew, A.V. Kavokin, T. Ostatnicky, M. Kaliteevski, I.A. Shelykh, R.A. Abram, Exciton-polariton integrated circuits. Phys. Rev. B **82**(3), 033302 (2010)
50. T.C.H. Liew, A.V. Kavokin, I.A. Shelykh, Optical circuits based on polariton neurons in semiconductor microcavities. Phys. Rev. Lett. **101**(1), 016402 (2008)
51. T.C.H. Liew, Y.G. Rubo, A.V. Kavokin, Generation and dynamics of vortex lattices in coherent exciton-polariton fields. Phys. Rev. Lett. **101**(18), 187401 (2008)
52. A.P.D. Love, D.N. Krizhanovskii, D.M. Whittaker, R. Bouchekioua, D. Sanvitto, S. Al Rizeiqi, R. Bradley, M.S. Skolnick, P.R. Eastham, R. André, Le Si Dang. Intrinsic decoherence mechanisms in the microcavity polariton condensate. Phys. Rev. Lett. **101**(6), 067404 (2008)
53. G. Malpuech, A. Kavokin, A. Di Carlo, J.J. Baumberg, Polariton lasing by exciton-electron scattering in semiconductor microcavities. Phys. Rev. B **65**(15), 153310 (2002)
54. C. Mora, Y. Castin, Extension of bogoliubov theory to quasicondensates. Phys. Rev. A **67**(5), 053615 (2003)
55. N. Na, Y. Yamamoto, Massive parallel generation of indistinguishable single photons via the polaritonic superfluid to mott-insulator quantum phase transition. New J. Phys. **12**(12), 123001 (2010)
56. G. Nardin, K.G. Lagoudakis, B. Pietka, F. Morier-Genoud, Y. Leger, B. Deveaud-Pledran, Selective photoexcitation of confined exciton-polariton vortices. Phys. Rev. B **82**(7), 073303 (2010)
57. T. Nikuni, E. Zaremba, A. Griffin, Two-fluid dynamics for a bose–einstein condensate out of local equilibrium with the noncondensate. Phys. Rev. Lett. **83**(1), 10–13 (1999)
58. T. Ostatnicky, I.A. Shelykh, A.V. Kavokin, Theory of polarization-controlled polariton logic gates. Phys. Rev. B **81**(12), 125319 (2010)
59. T.K. Paraiso, M. Wouters, Y. Leger, F. Morier-Genoud, B. Deveaud-Pledran, Multistability of a coherent spin ensemble in a semiconductor microcavity. Nat. Mater. **9**(8), 655–660 (2010)
60. S. Pigeon, I. Carusotto, C. Ciuti, Hydrodynamic nucleation of vortices and solitons in a resonantly excited polariton superfluid. Phys. Rev. B **83**(14), 144513 (2011)
61. S. Pilati, S. Giorgini, N. Prokof'ev, Critical temperature of interacting bose gases in two and three dimensions. Phys. Rev. Lett. **100**(14), 140405 (2008)
62. L. Pitaevskii, S. Stringari, *Bose–Einstein Condensation*. (Oxford University Press, Oxford, 2003)
63. D. Porras, C. Ciuti, J.J. Baumberg, C. Tejedor, Polariton dynamics and bose–einstein condensation in semiconductor microcavities. Phys. Rev. B **66**(8), 085304 (2002)
64. A. Posazhennikova, Colloquium: Weakly interacting, dilute bose gases in 2d. Rev. Mod. Phys. **78**(4), 1111–1134 (2006)
65. N. Prokof'ev, O. Ruebenacker, B. Svistunov, Critical point of a weakly interacting two-dimensional bose gas. Phys. Rev. Lett. **87**(27), 270402 (2001)
66. A. Recati, N. Pavloff, I. Carusotto, Bogoliubov theory of acoustic hawking radiation in bose–einstein condensates. Phys. Rev. A **80**(4), 043603 (2009)
67. G. Roumpos, M.D. Fraser, A. Loffler, S. Hofling, A. Forchel, Y. Yamamoto, Single vortex-antivortex pair in an exciton-polariton condensate. Nat. Phys. **7**(2), 129–133 (2011)

68. D. Sarchi, V. Savona, Long-range order in the Bose–Einstein condensation of polaritons. Phys. Rev. B **75**(11), 115326 (2007)
69. D. Sarkar, S.S. Gavrilov, M. Sich, J.H. Quilter, R.A. Bradley, N.A. Gippius, K. Guda, V.D. Kulakovskii, M.S. Skolnick, D.N. Krizhanovskii, Polarization bistability and resultant spin rings in semiconductor microcavities. Phys. Rev. Lett. **105**(21), 216402 (2010)
70. V. Savona, Effect of interface disorder on quantum well excitons and microcavity polaritons. J. Phys. Condens. Matter **19**(29), 295208 (2007)
71. M.O. Scully, M.S. Zubairy, *Quantum optics*. (Cambridge University Press, Cambridge, 1997)
72. I.A. Shelykh, R. Johne, D.D. Solnyshkov, G. Malpuech, Optically and electrically controlled polariton spin transistor. Phys. Rev. B **82**(15), 153303 (2010)
73. A. Sinatra, C. Lobo, Y. Castin, The truncated wigner method for bose-condensed gases: limits of validity and applications. J. Phys. B Atom. Mol. Opt. Phys. **35**(17), 3599 (2002)
74. D. Snoke, Polariton condensates - a feature rather than a bug. Nat. Phys. **4**(9), 674–675 (2008)
75. D. Snoke, P. Littlewood, Polariton condensates. Phys. Today **63**(8), 42–47 (2010)
76. R.M. Stevenson, V.N. Astratov, M.S. Skolnick, D.M. Whittaker, M. Emam-Ismail, A.I. Tartakovskii, P.G. Savvidis, J.J. Baumberg, J.S. Roberts, Continuous wave observation of massive polariton redistribution by stimulated scattering in semiconductor microcavities. Phys. Rev. Lett. **85**(17), 3680–3683 (2000)
77. M.H. Szymanska, J. Keeling, P.B. Littlewood, Nonequilibrium quantum condensation in an incoherently pumped dissipative system. Phys. Rev. Lett. **96**(23), 230602 (2006)
78. A.I. Tartakovskii, M. Emam-Ismail, R.M. Stevenson, M.S. Skolnick, V.N. Astratov, D.M. Whittaker, J.J. Baumberg, J.S. Roberts, Relaxation bottleneck and its suppression in semiconductor microcavities. Phys. Rev. B **62**(4), R2283–R2286 (2000)
79. F. Tassone, C. Piermarocchi, V. Savona, A. Quattropani, P. Schwendimann, Bottleneck effects in the relaxation and photoluminescence of microcavity polaritons. Phys. Rev. B **56**(12), 7554–7563 (1997)
80. F. Tassone, Y. Yamamoto, Exciton-exciton scattering dynamics in a semiconductor microcavity and stimulated scattering into polaritons. Phys. Rev. B **59**(16), 10830–10842 (1999)
81. M. Trujillo-Martinez, A. Posazhennikova, J. Kroha, Nonequilibrium josephson oscillations in bose–einstein condensates without dissipation. Phys. Rev. Lett. **103**(10), 105302 (2009)
82. S. Utsunomiya, L. Tian, G. Roumpos, C.W. Lai, N. Kumada, T. Fujisawa, M. Kuwata-Gonokami, A. Loeffler, S. Hoefling, A. Forchel, Y. Yamamoto, Observation of bogoliubov excitations in exciton-polariton condensates. Nat. Phys. **4**(9), 700–705 (2008)
83. D.F. Walls, G.J. Milburn, *Quantum Optics*. Springer-Verlag, Berlin (2008).
84. E. Wertz, L. Ferrier, D.D. Solnyshkov, R. Johne, D. Sanvitto, A. Lemaitre, I. Sagnes, R. Grousson, A.V. Kavokin, P. Senellart, G. Malpuech, J. Bloch, Spontaneous formation and optical manipulation of extended polariton condensates. Nat. Phys. **6**(11), 860–864 (2010)
85. E. Wertz, L. Ferrier, D.D. Solnyshkov, P. Senellart, D. Bajoni, A. Miard, A. Lemaitre, G. Malpuech, J. Bloch, Spontaneous formation of a polariton condensate in a planar gaas microcavity. Appl. Phys. Lett. **95**(5), 051108 (2009)
86. M. Wouters, T.C.H. Liew, V. Savona, Energy relaxation in one-dimensional polariton condensates. Phys. Rev. B **82**(24), 245315 (2010)
87. M. Wouters, I. Carusotto, Excitations in a nonequilibrium bose–einstein condensate of exciton polaritons. Phys. Rev. Lett. **99**(14), 140402 (2007)
88. M. Wouters, I. Carusotto, Superfluidity and critical velocities in nonequilibrium bose–einstein condensates. Phys. Rev. Lett. **105**(2), 020602 (2010)
89. M. Wouters, I. Carusotto, C. Ciuti, Spatial and spectral shape of inhomogeneous nonequilibrium exciton-polariton condensates. Phys. Rev. B **77**(11), 115340 (2008)
90. M. Wouters, V. Savona, Stochastic classical field model for polariton condensates. Phys. Rev. B **79**(16), (2009)
91. M. Wouters, V. Savona, Superfluidity of a nonequilibrium Bose–Einstein condensate of polaritons. Phys. Rev. B **81**(5), 054508 (2010)

Chapter 11
Exciton–Polariton Coupling with Acoustic Phonons

Edgar Cerda-Méndez, Dmitryi Krizhanovskii, Michiel Wouters, Klaus Biermann, Rudolf Hey, Maurice S. Skolnick, and Paulo V. Santos

Abstract Exciton–polariton are solid-state composite bosons with a high photonic character and low effective mass, which have been proven to undergo a thermodynamic phase transition to a macroscopically occupied state—a condensate—above a characteristic threshold density (Kasprzak et al. Nature 443:409, 2006; Balili et al. Science 316:1007, 2007). Full exploitation of the unique polariton properties requires dynamic processes for the dynamic confinement and control of the interaction between condensates. Here, we demonstrate a novel approach for the formation of arrays of interacting polariton condensates based on the spatial and temporal modulation by a coherent acoustic phonon. Analogous to the confinement of atomic Bose–Einstein condensates by optical lattices, the acoustic spatial modulation forms an array of polariton wires aligned with the phonon wavefronts. We show that the moving acoustic modulation controls both the energetic configuration and the spatial coherence length of the polariton condensates. Furthermore, the confinement potential moves with the acoustic velocity, thereby transporting the polariton wires. These moving acoustic confinement potentials provide, therefore, a powerful framework for manipulation and transport of solid-state condensates.

E. Cerda-Méndez (✉) · K. Biermann · R. Hey · P. Santos
Paul-Drude-Institut für Festkörperelektronik, Hausvogteiplatz 5–7, 10117 Berlin, Germany
e-mail: ecerda@pdi-berlin.de

D. Krizhanovskii · M.S. Skolnick
Department of Physics and Astronomy, University of Sheffield, Sheffield S3 7RH, United Kingdom

M. Wouters
Department of Physics, University of Antwerpen, Campus Drie Eiken D.N.018
Universiteitsplein 1, 2610, Edegem, Belgium

D. Sanvitto and V. Timofeev (eds.), *Exciton Polaritons in Microcavities*,
Springer Series in Solid-State Sciences 172, DOI 10.1007/978-3-642-24186-4_11,

11.1 Introduction

Microcavity polaritons (MPs) are quasiparticles arising from the strong coupling between excitons and photons confined in a semiconductor microcavity. We will consider here polaritons in planar microcavities (MCs), where they result from the interaction between two-dimensional excitons, which are confined in quantum wells (QW) embedded between Bragg mirror stacks. MPs are composite bosons which, due to their half photonic character have very low masses (on the order of 10^{-5} of the electron mass), and therefore, long spatial coherence lengths. As expected from the bosonic character, polaritons undergo a transition to a macroscopic quantum phase (which we will refer to as a condensate) at a critical particle density N_C. The low effective mass results in N_C values substantially smaller and de Broglie wavelengths much longer than for conventional Bose–Einstein condensates (BECs) such as cold atoms or liquid helium. Polariton condensation effects have been reported up to 30 K [6, 23] in GaAs system and up to room temperature in GaN MCs [7, 11, 11]. Unlike conventional BEC condensates, polariton condensates result from a dynamic balance between the injection of polaritons, and photon escape from the cavity and are, therefore, intrinsically metastable. They have been shown, however, to exhibit macroscopic quantum properties similar to those of equilibrium condensates including superfluidity [1, 2], vorticity [27], and extended spatial [23] and temporal coherence [30].

A full exploitation of the properties of polaritons and their condensates requires processes for the manipulation of these chargeless quasiparticles. In particular, the long spatial coherence (up to a few tens of μm) opens the way for the coherent control of their properties using μm-sized lateral potentials. These potentials can be created by patterning by directly patterning the sample [5, 13, 22, 28, 39]. Recently, the polariton modulation using periodical photonic potential has been demonstrated [28]. The previously listed patterning methods are static: they are established during the preparation of the sample and cannot be subsequently modified. Dynamic approaches for real-time control of polaritons have also been introduced. Polariton confinement potentials can be produced using spatially and polarization modulated light beams [3]. Here, one explores the different nonlinear interaction strengths for polariton width different polarization. Polariton can be also be confined by locally modifying the materials band gap via the application of strain fields [6, 15].

In this contribution, we will address an alternative process of the modulation of polaritons and polariton condensates based on the use of the microscopic strain field produced by a surface acoustic wave (SAW) [10]. SAWs are waves [4, 32, 35] are elastic vibrations propagating along the surface of a material. These waves become specially interesting in piezoelectric materials, where they can be electrically generated using interdigitated transducers (IDTs) deposited on the sample surface [40, 41]. A typical configuration for the generation of SAWs using IDTs is illustrated in Fig. 11.1a. The IDTs consist of a metal finger grating with a periodicity equal to the SAW wavelength λ_{SAW}. The application of a radio-frequency (rf)

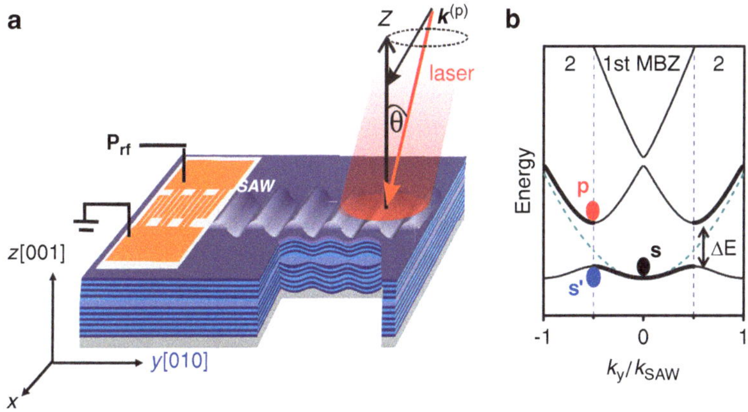

Fig. 11.1 Schematic diagram of the polariton modulation by surface acoustic waves (SAWs) generated by an interdigital transducer (IDT). (**b**) Dispersion of the lower polariton branch under the periodic modulation by a SAW displayed in the extended representation (*thick lines*) and folded within the first mini-Brillouin zone (MBZ). The dashed line is the corresponding dispersion in the absence of acoustic excitation [10] ΔE denotes the energy gap at the border of the first MBZ.

voltage with frequency $f_{SAW} = v_{SAW}/\lambda_{SAW}$ produces, via the inverse piezoelectric effect, a moving strain field that propagates to the region outside the transducer. Here, v_{SAW} and λ_{SAW} denote, respectively, the propagation velocity and wavelength of the acoustic mode. The SAW acoustic field typically penetrates the underlying material to depth corresponding to λ_{SAW}. By taking into account the acoustic velocity $v_{SAW} \approx 3{,}000\,\text{m/s} = 3\ \mu\text{m/ns}$ of (Al,Ga)As materials, one can easily estimate that the waves with frequency between 0.1 and 1 GHz (corresponding to λ_{SAW} between 30 and 3 μm) have penetration depths covering the typical depth of QWs within planar MCs.

The strain-induced modulation of the refractive index induced by a SAW has been exploited for different concepts for acousto-optic modulators [17, 24, 29, 38]. In particular, SAWs have been successfully applied for the modulation of the resonant energy of bare microcavities (i.e, those without embedded QWs) [14]. The modulation, in this case, relies in the changes not only of the refractive index but also of the cavity thickness induced by the SAW fields. The periodicity of the SAW field has a strong effect on the $\hbar\omega_c \times k$ light dispersion of the MC, which is approximately parabolic for small in-plane wave vectors k (cf. dashed lines in Fig. 11.1b). The application of a SAW along the surface direction y folds the dispersion along k_y within a mini-Brillouin zone (MBZ) with dimension $2\pi/\lambda_{SAW}$ (solid lines in Fig. 11.1b). In addition, energy gaps open up in the center and at the edges of the MBZ with amplitude depending on SAW intensity. Dispersion folding and the opening of gaps are a direct consequence of the formation of an artificial lattice under the periodic SAW modulation.

The use of SAWs for the modulation of microcavity polaritons was initially suggested by Ivanov and Littlewood [20], motivated by the relatively high sensitivity of the exciton energies to strain. Soon after that, Lima et al. [15] demonstrated that a SAW folds both the lower and upper polariton dispersion branches, thus proving that the quasiparticles have spatial coherence lengths exceeding several acoustic wavelengths. These experiments have been carried out for low polariton densities (i.e., in the linear regime) using piezoelectric surface acoustical wave (SAWs). More recently, we have demonstrated the coherent spatial modulation as well as the confinement of MC polariton condensates under SAW potentials [10]. These results show that SAWs can produce tunable and periodic confinement potentials for polaritons with μm dimensions. Apart from the tunability, one important advantage of the acoustic modulation is the fact that it does not require the structuring of the modulation region, which can result in degradation of the materials properties.

This chapter reviews the use of SAWs for the modulation of polaritons and polariton condensates. Section 11.2 addresses the modulation mechanisms as well as the issues associated with an efficient coupling of the SAW fields to the active region of the MCs. Section 11.3 reviews the effects of SAW fields of polaritons in the low particle density (Sect. 11.3.1) as well as in the high particle density regime, where polariton condensates are created with resonant (Sect. 11.3.2) excitation. Finally, Sect. 11.3.4 summarizes the main conclusions as well as future perspectives in this field.

11.2 Polariton Microcavities for Acoustic Modulation

The design of the layer structure of polariton MCs normally aims at the optimization of the coupling between the optical and excitonic fields as well as at achieving long photon lifetimes. The latter is normally achieved in semiconductor microcavities by using Bragg reflectors with many pairs, due to the relatively small contrast in the index of refraction between the mirror layers. In samples for acoustic modulation, it also becomes important to ensure a good coupling between the polariton field, which depends on the SAW strain-induced spacer region of the MC.

The effects of a SAW on the polariton field can be understood by considering the impact on the energy levels of its excitonic and photonic components. The SAW strain modulates the MC optical energy (with amplitude $\hbar\Delta\omega_c$) as well as the conduction (E_{CB}) and valence (E_{VB}) band edges of the QWs. Two conditions need to be satisfied to maximize the acoustic modulation efficiency: (1) ΔE_{CB} and ΔE_{VB} should be out-of-phase to produce a type-I band edge-modulation confining electrons and holes at the same spatial positions, thereby enhancing the excitonic oscillator strength (cf. Fig. 11.3d) and (2) $\hbar\Delta\omega_c$ and the band gap modulation $\Delta E_g = \Delta E_{\mathrm{CB}} - \Delta E_{\mathrm{VB}}$ should be in phase in order to maximize the light-matter coupling.

The determination of the acoustic modulation strength thus requires information about the depth distribution (z) of the SAW-strain field. The basic surface acoustic mode propagating on (001) cubic semiconductors consists of a Rayleigh wave with three strain components: a transverse component perpendicular to the surface

($\varepsilon_{zz}(z)$), a longitudinal component along the propagation direction l ($\varepsilon_{ll}(z)$), and a shear component $\varepsilon_{lz}(z)$. The modulation of the microcavity optical resonance $\hbar\Delta\omega_c$ is dictated by the strain-induced changes in the thickness (which are proportional to ε_{zz}) and refractive index of the spacer region [14]. The former normally dominates leading to relative energy changes $\Delta\omega_c/\omega_c \approx \varepsilon_{zz}$. The modulation of the band edges can be directly calculated from the strain using the appropriate deformation potentials.

A further important aspect regards piezoelectric effects associated with the strain field. Rayleigh SAWs on the (001) surface of III–V compounds are normally generated along a $l = x' = \{110\}$ (cf. Fig. 11.2b), where the shear strain ϵ_{lz} is coupled to a longitudinal piezoelectric field F_z. This piezoelectric coupling allows for the excitation of SAWs using IDTs deposited directly on the sample surface. In addition, F_z induces a type II modulation of the semiconductor band edges, which can reach hundreds of meV and is, thus, much stronger than the one induced by the strain field. The moving piezoelectric modulation provides the basis for the acoustic transport of electrons [19], electron–hole pairs [34], and spins [36, 37] by SAW fields. In the case of polaritons, however, this field has to be avoided since it destabilizes the quasi-particles by ionizing the excitons [33]. For efficient modulation, it becomes necessary to use nonpiezoelectric modes—this is the case for modes propagating along $l = x = [100]$ surface direction. These modes can still be electrically excited by placing the IDT over a piezoelectric island (consisting, for instance of a textured ZnO thin film), as indicated in Fig. 11.2a.

The previously listed constrains on $\hbar\omega_c$, ΔE_{CB}, and ΔE_{VB} can be satisfied by an appropriate design of the MC layer structures and selection of the acoustic

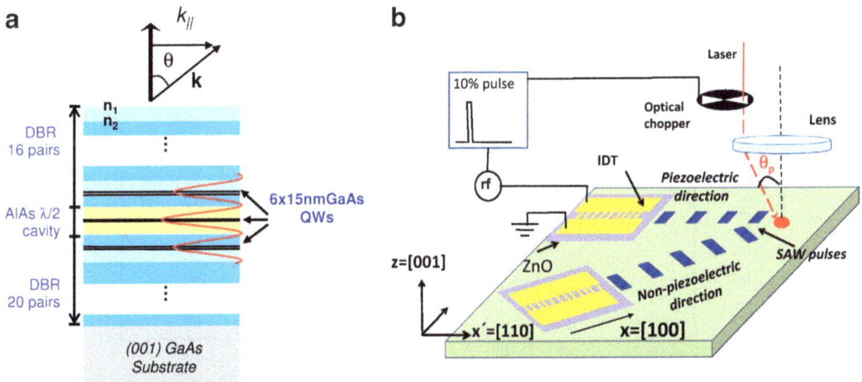

Fig. 11.2 (**a**) Layer structure of the microcavity samples for polariton modulation by SAWs. (**b**) Experimental setup for the modulation of the GaAs-based microcavities using SAWs generated by interdigital transducers (IDTs) oriented along piezoelectric ($x' = [110]$) and non-piezoelectric ($x = [100]$) directions. The IDTs were deposited on piezoelectric ZnO islands to allow SAW generation along the two directions. In the optical investigations, the polaritons were generated by laser pulses synchronized with the rf-pulses used to generate the SAW. θ_p is the angle of incidence of the exciting laser beam [9]

wavelength. We will consider here an MC with an $\lambda/2$ AlAs spacer containing 3 pairs of 15-nm GaAs QWs with 7-nm AlAs barriers illustrated in Fig. 11.2a. The QW pairs are located at the center of the cavity and at the interface of the adjacent DBR pairs, where the maxima of the electromagnetic field are located. The top and bottom DBRs consist of 16 and 23 pairs of $\lambda/4$ $Al_{0.15}Ga_{0.85}As/Al_{0.80}Ga_{0.20}As$ layers stacks, respectively. The AlGaAs pair layers contain Al concentrations of 15% and 80%. The whole structure was grown on GaAs (001) by molecular beam epitaxy.

On these samples, SAWs with a wavelength $\lambda_{SAW} = 8\ \mu m$ were excited both along the $x = [100]$ and the $x' = [110]$ directions (cf. Fig. 11.2) using IDTs deposited on a 600 nm thick ZnO island. This particular acoustic wavelength was selected to satisfy the modulation constraints listed above. The s_{11} power reflection coefficient displayed in Fig. 11.3a shows that SAWs with comparable strengths can be excited along the two directions. The lower resonance frequency for the [100] mode is attributed to the lower acoustic propagation velocity along this direction.

Figure 11.3 displays calculated profiles for the strain-induced lateral modulation of the (a) conduction band (ΔE_{CB}), (b) valence band ΔE_{VB}, and (d) band-gap modulation ($\Delta E_g = \Delta E_{CB} - \Delta E_{VB}$) of the QWs as a function of the SAW phase.

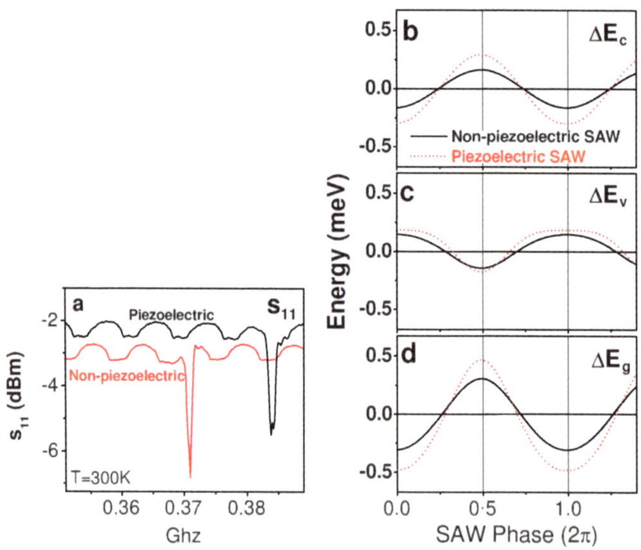

Fig. 11.3 (**a**) Measured rf-reflection coefficient s_{11} for IDTs along the x (*nonpiezoelectric, black line*) and x' (*piezoelectric, red line*) directions (*in relative units*). The dips correspond to the transduction range of the IDTs, where the reflected electric power reduces to the excitation of an acoustic mode. (**b**) Calculated profiles in the QW plane for the strain-induced (i.e., without including piezoelectric effects) modulation of the (**a**) conduction band (ΔE_{CB}), (**b**) valence band ΔE_{VB}, and (**d**) band-gap modulation ($\Delta E_g = \Delta E_{CB} - \Delta E_{VB}$). The horizontal axis is the SAW phase [9]. (*change labels!*)

The piezoelectric effects on the modulation of the band edges were not included in the plots for piezoelectric SAW along $x' = [110]$. Note that the strain-induced band gap modulation is of type-I for both SAW propagation directions, thus ensuring the confinement of electrons and holes at the same spatial positions. The peak-to-peak band gap modulation is of 0.6 and 1 meV for nonpiezoelectric and piezoelectric SAWs, respectively. The large band gap modulation for piezoelectric SAWs is due to the larger ΔE_{CB} for these SAWs. Finally, calculations of $\hbar\Delta\omega$ for the same structure presented in [9]) show that it is in phase with ΔE_g and has an amplitude approx. 50% lower.

11.3 Acoustic Modulation of Polaritons

11.3.1 Linear Regime

We examine in this section the effects of the acoustic modulation in the regime of low polariton densities, where the polariton dispersion is dominated by one-particle effects. The dispersion can be accessed by imaging the far-field photoluminescence (PL), as illustrated in Fig. 11.4a for a sample with the layer structure described in the previous section [9]. The polaritons in this sample have a coupling energy (the Rabi splitting) of $\hbar\Omega_R \approx 6$ meV. The full width on half maximum (FWHM) of the excitonic PL line was determined from measurements in control samples to be $\Gamma_{\mathrm{ex}} \sim 0.3$ meV, as determined from measurements in control samples. As will become clear below, these narrow linewidths are essential for the observation of SAW modulation for weak acoustic intensities.

The effects of the SAW on the PL are illustrated in Fig. 11.4b, which shows the dispersion measured under the same conditions as in Fig. 11.4a but under a non-piezoelectric SAW with $\lambda_{\mathrm{SAW}} = 8\ \mu$m. The SAW, in this case, was generated by applying a nominal rf-power $P_{\mathrm{rf}} = 28$ mW to the transducer (cf. Fig. 11.2b). In agreement with Fig. 11.1b, the dispersion shows up to four folded dispersion branches separated by energy gaps in the center ($k_l = 2m\pi/\lambda_{\mathrm{SAW}}$, $m = 0, 1, \dots$) and at the edges ($k_l = (2m + 1)\pi/\lambda_{\mathrm{SAW}}$) of the MBZs [15]. Note that in order to probe dispersion-folding effects, one needs to collect the emission from a sample region with dimensions of several acoustic wavelengths (a spot with diameter of approx 50 μm was used for PL excitation in Fig. 11.4). The observation of several dispersion branches indicates that spatial phase coherence of the polaritons exceeds several acoustic wavelengths. From the momentum uncertainty Δk for an LP branch energy at the center of the MBZ, we estimate a lower bound for the coherence length $\sim 2\pi/\Delta k \sim 11\mu$m, which exceeds the SAW period ($\lambda_{\mathrm{SAW}} = 8\ \mu$m).

For low acoustic powers, the width ΔE of the lowest energy gap at the edge of the MBZ (cf. Fig. 11.1b) and Fig. 11.3b–d becomes essentially equal the peak-to-peak amplitude of the effective modulation potential ϕ_0^{LP} acting on the lower polariton branch [16]. This approximation allows the direct determination of ϕ_0^{LP}.

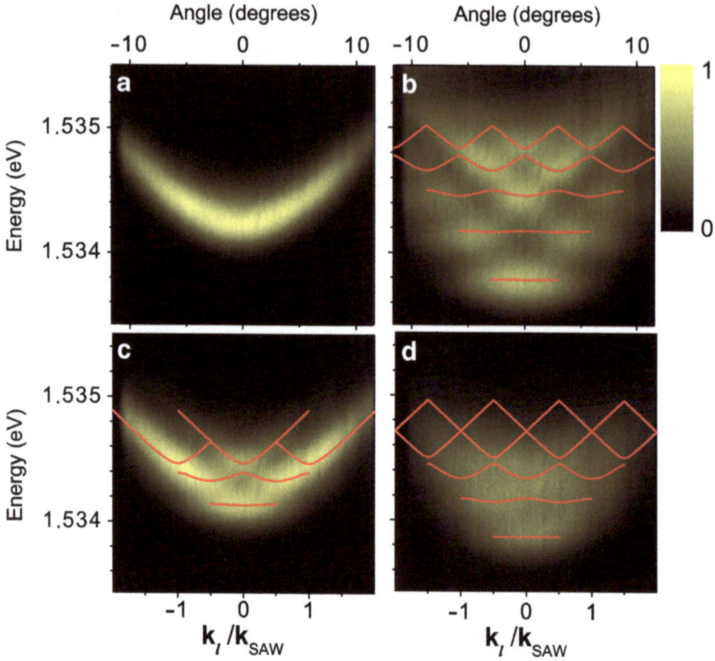

Fig. 11.4 Energy *vs.* wave vector polariton dispersion (**a**) without and (**b**) under a nonpiezoelectric SAW with $P_{rf} = 28\,\text{mW}$. The lower panels compare the dispersion under a (**c**) nonpiezoelectric and a (**d**) piezoelectric SAW excited using the same rf-power $P_{rf} = 2.2\,\text{mW}$. The horizontal axis corresponds to the wave vector k_l along the SAW propagation direction in units of $k_{SAW} = 2\pi/\lambda_{SAW}$. The measurements were carried out on a sample with Rabi splitting $\hbar\Omega_R \approx 6\,\text{meV}$ and a detuning $\delta \approx$ between the optical ($\hbar\omega_c$) and excitonic $\hbar\omega_x$ levels of approx. $\delta = \hbar\omega_c - \hbar\omega_x \approx -1\,\text{meV}$. The PL excitation was provided by a laser excitation beam impinging on the sample surface under an angle $\theta_p \sim 13°$ and with energy 1.4 meV above the bottom of the LP branch in (**a**) [9]

Figure 11.5a shows that ΔE can be tuned by changing the acoustic power P_{rf}. Note that $\phi_0^{LP} \propto \sqrt{P_{rf}}$: such a relationship is expected from the fact that amplitude of the SAW-strain field, which determine the changes in polariton energy, is proportional to $\sqrt{P_{rf}}$ [16].

The red lines superimposed on Fig. 11.4b displays the calculated polariton dispersion assuming a potential modulation ϕ_0^{LP} determined from ΔE. The calculations reproduce very well the measured data. In particular, they reproduce the flattening of the lowest dispersion branch with increasing modulation amplitude. The energetic width of the branches of W_{EB} (see Fig. 11.5b–d) is determined by the coupling between states localized in adjacent minima of the confinement potential created by the SAW. For high acoustic powers, this coupling reduces: the lower lying states become well-confined close to the minima of the SAW potential. These modes correspond, therefore, to an array of weakly coupled polariton wires allied parallel to the SAW wave fronts. Finally, the calculations also predict the observed red-shift

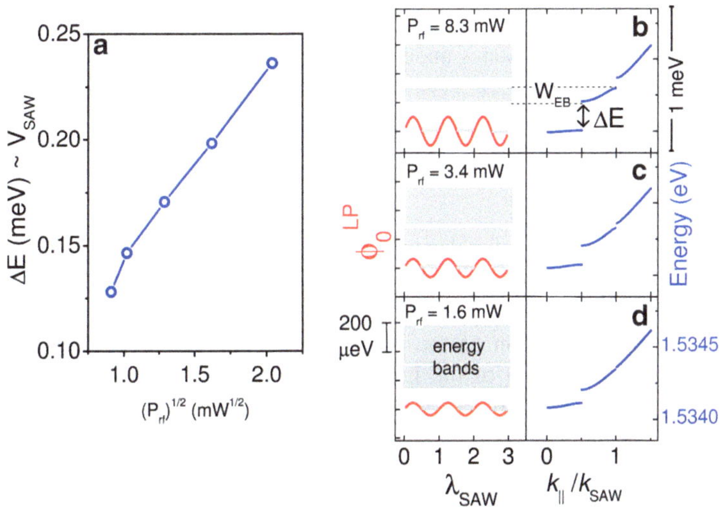

Fig. 11.5 (**a**) Dependence of first zone-edge energy gap Δ_E (at $|k_y| = \pi/\lambda_{\mathrm{SAW}} = k_{\mathrm{SAW}}/2$) on the nominal IDT excitation power $\sqrt{P_{\mathrm{rf}}}$ below condensation threshold. (**b-d**) Illustration of the process of flattening of the confined bottom dispersion band (right side panels) and formation of upper energy bands (left side panels) as the SAW potential increases

of the lower dispersion branch with increasing acoustic powers (cf. Figs. 11.4a,b). This shift, which exceeds the modulation amplitude, results from the energy renormalization of lower polariton branches due to SAW-induced coupling to higher branches [14].

The clear observation of up to three dispersion branches in Fig. 11.4 becomes possible because the polariton phase coherence, which is reflected in the PL linewidth which remains essentially constant when a nonpiezoelectric SAW is applied. By contrast, the application of a piezoelectric SAW leads to a considerable broadening and reduction in intensity of the PL lines. This effect is demonstrated in Fig. 11.4c and d, which compare the dispersion of nonpiezoelectric and piezoelectric SAWs, respectively, excited using the same rf-power $P_{\mathrm{rf}} = 2.2$ mW. For the nonpiezoelectric case (Fig. 11.4c), dispersion folding is again clearly observed. For piezoelectric SAWs, in contrast, the modulation seems to be larger but the dispersion becomes strongly blurred even for acoustic powers an order of magnitude lower than in Fig. 11.4b. The stronger modulation by the piezoelectric SAW is in agreement with the calculations in Fig. 11.3d. The blurring of the dispersion behavior, in contrast, is attributed to the dissociation of excitons by the piezoelectric field [44]. The latter is expected to impact the polariton linewidths when the field-induced exciton ionization time becomes shorter than $\hbar/\Gamma_{\mathrm{ex}}$. Using the expression for the exciton ionization time from [33], it can be shown that this condition becomes satisfied for fields exceeding a few kV/cm. These field levels are comparable to the ones generated by piezoelectric SAWs in the plane of the QWs [9]. The increase in linewidth (and, therefore, also in Δk) leads, in this case, to a reduction of the phase coherence of the LP states under a piezoelectric SAW.

11.3.2 Resonantly Pumped Condensate

The coherent modulation of polaritons can be done as well above the condensation threshold. We investigate condensates formed in the optical parametric oscillator (OPO) geometry, where the plane of the incident laser is perpendicular to the propagation direction of the SAW (cf. Fig 11.6). The excitation is done resonantly with the bottom of the LP branch by an in-plane pump wave vector $\mathbf{k}^{(\mathbf{p})} = (k_x^{(p)}, k_y^{(p)}) = (1.4, 0)\ \mu\text{m}^{-1} = (1.8, 0)k_{\text{SAW}}$.

Above threshold, self-organization sets in [25], forming a macroscopically occupied signal condensate at $k \sim 0$ containing 500–1,000 particles as well as a weaker idler at $k \sim 2k^{(p)}$ (cf. Fig. 11.6). We note that in the OPO excitation geometry, the SAW spatially modulates the pump as well as the signal polariton states, thus also creating weaker pump polaritons with wave vectors $(1.8, \pm 1)k_{\text{SAW}}$. Emission from these polaritons leads to diffracted beams with intensities measured to be $\leq 4\%$ of the main pump mode at $k^{(p)}$.

Figure 11.7a–d display E vs. k_y images for the signal state recorded under different SAW powers for a fixed optical excitation power density of $I_L = 2I_L^{\text{th}} = 1.2 \times 10^3$ W/cm^2 (where I_L^{th} is the condensation threshold power) distributed over a 50-μm-diameter spot. In the absence of a SAW, a condensate is formed at $k = 0$ and energy of 1.5345 eV, which is approximately 0.4 meV above the bottom of the LP branch below threshold. The blueshift arises from interparticle interactions between polaritons in the signal and the pump [12]. Figure 11.7e–g show the corresponding spatial interference patterns, from which the first-order spatial correlation functions $g^{(1)}(\mathbf{r}, -\mathbf{r})$ shown in Fig. 11.7h–j were extracted. The full width at $1/2.7$ of the maxima of $g^{(1)}(\mathbf{r}, -\mathbf{r})$ gives the corresponding coherence lengths L_x and L_y along the x and y directions, respectively.

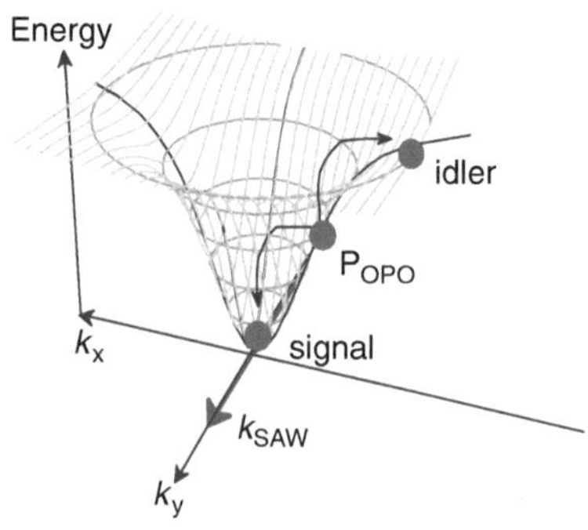

Fig. 11.6 Geometry of excitation for generation and modulation of OPO condensates with SAWs

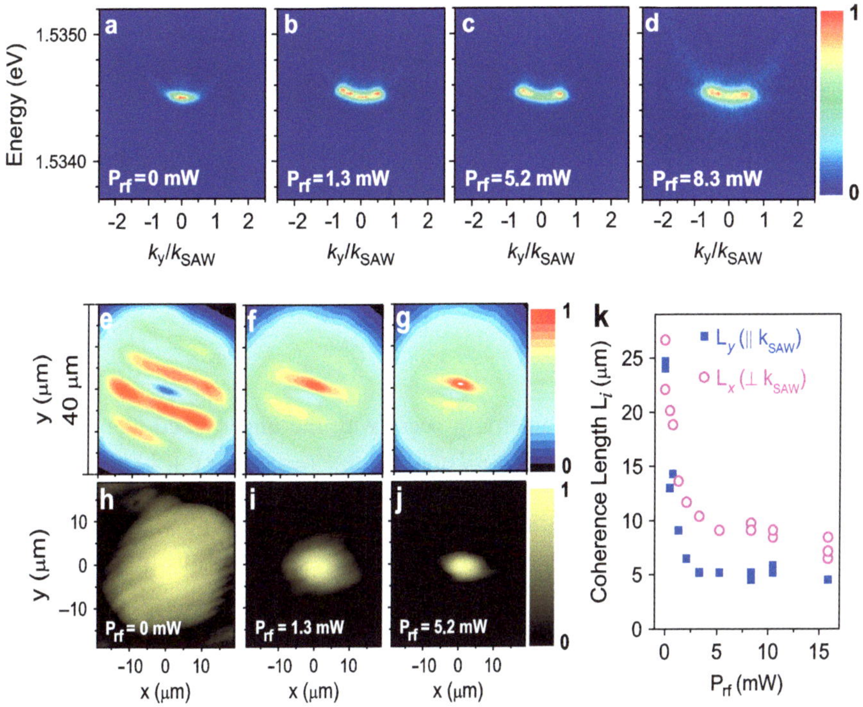

Fig. 11.7 Polariton condensates formed by resonant OPO excitation in a SAW-induced periodic potential. Panels (**a–d**) show E *vs.* k_y images of the condensates for a pump power $I_L = 1.2 \times 10^3$ W/cm^2 equal to twice the OPO condensation threshold I_L^{th} and different applied acoustic powers P_{rf}. (**e–g**) and (**h–j**) show the corresponding spatial interference patterns (**e–g**) and first order spatial correlation functions (**h–j**) of the condensates for the same acoustic powers. In the unperturbed state (i.e., no SAW applied), a condensate with extended spatial coherence forms at $k_y = 0$ (panel a). With a SAW, (i.e. no SAW appplied) the condensate becomes spatially confined in the form of wires aligned along the SAW wave fronts, leading to the spread of its momentum distribution in k-space (**b–d**). At the intermediate P_{rf} in (**c**), condensation takes place at the extrema $k_y = \pm k_{\mathrm{SAW}}/2$ of the first MBZ (corresponding to the s' states in Fig. 11.1b) with a coherence length $\sim \lambda_{\mathrm{SAW}}/2$. Panel (**k**) summarizes the dependence of the spatial coherence lengths L_x and L_y along the x and y directions as a function of P_{rf}

In the absence of acoustic excitation, the condensate is spatially homogeneous with equal coherence lengths $L_x = L_y = 25\ \mu$m (cf. Fig. 11.7d) along the x and y directions. When a weak SAW is applied ($P_{\mathrm{rf}} = 1.3$ mW, cf. Fig. 11.7b), the emission from the signal polaritons first spreads in momentum space while the spatial extent of $g^{(1)}$ (cf. Fig. 11.7i) reduces both along and perpendicular to the SAW propagation direction. Interestingly, at intermediate powers ($P_{\mathrm{rf}} = 5.24$ mW in Fig. 11.7c) the emission peaks not around $k_y = 0$ but at $k_y = \pm k_{\mathrm{SAW}}/2$, corresponding to the s' states in Fig. 11.1b at the edge of the lowest folded branch. This behavior is related to the scattering processes associated with the OPO. A likely explanation for their origin is that the diffraction replicas of the pump state with

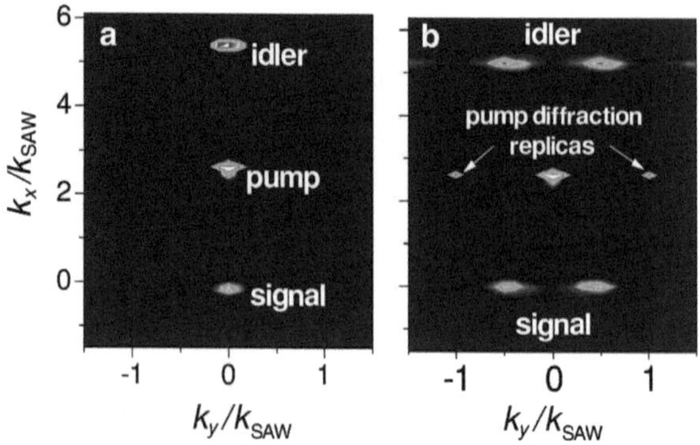

Fig. 11.8 Scattering mechanisms in the OPO excitation geometry without (panel (**a**)) and with a SAW (panel (**b**)) which give rise to excitation of the s'-states

$k_y = \pm k_{SAW}$ undergo a parametric process with the unperturbed pump state at $k_y = 0$, namely, $k_y^{(p)}(=0) + k_y^{(p)}(=k_{SAW}) \rightarrow k_y^{(s)}(=k_{SAW}/2) + k_y^{(i)}(=k_{SAW}/2)$ (see Fig. 11.8), thus providing the observed parametric gain at $k_y = \pm k_{SAW}/2$. Theoretical simulations based on a classical field model (see Fig. 11.8) reproduce this behavior. Finally, at higher SAW powers ($P_{rf} \geq 8.3\,mW$) the momentum of the condensate is no longer well defined due to the stronger confinement and, as a result, the emission pattern in momentum space spreads over the whole MBZ branch (Fig. 11.7d).

The dependence of the coherence lengths L_x and L_y on acoustic power summarized in Fig. 11.7k shows that they can be controllably reduced by a factor of five and three, respectively. The lower tunneling probability between adjacent potential minima at high SAW power reduces the phase coherence along y, resulting in the formation of polariton wires oriented along the SAW wavefronts with a coherence length $L_y = 5\ \mu m$ close to $\lambda_{SAW}/2 = 4\ \mu m$. The coherence reduction along x is at first sight more surprising: it is, however, consistent with the one-dimensional character of the wires, which makes the coherence very sensitive to phase fluctuations along the wire axis. This phenomenon is well known at thermodynamic equilibrium [31] and has also been predicted for nonequilibrium systems like one-dimensional parametric oscillators [43]. The acoustic field induces, therefore, a controlled transition from an extended condensate to one fragmented into an array of wires localized in the SAW potential minima. It is notable that over the whole range of SAW powers the linewidth of the total stimulated polariton emission (resolution limited to a FWHM of $\sim 60\ \mu eV$, cf. Fig. 11.7a–c) remains three times narrower than the one below threshold, indicating also persistence of the temporal coherence. Even though each condensate wire emits independently from the others at high acoustic powers, their spectral emission energies overlap within the measured linewidths, thus attesting to the high homogeneity of the modulation potential.

The degree of localization of the condensate due to the SAW may also be varied by increasing the polariton density. For condensates trapped at the minimum of the SAW potential, the blueshift [42] due to polariton interactions at high pumping

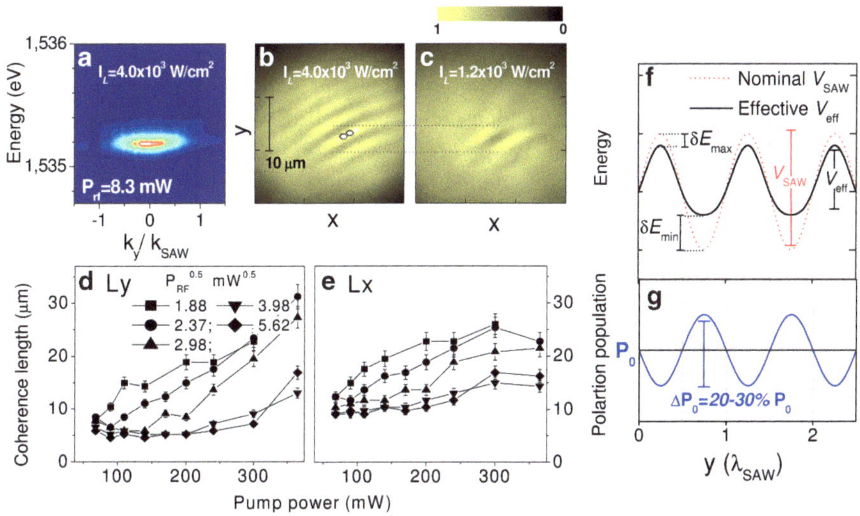

Fig. 11.9 Panel (**a**) shows the k-space image of an OPO condensate for pump and acoustic powers of $I_L = 6.7I_L^{th}$ and $P_{rf} = 8.3$ mW, respectively. In this case, the condensate energy blueshifts due to polariton-polariton interaction by 0.9 meV with respect to the bottom of the LP branch. Panels (**b**) and (**c**) show interference patterns of the signal condensate under an acoustic field of $P_{rf} =$ 8.3 mW for pump powers $I_L = 6.7I_L^{th}$ and $2I_L^{th}$, respectively. The increased coherence length for larger pump powers is attributed to the screening of the SAW potential at high polariton densities. Panels (**d**) and (**e**) show the dependence of the coherence lengths across and along the wires as a function of applied optical power and P_{rf}. Above the OPO threshold, the spatial modulation of the polariton energy by the SAW potential V_{SAW} (*red dotted line in panel* (**f**)) produces a modulation of the density of pump polaritons P_0 of about 20% above the OPO threshold, as seen in panel (**g**). As a result, the energy shifts δE_{min} at the minima of V_{SAW} are larger than those of the maxima (δE_{max}). The overall effect is a spatially dependent screening of V_{SAW}, which results in a smaller effective modulation potential V_{eff}, as shown by the solid black line in (**a**)

powers can raise the condensate energy level above the percolation limit in the SAW potential, thus partially reversing the confinement effects induced by the SAW. In this case, the energy of the pump laser needs to be slightly increased to account for the blueshift of the polariton dispersion.

Figure 11.9a shows E *vs.* k_y an image of the signal state under an acoustic power $P_{rf} = 8.3$ mW obtained using a pump beam detuned to the blue by $\delta_P = 1$ meV with respect to the LP branch below threshold. The pump power density $I_L = 6.7I_L^{th}$ is 3.3 times higher than that of Fig. 11.7. The higher optical excitation leads to a larger blue shift of the emission of 1 meV as well as to a fourfold increase in the signal polariton density per unit area, as determined by the intensity emitted into the signal beam [42].

Figure 11.9b,c compare the spatial interference pattern of condensates under the same applied SAW intensity for high ($I_L = 6.7I_L^{th}$) and low ($2I_L^{th}$), corresponding to dispersion image of Fig. 11.7d) pump powers, respectively. The latter yields coherence lengths L_y=5 and L_x=10 μm (cf. Fig. 11.7k). However, at high pump powers (Fig. 11.9b) the interference pattern extends over a wider region, yielding coherence lengths $L_y \sim 10$ and $L_x \sim 15\ \mu$m a factor of 1.5–2 greater than in Fig. 11.9c. The longer coherence lengths at equal P_{rf} demonstrate the increased coupling between neighboring condensate wires, as the SAW confinement potential becomes screened by the repulsive polariton-polariton interactions. The screening effect over can be monitored in detailed by scanning different applied optical and radiofrequency powers, as plotted in Fig. 11.9d,e.

The screening of the periodic potential at higher density can be understood as follows. We first consider the case of spatially unmodulated polaritons. A single-mode model [42] describing interacting pump polaritons excited with an external laser source and taking into account the detuning of the external laser frequency with respect to that of the LP branch, predicts a bistable behavior of the pump polariton population as a function of external laser power. In our case, the parameters of the system are such that the bistability threshold usually coincides with the OPO condensation threshold [25]. Following this model, at increased detuning of the pump frequency with respect to the LP branch, a stronger optical excitation power and hence a higher pump polariton density is required to bring the LP mode into resonance with the laser and to switch on the OPO. Once such a critical pump polariton density is created, it becomes impossible to inject more pump polaritons into the system, since a higher polariton population would blueshift the polariton resonance away from the laser frequency.

Polariton modulation by the SAW (Fig. 11.9) results in a spatial modulation of the detuning between the pump and the lower polariton branch. Following the discussion above, a higher pump polariton population in the SAW minima than that in the SAW maxima is created at powers above the OPO threshold, resulting in a spatial modulation of the pump polariton density $P_0(y)$ in antiphase with the SAW potential.

The dependence $P_0(y)$ in Fig. 11.9g, calculated using the bistability model of [42], shows a modulation amplitude $\Delta P_0 = P_0(max) - P_0(min)$ of about 20% of P_0 [42]. The blueshift due to signal-pump interactions $\delta E_{shift}(y) = G \times P_0(y)$ (G is the interaction constant) of the signal polaritons located in the SAW minima is stronger than that in the maxima. Therefore, the amplitude of the effective periodic potential (solid line in Fig. 11.9f) $V_{eff} = V_{SAW} - \delta E_{shift}$ experienced by the signal state reduces. $\delta E_{shift} = G \times \Delta P_0$ is the difference between the population induced energy shifts of polaritons in the minima and maxima of the SAW potential.

For an acoustic power $P_{rf} = 8.3$ mW, V_{SAW} is about 0.2 meV. From the blueshift of the condensate energy at high optical pump power ($6.7 I_{th}^{OPO}$), given by $G \times P_0 \approx 1$ meV, we estimate $\delta E_{shift} = G \times \Delta P_0 \sim 0.2$ meV and $V_{eff} \approx 0$. There is, therefore, a substantial screening of the SAW potential by polariton–polariton interactions. This behavior can be contrasted with the case of lower optical powers ($2 I_{th}^{OPO}$), where $G \times P_0 \approx 0.4$ meV and, therefore, $V_{eff} = V_{SAW} - G \times \Delta P_0(y) \approx 0.6 V_{SAW}$.

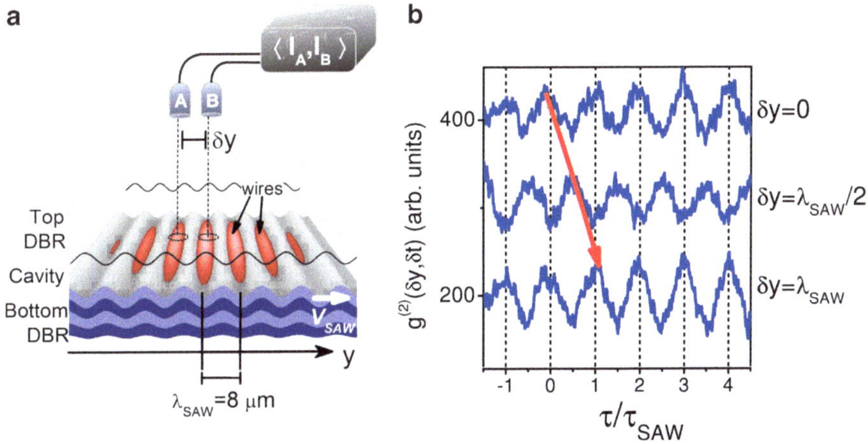

Fig. 11.10 Acoustic transport of confinement potentials at $P_{\rm rf} = 12.5\,$mW. Figure (**a**) is a schematic diagram of the experiments to determine the spatiotemporal correlation $g^{(2)}(\delta y, \delta\tau) = \langle I_{PL}(0,0), I_{PL}(\delta y, \delta\tau)\rangle$ between the intensities detected by the two avalanche photodiodes A and B for different separations δy. Panel (**b**) shows correlation histograms for different separations $\delta y = n\lambda_{\rm SAW}/2$ $(n = 0, 1, 2..)$ between the detection spots A and B. The oscillations have the SAW period $\tau_{\rm SAW} = 1/f_{\rm SAW} = 2.65\,$ns, and phase shift when the detectors are separated from one other. The phase is recovered when $\delta y = \lambda_{\rm SAW}$. The modulation amplitude of correlation histograms is about 20% and limited by the optical resolution of the setup

11.3.3 Transport of Polariton Lattice Potential

The confined condensates created by the SAW move with the acoustic propagation velocity $v_{\rm SAW} = 3\times10^3$ m/s. In order to demonstrate this dynamic character, we measured the spatiotemporal correlation function $g^{(2)}(\delta y, \delta t) = \langle I_{\rm PL}(0,0), I_{\rm PL}(\delta y, \delta t)\rangle$ of the PL intensity $I_{\rm PL}$ of OPO condensates using a Hanbury Brown and Twiss (HBT) setup. For that purpose, the emission from two spots within the condensate array, each 2 μm wide and separated by a distance δy, were collected and directed to the arms of the HBT setup (see Fig. 11.10a). Figure 11.10b displays time histograms for the probability for two-photon emission recorded for different spot separations δy. The time oscillations in the correlation function for $\delta y = 0$ (upper histogram) have the SAW repetition period and arise since the emission reaches a maximum when a confined condensate crosses the detection region. The phase of the oscillations changes when the spots are displaced by δy: the fact that the rate of change $\delta y/\delta\tau = v_{\rm SAW}$ proves that the condensate regions move with the acoustic velocity. We note, however, that while these results demonstrate that the confinement potentials move, they do not imply long-range transport of a coherent state, since the measured coherence time of unperturbed condensates of 200–500 ps [26, 30] is significantly shorter than the SAW period.

11.3.4 Conclusions

In conclusion, we have demonstrated the coherent spatial modulation of polariton condensates using propagating surface acoustic fields. The periodic acoustic potential creates a fragmented condensate in the form of wires aligned along the SAW wavefronts: the coherence length and the interaction between the wires can be controlled by intensity of the acoustic field. In dilute atomic gases, the dynamic manipulation of macroscopic quantum phases by tunable periodic potentials has allowed for the investigation of interesting many-body phenomena such as the superfluid-to-Mott insulator transition [18, 21] and the Josephson interaction between condensates [8]. The modulation by acoustic fields provides an interesting framework for similar studies on solid-state condensates.

Acknowledgements We acknowledge the technical support from A.-K. Bluhm, M. Höricke, S. Krauß, W. Seidel, H.-P. Schönherr, and E. Wiebicke in the fabrication of the samples.

References

1. A. Amo, D.B. Sanvitto, F. Laussy, E. del Valle, M. Martin, A. Lemaitre, J. Bloch, D. Krizhanovskii, M. Skolnick, C. Tejedor, L. Vina, Collective fluid dynamics of a polariton condensate in a semiconductor microcavity. Nature **457**, 291 (2009). DOI 10.1038/nature07640
2. A. Amo, J. Lefrère, S. Pigeon, C. Adrados, C. Ciuti, I. Carusotto, R. Houdr, E. Giacobino, A. Bramati,, Superfluidity of polaritons in semiconductor microcavities. Nat. Phys. **5**, 805 (2009). DOI 10.1038/nphys1364
3. A. Amo, S. Pigeon, C. Adrados, R. Houdré, E. Giacobino, C. Ciuti, A. Bramati, Light engineering of the polariton landscape in semiconductor microcavities. Phys. Rev. B **82**(8), 081, 301 (2010). DOI 10.1103/PhysRevB.82.081301
4. B.A. Auld, *Acoustic Fields and Waves in Solids*. (Robert E. Krieger Publishing Company, Inc, Malabar, Florida 1990)
5. D. Bajoni, P. Senellart, E. Wertz, I. Sagnes, A. Miard, A. Lemaître, J. Bloch, Polariton laser using single micropillar GaAs-GaAlAs semiconductor cavities. Phys. Rev. Lett. **100**, 047, 401 (2008)
6. R. Balili, V. Hartwell, D. Snoke, L. Peiffer, K. West, Bose-Einstein condensation of microcavity polaritons in a trap. Science **316**, 1007 (2007)
7. J.J. Baumberg, A.V. Kavokin, S. Christopoulos, A.J.D. Grundy, R. Butt, G. Christmann, D.D. Solnyshkov, G. Malpuech, G.B.H. von Högersthal, E. Feltin, J.F. Carlin, N. Grandjean, Spontaneous polarization buildup in a room-temperature polariton laser. Phys. Rev. Lett. **101**, 136,409 (2008). DOI 10.1103/PhysRevLett.101.136409. URL http://link.aps.org/doi/10.1103/PhysRevLett.101.136409
8. F.S. Cataliotti, S. Burger, C. Fort, P. Maddaloni, F. Minardi, A. Trombettoni, A. Smerzi, M. Inguscio, Josephson junction arrays with bose-einstein condensates. Science **293**(5531), 843–846 (2001). DOI 10.1126/science.1062612
9. E.A. Cerda-Méndez, D.N. Krizhanovskii, K. Biermann, R. Hey, P.V. Santos, M. Skolnick, Effects of the piezoelectric field in the modulation of exciton-polaritons by surface acoustic waves. Superlattices Microstruct. **49**(3), 233–240 (2010). DOI 10.1016/j.spmi.2010.06.006

10. E.A. Cerda-Méndez, D.N. Krizhanovskii, M. Wouters, R. Bradley, K. Biermann, K. Guda, R. Hey, P.V. Santos, D. Sarkar, M.S. Skolnick, Polariton condensation in dynamic acoustic lattices. Phys. Rev. Lett. **105**, 116, 402 (2010). DOI 10.1103/PhysRevLett.105.116402. URL http://link.aps.org/doi/10.1103/PhysRevLett.105.116402
11. S. Christopoulos, G.B.H. von Högersthal, A.J.D. Grundy, P.G. Lagoudakis, A.V. Kavokin, J.J. Baumberg, G. Christmann, R. Butté, E. Feltin, J.F. Carlin, N. Grandjean, Room-temperature polariton lasing in semiconductor microcavities. Phys. Rev. Lett. **98**(12), 126, 405 (2007). DOI 10.1103/PhysRevLett.98.126405
12. C. Ciuti, P. Schwendimann, A. Quattropani, Parametric luminescence of microcavity polaritons. Phys. Rev. B **63**(4), 041, 303 (2001). DOI 10.1103/PhysRevB.63.041303. URL http://link.aps.org/abstract/PRB/v63/e041303
13. G. Dasbach, M. Schwab, M. Bayer, D. Krizhanovskii, A. Forchel, Tailoring the polariton dispersion by optical confinement: Access to a manifold of elastic polariton pair scattering channels. Phys. Rev. B **66**(20), 201, 201 (2002). DOI 10.1103/PhysRevB.66.201201
14. Jr. M.M. de Lima, R. Hey, P.V. Santos, A. Cantarero, Phonon-induced optical superlattice. Phys. Rev. Lett. **94**, 126, 805 (2005)
15. Jr. M.M. de Lima, M. van der Poel, P.V. Santos, J.M. Hvam, Phonon-induced polariton superlattices. Phys. Rev. Lett. **97**, 045, 501 (2006)
16. M.M. de Lima Jr., P.V. Santos, Modulation of photonic structures by surface acoustic waves. Rep. Prog. Phys. **68**, 1639 (2005)
17. C. Gorecki, F. Chollet, E. Bonnotte, H. Kawakatsu, Silicon-based integrated interferometer with phase modulation driven by surface acoustic waves. Opt. Lett. **22**, 1784 (1997)
18. M. Greiner, O. Mandel, T. Esslinger, T.W. Hansch, I. Bloch, Quantum phase transition from a superfluid to a mott insulator in a gas of ultracold atoms. Nature **415**(6867), 39–44 (2002). DOI 10.1038/415039a
19. M.J. Hoskins, H. Morkoç, B.J. Hunsinger, Charge transport by surface acoustic waves in GaAs. Appl. Phys. Lett. **41**, 332 (1982). DOI 10.1063/1.93526. URL http://link.aip.org/link/?APL/41/332/1
20. A.L. Ivanov, P.B. Littlewood, Resonant acousto-optics of microcavity polaritons. Semicond. Sci. Technol. **18**, S428 (2003)
21. D. Jaksch, C. Bruder, J.I. Cirac, C.W. Gardiner, P. Zoller, Cold bosonic atoms in optical lattices. Phys. Rev. Lett. **81**(15), 3108 (1998). DOI 10.1103/PhysRevLett.81.3108
22. R.I. Kaitouni, O.E. Daif, A. Baas, M. Richard, T. Paraiso, P. Lugan, T. Guillet, F. Morier-Genoud, J.D. Ganiere, J.L. Staehli, V. Savona, B. Deveaud, Engineering the spatial confinement of exciton polaritons in semiconductors. Phys. Rev. B Condens. Matter Mater. Phys. **74**(15), 155311 (2006). DOI 10.1103/PhysRevB.74.155311. URL http://link.aps.org/abstract/PRB/v74/e155311
23. J. Kasprzak, M. Richard, S. Kundermann, A. Baas, P. Jeambrun, J.M.J. Keeling, F.M. Marchetti, M.H. Szymańska, R. André, J.L. Staehli, V. Savona, P.B. Littlewood, B. Deveaud, L.S. Dang, Bose-Einstein condensation of exciton polaritons. Nature **443**, 409 (2006)
24. A. Korpel, Acousto-Optics. (Marcel Dekker, Inc., New York 1997)
25. D.N. Krizhanovskii, S.S. Gavrilov, A.P.D. Love, D. Sanvitto, N.A. Gippius, S.G. Tikhodeev, V.D. Kulakovskii, D.M. Whittaker, M.S. Skolnick, J.S. Roberts, Self-organization of multiple polariton-polariton scattering in semiconductor microcavities. Phys. Rev. B **77**(11), 115336 (2008). DOI 10.1103/PhysRevB.77.115336. URL http://link.aps.org/abstract/PRB/v77/e115336
26. D.N. Krizhanovskii, D. Sanvitto, A.P.D. Love, M.S. Skolnick, D.M. Whittaker, J.S. Roberts, Dominant effect of polariton-polariton interactions on the coherence of the microcavity optical parametric oscillator. Phys. Rev. Lett. **97**, 097, 402 (2006)
27. K.G. Lagoudakis, M. Wouters, M. Richard, A. Baas, I.R.A. Carusotto, L.S. Dang, B. Deveaud-Plédran, Quantized vortices in an excitonpolariton condensate. Nat. Phys. **4**, 706 (2008). DOI 10.1038/nphys1051

28. C.W. Lai, N.Y. Kim, S. Utsunomiya, G. Roumpos, H. Deng, M.D. Fraser, T. Byrnes, P. Recher, N. Kumada, T. Fujisawa, Y. Yamamoto, Coherent zero-state and pi-state in an exciton-polariton condensate array. Nature **450**, 529 (2007)
29. M.M. de Lima Jr., M. Beck, R. Hey, P.V. Santos, Compact Mach-Zehnder acousto-optic modulator. Appl. Phys. Lett. **89**, 121, 104 (2006)
30. A.P.D. Love, D.N. Krizhanovskii, D.M. Whittaker, R. Bouchekioua, D. Sanvitto, S.A. Rizeiqi, R. Bradley, M.S. Skolnick, P.R. Eastham, R. André, L.S. Dang, Intrinsic decoherence mechanisms in the microcavity polariton condensate. Phys. Rev. Lett. **101**(6), 067404 (2008). DOI 10.1103/PhysRevLett.101.067404. URL http://link.aps.org/abstract/PRL/v101/e067404
31. L. Pitaevskii, S. Stringari, Bose-Einstein Condesation. (Clarendon Press, Oxford, 2003)
32. L. Rayleigh, On waves propagated along the plane surface of an elastic solid. Proc. London Math. Soc. **s1-17**(1), 4–11 (1885). DOI 10.1112/plms/s1-17.1.4. URL http://plms.oxfordjournals.org/content/s1-17/1/4.short
33. C. Rocke, O. Govorov, A. Wixforth, G. Böhm, G. Weimann, Exciton ionization in a quantum well studied by surface acoustic waves. Phys. Rev. B **57**, R6850 (1998)
34. C. Rocke, S. Zimmermann, A. Wixforth, J.P. Kotthaus, G. Böhm, G. Weimann, Acoustically driven storage of light in a quantum well. Phys. Rev. Lett. **78**, 4099 (1997)
35. D. Royer, E. Dieulesaint, *Elastic Waves in Solids.* (Springer, Heidelberg, 2000)
36. T. Sogawa, P.V. Santos, S.K. Zhang, S. Eshlaghi, A.D. Wieck, K.H. Ploog, Transport and lifetime enhancement of photoexcited spins in GaAs by surface acoustic waves. Phys. Rev. Lett. **87**, 276, 601–1 (2001)
37. J.A.H. Stotz, R. Hey, P.V. Santos, K.H. Ploog, Coherent spin transport via dynamic quantum dots. Nat. Mater. **4**, 585 (2005)
38. C.S. Tsai, *Guided-Wave Acousto-Optics.* (Springer, Berlin, 1990)
39. S. Utsunomiya, L. Tian, G. Roumpos, C.W. Lai, N. Kumada, T. Fujisawa, M. Kuwata-Gonokami, A. Löffler, S. Höfling, A. Forchel, Y. Yamamoto, Observation of bogoliubov excitations in exciton-polariton condensates. Nat. Phys. **4**, 700 (2008). DOI 10.1038/nphys1034
40. R.M. White, Surface elastic waves. In: Proceedings of the IEEE, vol. 58, (IEEE, New York, 1970) p. 1238
41. R.M. White, F.W. Vollmer, Direct piezoelectric coupling to surface elastic waves. Appl. Phys. Letts. **7**(12), 314 (1965)
42. D.M. Whittaker, Effects of polariton-energy renormalization in the microcavity optical parametric oscillator. Phys. Rev. B **71**(11), 115, 301 (2005). DOI 10.1103/PhysRevB.71.115301. URL http://link.aps.org/abstract/PRB/v71/e115301
43. M. Wouters, I. Carusotto, Absence of long-range coherence in the parametric emission of photonic wires. Phys. Rev. B **74**(24), 245316 (2006). DOI 10.1103/PhysRevB.74.245316. URL http://link.aps.org/abstract/PRB/v74/e245316
44. K.S. Zhuravlev, D.P. Petrov, Y.B. Bolkhovityanov, N.S. Rudaja, Effect of surface acoustic waves on low-temperature photoluminescence of GaAs. Appl. Phys. Lett. **70**, 3389 (1997)

Chapter 12
Polariton Condensation and Lasing

David Snoke

Abstract The similarities and differences between polariton condensation in microcavities and standard lasing in a semiconductor cavity structure are reviewed. The recent experiments on "photon condensation" are also reviewed.

Polariton condensation and lasing in a semiconductor vertical-cavity, surface-emitting laser (VCSEL) have many properties in common. Both emit coherent light normal to the plane of the cavity, both have an excitation density threshold above which there is optical gain, and both can have in-plane coherence and spontaneous polarization. Is there, then, any real difference between them? Should we drop the terminology of Bose–Einstein condensation (BEC) altogether and only talk of polariton lasing?

The best way to think of this is to view polariton condensation and standard lasing as two points on a continuum, just as Bose–Einstein pair condensation and Bardeen-Cooper-Schreiffer (BCS) superconductivity are two points on a continuum. In the BEC-BCS continuum, the parameter which is varied is the ratio of the pair correlation length to the average distance between the underlying fermions; BEC occurs when the pair correlation length, i.e., the size of a bound pair, is small compared to the distance between particles, while BCS superconductivity (which has the same mathematics as the excitonic insulator (EI) state, in the case of neutral electron–hole pairs) occurs when the pair correlation length is large compared to the distance between the particles. In the condensation-lasing continuum, the parameter which is varied is the ratio of the recombination time of the electron–hole pairs to the interaction time between pairs; in the exciton condensate limit, the recombination lifetime is essentially infinite compared to the time scale for interactions between the excitons, while in the case of lasing, the lifetime for recombination is short

D. Snoke (✉)
Department of Physics and Astronomy, University of Pittsburgh, Pittsburgh, PA 15260, USA

D. Sanvitto and V. Timofeev (eds.), *Exciton Polaritons in Microcavities*, Springer Series in Solid-State Sciences 172, DOI 10.1007/978-3-642-24186-4_12,

compared to the interaction time, and in fact, there may be no interaction between the pairs at all.

In these different limits, there are various properties of the BEC and standard lasing states that allow them to be distinguished. In this chapter, we will review the connections between the two states and also discuss the experiments that show the differences.

12.1 The State of Matter in Excitonic Condensation and Lasing

The fact that exciton condensation and lasing are fundamentally similar can be seen by looking at the basic mathematics of exciton condensation. In second quantization language, the creation operator for an exciton with center-of-mass momentum K is (see, e.g., [1], Sect. 11.2.1)

$$c_K^\dagger = \sum_k \phi(K/2 - k) b_{c,K-k}^\dagger b_{v,-k}, \tag{12.1}$$

where $b_{n,k}^\dagger$ is the fermion creation operator for an electron in band n with momentum k, and $\phi(k)$ is the orbital wave function for the electron–hole relative motion. In other words, creating an exciton corresponds to creating a superposition of electron–hole pairs in different k-states in the conduction band and valence bands. A Bose–Einstein condensate corresponds to a coherent state of excitons in a single center-of-mass K-state, which we will assume is $K = 0$:

$$|\alpha\rangle = \mathrm{e}^{-|\alpha|^2/2} \sum_n \frac{(\alpha c_0^\dagger)^n}{\sqrt{n!}} |0\rangle, \tag{12.2}$$

where $\alpha = \sqrt{N}\mathrm{e}^{\mathrm{i}\theta}$ is the phase of the condensate and N is the number of excitons in the condensate.

Let us look at this formalism in the limit of Frenkel excitons. Frenkel excitons correspond to excitons with electron and hole on the same lattice site, as opposed to Wannier excitons, in which the exciton and hole orbit each other at some distance. We write the fermion creation operator as the Fourier transform of a spatial creation operator for a single lattice site i:

$$b_{nk}^\dagger = \frac{1}{\sqrt{N_\mathrm{s}}} \sum_i \mathrm{e}^{-\mathrm{i}k \cdot r_i} b_{ni}^\dagger, \tag{12.3}$$

where N_s is the number of lattice sites. The creation operator (12.1) is then

$$c_K^\dagger = \sum_k \phi(K/2-k)\frac{1}{N_s}\sum_{i,j} e^{-i(K-k)\cdot r_i} b_{c,i}^\dagger e^{-ik\cdot r_j} b_{v,j}$$

$$= \frac{1}{N_s}\sum_{i,j} b_{c,i}^\dagger b_{v,j} e^{-iK\cdot r_i}\left(\sum_k \phi(K/2-k)e^{ik\cdot(r_i-r_j)}\right). \tag{12.4}$$

The Frenkel limit corresponds to picking $\phi(k) = \text{constant} = 1/\sqrt{N_s}$, in which case the term in the parentheses, which is the real-space wave function for the electron–hole relative motion, equals $\sqrt{N_s}\delta_{ij}$. Putting this into our definition of $c_K^\dagger$ gives us

$$c_K^\dagger = \frac{1}{\sqrt{N_s}}\sum_i e^{-iK\cdot r_i} b_{c,i}^\dagger b_{v,i}, \tag{12.5}$$

i.e., a superposition of excitations at all lattice sites.

Creating a macroscopic number of $K = 0$ excitons in one state involves terms of the form

$$(c_0^\dagger)^n = \left(\frac{1}{\sqrt{N_s}}\sum_i b_{c,i}^\dagger b_{v,i}\right)^n$$

$$= \frac{1}{N_s^{n/2}}\left(b_{c,1}^\dagger b_{v,1} + b_{c,2}^\dagger b_{v,2} + \ldots\right)^n. \tag{12.6}$$

This is a superposition of all possible ways to have N excitons in N_s lattice sites.

The expectation value $\langle\alpha|b_{c,i}^\dagger b_{c,i}|\alpha\rangle$ gives the probability of a lattice site i being in the excited state. If we pick site i, then in the product (12.6), there are n terms which will create an exciton at that site. For each of these terms, the number of ways of picking the remaining terms so as to not have double fermion creation operators (which vanish) is

$$(N_s-1)(N_s-2)\ldots(N_s-n) = \frac{(N_s-1)!}{(N_s-n)!}.$$

In the limit $N_s \gg n$, we ignore the possibility of double occupation and approximate

$$\frac{(N_s-1)!}{(N_s-n)!} \simeq N_s^{n-1}.$$

Each product term created in $|\alpha\rangle$ will have $n!$ matching terms in the term with the same order of n in $\langle\alpha|$. The expectation value is therefore

$$\begin{aligned}\langle\alpha|b^{\dagger}_{c,i}b_{c,i}|\alpha\rangle &= \mathrm{e}^{-N}\sum_n \frac{N^n}{(n!)^2}\frac{1}{N_\mathrm{s}^n}\frac{n}{N_\mathrm{s}}N_\mathrm{s}^n n! \\ &= \mathrm{e}^{-N}\sum_n \frac{N^n}{n!}\frac{n}{N_\mathrm{s}} \\ &= \frac{N}{N_\mathrm{s}}.\end{aligned} \tag{12.7}$$

In other words, the probability of a given lattice site being in the excited state is equal to the number of excitons in the condensate divided by the total number of lattice sites. If the number of excitons increases, at some point, the possibility of double occupation becomes significant and will cause this relation to become sublinear with N. This is known as "phase space filling."

The calculation for the dipole moment term $b^{\dagger}_{v,i}b_{c,i}$ is similar, except that this term changes the n term in the coherent state sum to the $(n-1)$ term. We therefore have

$$\begin{aligned}\langle\alpha|b^{\dagger}_{v,i}b_{c,i}|\alpha\rangle &= \mathrm{e}^{-N}\sum_n \frac{N^{n/2}}{n!}\frac{N^{(n-1)/2}}{(n-1)!}\frac{\mathrm{e}^{\mathrm{i}n\theta}\mathrm{e}^{\mathrm{i}(n-1)\theta}}{N_\mathrm{s}^{n/2}N_\mathrm{s}^{(n-1)/2}}\frac{n}{N_\mathrm{s}}N_\mathrm{s}^n(n-1)! \\ &= \sqrt{\frac{N_\mathrm{s}}{N}}\mathrm{e}^{-\mathrm{i}\theta}\mathrm{e}^{-N}\sum_n \frac{N^n}{n!}\frac{n}{N_\mathrm{s}} \\ &= \sqrt{\frac{N}{N_\mathrm{s}}}\mathrm{e}^{-\mathrm{i}\theta}.\end{aligned} \tag{12.8}$$

This is proportional to the amplitude of the condensate wave function. The total dipole moment of the solid is N_s times this. This corresponds to a macroscopic radiating dipole moment which will emit coherent light.

Let us now switch to thinking about a standard laser. In the theory of lasing in an ensemble of independent two-level oscillators, we start with a single Bloch oscillator coupled to a coherent electromagnetic wave:

$$(u_i + v_i b^{\dagger}_{c,i}b_{v,i})|0\rangle, \tag{12.9}$$

where $|u_i|^2 + |v_i|^2 = 1$. The coefficient v_i determines the degree of excitation. For an ensemble of identical oscillators, we repeatedly create identical states at all the different sites:

$$|l\rangle = \prod_i (u + v b^{\dagger}_{c,i}b_{v,i})|0\rangle. \tag{12.10}$$

The expectation value is $\langle l|b^{\dagger}_{c,i}b_{c,i}|l\rangle$ is $|v|^2$, which we can set equal to N/N_s.

As shown in [1], Sect. 11.2.3 (and discussed earlier in [2]), the state (12.10), which has the form of a BCS wave function, is equivalent to a coherent state of the form (12.2), where we equate

$$\alpha\phi = \frac{v}{u}. \tag{12.11}$$

If we choose $\phi = 1/\sqrt{N_s}$ and $|v|^2 = N/N_s$, in the limit where $u \simeq 1$ when phase space filling is negligible, we have exactly the same state as our Frenkel exciton example above, with probability of excitation of a single lattice site equal to N/N_s. Lasing occurs when phase space filling (inversion) is significant, in which case the state of the system resembles a standard BCS state.

The case of a coherent electromagnetic wave passing through a transparent medium with a dielectric constant (index of refraction) produces the same polarized state of matter. As with a laser, we model the system as an ensemble of two-level oscillators in phase with each other (see, e.g., Sects. 7.1 and 7.3 of [1]). In this case, the frequency of the wave is well below the electronic resonance so that the radiation from the polarized medium gives a phase shift to the electromagnetic wave, and a change in phase velocity, but no absorption or gain.

We thus have three physical scenarios which all lead to a state of matter which consists of an ensemble of electronic excitations in phase with each other: excitonic condensation, lasing, and linear refraction of a wave in a dielectric medium. The first two involve spontaneous coherence, while the last has electronic coherence because it is driven by an external electromagnetic field. Some people have used the term "driven condensate" to refer to coherence of a medium driven by an external field, but this terminology is not very helpful because it would have to also include every sound wave produced by a loudspeaker and every radio wave driven by an antenna. The key difference between lasing and excitonic condensation is that lasing involves *inversion*, which excitonic condensation does not. The expectation value (12.7) can be small in the case of a condensate, while in the case of lasing it must be near unity.

12.2 Condensation and Classical Waves

We will discuss the differences of exciton-polariton BEC and standard lasing below, but before we do that, let us consider another connection. As discussed above, every BEC state is a coherent state of bosons of the form (12.2). A general property of coherent states with large N is that they have exactly the same properties as classical wave states; in fact, in the modern understanding, all classical waves such as sound waves and electromagnetic waves are actually quantum mechanical coherent states with high occupation number of bosons (phonons and photons, respectively). In the case of a standard BEC, the behavior of the coherent condensate is described by the Gross–Pitaevskii equation:

$$i\hbar \frac{\partial \psi}{\partial t} = -\frac{\hbar^2}{2m} \nabla^2 \psi + U|\psi|^2 \psi, \tag{12.12}$$

which becomes the Ginzburg–Landau equation if the time derivative is replaced by $E\psi$. This is also called the nonlinear Schrödinger equation because it has exactly the same form as a single-particle Schrödinger equation with wave function ψ and a nonlinear potential energy that is proportional to the square of the amplitude of the wave function $|\psi|^2$. In the context of BEC, the wave function ψ is normalized by $(1/L^d) \int d^d x \; |\psi|^2 = N$, where N is the total number of particles in the condensate and d is the dimensionality. The Gross–Pitaevskii equation is often viewed as a heuristic assumption, but it can be justified fairly rigorously in terms of the underlying microsopics, as discussed in [1], Sect. 11.1.3. It is simply a consequence of the condensate being a coherent state, and the particles having a two-body interaction that leads to local potential energy proportional to the particle density.

The fact that a condensate can be described by a classical wave equation with a nonlinear term has also led to other qualms that a polariton BEC is no different from a coherent classical optical system. We can see that the two have a fundamental connection by showing that we can also derive the Gross–Pitaevskii equation starting with Maxwell equations.

We start with the Maxwell wave equation in a nonlinear isotropic medium,

$$\nabla^2 E = \frac{n^2}{c^2} \frac{\partial^2 E}{\partial t^2} + 4\mu_0 \chi^{(3)} \frac{\partial^2}{\partial t^2} |E|^2 E, \tag{12.13}$$

where $\chi^{(3)}$ is the standard nonlinear optical constant, and we ignore frequency-mixing terms in the general E^3 nonlinear response. We write a solution:

$$E = \psi e^{-i\omega t}, \tag{12.14}$$

where ψ is an amplitude which may vary in time and space. We write this envelope amplitude suggestively as ψ because we will see that it plays the same role as the matter wave ψ in the Gross–Pitaevskii equation.

Keeping only leading terms in frequency (known as the slowly varying envelope approximation), we have for the time derivative of E,

$$\frac{\partial^2 E}{\partial t^2} \simeq \left(-\omega^2 \psi - 2i\omega \frac{\partial \psi}{\partial t} \right) e^{-i\omega t}, \tag{12.15}$$

and for the time derivative of the nonlinear term,

$$\frac{\partial^2}{\partial t^2} |E|^2 E \simeq -\omega^2 |\psi|^2 \psi e^{-i\omega t}.$$

The standard polariton structure uses a planar or nearly planar cavity to give one confined direction of the optical mode. We therefore distinguish between the component of momentum k_z in the direction of the cavity confinement and the momentum $k_\parallel$ in the two-dimensional plane perpendicular to this direction. We therefore write

$$\psi = \psi(\vec{x})e^{i(k_\parallel\cdot\vec{x}+k_z z)}. \tag{12.16}$$

The full Maxwell wave (12.13) then becomes

$$(-(k_z^2 + k_\parallel^2)\psi + \nabla_\parallel^2\psi) = (n/c)^2\left(-\omega^2\psi - 2i\omega\frac{\partial\psi}{\partial t}\right) - 4\mu_0\chi^{(3)}\omega^2|\psi|^2\psi.$$

Since $\omega^2 = (c/n)^2(k_z^2 + k_\parallel^2)$, this becomes

$$\nabla_\parallel^2\psi = (n/c)^2\left(-2i\omega\frac{\partial\psi}{\partial t}\right) - 4\mu_0\chi^{(3)}\omega^2|\psi|^2\psi. \tag{12.17}$$

Near $k_\parallel = 0$, we can approximate

$$\hbar\omega = \hbar(c/n)\sqrt{k_z^2 + k_\parallel^2} \simeq \hbar(c/n)k_z\left(1 + \frac{k_\parallel^2}{2k_z^2}\right) \equiv \hbar\omega_0 + \frac{\hbar^2 k_\parallel^2}{2m}, \tag{12.18}$$

which gives an effective mass for the photon motion in the plane. This does not take into account the renormalization of the effective mass due to the coupling with the exciton states, which typically makes the effective mass about a factor of 2 heavier. To account for this, we would have to take into account the dependence of the index of refraction n on frequency due to the exciton resonance, as discussed in [1], Sect. 7.4. Neglecting this correction, for the first term on the right-hand side, we approximate

$$\omega \simeq \omega_0 = \frac{m(c/n)^2}{\hbar}, \tag{12.19}$$

so that we have

$$i\hbar\frac{\partial\psi}{\partial t} = -\frac{\hbar^2}{2m}\nabla_\parallel^2\psi - \frac{2\mu_0\chi^{(3)}(\hbar\omega)^2}{m}|\psi|^2\psi, \tag{12.20}$$

which we can rewrite as

$$i\hbar\frac{\partial\psi}{\partial t} = -\frac{\hbar^2}{2m}\nabla_\parallel^2\psi + U|\psi|^2\psi. \tag{12.21}$$

This is the Gross–Pitaevskii equation, or nonlinear Schrödinger equation. Note that although the Maxwell wave equation is second order in the time derivative, this equation is first order in time derivative, as in a typical Schrödinger equation.

One useful result of this derivation is that we can write the nonlinear $\chi^{(3)}$ value in terms of experimentally measured parameters. To do this, we first need to calibrate the electric field amplitude in terms of the density of particles. From basic photon theory (see [1], Sect. 4.4) the number of photons is related to the electric field amplitude by

$$N = \frac{\epsilon_0 V}{2\hbar\omega} E^2, \tag{12.22}$$

which for our two-dimensional system can be written as

$$\psi^2 = \frac{2\hbar\omega}{\epsilon_0 d} \frac{N}{A}, \tag{12.23}$$

where A is the area and d is the effective thickness. If we switch ψ in (12.21) from standing for electric field amplitude to standing for a number density wave function, the nonlinear term therefore corresponds to a potential energy linearly proportional to the density, with constant of proportionality equal to

$$U = \frac{2\mu_0|\chi^{(3)}|(\hbar\omega)^2}{m} \frac{2\hbar\omega}{\epsilon_0 d}.$$

We can use this result to equate the measured coefficient of linear shift of the energy of the polaritons, measured from the shift of the spectral line, with an effective $\chi^{(3)}$ value. Typical values for polariton condensates are 1 meV of shift for a density of $10^9\,\mathrm{cm}^{-2}$. Using a typical microcavity length of 400 nm and material constants for GaAs, we obtain

$$\frac{|\chi^{(3)}|}{\epsilon_0} = \frac{\mathrm{Umd}}{4\mu_0(\hbar\omega)^3} \tag{12.24}$$

$$\simeq 3 \times 10^{-14}\ \mathrm{m}^2/\mathrm{V}^2, \tag{12.25}$$

which is equivalent to 2×10^{-6} esu. By comparison, typical nonlinear coefficients for strongly nonlinear media are around three orders of magnitude lower [3]. This is one of the main appeals of polariton condensates as optical systems—they have world-record nonlinearities, allowing, e.g., optical parametric oscillator (OPO) behavior with continuous laser pump powers of a milliwatt or less, compared to standard broadband OPO systems which require pulsed lasers with instantaneous power at least six orders of magnitude higher. The trade-off is that polariton systems only work for a narrow range of wavelengths; the huge nonlinearity is obtained by operating near a sharp exciton resonance. This can be done with resonances of single atoms also, but to do that requires ultrahigh vacuum and extremely good laser frequency stability.

The fact that polariton condensate Gross–Pitaevskii equation maps to the nonlinear Maxwell wave equation in this system should not be cause to doubt the validity of the BEC description. Every condensate is described by a classical wave equation because the Gross–Pitaevskii equation is a classical wave equation. It does tell us, however, that some experimental results with polariton condensates do not intrinsically distinguish between condensation and lasing, or more generally, a coherent optical field. For example, quantized vortices are fundamentally a result of having a single-valued, coherent wave. Since this occurs also in the optical modes of a laser, quantized vortices can also be seen in a standard laser or VCSEL [4]. In the same way, in-plane phase coherence of the wave function is also seen in spatially extended laser modes from a VCSEL [5]. However, the central core radius of a vortex of a polariton BEC, in which the condensate is depleted, is determined by the healing length of the condensate ξ, which is determined by the interaction strength of the particles—stronger interaction gives shorter healing length (See [1], Sect. 11.1.4). In vortices of laser modes, the vortex structure is determined by the spatial dimensions of the laser modes.

12.3 Differences Between Polariton Condensation and Lasing

To see the difference between a polariton condensate coherent state and a standard laser, we start by examining how these states arise. Figure 12.1a shows the standard band picture of a semiconductor laser. Electrons from the valence band are pumped high up into the conduction band through what is typically an incoherent process, and then both the electrons and holes created by the pumping fall in energy to near the band edges (holes float upward as they lose energy, in this picture). This energy relaxation is an incoherent process so that even if a coherent pump was used to excite the carriers, by the time they arrive in the lowest energy states, they have lost all memory of the pump laser coherence. Fundamentally, the semiconductor laser is no different from an ensemble of isolated atoms, e.g., atoms in a gas or dopant atoms in a transparent matrix, with two occupied levels and two unoccupied levels. The pumping of the electrons into the upper states leads to *inversion*, i.e., higher probability of carriers in the upper state than in the lower state of the two

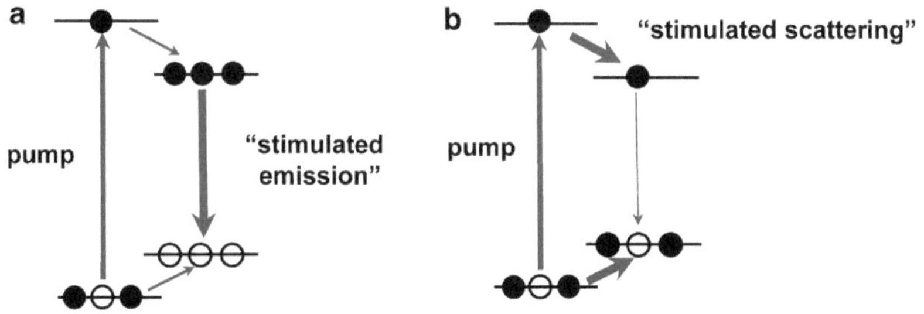

Fig. 12.1 (**a**) Pumping and emission scheme for a standard semiconductor laser. (**b**) Pumping and emission for an excitonic condensate. The emission rate can be low, with no optical gain in the emission process. Gain occurs in the redistribution of the excited carriers

levels involved in the light emitting transition (so-called negative temperature). This leads to a spectral range for electromagnetic waves with *gain*, e.g., amplification of coherent electromagnetic waves. As shown in [1], Sect. 11.3.1, a simple model of a laser as a homogeneous ensemble of pumped two-level atoms leads to the result

$$\frac{\partial \psi}{\partial t} = \frac{\omega}{2\epsilon}(A\psi - B|\psi|^2\psi), \tag{12.26}$$

where ϵ is the dielectric constant and A and B are two terms which depend on the pumping rate and oscillator strength; A is positive for net pumping rate exceeding the spontaneous emission rate, and B has the same sign as A. If the system is initially incoherent, but there is some small fluctuation that gives a coherent seed, this coherence will be amplified by the A term until there is a macroscopic coherence. The amplification will eventually be limited by the nonlinear B term, which arises from the physical constraint that the upper level of the two-level atoms cannot have higher than unity occupation, and Rabi flopping will depopulate the upper level if the coherent field grows too strong.

One can see from this that the onset of lasing is a type of spontaneous symmetry breaking since we cannot predict the exact value of the complex amplitude of the seed that will be amplified by (12.26). The same type of spontaneous symmetry breaking occurs in Bose–Einstein condensation. A calculation to be presented elsewhere [6] gives for the complex amplitude $\langle a_k \rangle$ of a condensate

$$\begin{aligned}\frac{\mathrm{d}}{\mathrm{d}t}\langle a_k \rangle = \langle a_k \rangle \frac{2\pi}{\hbar} 2U^2 \sum_{k_2,k_3} \Big[&\langle \hat{N}_{k_3} \rangle \langle \hat{N}_{k_4} \rangle (1 + \langle \hat{N}_{k_2} \rangle) \\ &- \langle \hat{N}_{k_2} \rangle (1 + \langle \hat{N}_{k_3} \rangle)(1 + \langle \hat{N}_{k_4} \rangle) \Big] \delta(E_k + E_{k_2} - E_{k_3} - E_{k_4}),\end{aligned} \tag{12.27}$$

where the $\langle \hat{N}_k \rangle$ are the average occupation numbers of the excited k-states, the same occupation numbers that go into the quantum Boltzmann equation for the increase of the population in the condensate (see [1], Sect. 4.8). The first term is normally small relative to the second since it involves a product of two occupation numbers while the second involves only one, but when the number of particles in low-energy states becomes large, it can lead to gain of the coherent amplitude of the condensate. (The topic of the onset of phase coherence in BEC has been a long-standing problem in the theory of BEC, addressed by numerous authors (e.g., [7–9].) It is not essential that a collisional interaction be the means of establishing condensation; condensation can also occur by emission and reabsorption of phonons (see [10], Sect. 8.2.3). But in every case, the stimulated scattering $(1 + \langle \hat{N}_k \rangle)$ factors, which connect the occupation of one particle state to another, drive the onset of the phase coherence. As illustrated in Fig. 12.1b, the gain in the system occurs in the coupling of the continuum of excited states to the emitting states, not in the emission process itself.

These two different relations, (12.26) and (12.27), tell us a lot about both the similarities and the differences between lasing and excitonic condensation. We see that both are examples of spontaneous coherence. In the spontaneous symmetry breaking of a laser, the control parameter is not the temperature T, as the transition is not a thermodynamic phase transition; instead, the control parameter is the incoherent pumping rate which goes into the gain constant A. Nevertheless, spontaneous symmetry breaking is not unique to BEC.

The *onset* and *stability* of standard lasing arise from the net gain from incoherent pumping, and the coherence arises from the stimulated emission of photons, which is taken into account in (12.26) by the $A\psi$ term. This implies that a standard laser must have *inversion*, and the stimulated emission occurs *only in the coherent state*—whatever state has many photons will be amplified if there is gain. In a standard laser, that photon state is defined by the mirrors of the cavity, which recirculates photons to suppress the loss rate. Electronic excitations not resonant with that state will undergo transitions to pump that state only incoherently.

In a polariton condensate, as with any condensate, the onset and stability of the coherent amplitude come fundamentally from the *interaction of the particles in nearby states*. The two terms which amplify the coherence in (12.27) come from the occupation numbers of the noncondensed states. If all of the occupation numbers $\langle \hat{N}_{k_2} \rangle$, $\langle \hat{N}_{k_3} \rangle$, and $\langle \hat{N}_{k_4} \rangle$ are all small compared to unity, then there will always be net dephasing. If $\langle \hat{N}_{k_3} \rangle$ and $\langle \hat{N}_{k_4} \rangle$ are large compared to unity, however, while $\langle \hat{N}_{k_2} \rangle$ is small, then there will be amplification. Onset of condensation therefore requires not just that the condensate state itself is highly occupied but that nearby states also have large (though not necessarily macroscopic) occupation. The coherence of the system is shared as a collective property of particles in many states. Figure 12.2 shows how the states near to the condensate state in a polariton system are also highly occupied.

In a condensate of particles with finite lifetime, there is a pump that incoherently creates new particles to compensate for the loss of the particles, but the stimulated process that gives rise to coherence is in the scattering of these incoherently created particles with each other, not in gain in the recombination process. Inversion is not only not necessary for polariton condensation, it actually kills polariton condensation because inversion corresponds to Pauli state filling of the bands, which prevents pairing of the carriers into bosonic excitons and thus shuts off the process of bosonic stimulated scattering which leads to condensation.

One can think of the two limits of the BEC-lasing continuum in terms of what drives the coherence. In the case of lasing, we can imagine starting with a very small coherent electromagnetic field which drives the two-level oscillators, similar to the way that an electromagnetic wave drives the oscillators which give the dielectric response of the medium. This small coherent electromagnetic wave drives a small coherent polarization of the medium, which then radiates, and this radiation is amplified by the gain due to the inversion.

In an excitonic BEC, the coherent polarization of the medium drives the electromagnetic radiation. The coherence in the polarization of the medium leads to

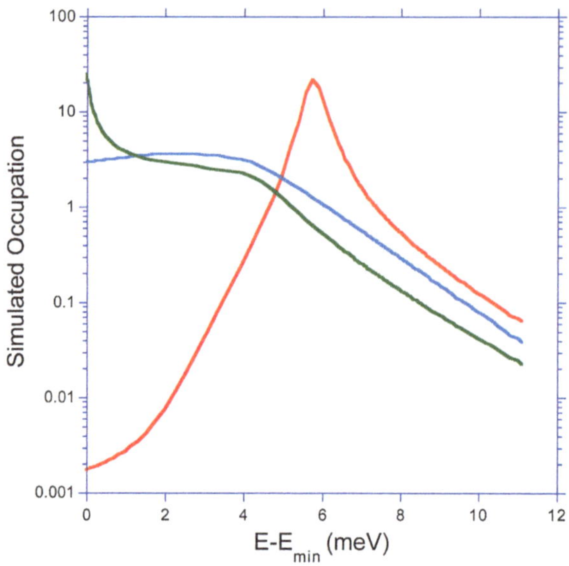

Fig. 12.2 Energy distribution of polaritons in a GaAs microcavity structure from numerical solutions of the quantum Boltzmann equation in steady state for the parameters of the experimental data. *Red, lower curve*: when only interactions with phonons are accounted for. Above 6 meV, the states are essentially exciton states, with a Maxwell–Boltzmann distribution. The "polariton bottleneck" at 6 meV occurs because the phonons cannot efficiently couple exciton states to polariton states. *Blue curve*: polariton-polariton interactions included, but final-states $(1 + N_k)$ factors not included. *Green curve*: full calculation including final states $(1 + N_k)$ factors. There is not only a peak ak $E = 0$, there is also a pileup of particles in low-momentum states. From [11]

emission of coherent radiation but would still occur even if there were no radiation, which can be the case, e.g., when the emission of radiation is forbidden by the symmetry of the semiconductor bands [12] or by a barrier between the electrons and holes, as in the case of excitons in double quantum wells [13]. We imagine starting with a small coherent polarization of the medium, which then is amplified by pulling in carriers from the incoherent excited population of the medium, and then this amplified coherent polarization can emit radiation.

In other words, in both systems, there are electromagnetic degrees of freedom and electronic degrees of freedom, but in a laser, the coherence of the electronic states of the medium is driven by the electromagnetic coherence, as in a linear coherent wave passing through a dielectric medium. In an excitonic condensate, the coherence of the medium arises first, and coherent radiation is emitted by this polarization as a by-product in some cases.

The difference between the two cases can be seen by thinking of what happens when a new photon is introduced into the system with energy resonant with the condensate. In the case of a laser, the system has gain, so the most likely event is the stimulated emission of another photon. In the case of an excitonic condensate, the most likely event is absorption of the additional photon. The high occupation number of the condensate implies that the absorption process will be essentially deterministic. This property may have important technological applications as a coherent absorber.

12.4 Experimental Differences Between Condensation and Lasing

A standard exciton-polariton microcavity structure used for condensation is different from a laser in having *strong coupling* between exciton and photon states. Strong coupling is defined as a Rabi splitting between the upper and lower polariton states which is large compared to the line width of either line. (For a review of the basic properties of microcavity polaritons, see [14].) The term "Rabi frequency" for the splitting between the states is no accident—it is equal to the standard Rabi frequency $\omega_R = (e\langle p\rangle/m\hbar\omega_0)E_0$ in a two-level oscillator, where $\langle p\rangle$ is the dipole matrix element coupling the two states, when E_0 is the electric field amplitude of a single photon of frequency ω_0 in the cavity and ω_0 is the exciton resonance.

When the system is in strong coupling, the polariton is the proper eigenstate, with an equal, or near-equal, superposition of an exciton and a photon. This allows interactions between the polaritons to occur without dephasing via the elastic exciton-exciton collision process. This gives the collision terms which appear in (12.27) which allow onset of coherence without the gain term of (12.26).

In many experiments, the system may be in strong coupling at low excitation density but revert to weak coupling at high density. The reason is that the dipole matrix element $\langle p\rangle$ is in general a function of carrier density and can become greatly reduced at high density due to Pauli phase space filling and carrier screening; essentially, the excitons no longer are well-defined bound states at high density so that the system becomes a gas of uncorrelated electrons and holes, i.e., a plasma, and the dipole coupling between photons and free electrons and holes is much less than the coupling of photons to excitons. The transition from excitons to free carriers is sometimes called the "Mott transition," although the nature of this exciton-plasma transition in semiconductors is of quite different nature from that considered in cold atom gases and in doped solids [15, 16].

If the system reverts to weak coupling, this will result in closing of the gap between the upper and lower polaritons at the zero detuning point (the point at which the excitons and photons have exactly the same energy). It is not always easy to know when this has happened, however. If the bare exciton and photon states are not perfectly equal in energy, then there will be a splitting between them just due to the energy difference of the bare states. This energy gap will not close down when the system goes into weak coupling. A better test of strong coupling is to measure the effective mass of the lower polaritons via angle-resolved far-field emission (this has been done in several experiments). In strong coupling, the effective mass of the polaritons will be twice the mass of the bare cavity photon mode [18].

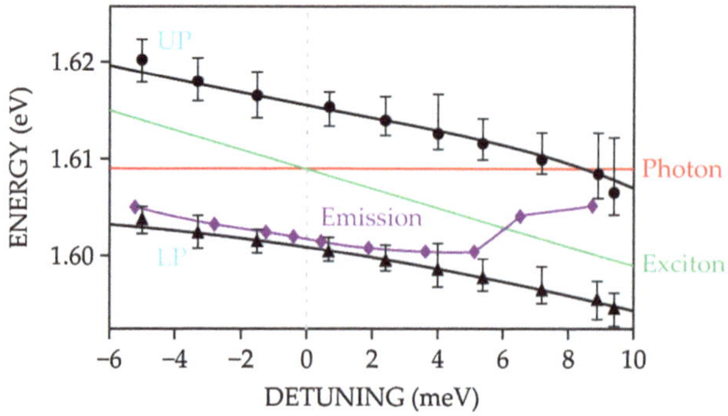

Fig. 12.3 *Black circles*: energy of the upper polariton (UP) and lower polariton (LP) from reflectivity data in a GaAs microcavity, as a function of detuning as the exciton state is shifted by applied stress. *Purple diamonds*: the energy of the coherent emission from the sample as a function of stress. At detuning less than +4 meV, the coherent emission follows the exciton shift, showing the system is still in the strong coupling limit. The green and red lines are the bare exciton and photon modes, determined by the fit to the reflectivity data shown as the black lines. The line connecting the purple diamonds is a guide to the eye. (From [21])

One way to allow higher polariton density without reverting to weak coupling, which has been adopted by many groups, is to use multiple quantum wells in a cavity instead of just one. This greatly reduces the amount of phase space filling since the exciton component of any one polariton is shared among many wells. Typical samples now use 12 or 16 wells placed at antinodes of the confined cavity photon mode [17, 19, 20].

In such samples, it is possible to observe a clear distinction between polariton condensation and lasing. Figure 12.3 shows the emission energy from polaritons in a microcavity in which the exciton energy was shifted using stress to tune the exciton resonance frequency while leaving the photon cavity frequency essentially unchanged. As seen in the reflectivity data, the lower polariton mode follows the exciton shift with stress since it has an excitonic component when there is strong coupling. At high density, when the polariton condensate appears, the coherent emission still shifts with stress to follow the exciton mode. If the system were in a lasing mode with weak coupling, the emission would occur at the cavity photon mode, indicated by the dashed line in Fig. 12.3. (Actually, at high density, this cavity mode is slightly redshifted from its low-density value due to a density-dependent change of the index of refraction [22], but this redshift is unaffected by the stress if the system is in weak coupling.) As seen in this figure, at very large detuning, the coherent emission finally jumps up to near the bare cavity photon frequency, as the system reverts to weak coupling, and the system is at the point of standard VCSEL lasing mode.

The excitation density required to get coherent emission is not the same in the polariton condensate state and the lasing state. The polariton condensate in general requires much lower excitation power because it does not require inversion, as discussed above. In the sample used for the data of Fig. 12.3, the threshold for

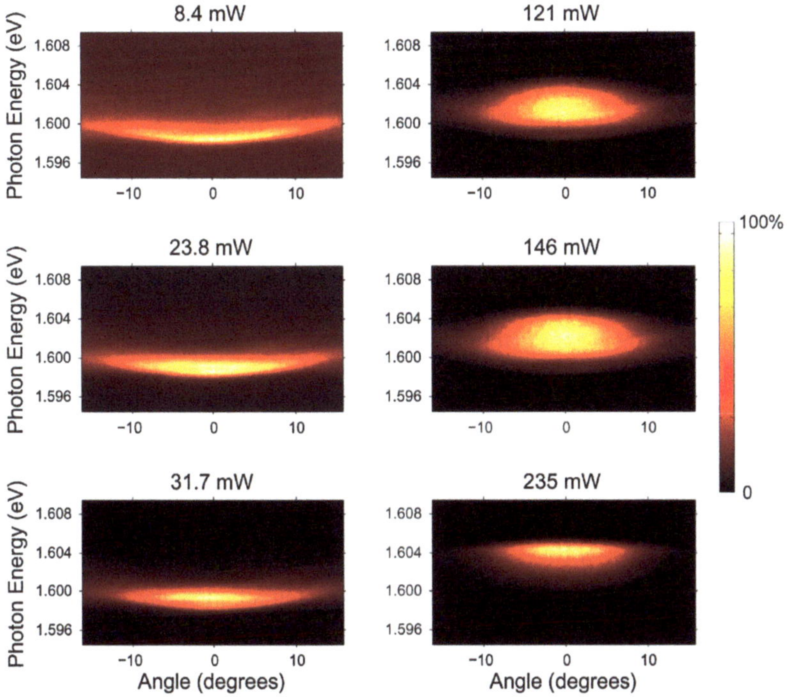

Fig. 12.4 Angle-resolved photon emission from a GaAs microcavity sample (which gives the momentum and energy distribution, i.e., spectral function times the occupation number) for several excitation pump intensities. Polariton condensation is at 31.7 mW, with energy and momentum narrowing; standard lasing is seen at 235 mW. (From [25])

polariton condensation in the strong coupling limit is about six times smaller than the threshold for lasing at large positive detuning at the same place on the same (recall that via stress tuning, the detuning can be varied for the same spot of the sample.) Ref. [20] showed that polariton condensation could have a threshold as much as a factor of 20 lower than the lasing threshold in the same sample when the detuning was changed by moving the laser spot to a different place in the sample with different cavity thickness.

Since the lasing threshold is higher in power, it is possible to push the system from polariton condensation to standard lasing simply by turning up the excitation power. Figure 12.4 shows a sequence of images which give the angle-resolved spectrum of the polariton emission from the same microcavity sample used for Fig. 12.3 when the stress is chosen to give nearly zero detuning, and the excitation power is increased. The angle of emission of the externally emitted photons maps directly to the momentum of the polaritons in the plane of their motion.

Comparing the 23.8-mW data to the 8.4-mW data, one sees that at first, as the power is increased, the polariton spectrum shifts to blue and broadens. Both of these effects are expected for particles with repulsive interactions. Because these spectra are time averaged, there is also a contribution to the broadening if the blueshift fluctuates in time due to laser power fluctuations. The 31.7-mW data shows the spectrum near the polariton condensation threshold. The spectrum has narrowed in energy and also has become narrower in momentum space. The energy width here is limited by instrumental resolution; Love et al. [23] have recently showed that extreme narrowing, corresponding to coherence times of more than 200 ps, can be measured in this type of system when a very stable laser with low intensity fluctuations is used.

If the power is increased further, the spectrum broadens again, and shifts even more strongly to the blue. Both of these effects are related to phase space filling at high density. As discussed above, phase space filling reduces the coupling and therefore reduces the Rabi splitting between the upper and lower polaritons, which shifts the lower polariton up. The increased broadening is also likely to do strong dephasing due to free electrons and holes in the system. When the system reaches 235 mW, however, a second spectral narrowing is seen. This transition corresponds to standard lasing, and the energy of the photons corresponds to the bare cavity photon energy. At all high powers, the coherent emission stays at this energy, pinned at the cavity photon energy as expected for a laser.

We thus see that the two different transitions, polariton condensation and standard lasing, are easily distinguishable experimentally, leading to two separate line narrowings at different excitation powers in the same place on the sample. Bloch and coworkers [24] have also seen similar behavior by varying the laser power. In general, in all these experiments, it is quite easy to see when the system has reverted to lasing because the emission energy pops up to the bare cavity photon energy.

12.5 Josephson Junctions, Phase Locking, Solitons, and Vortices

In the past few years, the number of polariton condensate experiments has exploded, with a number of demonstrations of effects analogous to those of atomic condensates, superconductors, or liquid helium. In one sense, it could be said that the main interest of these effects in atomic condensates, superconductors, and helium is to show wavelike behavior of matter, i.e., these effects all follow from being able to write down a Gross–Pitaevskii equation which treats the condensate as a classical wave. In the case of polaritons, the wavelike behavior is perhaps less surprising because light is already seen as a wave. But unless one disbelieves quantum mechanics, all matter is described as waves. The significance of condensation in all these systems, e.g., atoms, Cooper pairs, and polaritons, is that a system e.g., atoms, Cooper pairs, and polaritons, is that a system which normally has strong dephasing can spontaneously acquire a macroscopic coherent amplitude due to the collective effects. Polariton systems, like atomic systems, have a threshold of low temperature and high density at which the coherence due to interparticle interaction defeats the dephasing processes.

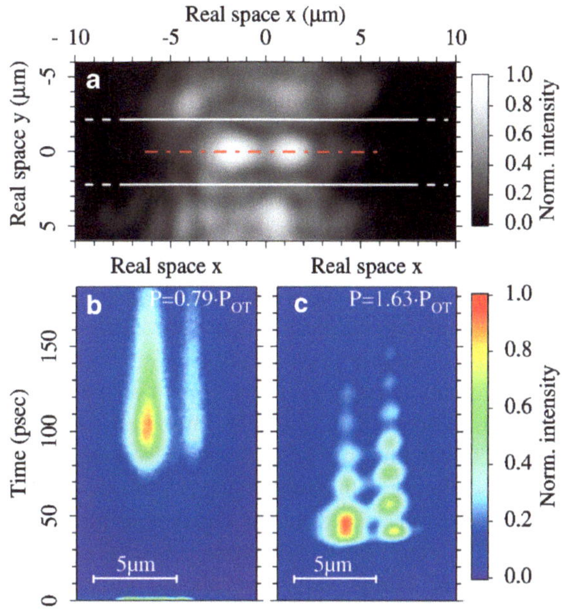

Fig. 12.5 Polariton condensate Josephson junction experiment (from [26]). (**a**) Spatial intensity distribution for two adjacent traps. (**b**) Time- and space-resolved intensity showing the population of the polaritons in the traps, below the critical density threshold for condensation. (**c**) Time- and space-resolved intensity when the two traps hold polariton condensates. The experiment gives an AC current for a DC potential difference between the two traps

Some of these effects with polaritons have analogs in lasing systems, but their behavior is different. Figure 12.5 shows an example of oscillations in the emission of two traps containing polariton condensates connected by a thin tunneling barrier. Similar results have been seen by Bloch and coworkers [27]. The period of the oscillations depends on the chemical potential difference between the two traps, as in a standard Josephson junction experiment.

These oscillations look superficially like the beating of two laser modes. But in the case of laser mode beating, the two states which interfere have discrete energies well separated in frequency from other laser modes. The stability of the oscillations comes from the fact that there are no other nearby states which have gain. In the case of a polariton Josephson junction, as discussed above in Sect. 12.3, the stability comes from the interaction of the particles in a continuum of k-states of the polaritons.

At first, it may seem that the Josephson junction effect in a superconductor involves very different physics from a Josephson junction of polaritons. The superconductor case involves two Fermi levels of the electrons, and the oscillation frequency depends on the difference between these Fermi levels. But recall that a BCS superconductor is just the high-density limit of BEC (see, e.g., [1], Sect. 11.2.3). The Fermi level in a BCS superconductor is determined by the pair wave function of the condensed pairs. A condensate of excitons or polaritons also has a spread of k-states which is determined by the pair wave function, but the density is low enough that there is no phase space filling which produces a Fermi level in this case. In both cases, once a condensate of pairs is formed, the oscillation frequency of a Josephson junction depends on the chemical potential between the two sides, which depends on the sum of any externally applied potential plus the potential energy due to interactions of the particles.

Because the chemical potential difference between the two sides depends not just on the static frequency difference between the two traps but also on the polariton-polariton interactions and on interactions of polaritons with excitons in higher energy states, the period of the oscillations changes in time for the polariton condensate. As the density drops due to recombination, the strength of these interactions drops, and the frequency difference between the traps is shifted. Again, one could argue that a laser system with a strong nonlinear shift of the index of refraction could also show this effect, but as discussed in Sect. 12.2, the exceedingly high value of the nonlinear term in the condensate Gross–Pitaevskii equation is what makes the polariton system special. This high value of the nonlinear term comes from dressing the photons in the system with an excitonic part, which gives a strong interaction between the particles through the long-range Coulomb interaction. The cost of this is that the excitonic part also leads to much higher intrinsic dephasing due to scattering with phonons and impurities. But the bosonic stimulated scattering of the polaritons can overcome this dephasing, leading to phase coherence which is like that of a laser but which shows nonlinear effects at much lower carrier densities.

There are many other experiments with polaritons that can be described by the formalism of condensation, including vortices [28, 29], soliton propagation over long distances [30], and propagation without scattering in wires (channels) [31]. The bottom line is that these systems are easily described by the language of condensation, especially when taking into account the interplay between the condensate and excited states in the continuum [32], and insisting on the language of laser optics only adds difficulty. The condensation paradigm has become dominant because it is successful in describing the experiments.

When the polariton condensate is pumped resonantly by direct coupling to an external laser beam, the choice of language is more ambiguous. As we have seen, once a condensate is formed, it can be described by a classical wave equation, the Gross–Pitaevskii equation. One can also create such a state by direct pumping with a laser at the same frequency as a polariton state. In this case, the system can be far from equilibrium, and the contribution to the optical effects from particles in incoherent excited states can be unimportant. In general, many of the characteristics of an equilibrium condensate, such as the linear Bogoliubov excitations, persist even as the system moves into a not-fully-equilibrated state [33].

12.6 Photon Condensation

Recently, experiments on "Bose condensation of photons" have been reported [34]. What the authors mean by this is that they created a system with approximate number conservation of photons in a cavity over a short time, similar to the case of a polariton condensate in a microcavity, but in the limit of weak coupling, so that the particles were purely photons. In this limit, the photons thermalized not by

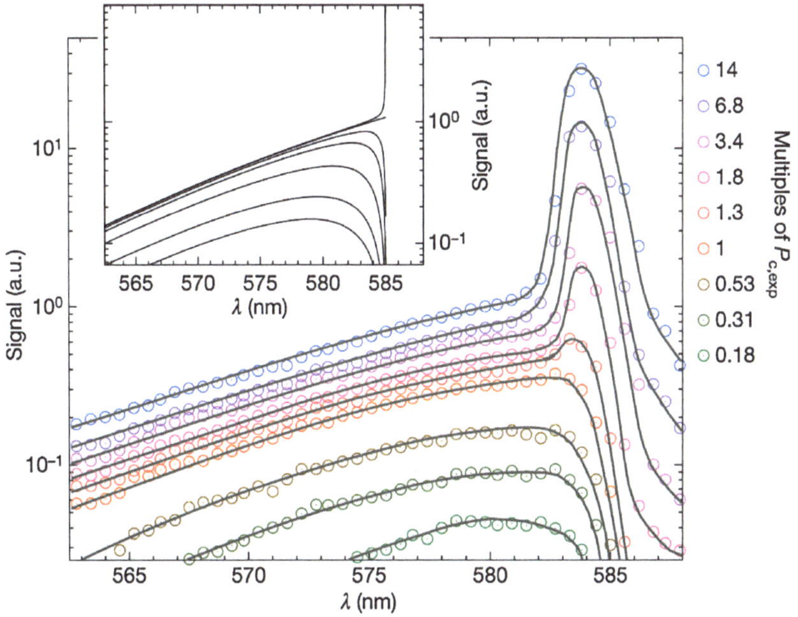

Fig. 12.6 Emission energy spectra from a photon gas in weak coupling as a function of density, with a "photon condensate" at high density (from [34]). *Inset*: the prediction for an ideal Bose gas at the same densities

collisions with each other, since they are effectively non-interacting, but instead by being absorbed and then reemitted by dye molecules in the cavity. The dye was chosen such that one photon absorbed led to one photon emitted at a different energy, with very little nonradiative loss; the energy of the emission depended on the temperature of the dye. As in the case of strong coupling, the photons have an effective mass determined by the cavity properties. At high photon density, the energy distribution of the photons thermalized and then showed a peak at $k = 0$ as in the case of a polariton condensate, as shown in Fig. 12.6.

This system is not the same as standard lasing because the number of photons was approximately conserved. However, it is not really the same as a standard condensate either. As discussed above, one of the main properties of standard condensation is spontaneous phase coherence of the condensate, which comes about via interactions of the particles. In the case of the "photon condensate," there is no interaction of the particles which could lead to phase locking, and the absorption and reemission of the photons by the dye molecules introduces random dephasing every time a photon undergoes this process. One thus has the situation that the number conservation at a given temperature makes the particles want to occupy the lowest available state, but the constant dephasing prevents them from actually becoming a phase-coherent condensate. This is seen in Fig. 12.6 in the relatively broad width of the peak at $k = 0$, with a full width at half maximum of around 2 nm, or 7 meV, much greater than the spectral resolution. This spectral width corresponds to a coherence time of about 100 fs. By contrast, coherence times of hundreds of picoseconds have been observed in polariton condensates [23].

The experiments with photons in weak coupling are therefore an interesting intermediate case in which the system is prevented from acquiring optical coherence either by lasing or polariton condensation in strong coupling. The system acts in many ways, like an ideal Bose gas, which, as is well known, cannot undergo true Bose–Einstein condensation [35].

12.7 Conclusions

Both standard lasing and polariton condensation are states in which there is a coherent polarization of the electrons in the matter, leading to spontaneous phase coherence of the electron polarization and the light emission. In the case of a condensate, the spontaneous coherence comes about due to a thermodynamic phase transition (even when not fully equilibrated) in which temperature is the controlling parameter, while in the case of a laser, the spontaneous coherence comes about far from equilibrium in a transition in which the pump power is the controlling parameter; a laser requires population inversion, while polariton condensation is inhibited by population inversion. Experiments have been performed in which each of these transitions can be seen separately in the same physical system, with very different behavior.

The recent experiments on photon condensation in the weak coupling limit do not exhibit phase coherence; in many ways this system acts like a polariton condensate, with an effective mass that gives a non-Planckian energy distribution, but the nearly noninteracting nature of the photons in this case means that they can only thermalize via incoherent interactions with a dye medium, which leads to strong dephasing.

Acknowledgements This work has been supported by the National Science Foundation through Grant DMR-0706331.

References

1. D.W. Snoke, *Solid State Physics: Essential Concepts*, (Pearson, USA, 2009)
2. J.M. Blatt, *Theory of Superconductivity*, (Academic Press, New York, 1964)
3. E.g.F. Kalzar, S. Etemad, G.L. Baker, Z. Mesaer, Solid State Comm. **63**, 1113 (1987)
4. M. Vaupel, K. Staliunas, C.O. Weiss, Phys. Rev. A **54**, 880 (1996)
5. B. Zhang, D.W. Snoke, and A.P. Heberle, in press
6. D.W. Snoke, G. Liu, and S.M. Girvin, Annals of Physics, in press.
7. H.T.C. Stoof, *Bose-Einstein Condensation*, A. Griffin, D.W. Snoke, S. Stringari, eds. (Cambridge University Press, Cambridge, 1995), p. 226; Phys. Rev. A **45**, 8398 (1992)
8. M.O. Scully, Phys. Rev. Lett. **82**, 3927 (1999)

9. C.W. Gardiner, P. Zoller, Phys. Rev. A **55**, 2902 (1997)
10. S.A. Moskalenko, D.W. Snoke, *Bose-Einstein Condensation of Excitons and Biexcitons and Coherent Nonlinear Optics with Excitons*, (Cambridge University Press, Cambridge, 2000)
11. V.E. Hartwell, D.W. Snoke, Phys. Rev. B **82**, 075307 (2010)
12. K. Yoshioka, T. Ideguchi, A. Mysyrowicz, M. Kuwata-Gonokami, Phys. Rev. B **82**, 041201 (2010)
13. For a review see D.W. Snoke, Dipole excitons in coupled quantum wells: toward an equilibrium exciton condensate, in *Quantum Gases: Finite Temperature and Nonequilibrium Dynamics*, N. Proukakis, S. Gardiner, and M. Davis, eds., (World Scientific Publishing), in press.
14. A.V. Kavokin, J.J. Baumberg, G. Malpuech, G.P. Laussy, *Microcavities*, (Oxford University Press, Oxford, 2007)
15. D.W. Snoke, Solid State Comm. **146**, 73 (2008)
16. D. Semkat, F. Richter, D. Kremp, G. Manzke, W.-D. Kraeft, K. Henneberger, Phys. Rev. B **80**, 155201 (2009)
17. R. Balili, V. Hartwell, D.W. Snoke, L. Pfeiffer, K. West, Science **316**, 1007 (2007)
18. O.L. Berman, Y.E. Lozovik, D.W. Snoke, Phys. Rev. B **77**, 155317 (2008)
19. H. Deng, G. Weihs, C. Santori, J. Bloch, Y. Yamamoto, Science **298**, 199 (2002)
20. H. Deng, G. Weihs, D. Snoke, J. Bloch, Y. Yamamoto, Proc. Nat. Acad. Sci. **100**, 15318 (2003)
21. R. Balili, B. Nelsen, D.W. Snoke, L. Pfeiffer, K. West, Phys. Rev. B **79**, 075319 (2009)
22. D. Bajoni, P. Senellart, A. Lemaitre, J. Bloch, Phys. Rev. B **76**, 201305R (2007)
23. A.P.D. Love, D.N. Krizhanovskii, D.M. Whittaker, R. Bouchekioua, D. Sanvitto, S. Al Rizeiqi, R. Bradley, M.S. Skolnick, P.R. Eastham, R. André, Le Si Dang, Phys. Rev. Lett. **101**, 067404 (2008)
24. D. Bajoni, P. Senellart, E. Wertz, I. Sagnes, A. Miard, A. Lemaître, Jacqueline Bloch, Phys. Rev. Lett. **100**, 047401 (2008)
25. B. Nelsen, R. Balili, D.W. Snoke, L. Pfeiffer, K. West, J. Applied Phys. **105**, 122414 (2009)
26. K.G. Lagoudakis, B. Pietka, M. Wouters, R. André, B. Deveaud-Plédran, Phys. Rev. Lett. **105**, 120403 (2010)
27. L. Ferrier, A. Amo, J. Bloch, private communication.
28. K.G. Lagoudakis, T. Ostatnicky, A.V. Kavokin, Y.G. Rubo, R. André, B. Deveaud-Plédran, Science **326**, 974 (2009)
29. D.N. Krizhanovskii, D.M. Whittaker, R.A. Bradley, K. Guda, D. Sarkar, D. Sanvitto, L. Viña, E. Cerda, P. Santos, K. Biermann, R. Hey, M.S. Skolnick, Phys. Rev. Lett. **104**, 126402 (2010)
30. A. Amo, D. Sanvitto, F.P. Laussy, D. Ballarini, E. del Valle, M.D. Martin, A. Lemâitre, J. Bloch, D.N. Krizhanovskii, M.S. Skolnick, C. Tejedor, L. Viña, Nature **457**, 291 (2009)
31. E.Wertz, L. Ferrier, D.D. Solnyshkov, R. Johne, D. Sanvitto, A. Lemaître, I. Sagnes, R. Grousson, A.V. Kavokin, P. Senellart, G. Malpuech, J. Bloch, Spontaneous formation and optical manipulation of extended polariton condensates, Nat. Phys. **6**, 860 (2010)
32. H. Deng, H. Haug, Y. Yamamoto, Rev. Mod. Phys. **82**, 1489 (2010)
33. M. Assmann, J.-S. Tempela, F. Veita, M. Bayer, A. Rahimi-Iman, A. Löffler, S. Höfling, S. Reitzenstein, L. Worschech, A. Forchel, Proc. Nat. Acad. Sci. **108**, 1804 (2011)
34. J. Klaers, J. Schmitt, F. Vewinger, M. Weitz, Nature **468**, 545 (2010)
35. M. Combescot, D.W. Snoke, Phys. Rev. B **78**, 144303 (2008); and references therein

Chapter 13
The Future Prospects of Room-Temperature Polariton Lasers

Gabriel Christmann and Jeremy J. Baumberg

Abstract Polariton lasers offer great opportunities for a new generation of low-threshold coherent light emitters and sensors. However, operation at room temperature is challenging as thermal broadening is detrimental to the strong coupling regime. In this chapter, we present different approaches for the realization of room-temperature polariton lasers. We discuss the importance of material choices and sample design and present successful demonstration of polariton lasing using III-nitride semiconductors. Finally, we present some perspectives for the improvement of current structures and discuss issues related to electrical injection.

13.1 Introduction

Since the first demonstration of the semiconductor laser [1–4], such devices have become widespread in everyday life thanks to their small size, high efficiency, and low cost. Success in their many applications, which include high-bit-rate optical communications, optical data storage, high-resolution printing and more, has been enabled by continuous developments in improving performance and, in particular, by reducing the lasing threshold. From the required population inversion at the heart of lasing, one natural approach has been to reduce the electronic density of states and the sample volume by utilising nanostructures. After the original bulk *p-n* junctions were demonstrated, quantum well and then quantum dot lasers have been developed, allowing dramatic decreases in the device threshold current. A different focus for laser diodes (LDs) is in optimizing their optical cavity. Considerable work thus focussed on the optimization of the waveguide cavities of edge-emitting LDs. A breakthrough was achieved with the development of distributed Bragg reflector

G. Christmann (✉) · J.J. Baumberg
Nanophotonics Centre, Cavendish Laboratory, University of Cambridge, Cambridge CB3 0HE, UK
e-mail: gprmc2@cam.ac.uk; jjb12@cam.ac.uk

D. Sanvitto and V. Timofeev (eds.), *Exciton Polaritons in Microcavities*, Springer Series in Solid-State Sciences 172, DOI 10.1007/978-3-642-24186-4_13,

(DBR) microcavities (MCs) which allowed for much stronger optical confinement and very low optical losses. This finally enabled the realization of vertical cavity surface emitting lasers (VCSELs) [5]. In addition to a lower threshold, VCSELs posses other advantages when compared to edge-emitting LDs, such as a better mode quality, a more compact design, and vertical emission which allows their use in arrays. However, possibly the strongest commercial advantage was the ability to test large numbers of devices at an early stage on chip before packaging.

In spite of all these improvements, LD and VCSEL thresholds are intrinsically limited by the population inversion/stimulated emission mechanism. In this context, Imamoğlu and coworkers proposed a new mechanism to realize an inversionless laser, the polariton laser [6]. In such a device, the idea is to move away from the Fermi statistics that governs the lasing threshold of a conventional laser and use the bosonic nature of polaritons. The consequence is that the relaxation of polaritons to the ground state is stimulated by the ground state occupation, and when the corresponding rate becomes faster than the loss rate (mainly through polariton emission), the system builds a up a massive ground state population and acquires coherence. Hence, the emission properties are very similar to those of a VCSEL: coherent and directional light emission, although the coherence exists in both the matter and light parts of the system.

Optically pumped working devices have been demonstrated in a whole variety of systems now, including II-tellurides [7,8], III-arsenides [9–11], III-nitrides [12,13], and organic MCs [14]. In the case of II-tellurides and III-arsenides, the lasing phenomenon was only observed at low temperature, whereas the higher exciton energy of III-nitrides and organic semiconductors allows for polariton lasing at room temperature. In parallel, the recent development of III-arsenide polariton diodes opens very interesting potential for the realization of electrically pumped devices [15–17]. Such devices are discussed in the chapter by Tsintzos and Savvidis.

13.2 Strong Coupling Regime at Room Temperature

13.2.1 Generalities on the Strong Coupling Regime

In this section some general features of the strong coupling regime will be briefly presented, in particular, how the MC design affects the characteristics of the strong coupling regime. For more details, the reader is referred to a recent book [18]. In an MC in the strong coupling regime, an excitonic transition is coupled to the Fabry-Perot mode of an optical cavity. The system is designed so that an emitted photon is reabsorbed before exiting the optical cavity. As a result, the resonantly coupled exciton and confined optical mode periodically exchange energy [19]. The new quasiparticles resulting from this coupling, called cavity polaritons, are a quantum superposition of the cavity photon and the exciton. They obey Bose statistics and their dispersion curves, the so-called lower (LPB) and upper (UPB) polariton

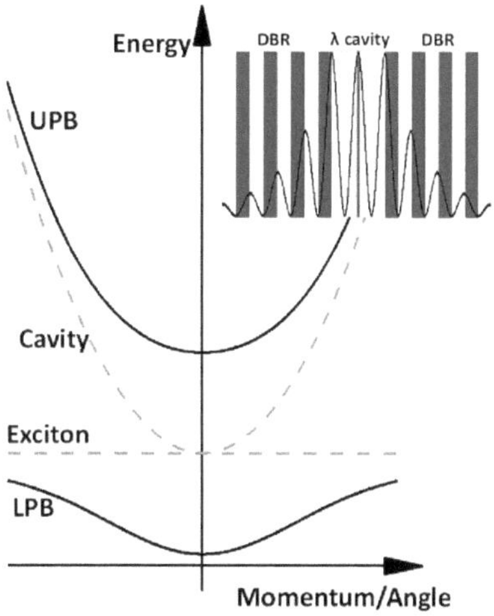

Fig. 13.1 Typical lower (LPB) and upper (UPB) polariton branch dispersions (*black lines*), the uncoupled exciton and cavity dispersions are also displayed (*gray dashed lines*). Inset: typical MC design sketch of a λ MC with one QW at the center. Black line displays the electric field intensity

branches, arise from the mixing of the near-parabolic Fabry-Perot mode and the almost dispersionless exciton mode (Fig. 13.1a). The typical MC design (inset of Fig. 13.1) is composed of a cavity spacer layer whose thickness is a multiple of half wavelengths, surrounded by DBRs. The exciton-containing structure is generally quantum wells which are placed at the antinodes of the cavity light field to maximize the exciton-photon interaction, but bulk layers can also be used [20].

The vacuum Rabi splitting (Ω_{VRS}) for a QW MC is given by [21]:

$$\hbar\Omega_{\mathrm{VRS}} \approx 2\hbar\left(\frac{2\Gamma_0 c N_{\mathrm{QW}}}{n_{\mathrm{c}} L_{\mathrm{eff}}}\right)^{(1/2)}, \tag{13.1}$$

where N_{QW} is the number of QWs in the cavity; $\hbar\Gamma_0$ is the radiative width of a free exciton, which can be expressed in terms of the QW exciton oscillator strength per unit area [21]; L_{eff} is the cavity effective length (including penetration in the DBRs [22]); and n_{c} is the cavity refractive index. From this equation, designs favoring large Rabi splitting require a large number of QWs with strong oscillator strength while keeping the optical length of the cavity as short as possible.

Important parameters in this mixing are the linewidths of the uncoupled cavity mode and the uncoupled exciton. Broadened transitions are detrimental to the strong coupling regime (SCR). Indeed, if the polariton linewidths become of the order of the Rabi splitting, this regime is washed out. In the case of homogeneous (Lorentzian) lineshapes, it can be shown that the resulting broadening for the polaritons is the average between the cavity mode and the exciton linewidth [23]. The case of inhomogeneous broadening is slightly different because of the different lineshape, which is generally considered to be Gaussian [24]. In consequence of its rapid exponential decay, the effect on the Rabi splitting from the inhomogeneous broadening is nullified [24]. The inhomogeneous broadening will however seriously impact the SCR if it becomes of the order of the Rabi splitting.

13.2.2 Designs for Room Temperature

Before considering polariton lasing at room temperature, it is necessary to achieve the strong coupling regime at room temperature. Indeed, when raising the temperature, the resulting thermal energy is balanced against the exciton binding energy. It is therefore necessary to use a material with a strong exciton binding energy. It is also important that the total broadening of the excitonic transition remains small enough compared to the Rabi splitting in order to remain in the strong coupling regime. The thermal energy at room temperature will lead to a significant increase of the homogeneous broadening. The QW inhomogeneous broadening can also become an important issue in some cases for materials with strong exciton binding energies, and one has to be careful that the gains in oscillator strength and exciton binding energy are nullified by such a broadening resulting from material growth issues.

Table 13.1 groups some properties of inorganic semiconductor materials that are commonly used in strong coupling MCs: III-arsenides II-tellurides, III-nitrides, and ZnO. Although III-arsenide QWs and II-telluride semiconductors can exhibit excitonic effects at room temperature, it appears that only GaN and ZnO have bulk exciton binding energy above the thermal energy at room temperature, which make them the current materials of choice for SCR nonlinearities at room temperature [35]. Their large oscillator strength also help in increasing the important Rabi splitting [36, 37]. Furthermore, their smaller Bohr radius means a higher saturation density, which favors the observation of polariton lasing. This also allows use of thinner QWs, and thus putting more of them in a given thickness, which increases the Rabi splitting.

On the other hand, these two materials create detrimental issues compared to III-arsenides. First, they exhibit a wurtzite crystalline structure. As a result, heterostructures grown along the *c* axis exhibit strong spontaneous and piezoelectric fields that lead to quantum confined Stark effects (QCSE) [30, 38]. This leads to a dramatic reduction of the exciton binding energy and oscillator strength in wide wells and insists on the use of narrow wells which are a lot more sensitive to interface and alloy composition disorder. The growth of heterostructures on nonpolar facets allows potential addressing of this problem but is challenging and

Table 13.1 Summary of III-nitride and other semiconductor compound excitonic properties

	III-Arsenides	II-Tellurides	III-Nitrides[a]	ZnO[a]
Binding energy (meV)				
Bulk:	4.9 [25]	11 [25]	25 − 27 [28]	60 [31]
Quantum well:	7 − 9 [26]	~22 [27]	~50 [29]	~90 [30]
Bohr radius (nm)				
Bulk:	11.2 [25]	12.2 [25]	2.8	2.3 [31]
Quantum well:	~10 [32]	NA	~1.7 [29]	~2 [30]
Quantum well oscillator strength (cm^{-2})	4.8×10^{12} [33]	2.3×10^{13} [27]	4.8×10^{13} [34]	NA

[a] QW values are for *c* plane (polar) growth

the resulting structures show strong anisotropy in the optical properties due to the anisotropy of the unit cell.

Another issue comes from the technological challenges in the growth of such materials due to the reduced availability of bulk substrates. Thus, the fabrication generally relies on heteroepitaxy on foreign substrates such as sapphire. The lattice mismatch results in an increased defect density which is, as an illustration, typically between 10^8 and 10^9 cm^{-2} for a 2 μm thick GaN buffer grown on sapphire. Defects are important not only for the excitonic properties but also for the cavity uniformity.

Another source of defects in the case of MC devices is the growth of DBRs below the active region. In the case of III-arsenide semiconductors, the nearly lattice-matched AlAs/GaAs system offers a very efficient system to grow high-reflectivity DBRs. However, for III-nitrides, the equivalent AlN/GaN system exhibits a 2.4% lattice mismatch which leads to built-in strain and thus dislocations and cracks during the growth. Working with AlGaN alloys can reduce the built in strain, but it is at the cost of the refractive index contrast. An alternative approach, which has proven the most successful for epitaxial III-nitride DBRs, is to use $Al_{0.93}In_{0.17}N$ lattice matched to GaN, which combines both significant refractive index contrast and lattice matching [39]. More details on the strategy to grow III-nitride DBRs and MCs can be found in the review article [40]. The case of ZnO is more problematic: the best quality epitaxial ZnO is grown by molecular beam epitaxy (MBE), but MBE grown DBRs using ZnO as one of the materials have not been reported until very recently [41]. Consequently, MBE-grown ZnO MCs generally use a III-nitride bottom DBR and a dielectric top one [42,43]. One can also use other growth techniques such as pulsed laser deposition or RF sputtering to grow the DBRs and the active ZnO regions, at the cost of reduced material quality. ZnO MCs grown with such techniques with sufficiently good excitonic properties for SCR were, however, reported [44,45].

Alternatively, it is possible to avoid the problems caused by the epitaxy of DBRs by using two dielectric mirrors. This requires growth of a cavity layer above a sacrificial layer, deposition of a top dielectric DBR, removal of the sacrificial layer to release the cavity layer, and finally, deposition of the second DBR underneath. Such a process is technologically challenging and does not completely suppress the material issues. Indeed, the growth of the sacrificial layer can also add defects to the structure, and the liftoff process generally creates surface roughness issues. This full dielectric approach has been successfully employed to produce high-quality III-nitride based MCs [46].

All the above-mentioned issues affect the epitaxial quality of the cavity layer and lead to inhomogenous broadening. Indeed, although in very specific cases QWs with low temperature emission linewidths of a few meV can be grown, both of the GaN and ZnO [47, 48] heterostructures in realistic cavity layers have inhomogeneous broadenings of at least 20 meV [29, 34]. In this context, to avoid important significant inhomogeneous broadening, which is detrimental to the strong coupling regime, it has been profitable to use a bulk active region in which the inhomogeneous broadening is less sensitive to defects.

13.2.3 Experimental Realizations

13.2.3.1 III-Nitrides

Bulk Active Region

Due to the difficulty in growth of structures with high-reflectivity DBRs and high-quality QWs, early attempts to demonstrate strong coupling in III-nitrides used low quality (Q) factor MCs and bulk active regions. In 2003, Antoine-Vincent and coworkers demonstrated the strong coupling regime at cryogenic temperatures in a $\lambda/2$ bulk GaN MC [49]. The sample used the reflectivity of the silicon substrate topped with a 1.5 pair AlN/AlGaN DBR as a bottom mirror and a 4 pair SiO_2/Si_3N_4 DBR as a top one. The same group finally succeeded in observing the SCR at room temperature 2 years later in a very similar structure using an Al top mirror [50]. These samples illustrated that the large oscillator strength of GaN do allow one to reach the strong coupling regime even in samples exhibiting very low Q factors and significant broadening. However, to observe nonlinear effects such as polariton lasing, it is necessary to have a high Q factor MC. Indeed, one needs a long polariton lifetime in order to get polariton relaxation faster than the radiative decay. Subsequent efforts by this group to fabricate higher Q factor MCs in the strong coupling regime [51, 52] remained somewhat limited to fairly low Q values (190). Others have fabricated similar hybrid bulk GaN MC structures and observed SCR at room temperature, but here as well the Q factor was low (160) [53]. As a result, to date, no polariton lasing has been observed from such low Q factor samples.

To realize our bulk GaN MC for SCR at room temperature and polariton lasing, we aimed to produce structures with high-reflectivity DBRs to get a Q factor as high as possible. The samples were grown by metalorganic vapor phase epitaxy (MOVPE) and consist first of a standard 3-μm thick GaN buffer layer, followed by a strain-relieving layer made of a 10-pair AlN/GaN superlattice (3–4 nm), and a 1-μm thick AlGaN layer. The microcavity itself is made of a $\frac{3\lambda}{2}$ GaN cavity layer ($\lambda \sim 360$ nm) sandwiched between a 35-pair lattice-matched $Al_{0.85}In_{0.15}N/Al_{0.2}Ga_{0.8}N$ DBR and a 10-pair SiO_2/Si_3N_4 Bragg mirror [54]. The use of such lattice-matched $Al_{0.85}In_{0.15}N/Al_{0.2}Ga_{0.8}N$ material systems has allowed the growth of high-reflectivity (>99%) mirrors [55] and preserves good material quality for the growth of the bulk active region. Furthermore, MCs using similar lattice-matched $Al_{0.83}In_{0.17}N/GaN$ have been shown to exhibit very high Q factors in the range of 2,800 up to 6,400 when measured locally [56, 57].

To check that the sample is in the strong coupling regime, angle-resolved photoluminescence (PL) is performed at room temperature under transverse electric polarization (Fig. 13.2a). Two branches (bold lines) are clearly observed and exhibit an anticrossing. The low energy PL line exhibit an asymptotic approach toward the exciton X_A at increasing angle, while the second stays above X_C and goes to high energy at large angles. This behavior is characteristic of the strong coupling regime,

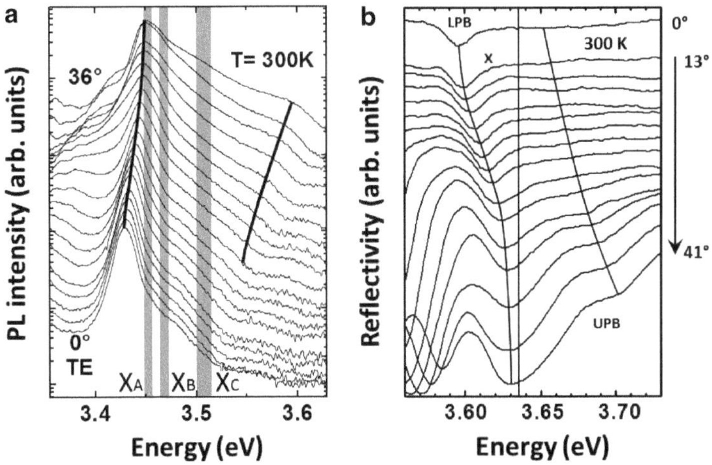

Fig. 13.2 (**a**) Angle-resolved PL spectra acquired on the bulk GaN MC at room temperature. Black curves are guides to the eye. Gray vertical bars display the A, B, and C exciton energies. (**b**) Angle-resolved reflectivity spectra acquired on an $Al_{0.17}Ga_{0.83}N$/GaN MQW MC. Black curves show the lower (LPB) and upper (UPB) polariton branches. Vertical line displays the uncoupled exciton energy. *Results from EPFL*

and in fact, in the present case, a more detailed analysis of all the shoulder features in the PL spectra between the X_A and X_C analysis reveals that strong coupling between the cavity mode and the three X_A, X_B, X_C excitons should be considered [54].

To sum up, such bulk GaN MC structures combine strong coupling at room temperature and potentially very high Q factors. Such samples are therefore very promising for the observation of polariton lasing at room temperature.

QW Active Region

Using QWs as an active region provides some advantages because the quantum confinement in such quantum wells increases the exciton binding energy and the oscillator strength. The first report on the room-temperature strong coupling regime in III-nitride systems was achieved by Tawara and coworkers [58]. Their MC structure used a wafer bonding technique with two SiO_2/ZrO_2 DBRs, which allowed them to obtain a fairly high Q factor of 400. One puzzling aspect to this report is that SCR was claimed to be found with InGaN/GaN QWs which are known to exhibit very strong inhomogeneous broadening due to In composition inhomogeneities. Indeed, the authors report a 200-meV QW emission linewidth, a value far larger than the reported 17-meV Rabi splitting. Such a broadening should prevent the observation of the SCR according to theory [24]. Investigation of the SCR was attempted by moving along the sample growth gradient, and the two signatures were only visible very close to zero detuning. No angle-resolved measurements were done to check that the system is in the SCR. It is therefore not clear whether the sample is indeed in the strong coupling regime or the observations comes from another effect, possibly photonic disorder in the cavity [56].

Our own design approach for observing room-temperature SCR in QW MCs was first to chose appropriate QWs, which means with reasonably low-emission linewidths. The InGaN/GaN system is the most commonly employed material system but is characterized with very strong inhomogeneous broadening. On the other hand, the GaN/AlGaN system allows narrow emission linewidths (down to 5 meV at low temperature) [55]. We thus chose to use GaN/$Al_{0.2}Ga_{0.8}N$ QWs which combine a room-temperature emission linewidth of 42 meV and a sufficiently high barrier to ensure good confinement and a limited contribution of barrier absorption to the upper polariton. The samples we studied used a 35-pair lattice-matched $Al_{0.85}In_{0.15}N/Al_{0.2}Ga_{0.8}N$ DBR, an $Al_{0.2}Ga_{0.8}N$ cavity layer, and finally, a 13-pair SiO_2/Si_3N_4 dielectric DBR. The structures are grown on a strain-relieving template using two superlattices in order to avoid the formation of cracks. Using a 67 period GaN (1.2 nm)/$Al_{0.2}Ga_{0.8}N$ (3.6 nm) multiple QW (MQW) active region allowed a significant increase in the light-matter interaction compared to previous results with only two sets of three quantum wells [13, 34]. The resulting sample was characterized by means of angle-resolved reflectivity at room temperature (Fig. 13.2b). A clear anticrossing is observed, exhibiting a Rabi splitting of 56 meV, a value almost double of previous samples using two sets of three QWs [34].

13.2.3.2 ZnO

Several groups have reported SCR at room temperature (RT) in ZnO materials. First, Shimada et al. reported a hybrid structure grown by MBE using an $Al_{0.5}Ga_{0.5}N$/GaN bottom DBR, a bulk λ ZnO cavity, and a top SiO_2/Si_3N_4 DBR. A Rabi splitting of 50 meV was extracted, comparable to III-nitride MC values [42]. Faure et al. also reported SCR at RT in a very similar structure: an MBE-grown hybrid MC with a bottom $Al_{0.3}Ga_{0.7}N$/AlN DBR, a $\lambda/4$ ZnO active region, and a top SiO_2/Si_3N_4 DBR. The Rabi splitting value is smaller (30 meV) due to the thinner active region [59]. Finally, Nakayama et al. observed the SCR regime at low temperature in a slightly different structure using two dielectric HfO_2/SiO_2 DBRs and an active region of ZnO deposited by pulsed laser deposition (PLD). In spite of the lower quality of epitaxial films grown by PLD compared with MBE, they were still able to observe a Rabi splitting of 80 meV at 10 K [44]. However, they did not report any SCR at room temperature. All these results exemplify the great potential of ZnO for SCR applications at room temperature. It is, however, worth pointing out that the samples presented here have low Q factors, which would be problematic for polariton lasing observation.

13.3 Room-Temperature Polariton Lasing

To the best of our knowledge at the present time, room-temperature polariton lasing has been only observed in GaN-based systems and organic systems. As this chapter discusses inorganic systems, only the results on III-nitride systems will be presented. Polariton lasing in organic systems is thoroughly discussed in another chapter of this book.

13.3.1 Bulk Microcavities

The nonlinear emission properties of the GaN bulk microcavity were characterized by nonresonant photoluminescence experiments. The excitation source used was tuneable frequency-doubled visible emission from a 250-kHz optical parametric amplifier providing 150-fs long pulses. The sample is excited at a 20° incidence angle, the spot size is 60 μm, and the pumping energy is tuned in order to match the first dielectric distributed Bragg reflector reflectivity minimum (this corresponds to an energy of 4.14 eV). The photoluminescence is collected at specific angles with a 3° selection and focused onto a UV collection multimode fiber. The photoluminescence is sent to a 0.5 m monochromator combined with liquid nitrogen cooled charge-coupled device camera. All the measurements are performed at room temperature.

The integrated output intensity collected at normal incidence for these conditions is shown as a function of pump intensity in Fig. 13.3a with slightly positive detuning. A clear nonlinear behavior is found for the emission at $\lambda \sim 365$ nm, with an increase of over 10^3 at the critical threshold (average pump power at threshold, 1.0 mW), as seen in Fig. 13.3b. A careful estimation of the three-dimensional carrier density yields $N_{3D} \sim 2.2 \times 10^{18}$ cm^{-3}, which is an order of magnitude below both the Mott density ($\sim 1-2 \times 10^{19}$ cm^{-3} in GaN at room temperature [60]), and the transparency density for bulk GaN, which provides the lower limit for conventional lasing.

Commensurate with the nonlinear power increase is a sudden collapse in emission linewidth at threshold down to the resolution limit of the spectrometer (1.7 meV). Note that when increasing further the excitation power, the emission linewidth starts to increase again. This observation results from decoherence induced by polariton–polariton interactions. Note also that the precise peak energy position above threshold is extremely sensitive to the spatial alignment on the sample. Furthermore, spectra often show multiple line emission, for example, L_{1-3} in Fig. 13.3a with the spacing between these sharp emission peaks being irregular and varying over both time and sample position. These observations originate from the photonic disorder that localizes the polariton condensate laterally, with the different modes first attaining unity occupancy on different laser shots statistically. As the pump power increases, the peak emission energy blue shifts below threshold (Fig. 13.3d). Above threshold, the situation is different as some lines blueshift while

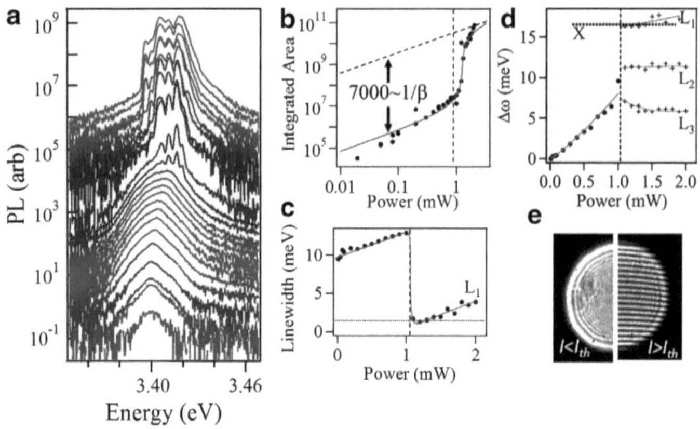

Fig. 13.3 (**a**) Emission spectra at average pump powers from 20 μW to 2 mW at 0°, shifted vertically for clarity. (**b**) Integrated intensity versus pump power (*solid points*), with fit. (**c**) Spectral linewidth and (**d**) energy shift of peak emission with pump power (for different modes L_{1-3}). (**e**) Interference of far-field emission cone through a slightly misaligned Michelson interferometer below and above threshold. (*Results from University of Southampton*)

some other stays at constant energy. Such a blue shift is expected from polariton–polariton interactions. Near-field imaging of the emission shows a strong spatial collapse of the emission above threshold due to photonic disorder that leads to polariton lasing only on the highest-Q-factor areas within the excitation spot [56]. Finally, measurements using a Michelson interferometer demonstrate the onset of first-order coherence at threshold as exemplified in Fig. 13.3e. Interference fringes appear above threshold, and they can be observed for time delays up to 700 fs after the excitation.

To summarize, a nonlinear emission threshold of the lower polariton emission under nonresonant excitation has been observed from the bulk GaN microcavity. In addition, the appearance of first-order coherence has been reported. These features demonstrated polariton lasing at room temperature for the first time [12].

A surprising feature observed during the measurements on this bulk microcavity is that, above threshold, no signature of polarization is observed. This contrasts with other measurements performed on II-VI and III-arsenide macrocavities where a linear polarization pinned along the crystallographic axes by strain is reported [8, 10]. The absence of a polarization in the coherent regime of such a polariton system is at first sight surprising. As the experiment is performed under pulsed excitation, the polarization measurements are averaged over many experimental realizations. It is thus interesting to measure the polarization properties for individual experimental realizations.

There is much deeper underlying physics in the way polariton condensates gain their polarization. Phase transitions are characterized by a spontaneous symmetry breaking, which is characterized by an order parameter. Its value is zero for the noncondensed (symmetric) phase and becomes nonzero for the condensed (broken symmetry) phase. Polaritons are spinor quasiparticles with two possible spin projections: up (and down) corresponding to right (and left) circular polarizations of the emitted light, respectively. Therefore, the order parameter $\psi(\mathbf{r})$ for exciton-polariton Bose-Einstein condensation possesses two components: $\psi(\mathbf{r}) = \begin{bmatrix} \psi_\uparrow(\mathbf{r}) \\ \psi_\downarrow(\mathbf{r}) \end{bmatrix}$, where $\psi_\uparrow(\mathbf{r})$ and $\psi_\downarrow(\mathbf{r})$ are the complex spin-up and spin-down wave functions, respectively. The three-dimensional polarization vector, $\mathbf{S}$ (termed the Stokes vector in classical optics, or pseudospin in quantum mechanics), is linked to these wave functions through $S_x = \mathrm{Re}\left(\psi_\uparrow^* \psi_\downarrow\right)$, $S_y = \mathrm{Im}\left(\psi_\uparrow^* \psi_\downarrow\right)$, and $S_z = \frac{1}{2}\left(|\psi_\uparrow|^2 - |\psi_\downarrow|^2\right)$. The absolute polarization degree of the condensate is $\rho = \left(|\psi_\uparrow|^2 + |\psi_\downarrow|^2\right)/N_0$, where N_0 is the occupation number of the condensate. It is linked with linear, diagonal, and circular polarization degrees (ρ_l, ρ_d, and ρ_c) by $\rho = \sqrt{\rho_\mathrm{l}^2 + \rho_\mathrm{d}^2 + \rho_\mathrm{c}^2}$, where $\rho_{\mathrm{l,d,c}} = 2S_{x,y,z}/N_0$.

The experimental conditions used are quite similar to the previous experiments. However, this time the sample is pumped at an angle of 30° and negatively detuned (by < -30 meV). To resolve the polarization of the emission for each experimental realization of the condensate arising from each excitation pulse, the emission is polarization split and both orthogonal components (linear, diagonal, or circular) are focused either onto the input slit of a streak camera operated in single-shot mode or onto the cathodes of balanced photomultipliers. Previous experiments have shown that the maximum lifetime of electronic excitations in the sample is 35 ps (with an exponential decay time) [12]. Hence, between pulses, the sample completely recovers (successive pulses are separated by 4.2 μs), and each pulse forms a new experiment. For streak camera measurements, the polarization-split emission is focused onto different positions on the input slit of the streak camera, and their optical paths made equal so that the two polarization components of each pulse appear at the same time position on the streak image. The streak camera is manually triggered in single-shot mode, with a gain sufficient to record the relatively weak emission, allowing extraction of each polarization component averaged over each laser shot. Equivalently, for measurements with photomultipliers, the overall time resolution of the system is $< 1\,\mu$s, which allows extraction of the intensity of each polarization component for each pulse over thousands of successive shots. The relative intensities collected on each photomultiplier allows extraction of the polarization degree of the considered basis. Observations in Fig. 13.4a (for horizontal/vertical polarization) and Fig. 13.4b (for right/left circular polarization) present what is recorded using the streak camera method. Bright spots of varying intensity are observed on the streak camera, and dots at the same horizontal position correspond to the two intensity components along the considered basis. The polarization degree can be extracted from these measurements using, for example, $\rho_\mathrm{l} = \frac{I_\mathrm{H} - I_\mathrm{V}}{I_\mathrm{H} + I_\mathrm{V}}$ for linear polarization, where I_H and I_V are the intensities of the vertically and horizontally polarized emissions, respectively. Figure 13.4c shows the typical behavior of the linear polarization degree over several experiments

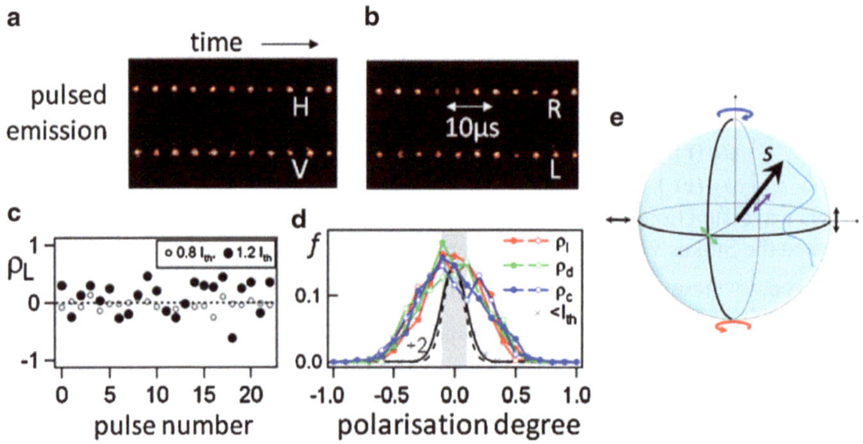

Fig. 13.4 Above-threshold polarization-resolved emission when analyzing along (**a**) horizontal/vertical and (**b**) right/left circular bases. (**c**) Extracted linear polarization degree showing stochastic variation from pulse to pulse above threshold. (**d**) Histogram of the fraction of each polarization state along linear, diagonal, and circular bases of nearly 2,000 polariton condensates. Open and closed circles show repeated measurement with reversed polarization split (e.g., H/V and L/R), while crosses show below-threshold unpolarized emission statistics (curve divided by 2 to fit on scale, within detection sensitivity shaded gray and scaled dashed line). (**e**) Diffusion of the pseudospin, **S** (*Stokes polarization vector*) of a polariton condensate on the Poincaré sphere. *Results from University of Cambridge*

below and above threshold. It is seen that below threshold, the polarization degree remains close to zero, while above threshold, from one experiment to another, this polarization degree exhibits nonzero values which randomly change from one pulse to the other. Similar results are obtained for diagonal and circular bases. Repeating the experiment for several hundred pulses in each polarization basis allows the construction of a polarization histogram (Fig. 13.4d). Below threshold, the microcavity emission is unpolarized, exhibiting thermal noise around zero absolute polarization. Above threshold, the emission is found to be instantaneously polarized but with no preferential orientation. The magnitude of the mean polarization is $\sim$25% for each basis (given by the standard deviations in Fig. 13.4d), giving a total mean polarization of $\sim$43% just above threshold. The mean polarization measured for a single shot is $<$ 100% due to the random walk of the polarization vector on a time scale given by the coherence time of the condensate (Fig. 13.4e).

To summarize, these results show that the bosonic exciton polaritons in GaN form a phase coherent state above a characteristic density and exhibit spontaneous symmetry breaking at room temperature. This spontaneous symmetry breaking has not been observed for Bose-Einstein condensates in II–VI and III-arsenide microcavities. The anisotropy of these latter systems forces them to choose always the same linear polarization, aligned with a crystallographic axis. In contrast, the use of this isotropic bulk GaN microcavity allowed observation of this spontaneous symmetry breaking phenomenon which is at the base of the theory of first-order phase transitions. The spontaneous symmetry breaking associated with the buildup of an order parameter is a central feature of Bose–Einstein condensation.

13.3.2 Multiple Quantum Well Microcavities

Room-temperature nonlinear properties of the $GaN/Al_{0.2}Ga_{0.8}N$ MQW sample have also been investigated using a frequency quadrupled Nd:YAG laser (λ_{laser} = 266 nm). The excitation is thus strongly nonresonant (more than 1 eV above the QW energy). The repetition rate is 8.52 kHz, and the pulse duration is 500 ps. The excitation angle is 44°, and the spot size is ~50 μm. Measurements were performed on regions exhibiting relatively small negative detunings ($\delta = -15$ to -20 meV) compared to the vacuum field Rabi splitting ($\Omega_{VRS} = 56$ meV). Choosing small detunings helps to discriminate emission from polariton states and from uncoupled states. A series of emission spectra measured at normal incidence at average pump power densities ranging from 0.16 to 28.8 Wcm^{-2} is displayed in Fig. 13.5a. A transition toward a nonlinear emission regime is clearly observed at a relatively low threshold pump power density P_{thr} = 17.7 Wcm^{-2}. This nonlinear emission is accompanied by a sharp increase in the integrated emission intensity (Fig. 13.5b) and a strong emission line narrowing as the measured emission linewidth decreases from ~15 meV just below threshold down to 460 μeV (single mode emission peak, see Fig. 13.5d). The latter is a signature of the significant increase in the coherence time. In addition, in Fig. 13.5c, it is seen that the emission line exhibits

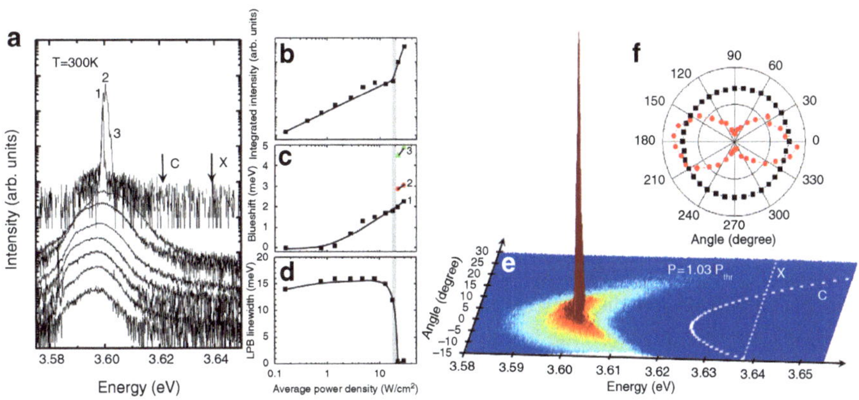

Fig. 13.5 (**a**) Semilogarithmic plot displaying room-temperature emission spectra at average pump power densities ranging from 0.16 to 28.8 Wcm^{-2} at $k_{\parallel} = 0$. The uncoupled cavity (C) and exciton (X) modes are also reported (*arrows*). (**b**) Integrated output intensity. (**c**) Blueshift of the emission. The three lines above threshold correspond to the three peaks labelled in (**a**). (**d**) Emission linewidth. In (**b**), (**c**), and (**d**), the lines are guides for the eyes. (**e**) 3D representation of the angle-resolved emission pattern just above threshold. Dashed white lines display the uncoupled cavity (C) and exciton (X) modes. (**f**) Integrated intensity versus analyzer angle below (*black*) and above (*red*) threshold. *Results from EPFL*

a slight blueshift ($E_{\text{shift}} = 1.8\,\text{meV}$ at threshold) ascribed to polariton–polariton interactions. Note also that above threshold, the nonlinear emission peak remains far from the position of the uncoupled cavity mode (arrow on Fig. 13.5a), which indicates that the nonlinear emission comes from polariton states. This clearly demonstrates polariton lasing at room temperature in this MQW MC. It is worth pointing out that the threshold value is strongly detuning dependent because of a competition between the thermodynamic and kinetic regimes [61].

The angular emission pattern of the microcavity above threshold is reported in Fig. 13.5e, which is reconstructed from individual angle-resolved measurements. It is seen that the above-threshold emission occurs at $k_{\parallel} = 0$, as in II-VI and III-arsenide microcavities [8, 10]. The angular aperture is estimated to be around 6°. Similar to what is observed in II-VI and III-arsenide QW microcavities, the emission is linearly polarized above threshold (polarization degree $\sim$80%) while it is unpolarized below threshold (Fig. 13.5f). The linear polarization is pinned on the crystallographic axes ($[1\bar{1}00]$ and $[2\bar{1}\bar{1}0]$). This observation differs from the random polarization of condensates in the bulk microcavity, probably because of the anisotropy added by the MQW growth. Further studies show, however, that when the excitation power is raised, a depinning of the polarization is observed, and a situation similar to the bulk microcavity case is observed [62].

13.4 Future Prospects

13.4.1 Novel Sample Designs

13.4.1.1 Air-Spaced DBRs

One recent approach that we have adopted to increase the Rabi splitting for room-temperature operation uses a new type of microcavity design that borrows concepts from VCSELs and plasmonics. By using air-spaced DBRs, reflectivities >99% can be obtained with only two or three mirror pairs, due to the high refractive index contrast at each interface. Combining this as a lower DBR mirror in a III-V microcavity together with a top mirror of 35 nm Au produces a Tamm plasmon microcavity in which the optical field is very large close to the Au layer where the QWs are then placed. Due to the high-efficiency reflectors, the effective cavity length becomes very short, and combined with the larger field density at the QW well layers, increases the Rabi splitting. In this way, such etched devices can show strong coupling at room temperature [63]. Such structures have a further series of advantages since the spacing between the mirror pairs can be electrically tuned using the top Au contact and the substrate. This produces a new generation of polaritonic MEMs devices. Additionally, this vertical bias field can be used to tune the exciton energy via the quantum-confined Stark effect hence allowing selective tuning of the polariton resonance. Finally, selective Au top contact act as polariton waveguides and can provide lateral forces on polaritons that allow them to be moved around and trapped. All these features look promising for future generations of polariton devices, both at low temperature and room temperature.

13.4.1.2 Oriented Polaritons

The results on polariton lasing presented here are very promising for new generations of coherent light emitters; however, currently, the threshold of these devices is limited by the rate of polariton-polariton scattering. This is illustrated by the large number of quantum wells (QWs) generally inserted in the MCs for polariton lasing [7, 11, 13], in comparison with equivalent VCSEL structures [64, 73]. This design is needed to allow high-enough polariton densities so that polariton-polariton scattering becomes faster than polariton decay. Reducing the decay rate by using high-Q cavities has led to polariton localization in the photonic disorder potential. The desire to reduce the minimum threshold for polariton lasing/condensation has thus led to various proposals to enhance polariton relaxation. One suggestion has been to use scattering by free electrons [66, 67]; however, to date, this has not been effective.

We recently proposed a new technique to improve the polariton–polariton scattering rate using electrically biased asymmetric double QWs (ADQWs), using III-arsenide semiconductors [68]. We use a polariton *p*-i-*n* structure comprising four sets of $In_{0.1}Ga_{0.9}As/GaAs/In_{0.08}Ga_{0.92}As$ (10 nm/4 nm/10 nm) ADQWs placed at the antinodes of the optical field (Fig. 13.6a). By applying an electric field to the structure, it is possible to bring the electron levels of neighboring QWs into resonance (Fig. 13.6b), while the hole levels remain uncoupled. In this resonant condition, the spatially direct (DX) and indirect (IX) excitons become coupled, so they share the strong oscillator strength of the optically allowed transition together with the strong static dipole moment in the growth direction of the indirect exciton. This results in a strong excitonic dipole moment at resonance (Fig. 13.6c).

The resulting *oriented polaritons* (OP), which show unusual dispersion features [68], are predicted to to exhibit stronger polariton–polariton interactions. This is

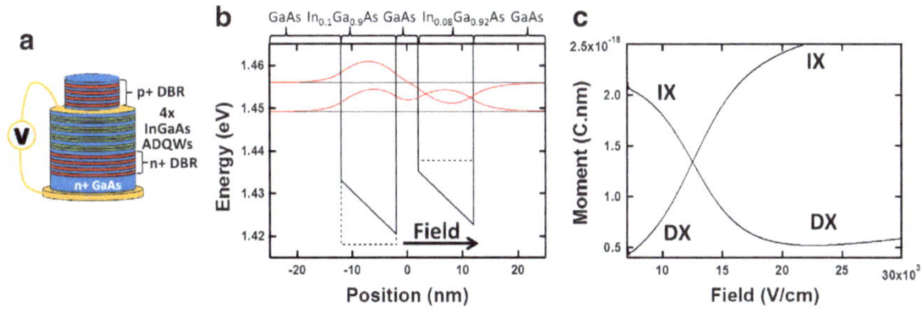

Fig. 13.6 (**a**) ADQW polariton mesa structure. (**b**) Conduction bands of one ADQW period (*black line at resonance, dashed at flat-band*) and corresponding electron wave functions (*red lines*). (**c**) Calculated dipole moment corresponding to DX and IX for a single electron-hole pair. *Results from University of Cambridge*

because this interaction is based on exciton dipole-dipole coupling and for excitons with dipole moments in the plane, this interaction is relatively weak. When the oscillating induced dipoles are combined with vertically oriented static dipole moments perpendicular to the QW planes, the repulsive interaction is greatly enhanced (the electrons are close to each other, as are the holes, and their attraction is at a minimum). Hence, it is of great interest to develop polariton lasers based on OPs because they should exhibit significantly lower polariton lasing thresholds. For room-temperature operation, it is also worth pointing out that the OP concept can easily be transferred to III-nitride semiconductors. The spontaneous and piezoelectric polarization present in III-nitride heterostructures could indeed enable OPs without the need of externally applied electric field.

13.4.1.3 Alternative Materials

At the moment, room-temperature polariton lasing has been demonstrated only in III-nitride and organic based MCs. However, several other materials are promising candidates. As previously discussed in this chapter, ZnO seems to be really advancing more rapidly and it has similar physical properties in some respects to GaN. The situation is currently similar to that of GaN a few years ago: several groups have reported the SCR; however, samples combining high-Q-factor MCs and low emission linewidth active region are still missing. The recent demonstration of VCSEL action in a ZnO-based MC is promising [41], and indeed a room-temperature polariton laser using this material has just been observed [69].

Among other material which have significant potential we pick out hybrid organic-inorganic system in which the SCR has been recently demonstrated at RT [70, 71]. Such organic-inorganic perovskites, having chemical formulas of the form $(R\text{-}NH_3)_2MX_4$ where R is an organic chain, M is a metal and X an halogen, self-assemble readily into organic-inorganic layered systems which are composed of hybrid quantum wells with extremely high oscillator strength and narrow linewidth. MCs using this system exhibit Rabi splittings in excess of 100 meV and present several advantages over J-aggregates MCs, including a better thermal insensitivity and significant photooxidation resistance. They are therefore good candidates for polariton lasing at room temperature.

13.4.2 Electrical Injection

13.4.2.1 Overview

A key requirement for the practical use of polariton laser devices is electrical injection. As mentioned in the introduction, recently, several groups have managed to fabricate polariton LEDs in the III-arsenide system [15–17]. Electrically driven polariton emission has been even reported at room temperature [65]. Such results are very encouraging as they prove that it is possible to electrically inject electrons and holes that form polaritons without destroying the SCR; however, to date, electrically injected polariton lasing is still to be reported. Furthermore, while some advanced III-arsenide-based MC designs exhibit SCR at room temperature, polariton lasing still remains to be demonstrated in these conditions. We will therefore discuss here mostly electrical injection in wide bandgap semiconductors. Furthermore, due to the less mature technology for ZnO optoelectronic devices (for a review, see [72]), the discussion will be focused on the III-nitride case.

13.4.2.2 III-Nitride Case

The widespread availability of III-nitride optoelectronic devices such as LEDs and laser diodes demonstrate that efficient electrical injection technologies have been developed. The case of a polariton light emitter is however a tricky one. As mentioned earlier in this chapter, the growth of high-Q-factor MCs is challenging. Adding to that, the process complexity of an electrically injected vertical cavity design makes the realization of a III-nitride polariton laser a considerable tour de force. A good illustration is that the first electrically injected III-arsenide/phosphide VCSEL was demonstrated in 1979 [5], while its III-nitride counterpart was only realized in 2008 [73]. This result is nevertheless really encouraging as it shows that it is possible to develop efficient injection processes.

The peculiar design of the SCR MCs studied in this chapter causes some problems when adapted for electrical injection. In particular, converting III-nitride MCs into polariton emitting devices will cause some issues. Doping the epitaxial layers is known to create defects that lead to significant inhomogeneous broadening, detrimental to the SCR. The designs used for our polariton laser structures have thick active regions, both bulk and MQW. A bulk active region is hard to electrically inject into. While the MQW sample seems better suited, its large thickness also makes it difficult to efficiently and uniformly inject carriers into. A possible solution would be to attempt intracavity contacting in order to electrically pump only a small portion of the QW: the SCR then optically spreads carriers throughout the whole active region [74]. Another problem with such GaN/AlGaN samples is that *p*-type doping of AlGaN layers is extremely challenging due to the increase of the activation energy of Mg acceptor levels [75]. Using InGaN/GaN QWs would therefore be preferable for an electrically injected polariton laser, but it brings back the problem of inhomogeneous broadening, as previously discussed [34]. This approach could become realistic provided that adequate efforts to develop narrow emission linewidth InGaN/GaN QWs, possibly using small indium content are successful.

To conclude, we have discussed the challenges toward electrical injection in II-nitride-based polariton lasers. While promising paths are outlined, it is clear that a lot of development work remains to be tackled, and some intense activity is needed for successful progress.

Acknowledgements We would like to acknowledge our many generous collaborators at the University of Southampton, University of Cambridge, University of Crete University Blaise Pascal, and École Polytechnique Fédérale de Lausanne who worked with us on this project.

References

1. R.N. Hall, G.E. Fenner, J.D. Kingsley, T.J. Soltys, R.O. Carlson, Phys. Rev. Lett. **9**, 366 (1962)
2. M.I. Nathan, W.P. Dumke, G. Burns, F.H. Dill Jr., G. Lasher, Appl. Phys. Lett. **1**, 62 (1962)
3. N. Holonyak Jr., S.F. Bevacqua, Appl. Phys. Lett. **1**, 82 (1962)
4. T.M. Quist, R.H. Rediker, R.J. Keyes, W.E. Krag, B. Lax, A.L. McWhorter, H.J. Zeigler, Appl. Phys. Lett. **1**, 91 (1962)
5. H. Soda, K. Iga, C. Kitahara, Y. Suematsu, Jpn. J. Appl. Phys. **18**, 2329 (1979)
6. A. Imamoğlu, R.J. Ram, S. Pau, Y. Yamamoto, Phys. Rev. A **53**, 4250 (1996)
7. Le Si Dang, D. Heger, R. André, F. Boeuf, R. Romestain, Phys. Rev. Lett. **81**, 3920 (1998)
8. J. Kasprzak, M. Richard, S. Kundermann, A. Baas, P. Jeambrun, J.M.J. Keeling, F.M. Marchetti, M.H. Szymaska, R. André, J.L. Staehli, V. Savona, P.B. Littlewood, B. Deveaud, Le Si Dang, Nature **443**, 409 (2006)
9. H. Deng, G. Weihs, C. Santori, J. Bloch, Y. Yamamoto, Science **298**, 199 (2002)
10. R. Balili, V. Hartwell, D. Snoke, L. Pfeiffer, K. West, Science **316**, 1007 (2007)
11. D. Bajoni, P. Senellart, E. Wertz, I. Sagnes, A. Miard, A. Lemaître, J. Bloch, Phys. Rev. Lett. **100**, 047401 (2008)
12. S. Christopoulos, G.B.H. von Hgersthal, A.J.D. Grundy, P.G. Lagoudakis, A.V. Kavokin, J.J. Baumberg, G. Christmann, R. Butt, E. Feltin, J.-F. Carlin, N. Grandjean, Phys. Rev. Lett. **98**, 126405 (2007)
13. G. Christmann, R. Butt, E. Feltin, J.-F. Carlin, N. Grandjean, Appl. Phys. Lett. **93**, 051102 (2008)
14. S. Kéna-Cohen, S.R. Forrest, Nat. Photon. **4**, 371 (2010)
15. D. Bajoni, E. Semenova, A. Lemaître, S. Bouchoule, E. Wertz, P. Senellart, J. Bloch, Phys. Rev. B **77**, 113303 (2008)
16. A.A. Khalifa, A.P.D. Love, D.N. Krizhanovskii, M.S. Skolnick, J.S. Roberts, Appl. Phys. Lett. **92**, 061107 (2008)
17. S.I. Tsintzos, N.T. Pelekanos, G. Konstantinidis, Z. Hatzopoulos, P.G. Savvidis, Nature **453**, 372 (2008)
18. A.V. Kavokin, J.J. Baumberg, G. Malpuech, F.P. Laussy, *Microcavities*, (Oxford University Press, Oxford, 1997)
19. C. Weisbuch, M. Nishioka, A. Ishikawa, Y. Arakawa, Phys. Rev. Lett. **69**, 3314 (1992)
20. A. Tredicucci, Y. Chen, V. Pellegrini, M. Börger, L. Sorba, F. Beltram, F. Bassani, Phys. Rev. Lett. **75**, 3906 (1995)
21. M.S. Skolnick, T.A Fisher, D.M. Whittaker, Semicond. Sci. Technol. **13**, 645 (1998)
22. D.I. Babic, S.W. Corzine, IEEE J. Quant. Electron. **28**, 514 (1992)
23. V. Savona, L.C. Andreani, P. Schwendimann, A. Quattropani, Solid State Commun. **93**, 733 (1995)
24. R. Houdré, R.P. Stanley, M. Ilegems, Phys. Rev. A **53**, 2711 (1996)
25. P.Y. Yu, M. Cardona, *Fundamentals of Semiconductors*, 2nd edn. (Springer, Berlin, 1999)
26. R. Atanasov, F. Bassani, A. DAndrea, N. Tomassini, Phys. Rev. B **50**, 14381 (1994)
27. R. André, D. Heger, Le Si Dang, Y. Merle dAubigné, J. Cryst. Growth **184–185**, 758 (1998)
28. K. Kornitzer, T. Ebner, K. Thonke, R. Sauer, C. Kirchner, V. Schwegler, M. Kamp, M. Leszczynski, I. Grzegory, S. Porowski, Phys. Rev. B **60**, 1471 (1999)
29. G. Christmann, R. Butté, E. Feltin, A. Mouti, P.A. Stadelmann, A. Castiglia, J.-F. Carlin, N. Grandjean, Phys. Rev. B **77**, 085310 (2008)

30. C. Morhain, T. Bretagnon, P. Lefebvre, X. Tang, P. Valvin, T. Guillet, B. Gil, T. Taliercio, M. Teisseire-Doninelli, B. Vinter, C. Deparis, Phys. Rev. B **72**, 241305(R) (2005)
31. R.T. Senger, K.K. Baja, Phys. Rev. B **68**, 045313 (2003)
32. P. Senellart, Ann. Phys. **28**, 1 (2003)
33. R. Houdré, R.P. Stanley, U. Oesterle, M. Ilegems, C. Weisbuch, J. Phys. IV **03**, C551 (1993)
34. G. Christmann, R. Butté, E. Feltin, J.-F. Carlin, N. Grandjean, Phys. Rev. B **73**, 153305 (2006)
35. M. Saba, C. Ciuti, J. Bloch, V. Thierry-Mieg, R. André, Le Si Dang, S. Kundermann, A. Mura, G. Bongiovanni, J.L. Staehli, B. Deveaud, Nature **414**, 731 (2001)
36. G. Malpuech, A. Di Carlo, A. Kavokin, J.J. Baumberg, M. Zamfirescu, P. Lugli, Appl. Phys. Lett. **81**, 412 (2002)
37. M. Zamfirescu, A. Kavokin, B. Gil, G. Malpuech, M. Kaliteevski, Phys. Rev. B **65**, 161205 (2002)
38. M. Leroux, N. Grandjean, M. Laügt, J. Massies B. Gil, P. Lefebvre, P. Bigenwald, Phys. Rev. B **58**, 13371 (1998)
39. J.-F. Carlin, M. Ilegems, Appl. Phys. Lett. **83**, 668 (2003)
40. R. Butté, E. Feltin, J. Dorsaz, G. Christmann, J.-F. Carlin, N. Grandjean, M. Ilegems, Jpn. J. Appl. Phys. **44**, 7207 (2005)
41. S. Kalusniak, S. Sadofev, S. Halm, F. Henneberge, Appl. Phys. Lett. **98**, 011101 (2011)
42. R. Shimada, J. Xie, V. Avrutin, Ü. Özgür, H. Morkoč, Appl. Phys. Lett. **92**, 011127 (2008)
43. S. Faure, T. Guillet, P. Lefebvre, T. Bretagnon, B. Gil, Phys. Rev. B **78**, 235323 (2008)
44. M. Nakayama, S. Komura, T. Kawase, D. Kim, J. Phys. Soc. Jpn. **77**, 093705 (2008)
45. C. Sturm, H. Hilmer, R. Schmidt-Grund, M. Grundman, New J. Phys. **11**, 073044 (2009)
46. T. Tawara, H. Gotoh, T. Akasaka, N. Kobayashi, T. Saitoh, Appl. Phys. Lett. **83**, 830 (2003)
47. E. Feltin, D. Simeonov, J.-F. Carlin, R. Butté, N. Grandjean, Appl. Phys. Lett. **90**, 021905 (2007)
48. J.-M. Chauveau, M. Teisseire, H. Kim-Chauveau, C. Deparis, C. Morhain, B. Vinter, Appl. Phys. Lett. **97**, 081903 (2010)
49. N. Antoine-Vincent, F. Natali, D. Byrne, A. Vasson, P. Disseix, J. Leymarie, M. Leroux, F. Semond, J. Massies, Phys. Rev. B **68**, 153313 (2003)
50. F. Semond, I.R. Sellers, F. Natali, D. Byrne, M. Leroux, J. Massies, N. Ollier, J. Leymarie, P. Disseix, A. Vasson, Appl. Phys. Lett. **87**, 021102 (2005)
51. K. Bejtka, F. Réveret, R.W. Martin, P.R. Edwards, A. Vasson, J. Leymarie, I.R. Sellers, J.Y. Duboz, M. Leroux, F. Semond, Appl. Phys. Lett. **92**, 241105 (2008)
52. F. Réveret, P. Disseix, J. Leymarie, A. Vasson, F. Semond, M. Leroux, J. Massies, Phys. Rev. B **77**, 195303 (2008)
53. A. Alyamani, D. Sanvitto, A.A. Khalifa, M.S. Skolnick, T. Wang, F. Ranalli, P.J. Parbrook, A. Tahraoui, R. Airey, J. Appl. Phys. **101**, 093110 (2007)
54. R. Butté, G. Christmann, E. Feltin, J.-F. Carlin, M. Mosca, M. Ilegems, N. Grandjean, Phys. Rev. B **73**, 033315 (2006)
55. E. Feltin, J.-F. Carlin, J. Dorsaz, G. Christmann, R. Butté, M. Laügt, M. Ilegems, N. Grandjean, Appl. Phys. Lett. **88**, 051108 (2006)
56. G. Christmann, D. Simeonov, R. Butté, E. Feltin, J.-F. Carlin, N. Grandjean, Appl. Phys. Lett. **89**, 261101 (2006)
57. R. Butté, G. Christmann, E. Feltin, A. Castiglia, J. Levrat, G. Cosendey, A. Altoukhov, J.-F. Carlin, N. Grandjean, Proc. SPIE **7216**, 721619 (2009)
58. T. Tawara, H. Gotoh, T. Akasaka, N. Kobayashi, T. Saitoh, Phys. Rev. Lett. **92**, 256402 (2004)
59. S. Faure, C. Brimont, T. Guillet, T. Bretagnon, B. Gil1, F. Médard, D. Lagarde, P. Disseix, J. Leymarie, J. Zúñiga-Pérez, M. Leroux, E. Frayssinet, J.C. Moreno, F. Semond, S. Bouchoule, Appl. Phys. Lett. **95**, 121102 (2009)
60. F. Binet, J.Y. Duboz, J. Off, F. Scholz, Phys. Rev. B **60**, 4715 (1999)
61. R. Butté, J. Levrat, G. Christmann, E. Feltin, J.-F. Carlin, N. Grandjean, Phys. Rev. B 80, 233301 (2009)
62. J. Levrat, R. Butté, T. Christian, M. Glauser, E. Feltin, J.-F. Carlin, N. Grandjean, D. Read, A.V. Kavokin, Y.G. Rubo, Phys. Rev. Lett. **104**, 166402 (2010)

63. C. Grossman, G. Christmann, J.J. Baumberg, Appl. Phys. Lett. (2011)
64. J.L. Jewell, J.P. Harbison, A. Scherer, Y.H. Lee, L.T. Florez, IEEE J. Quant. Electron. **27**, 1332 (1991)
65. S.I. Tsintzos, P.G. Savvidis, G. Deligeorgis, Z. Hatzopoulos, N.T. Pelekanos, Appl. Phys. Lett. **94**, 071109 (2009)
66. G. Malpuech, A. Kavokin, A. Di Carlo, J.J. Baumberg, Phys. Rev. B **65**, 153310 (2002)
67. P.G. Lagoudakis, M.D. Martin, J.J. Baumberg, A. Qarry, E. Cohen, L.N. Pfeiffer, Phys. Rev. Lett. **90**, 206401 (2003)
68. G. Christmann, A. Askitopoulos, G. Deligeorgis, Z. Hatzopoulos, S.I. Tsintzos, P.G. Savvidis, J.J. Baumberg, Appl. Phys. Lett. **98**, 081111 (2011)
69. T. Guillet, M. Mexis, J. Levrat, G. Rossbach, C. Brimont, T. Bretagnon, B. Gil, R. Butté, N. Grandjean, L. Orosz, F. Réveret, J. Leymarie, J. Zúniga-Pérez, M. Leroux, F. Semond, and S. Bouchoule, Appl. Phys. Lett. **99**, 161104 (2011)
70. G. Lanty, J.S. Lauret, E. Deleporte, S. Bouchoule, X. Lafosse, Appl. Phys. Lett. **93**, 081101 (2008)
71. K. Pradeesh, J.J. Baumberg, G. Vijaya Prakash, Opt. Express **17**, 22171 (2009)
72. Ü. Özgür, Ya. I. Alivov, C. Liu, A. Teke, M.A. Reshchikov, S. Dogan, V. Avrutin, S.-J. Cho, H. Morkoç J. Appl. Phys. **98**, 041301 (2005)
73. T.-C. Lu, C.-C. Kao, H.-C. Kuo, G.-S. Huang, S.-C. Wang, Appl. Phys. Lett. **92**, 141102 (2008)
74. R. Butté, N. Grandjean, Semicond. Sci. Technol. **26**, 014030 (2011)
75. M. Suzuki, J. Nishio, M. Onomura, C. Hongo, J. Cryst. Growth **189/190**, 511 (1998)

Chapter 14
Exciton–Polaritons in Organic Semiconductor Optical Microcavities

Stéphane Kéna-Cohen and Stephen R. Forrest

Abstract The large exciton binding energies and oscillator strengths of organic semiconductors have allowed the realization of strong exciton–photon coupling at room temperature in microcavities containing a wide variety of materials. The first part of this chapter reviews the physics of organic semiconductors and initial observations of cavity polaritons in this class of materials. In the second part, the linear optical properties of crystalline organic microcavities are discussed and contrasted to previous organic and inorganic microcavity results. The chapter concludes with a discussion of recent organic polariton lasing results and future prospects for realizing nonlinear behavior using organic polaritons.

14.1 Introduction

Organic semiconductors have both electrical and structural properties that are inherently different from their inorganic counterparts. Most importantly, the bonds between organic molecules that comprise their bulk properties are the result of intermolecular van der Waals bonds that make them far softer and more ductile than inorganic semiconductors that are primarily bonded by covalent and, sometimes, ionic forces [1, 2]. In addition to determining the unique mechanical properties of these materials, van der Waals bonds have many fundamental consequences to their optical and electronic nature as well. These weak and short-range intermolecular forces result in a very low polarizability of the solid. This results in a relatively

S. Kéna-Cohen
Department of Physics, Imperial College, London, UK
e-mail: s.kena-cohen@imperial.ac.uk

S.R. Forrest (✉)
Departments of Physics and Electrical Engineering and Computer Science, University of Michigan, Ann Arbor, MI 48109, USA
e-mail: stevefor@umich.edu

D. Sanvitto and V. Timofeev (eds.), *Exciton Polaritons in Microcavities*,
Springer Series in Solid-State Sciences 172, DOI 10.1007/978-3-642-24186-4_14,

low dielectric constant ($\varepsilon \approx 3$) compared to inorganic semiconductors (where $\varepsilon \approx 10$–15).

The consequences of such a low dielectric constant are profound and determine many of the optical and electronic properties of this very large class of materials. Most importantly, the excited state of organic materials is typically characterized by tightly bound electron–hole pairs or excitons. Hence, on the absorption of a photon, the mobile molecular excited state is generated rather than a free electron and hole as is the case in most inorganic semiconductors. The exciton binding energy, which scales as $1/\varepsilon$, is typically between 0.1 eV and 1 eV, compared to ~10 meV for materials such as GaAs or InP. It is the bound excitonic state, rather than the free carrier state, therefore, that determines the optical and optoelectronic properties of organics at room temperature, and hence they are often called "excitonic materials." This tightly bound state is then confined to only a single molecule (forming a so-called Frenkel exciton), or in the case of more strongly bonded molecular species such as 3,4,9,10-perylenetetracarboxylic dianhydride (PTCDA), the exciton can spread over only a few neighboring molecules, forming a charge transfer exciton. This latter state is of particular interest in the study of exciton phenomena as it is able to "sample" the dielectric environment of the more macroscopic crystal structure [3], approaching the conditions observed in inorganic semiconductors that are characterized by very loosely bound and large (typically spread over 10–100 lattice constants) Wannier–Mott excitons.

In contrast to the broad, parabolic density of states of inorganic materials, the excitonic density of states is narrow, as is characteristic of vibrationally broadened molecular energy levels. The broadening, both homogenous and inhomogenous, arises from the progression of vibronic energies that range from ~ 0.2 eV to only a few meV, depending on whether their origin is from strong, covalent bonds within a molecule or due to soft van der Waals bonds between molecules comprising the solid.

A further outcome of the weak van der Waals binding between molecules is that the materials are naturally "soft". Hence, the growth of single crystals, as opposed to amorphous or highly polycrystalline thin films, has presented a challenge. Coupled with the lack of substrates that provide an epitaxial template whose lattice constant matches that of the organic film grown on its surface, most growth techniques (including liquid, vacuum, and vapor phase methods) have resulted in a high degree of disorder that makes it difficult to observe many crystal-structure-dependent optical and electronic properties. Indeed, the typically monoclinic or triclinic unit cells that characterize small-molecular-weight organics, combined with the highly asymmetric molecular structure itself, often give rise to highly anisotropic charge mobilities, indicies of refraction, a Davydov splitting of the excitonic band along different crystal axes, etc., properties that can lead to the observation of complex phenomena associated with polaritons [4].

However, while the weak bond strength often can prohibit formation of macroscopic crystalline structure, at the same time it can significantly reduce stress in films grown on non-lattice-matched substrates. Therefore, under an appropriate set of conditions, highly ordered growth can be obtained on substrates that introduce significant strain with the overlying film [5] or even on completely amorphous substrates such as glass [6]. Indeed, some of the work reported in this chapter is based on single-crystalline films of anthracene grown from the liquid phase into cavities where there is no crystalline structural matching available to force the desired growth. Instead, it is the nucleation at surfaces and edges, coupled with the intermolecular van der Waals bonds that ultimately determine the resulting structures observed.

Finally, we note that the strength of the excitonic bond leads to the formation of polaritons that are also stable at room temperature. For this reason, they are of considerable interest for investigations of the strong-coupling regime. The stable exciton, coupled to a photon in a high quality factor (Q) microcavity, results in a robust polaritonic state that is easily observed at room temperature, making it interesting to study for the insights that follow from its fundamental properties. The stability of the polaritonic state also suggests that it may be useful in many practical applications. Indeed, the low thresholds achievable for room-temperature polariton lasing, as we will discuss in this chapter, may eventually lead to a very long sought-after electrically pumped organic laser that has potential applications in sensing, metrology, and quantum optics to name only a few examples.

14.2 Strong Exciton–Photon Coupling in Organic Semiconductor Microcavities

The small exciton binding energies typical of inorganic semiconductors have, until recently, limited the use of inorganic microcavities to low temperatures. There have been, however, recent advances in the growth of wide bandgap semiconductors allowing the realization of strong exciton–photon coupling in GaN and ZnO microcavities at room temperature [7, 8]. While in those reports, the thermally broadened polariton linewidths remain comparable to the Rabi splitting, polariton lasing has been demonstrated in GaN [9] and AlGaN/GaN quantum wells [10], suggesting that these new material sets may provide the means to achieve room-temperature electrically pumped polariton lasing.

An alternative means of achieving room-temperature polariton emission is to use organic semiconductors within the microcavity. As we have seen, organic excitons possess large binding energies, making them highly stable at room temperature. Moreover, their oscillator strengths are far greater than those of their inorganic counterparts which results in extremely large Rabi splittings.

The first observation of organic cavity polaritons was made by Lidzey et al. in 1998 [11]. In that sample, a blend of tetra-(2,6-*t*-butyl)phenol-porphyrin zinc (4TBPPZn) in an inert polystyrene (PS) matrix was used as the active material, and the microcavity was composed of a bottom SiO_2/SiN_x DBR and a top silver mirror. By measuring angle-resolved reflectivity, he observed the characteristic anticrossing of both polaritonic resonances and extracted a Rabi splitting of $\Omega \approx 160\,\text{meV}$. This was more than an order of magnitude higher than the largest Rabi splitting observed

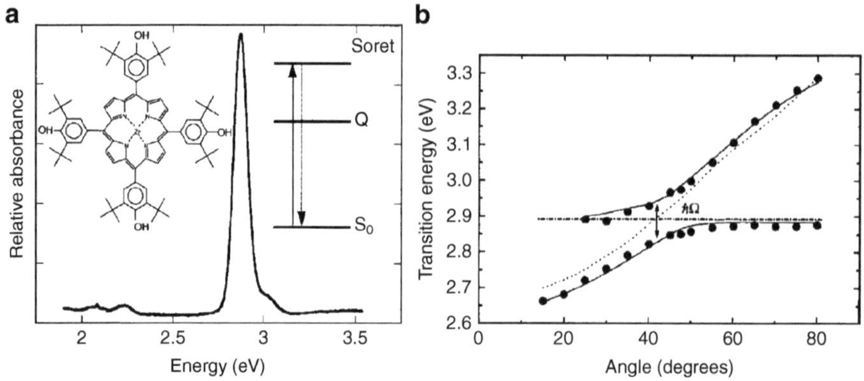

Fig. 14.1 (**a**) Absorption spectrum of the doped 4TBPPZn film used by Lidzey et al. [11]. The *inset* on the left shows the molecular structure of 4TBPPZn, while the inset on the right shows the energy level structure. Absorption corresponding to the strong 2.9 eV transition occurs from the ground state to the Soret band. (**b**) Dispersion relation extracted from angle-resolved reflectivity. The Rabi splitting is indicated by an *arrow*. Reprinted with permission from [11]. Copyright (1998) Nature Publishing Group

in an inorganic microcavity at the time. The absorption spectrum of 4TBPPZn and the extracted dispersion curve are reproduced in Fig. 14.1. The excitonic transition used in this demonstration is called the Soret band and corresponds to the second excited electronic energy level of the molecule. Because of fast relaxation, however, luminescence in organic molecules occurs almost exclusively from the lowest excited state (Kasha's rule). Polariton emission, as a result, was unlikely to be possible in this structure.

Room-temperature polariton emission was observed soon after in J-aggregates of the cyanine dye 2,2′-dimethyl-8-phenyl-5,6,5′,6′-dibenzothiacarbocyanine chloride [12]. This dye molecule possesses a net charge which causes the molecules to form longchains with their dipole moments aligned. The resulting structure is called a J-aggregate, and its excitations are effectively 1D Frenkel excitons [13]. The absorption spectrum of J-aggregates is red-shifted and narrowed with respect to the monomer due to interactions between the molecules. Optical transitions are only possible to the lowest lying Frenkel state, and its oscillator strength is $\propto N$, the number of monomers in the chain. Figure 14.2 shows the absorption spectrum of the J-aggregates used by Lidzey et al., as well as the photoluminescence (PL) from the microcavity structure. An interesting observation is that in contrast to inorganic semiconductors, no polariton emission was detected from the upper polariton branch. This absence of upper branch emission is now understood to result from rapid relaxation to the large number of final states available in the exciton reservoir [14–16].

In organic films, the different intramolecular vibrational modes of the molecular nuclei result in transition dipole moments that correspond to absorption from the ground to the first electronic state accompanied by the creation of one or more of vibrational modes. These transition dipole moments are scaled

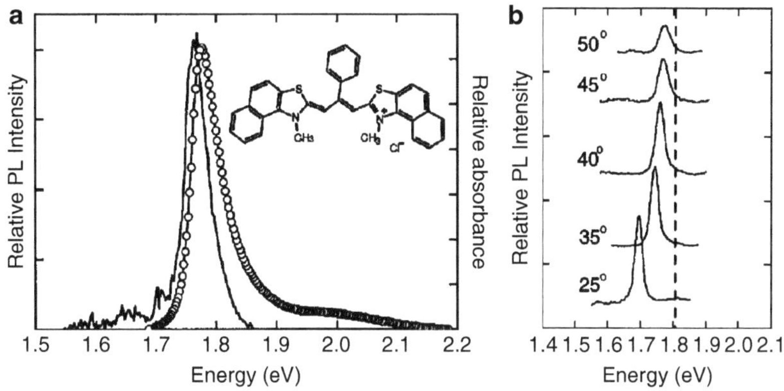

Fig. 14.2 (**a**) Absorption (*circles*) and PL spectrum (*solid*) of the J-aggregate used for the observation of polariton emission [12]. The inset on the left shows the molecular structure of the dye. (**b**) Microcavity PL for a range of selected angles. Photoluminescence is observed only from the lower branch and does not cross the bare exciton energy indicated by the dashed line. Reprinted with permission from [12]. Copyright (1999) by the American Physical Society

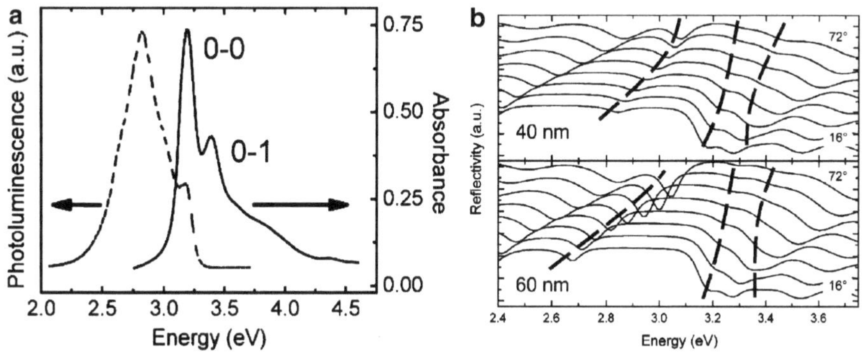

Fig. 14.3 (**a**) Absorption spectrum (*solid*) and PL (*dashed*) of the NTCDA films used by Holmes et al. [18]. Note the presence of the two intramolecular vibronics in the absorption spectrum. The emission spectrum, which does not clearly show the vibronics, is most likely dominated by defect emission. (**b**) Angle-resolved reflectivity from 40-nm to 60-nm-thick microcavities. The positions of the reflectivity minima trace out the dispersion relation and are indicated by *dashed lines*. The middle branch corresponds to a hybridized mixture of the 0–0 and 0–1 vibronic transitions. Reprinted with permission from [18]. Copyright (2004) by the American Physical Society

by the Franck–Condon factors, but when summed over all modes, the purely electronic oscillator strength is recovered [2, 17]. Holmes et al. showed that when the organic 3,4,7,8-napthalenetetracarboxylic dianhydride (NTCDA), which possesses two strong intramolecular vibronic transition, is placed within an optical microcavity, new hybrid polaritons emerge which correspond to a mixture of both vibronic resonances with the cavity photon [18]. The optical absorption of NTCDA and the microcavity reflectivity are shown in Fig. 14.3. In this case, an anticrossing of the polaritonic resonances, indicated by the dashed lines, was observed around both vibronic transitions. This data was interpreted in terms of a three-body coupled harmonic oscillator (CHO) model. It can be shown that indeed, the semiclassical result for the microcavity dispersion reduces to this model when a second excitonic resonance is added and A quantum theory describing this hybridization has recently been formulated [19].

In organic cavities, bulk layers of the resonant excitonic material are typically used. Moreover, since nonlocal polarization effects can be ignored in this case, we can derive an expression for the polariton dispersion using the semiclassical approach from [20]. If we neglect spatial dispersion, a general expression for the transverse dielectric constant is given by:

$$\varepsilon(\omega) = \varepsilon_{\mathrm{B}} \left(1 + \frac{\omega_{\mathrm{LT}}}{\omega_{\mathrm{ex}} - \omega - i\,\gamma_{\mathrm{ex}}} \right) \tag{14.1}$$

where ε_{B} is the background dielectric constant, ω_{LT} is the longitudinal-transverse splitting, ω_{ex} is the bare exciton resonant energy, and γ_{ex} is an empirical damping parameter.

The approximate expression for the dispersion in a cavity of thickness L_{c} is [20]:

$$\begin{aligned} \omega = & \frac{\omega_{\mathrm{c}} + \omega_{\mathrm{ex}}}{2} - i\,\frac{\gamma_{\mathrm{ex}} + \gamma_{\mathrm{c}}}{2} \\ & \pm \sqrt{\frac{(\omega_{\mathrm{ex}} - \omega_{\mathrm{c}})^2}{4} + V^2 - \frac{(\gamma_{\mathrm{ex}} - \gamma_{\mathrm{c}})^2}{4} + i\,\frac{(\omega_{\mathrm{ex}} - \omega_{\mathrm{c}})(\gamma_{\mathrm{c}} - \gamma_{\mathrm{ex}})}{2}} \end{aligned} \tag{14.2}$$

where ω_{c} is the bare cavity resonance energy, γ_{c} is the cavity damping, and the interaction potential is:

$$V = \sqrt{\frac{2\omega_{\mathrm{ex}}\omega_{\mathrm{LT}}L_{\mathrm{c}}}{L_{\mathrm{DBR}} + L_{\mathrm{c}}}} \tag{14.3}$$

where L_{DBR} is the DBR penetration depth [21]. We see from this expression that the Rabi splitting can simply be increased by increasing the cavity length and that its limiting value is $2V = 2\sqrt{2\omega_{\mathrm{ex}}\omega_{\mathrm{LT}}}$. For organic microcavities, an advantageous situation is obtained when $L_{\mathrm{c}} > L_{\mathrm{DBR}}$, but this fact is rarely appreciated.

In amorphous or nanocrystalline structures, it is clear that the picture of Frenkel excitons possessing a definite $\boldsymbol{k}$ is inappropriate. The excitons in those cases resemble localized molecular excitations. All strongly coupled organic microcavities presented in this section fall in this category. In this case, the cavity photon couples the different excitations together and does create some wavelike polaritonic states. It was shown by Agranovich, however, that a large number of the states remain localized [22]. Agranovich called these localized excitations incoherent states. The presence of these states can be easily understood based on the argument that wavelike polaritons can only exist when the uncertainty in momentum is small compared to the absolute momentum. Since a large number of incoherent states exist, they inevitably play an important role in the dynamics. Whether these states play a role in preventing polariton lasing is still an open question. Michetti has shown numerically the existence of incoherent states by directly diagonalizing the 1D light-matter Hamiltonian, while including energetic and positional disorder [23].

14.3 Single Crystalline Organic Microcavities

The first observation of strong exciton–photon coupling in an organic material possessing long-range order (on a scale larger than an optical wavelength) occurred in polycrystalline films of thermally evaporated tetracene [24]. In organic crystals, the periodicity and resonance intermolecular Coulomb interaction give rise to an excitonic band, and for crystals with more than one molecule per unit cell, the excitonic band is split into as many Davydov components as there are inequivalent molecules in the unit cell [4, 25]. As a result, the excitonic resonance interacts with both photon polarizations, and the polariton dispersion will depend on the azimuthal orientation of the sample [26, 27]. This is in contrast to the case of inorganic QWs where TE polarized photons interact with T-excitons, while TM-polarized photons interact with L- and Z-excitons. The situation is further complicated in the case of crystals with two molecules per unit cell [26,27]. In that case, the dielectric tensor is biaxial, and both Davydov components interact with the two photon polarizations. This gives rise to a dispersion possessing four polariton branches: two lower and two upper branches where each pair corresponds approximately to one excitonic resonance.

In the case of the polycrystalline tetracene microcavity, the most striking effect of crystallinity is the appearance of this second lower polariton branch in reflectivity (the second upper polariton branches spectrally overlap). In that work, the crystalline domains, shown in Fig. 14.4, were on the order of 5–10 μm which is larger than the experimental spot size. The observed spectra thus consisted of an incoherent sum of the reflectivity of the randomly oriented underlying crystallites. The calculated reflectivity for random crystalline orientations and the measured dispersion for a 160-nm-thick microcavity are shown in Fig. 14.4b. Tetracene, however, is not ideal for the observation of polariton lasing. Its high surface roughness complicates the realization of high quality factor microcavities. In addition, it possesses high nonradiative losses resulting from an efficient singlet fission mechanism [2].

A much better suited material for lasing is the archetypical molecular crystal anthracene (Fig. 14.5a). This material possesses a high quantum yield (0.99) and damage threshold, which are both advantageous. Moreover, a growth method was developed which avoided damage caused by direct deposition of dielectric mirrors on the soft organic material resulting in high Q cavities [28].

The fabricated structure, shown in Fig. 14.5b, was composed of 30-mm-long $\times$ 0.5-mm-wide $\times$ 140-nm-thick anthracene single-crystal channels sandwiched between two DBRs composed of 12, alternating $\lambda/4$ pairs (where λ is the wavelength) of SiO_2/SiN_x. The microcavity was fabricated in three steps. First, the DBRs were grown on quartz substrates by plasma-enhanced chemical vapor

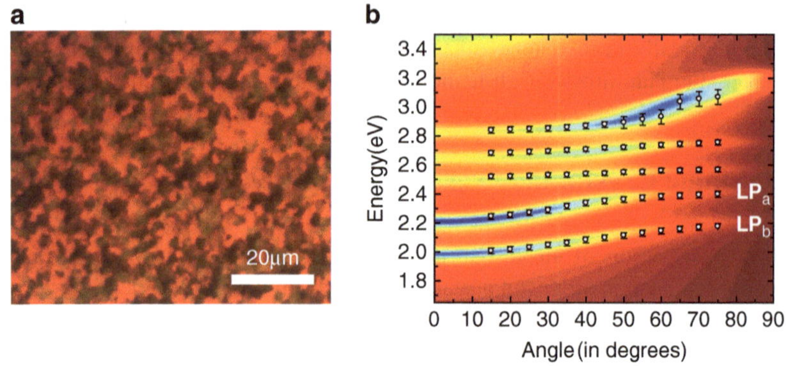

Fig. 14.4 (**a**) Differential interference contrast micrograph of a 115-nm-thick polycrystalline tetracene film from [24] grown on a silver substrate. The scale bar is indicated, and the crystalline domains are easily identified. (**b**) The contour map shows the calculated dispersion assuming an incoherent sum over randomly oriented crystalline microcavities [24]. Note the appearance of two lower branches associated with the **a**- and **b**-polarized excitonic resonances. The *circles* correspond to the experimentally determined dispersion from reflectivity

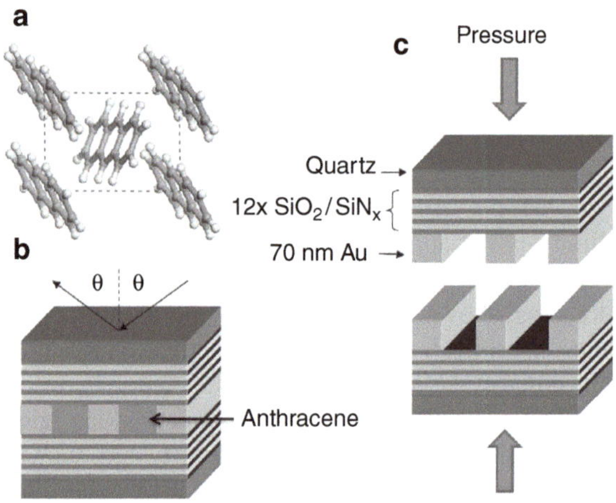

Fig. 14.5 (**a**) Crystal structure of anthracene as viewed in the [001] stacking direction. The molecular structure is also shown with white atoms corresponding to hydrogen and gray atoms to carbon. (**b**) Schematic of the complete microcavity structure from [28] showing anthracene channels enclosed between two opposing DBRs. (**c**) Schematic representation of the bonding process. Here two half structures are bonded together using mechanical pressure to form an empty microcavity. The air gaps are subsequently filled with anthracene from the melt

deposition (PECVD). Then, 70-nm-thick, 0.5-mm-wide gold stripes, separated by 0.5-mm-wide gaps, were defined along the length of the substrate using photolithography and lift-off. Two substrates were then put together such that the DBRs were positioned face-to-face with their gold stripes aligned. Application of pressure caused portions of the gold stripes to cold weld [29], holding the substrates together, as depicted schematically in Fig. 14.5c. The cold welding process was then completed by applying 75 MPa of pressure with a mechanical press for 30 s. This resulted in approximately 140-nm-thick, "empty" microcavity channels. The bonded sample was placed in a pyrex Petri dish partially filled with anthracene, arranged such that channel openings on only one side were covered by the powder. Upon melting in a nitrogen atmosphere at 240 °C, liquid anthracene penetrated the channels by capillary action and was crystallized by decreasing the temperature at a rate of $\sim$ 1 °C/min. The crystallization process was found to depend heavily on the gap thickness and cooling rate. For small gaps ($< 3\,\mu$m thick) and slow cooling ($< 1.5\,°$C/min), the process reproducibly yielded larger than cm-sized anthracene single crystals. Figure 14.6a shows a phase-contrast micrograph of the channel area taken from the finished structure. No contrast from multiple domains was observed within the channel region. In addition, the single-crystal nature of the films was confirmed by the rotational dependence of the electron diffraction patterns and the appearance of Kikuchi lines in the reflection high-energy electron diffraction patterns of the films [31]. X-ray diffraction on separated slides confirmed that, in all cases, the crystals were found to have **a** and **b** crystal axes lying in plane. The melting process itself has been described in more detail elsewhere [32, 33] but, in those cases, was found to result in uncontrolled thicknesses in the μm range.

The (inverted) transmission spectrum of a 120-nm-thick anthracene crystal grown from the melt is shown in Fig. 14.6c, with the electric field vector **E**

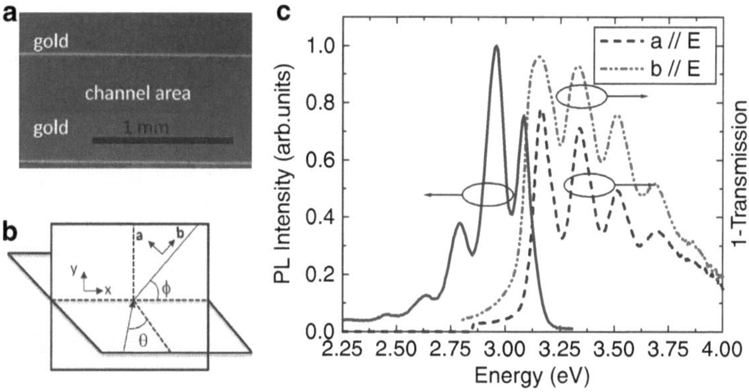

Fig. 14.6 (**a**) Differential interference contrast micrograph of the channel area taken from a complete microcavity [28]. No contrast is observed within the channel area at any magnification up to 50×. (**b**) Schematic representation showing the laboratory coordinate system (x, y), crystal axes **a** and **b**, incidence angle (θ), and the azimuthal angle (ϕ) [30]. (**c**) *Right*: inverse transmission $(1-T)$ spectrum of a 120-nm-thick anthracene single-crystal channel for electric field polarizations along the **a** and **b** crystal axes [28]. Because of the large oscillator strength of the first singlet transition in anthracene, there is a significant reflection at resonance, and the transmission cannot be taken as a direct measure of the absorption. *Left*: photoluminescence from the same film (uncorrected for self-absorption). The vibronic replicas mirror the absorption spectrum, with the highest energy peak distorted by self-absorption. Reprinted with permission from [28, 30]. Copyright (2008) by the American Physical Society

polarized along both the **a** and **b** crystal axes. Peaks in the spectra correspond to transitions from the ground state to the intermolecular vibronics of the first excited singlet state. From the transmission spectrum, the 0–0 transition energy along the **b** crystal axis is seen to be slightly red-shifted compared to that along the **a** crystal axis. At room temperature, the Davydov splitting between these two transitions (approximately 200 cm^{-1}) [34] is smaller than their inhomogenous linewidth. As a result, the redshift is difficult to resolve in transmission. Most of the oscillator strength, however, is located along the **b**-axis [34] which will be seen to result in a higher interaction potential for this excitonic component.

Angle-resolved reflectivity spectra of the microcavity structure are shown for TM-polarized light in Fig. 14.7. Spectra are shown for two different sample orientations, one with **a//E** and the other with the sample rotated 90° degrees such that **b//E**. The positions of the uncoupled excitonic resonances are indicated by gray bars. In both cases, three peaks are found to anticross around the 0–0, 0–1 and 0–2 vibronic resonances with increasing incident angle. These peaks are identified by arrows and are associated with the lower polariton (LP) and middle polariton (MP1 and MP2) branches. In contrast to the bare exciton Davydov components, LP_a and LP_b are easily distinguished. They are separated by more than 70 meV due mostly to the difference in background refractive index along each crystal axis. The dispersion extracted from the peak positions is shown in Fig. 14.8 for both TE and TM polarization with the sample oriented along either the **a** or the **b** crystal axis. The position of the TM-polarized DBR sideband is also shown.

Large differences were observed in the dispersion relation for TE- and TM-polarized light as well as for the two different sample orientations. Differences in the TE and TM dispersions occur due to the anisotropic polariton dispersion and the polarization splitting of the cavity mode. Even in the absence of anisotropy,

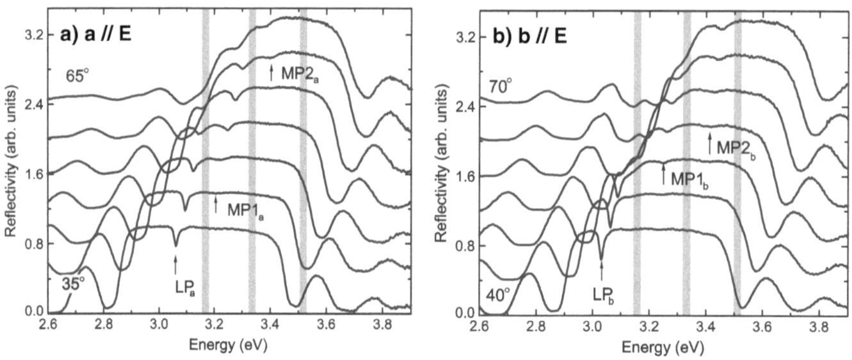

Fig. 14.7 (**a**) Angle-dependent TM-polarized reflectivity spectra taken for a sample orientation such that the propagation direction and polarization are (**a**) parallel to **a** (**b**) parallel to **b**. The lower polariton (LP) and middle polariton branches (MP1, MP2) are indicated by *arrows*. The *gray bars* indicate the energies of the uncoupled excitonic resonances. The dispersion can be extracted from the dips in reflectivity. Reprinted with permission from [28]. Copyright (2008) by the American Physical Society

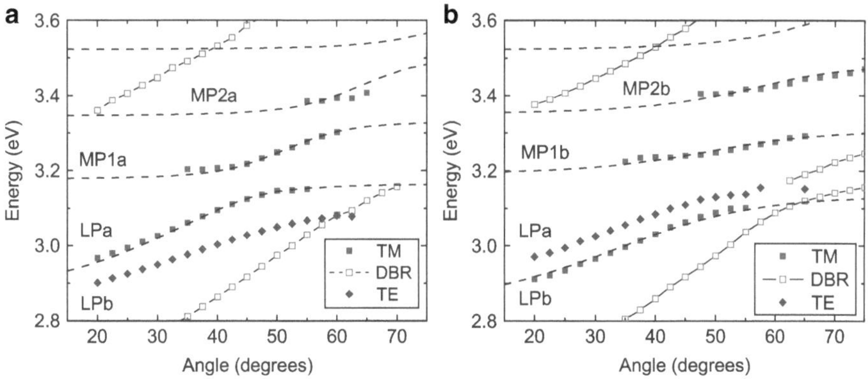

Fig. 14.8 (**a**) Dispersion extracted from the dip positions in reflectivity for a sample orientation such that the propagation direction is (**a**) parallel to **a** (**b**) parallel to **b** [28]. From the plot, we see that beyond $\theta = 60°$, both mirrors become transparent at the LP energy, and the LP ceases to exist. The *dashed lines* are a fit to a 4-body coupled harmonic oscillator Hamiltonian. For the case of TE polarization, only the dips corresponding to the LP branch are well resolved

a large polarization splitting can occur if the Fabry–Perot resonance defined by the cavity length differs significantly from the center of the DBR stop band. As the incidence angle increases, the DBR penetration depth increases for TM-polarized light but decreases for TE polarized light, leading to a polarization splitting of the photon component of the cavity [21]. In this case, the polarization splitting was calculated to be $\Delta E_{\text{TM-TE}} = 0.11 \pm 0.02\,\text{eV}$ at 60 °. At small angles, where the DBR penetration depths are nearly identical, differences between the TE and TM reflectivity are due solely to anisotropy in the crystalline cavity material. Differences between Fig. 14.8a and b, which result from rotating the sample, are also a result of the in-plane anisotropy.

When the electric field is polarized along one of the crystal axes, the dispersion is that of a cavity polariton in an isotropic material, with the appropriate excitonic resonances and cavity photon dispersion [26]. As a result, the polariton dispersion for TM-polarized light with **E**//to **a** or **b** can be modeled by a CHO Hamiltonian:

$$\mathbf{H} = \begin{pmatrix} E_{\text{ph}}(\theta) & V_1 & V_2 & V_3 \\ V_1 & E_{\text{ex0-0}} & 0 & 0 \\ V_2 & 0 & E_{\text{ex0-1}} & 0 \\ V_3 & 0 & 0 & E_{\text{ex0-2}} \end{pmatrix} \tag{14.4}$$

Interaction potentials (V_1, V_2, V_3) and uncoupled exciton energies ($E_{\text{ex0-0}}$, $E_{\text{ex0-1}}$, $E_{\text{ex0-2}}$) are included for the three lowest intermolecular vibronics [18]. The cavity photon dispersion is given by:

$$E_{\text{ph}}(\theta) = E_{\text{ph}}(0)\left(1 - \frac{\sin^2\theta}{{n_{\text{eff}}}^2}\right)^{-1/2} \tag{14.5}$$

Table 14.1 Harmonic oscillator fit parameters[a]

Orientation	$E_{ph}(0)$ (eV)	n_{eff}	V_1 (meV)	V_2 (meV)	V_3 (meV)
a//E	2.918 ± 0.006	1.73 ± 0.05	47 ± 5	43 ± 4	50 ± 30
b//E	2.939 ± 0.006	1.74 ± 0.03	108 ± 2	94 ± 4	80 ± 10

[a]The uncoupled exciton energies that were kept fixed at $E_{0\text{–}0} = 3.17\,\text{eV}$, $E_{0\text{–}1} = 3.34\,\text{eV}$, $E_{0\text{–}2} = 3.52\,\text{eV}$ for the **a** exciton, and $E_{0\text{–}0} = 3.16\,\text{eV}$, $E_{0\text{–}1} = 3.33\,\text{eV}$, $E_{0\text{–}2} = 3.51\,\text{eV}$ for the **b** exciton.

The fit which results from diagonalizing this Hamiltonian is shown in Fig. 14.8. The parameters used are summarized in Table 14.1. Damping is ignored since it results in negligible changes in the fit parameters.

The most striking differences between the two sets of parameters are the larger interaction potentials for the **b**-exciton due to its larger oscillator strength. This is also evident from the data. It should be emphasized the CHO description used here is only appropriate for propagation along the crystal axis indicated [35]. In general, we find that the dispersion depends on the *direction* of propagation.

Resonant Rayleigh scattering has been used to probe disorder in inorganic semiconductor and, more recently, to visualize the dispersion of inorganic semiconductor microcavities [36,37]. This method becomes especially useful in the case of anisotropic organic cavities where a single measurement allows all azimuthal directions to be probed.

Upon resonant excitation of a semiconductor by laser light, secondary emission at the excitation wavelength can occur either due to RRS or due to resonant PL [38]. In the absence of disorder, coherent reemission can only occur in the specular transmission and reflection directions. However, phase-preserving elastic scattering, such as that from static disorder, leads to reemission by RRS in other directions. Hence, RRS remains coherent with the excited interband polarization and, for single-scattering processes, retains the polarization direction of the exciting laser light.

The elastic scattering contours in an isotropic semiconductor microcavity form circular rings in $\boldsymbol{k}$-space as dictated by conservation of in-plane momentum, and these can be directly observed in the RRS profile [36,37,39]. The ring radius is given by the resonance condition, while their width is determined by the broadening of the polariton dispersion in $\boldsymbol{k}$-space. In an anisotropic crystal such as anthracene, a more complicated dispersion relation is thus expected, depending on both the incidence (θ) and the azimuthal (ϕ) angles, as shown in Fig. 14.6b. By mapping the RRS pattern, the full dispersion can be directly visualized, and the effects of anisotropy can be gauged by observing the shape of the elastic scattering contours.

For the RRS measurements, a tunable Ti:Sapphire laser producing 200 fs pulses with a repetition rate of 76 MHz was used for excitation. The laser output was frequency doubled by a 1-mm-thick β-BaB_2O_4 crystal and then filtered using a dichroic mirror. This generated tunable pulses spanning the entire range of energies of the lower polariton branch. The RRS pattern was projected onto a screen oriented normal to the sample surface in a transmission geometry and captured using a CCD camera. The half-angle spread in the angle of incidence was $\sim 0.5°$ for θ and ϕ.

The lower branches were resonantly excited with in-plane momentum $\boldsymbol{k}_{//} = (k_x, 0)$ (corresponding to $\phi = 0$) by selecting the appropriate polarization (TE for LP_a and TM for LP_b), energy, and angle θ. The RRS pattern obtained for particular angles and an energy of 2.950 eV is shown in Fig. 14.9. The speckle, which is observed in all patterns, suggests that the emission is mostly coherent, thereby justifying the assignment to RRS. We note that the polarized resonant PL and the RRS make it impossible to clearly distinguish between the two in a time-averaged measurement based solely on the degree of polarization with respect to the exciting laser. Indeed, regardless of the scattering direction, the emission is always polarized along the crystal axis corresponding to the excited polariton. From Fig. 14.9, it is observed that the RRS pattern corresponding to LP_a is nearly circular, while that corresponding to LP_b possesses significant asymmetry leading to an elliptical shape. This asymmetry increases as the incidence angle (and energy) is increased.

Upon laser excitation, a coherent polarization $P_{kx0,ky0}$ is created. Polarization transfer from disorder scattering can occur to $P_{kx,ky}$ if the final momentum vectors of the scattered polarization lie on the elastic scattering surface determined by the dispersion relation. The dispersion relation of crystalline organic microcavities has previously been derived for the case of perfect mirrors and two molecules per unit cells [26, 27]. It was shown that some weak mixing on the order of $(k_{//}/Q)$ [4] occurs between the two Davydov components, where $k_{//}$ is the in-plane wavevector and Q is the total wave vector. Neglecting mixing, the anisotropy effectively results in azimuthal angle–dependent interaction potentials. It was also pointed out [26] that the polaritons are composed of a combination of TE and TM modes such that the LP_a and LP_b polaritons are almost completely polarized along the **a** and **b**

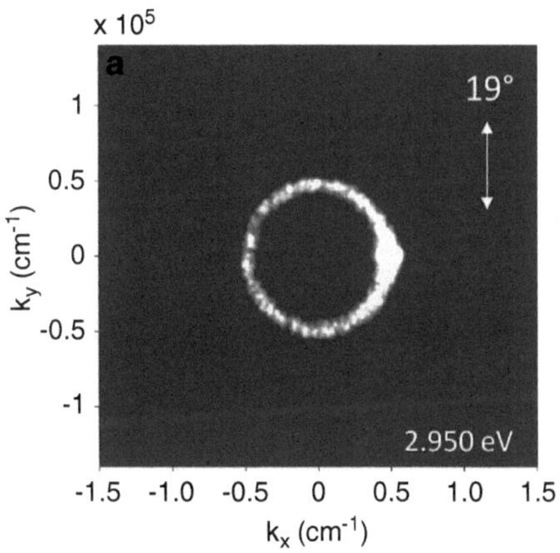

Fig. 14.9 (**a**) Resonant Rayleigh scattering (RRS) images of obtained by projecting the pattern onto a screen [30]. In both cases, the polarization direction, energy, and incident angle (θ) are indicated. (**a**) RRS pattern for excitation of LPa. (**b**) RRS pattern for excitation of LPb. The direction of excitation was chosen such that $k_y = 0$. Note the ellipticity of the LPb pattern. The bright spot is the specular transmission of the exciting laser beam. The $(k_x,\ k_y)$ axes range from (0,0) to (1.5, 1.4) $\times\ 10^5\ \text{cm}^{-1}$

crystal axes, respectively, which is consistent with our observations. Experimentally, however, the ellipticity observed is much stronger than that which was theoretically predicted.

The differences occur due to anisotropies in the photonic component of the cavity arising from the different background (i.e., nonresonant) refractive index along each crystal axis and the polarization-dependent penetration depth of the DBR mirrors. The latter accounts for the largest fraction of the ellipticity of the RRS contours by leading to an elliptical dispersion relation for the Fabry–Pérot cavity mode and a polarization-dependent interaction potential.

We have calculated the dispersion relation of our structure based on an anisotropic transfer matrix that consists of propagating the in-plane components of the electric and magnetic fields along the various interfaces according to first-order Maxwell equations [35, 40]. An orthorhombic dielectric tensor was used with $\varepsilon_{z0} = 2.9$, while $\varepsilon_{y0} \equiv \varepsilon_{\alpha}(\omega)$ and $\varepsilon_{x0} \equiv \varepsilon_{b}(\omega)$ taken from [41]. An exact calculation accounting for the monoclinic structure would require knowledge of the frequency-dependent Euler angles. Nominal measured film thicknesses were used, and no parameters were fitted. The resulting dispersion is in good agreement with experiment and is shown in Fig. 14.10. The method fully accounts for the response of the (isotropic) DBRs and, in particular, the polarization dependence of the penetration depth.

In the linear regime, angle-resolved PL from the single-crystalline microcavities was measured using continuous-wave excitation from the $\lambda = 325$ nm line of a HeCd laser. The sample was excited over a 100 μm spot at a power < 25 μW. As is often the case for high Q organic microcavities, photoluminescence is only observed from the lower branches. It was found that LP_a and LP_b emission is always linearly polarized along the **a**- and **b**-axes, respectively, regardless of the sample orientation with respect to the plane of incidence. Moreover, the polarization dependence of the PL at normal incidence, which is displayed in Fig. 14.11a, provides clear evidence of the single-crystalline nature of the cavity material within the spot size.

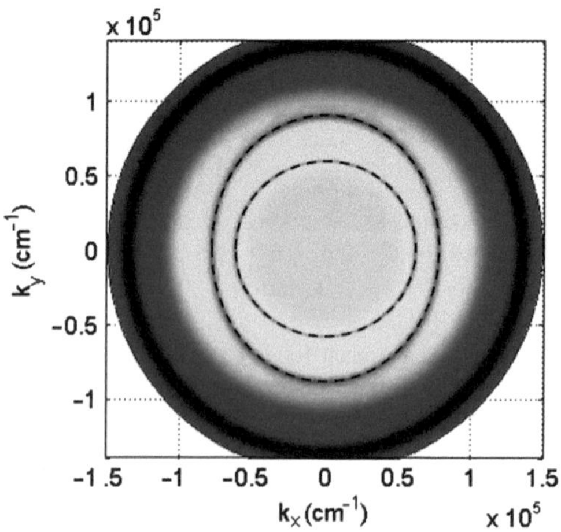

Fig. 14.10 (**a**) Calculated resonance condition for the two branches at an excitation energy of 2.950 eV [30]. The contours are plotted using a logarithmic color scale to reduce the contrast. The dark region identified by *dashed lines* corresponds to the poles of the system transfer matrix [30]. The *outer circle* corresponds to the edge of the light cone. Reprinted with permission from [30]. Copyright (2008) by the American Physical Society

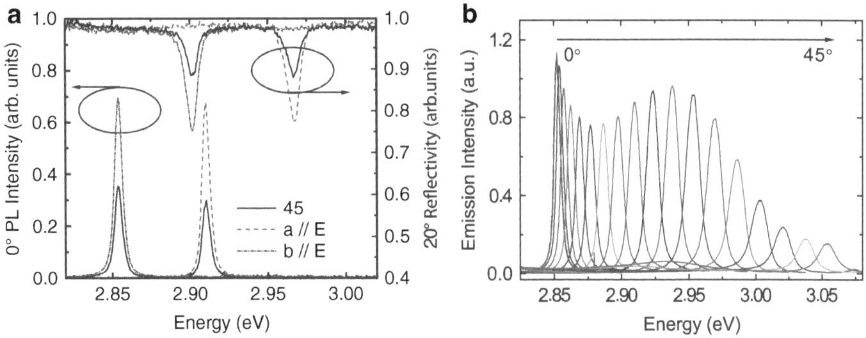

Fig. 14.11 (**a**) *Left*: normal incidence photoluminescence and *right*: 20° TM-polarized reflectivity shown for three different polarizer orientations. The sample was oriented such that **a**//**E** ($\phi = 90°$). The complete extinction of each polariton branch at normal incidence highlights the single-crystal nature of the microcavity. Reprinted with permission from [28]. Copyright (2008) by the American Physical Society. (**b**) Angle-resolved PL from the microcavity measured under along $\phi = 0°$ [30]. Only LP_b is shown. Note the peak in intensity at ~ 2.94 eV

The angle-resolved PL (shown only for LP_b) displayed in Fig. 14.11b possesses an unusual maximum at ~ 2.94 eV whose energy corresponds to a transition from the first electronically excited state to the first vibronic of the electronic ground state. This feature is present for both LP_b and LP_a in all cavities examined, regardless of the initial detuning. We will see that it plays an important role in the dynamics and provides the mechanism for polariton lasing.

Finally, the cavity quality factor, Q, can be extracted from the LP linewidths at shallow angles, when the LP is mostly photon-like. Photoluminescence measurements yield linewidths of $\sim$ 3.5–5.5 meV, corresponding to a lower bound of $Q \approx 800$. The measured linewidths, however, are limited by averaging over the spot size. Indeed, by probing different spots on the sample, we find that the detuning can be varied by approximately 50 meV due to intrinsic variations in cavity thickness. As an upper bound estimate, we note that the calculated Q factor for this structure is $Q \approx 6,000$, assuming perfectly flat DBR interfaces.

14.4 Polariton Lasing in Single Crystalline Organic Microcavities

Coherent, laser-like emission from polaritons has been predicted to occur when the ground-state occupancy of polaritons $\langle n_{gs} \rangle$ reaches 1 ([42]). This process, known as polariton lasing, can occur at thresholds much lower than required for conventional lasing. Polariton lasing and Bose–Einstein condensation (BEC) of polaritons have both recently been observed in inorganic semiconductors [9,43,44]. In this section, we review polariton lasing in crystalline anthracene microcavities, which occurs at room temperature and at a threshold below that of our best case estimate for conventional photon lasing.

The microcavity used in this study, shown schematically in Fig. 14.5b, was similar to that studied in the previous section with the exception that it contained a nominally 120-nm-thick anthracene layer such that the polariton ground state is located at $E_{\mathrm{LPB}} \approx 2.94\,\mathrm{eV}$, the same energy where a maximum in PL was observed. The full dispersion of the **b**-polarized polariton from this structure, including coupling with the higher lying vibronic replicas, is shown in Fig. 14.12.

To identify a threshold in the polariton ground-state occupation, the emission spectrum at normal incidence was measured as a function of the pump fluence, varied using a series of metallic neutral density filters (Fig. 14.13a). The sample was pumped nonresonantly with 150 fs, $\lambda = 360\,\mathrm{nm}$ pulses generated by an optical parametric amplifier (OPA) pumped by a 1 kHz Ti:Sapphire regenerative amplifier. These were focused to a 220-μm-diameter FWHM spot and incident at an angle of 10° from normal incidence. The $\lambda = 360\,\mathrm{nm}$ pump wavelength was chosen to correspond approximately to the first transmission maximum beyond the edge of the DBR stop band. The inset of Fig. 14.13b shows the integrated area of the **b**-polarized lower polariton ($\mathrm{LP_b}$) emission peak as a function of pump fluence. This emission transitions from sublinear at low fluence and dominated by bimolecular quenching, to superlinear above threshold ($P_{\mathrm{th}} = 120\,\mathrm{nJ}$). Immediately below and above

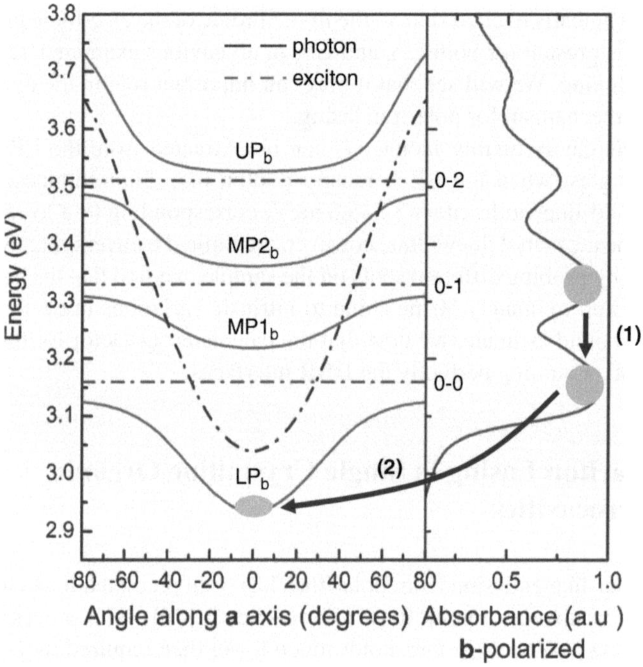

Fig. 14.12 Calculated dispersion of the **b**-polarized polariton along the **a** crystal axis ($\phi = 90°$). The lower polariton (LP), middle polariton (MP1, MP2), and upper polariton (UP) branches resulting from strong coupling of the cavity photon to the three strongest intramolecular vibronics are shown (*solid lines*). The positions of the bare exciton (*dash-dot*) and cavity photon dispersions (*dashed*) are indicated. Reprinted with permission from [45]

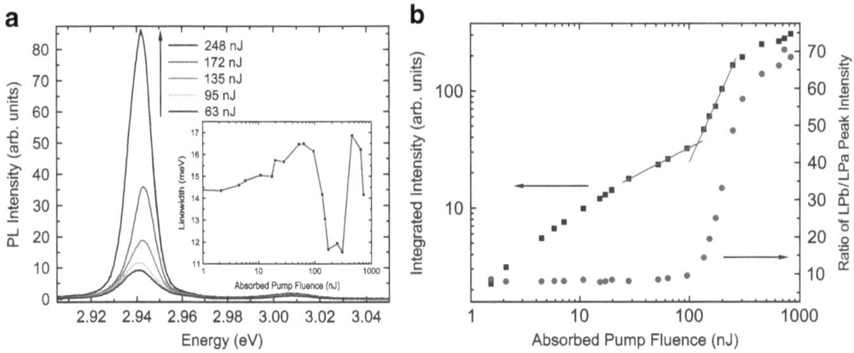

Fig. 14.13 (**a**) Photoluminescence measured at normal incidence as a function of increasing pump fluence. The peak at ∼2.94 eV corresponds to LP_b, while that at ∼3.01 eV to LP_a. *Inset*: linewidth (FWHM) of LP_b as a function of absorbed pump fluence. Upon reaching threshold, a collapse of the emission linewidth is observed. Reprinted with permission from from [45]. (**b**) Integrated intensity (*left*) and ratio of LP_b/LP_a intensity (*right*) as a function of increasing pump fluence. The lasing threshold is defined as the intersection between the sublinear regime, dominated by bimolecular quenching and the superlinear regime [45]

threshold, the intensity dependence is fit to a power law x^p with $p = 0.49 \pm 0.02$ and $p = 2.10 \pm 0.07$ respectively. At the highest fluences, the emission intensity saturates. On the other hand, emission from the **a**-polarized lower polariton (LP_a) remains in the sublinear regime both below and above threshold. A narrowing of the emission linewidth from 16.5 meV immediately below threshold to 11.5 meV above was observed. Note that because the cavity thickness is intentionally varied over the sample area, the observed linewidth is broadened here due to spatial averaging [45]. Indeed, for a second cavity probed with a smaller spot size, the emission line shape suggests the presence of multiple emission lines with a mode spacing on the order of $\Delta E \approx 1.4$ meV [45].

The absolute output power was measured at threshold with a high-sensitivity Joulemeter (3.224×10^8 V/J) to estimate the ground-state occupancy. We find an output fluence of 0.2 ± 0.1 pJ. Ignoring the weak anisotropy, the total number of states within the detection cone is calculated as [44]:

$$M = \frac{\pi D^2/4}{4\pi^2} \pi \left(k_0 \Delta\theta\right)^2, \tag{14.6}$$

where D is the spot diameter, $k_0 = 2\pi/\lambda$ and $\Delta\theta$ is the detection half angle. We obtain an average occupancy $\langle n \rangle \approx 238$/state. This value is well above the $\langle n_{gs} \rangle \approx 1$ threshold required for polariton lasing supporting the interpretation of final state stimulation.

A small blue shift of the polariton is observed above threshold, but this does not appear to be related to exciton–exciton interactions as in the case of inorganic semiconductors. Depending on the detuning between the excitonic resonance and the cavity mode, the polariton will blue or redshift slightly toward 2.94 eV upon reaching threshold. By measuring spots with different detunings, we have observed nonlinear emission from both the **a**- and **b**-polarized lower polaritons (LP_a and LP_b), with the laser emission polarized accordingly. However, lasing is only observed when the detuning is such that the *polariton ground state* is located near 2.94 eV. This corresponds to the energetic separation between the exciton reservoir and the first vibronic sublevel of the *molecular ground* state (E_{vib}). This suggests that the polariton ground state is directly populated from the reservoir.

To investigate the polariton population, angle-resolved PL was measured along the **a**-axis of the crystal with a detection cone of half angle, $\Delta\theta = 4°$. Photoluminescence below threshold, at a pump fluence of $0.003P_{th}$, is shown in Fig. 14.14a. The observed peak positions are directly associated with the polariton dispersion, $E_{LP}(k_a, k_b = 0)$, where k_a and k_b are the wave vectors along the **a** and **b** crystal axes respectively. Emission from LP_b is strongest and is TE polarized, while emission from LP_a is TM polarized. In both cases, emission is strongest at $\theta = 0°$ (corresponding to zero in-plane wave vector) and decreases monotonically with increasing angle. This reflects the decreasing polariton population and increasing lifetime with increasing angle along the branch. The emission spot below threshold is shown in Fig. 14.15a. It is circular and Gaussian in intensity.

The angle-resolved PL above threshold at a pump fluence of $1.3P_{th}$ shown in Fig. 14.14b indicates that the dispersion is unchanged, confirming that the structure is still in the strong-coupling regime. However, the emission occurs almost exclusively from LP_b and becomes localized near the bottom of the branch. In addition, the emission spot changes drastically, acquiring Hermite–Gaussian TEM_{01} modal structure. In regions with a detuning where lasing does not occur, neither modal structure nor bright spots are seen at the same pump intensity. The bright

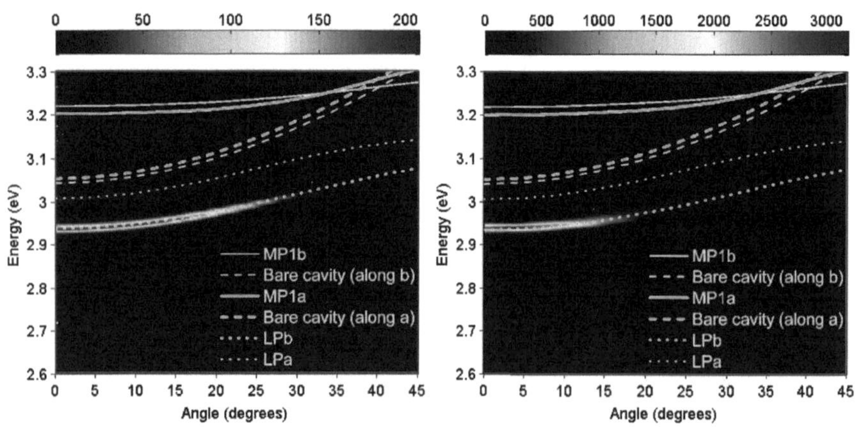

Fig. 14.14 (**a**) Angle-resolved photoluminescence measured below threshold. Emission is observed from both LP_a (weak) and LP_b. *Dashed lines* corresponding to the position of the "uncoupled" cavity modes for each polarization are shown for reference. The *solid line* shows the position of MP1 which is not observed in PL. (**b**) Same as in (**a**) but above threshold. Emission is seen to still occur from the lower polariton branches and not the "uncoupled" cavity, providing strong evidence for polariton lasing. Reprinted with permission from [45]

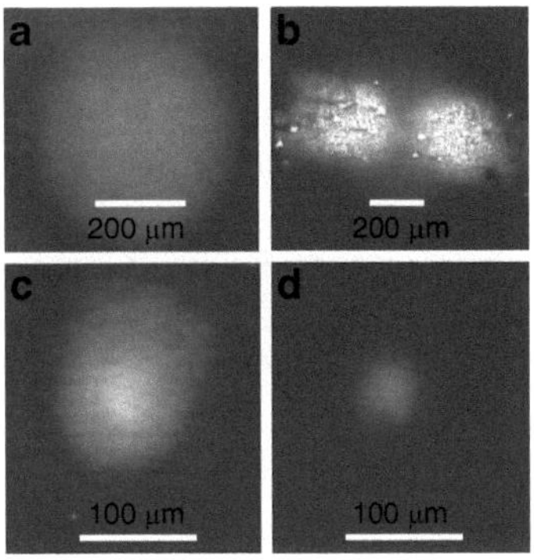

Fig. 14.15 Emission spot imaged in the near field using a charge-coupled device camera. The images are shown for a 120-nm-thick cavity (**a**) below and (**b**) above threshold [45] as well as a 230-nm-thick cavity (**c**) below and (**d**) above threshold. A different excitation spot size was used for the two different thickness microcavities as can be observed in the images

emission spots observed above threshold appear to be localized near structural inhomogeneities on the film, while the two modal lobes retain the same shape on various locations on the sample where lasing occurs. Similar transverse mode profiles have resulted from anisotropy in vertical-cavity surface-emitting lasers [46]. Note that for a thicker 230-nm cavity that was probed with a smaller spot size, a simple collapse of the emission spot was observed above threshold [45].

To determine the nature of the scattering mechanism responsible for populating the lower branch, we can calculate the scattering rate of a reservoir exciton to the polariton ground state via the nonradiative emission of an optical phonon of energy E_{vib} using the model proposed by Litinskaya and Agranovich [14]. In this model, the scattering rate from state $\boldsymbol{k}$ to $\boldsymbol{k}'$ is given by:

$$W_{k\to k'} = \frac{2\pi}{\hbar}\frac{gE_{\mathrm{vib}}}{N}\left|\beta_k\right|^2\left|\beta_{k'}\right|^2\delta\left(E\left(k\right)-E\left(k'\right)-E_{\mathrm{vib}}\right) \tag{14.7}$$

where N is the total number of molecules, β_k is the excitonic component of the state at $\boldsymbol{k}$, and g is the exciton–phonon coupling constant. The total scattering rate was calculated by summing numerically over the exact dispersion using the molecular density $\rho = 2/476.5\,\text{Å}^3$ and $g = 0.85$. We find $W_{\mathrm{total}} \approx (100\,\mathrm{ns})^{-1}$ which is much slower than the radiative lifetime of anthracene.

This suggests that instead, radiative scattering, mediated by the dipole interaction, is dominant [47]. This interaction results from the transition dipole moment that mixes the ground state of S_1 to the first excited vibronic of S_0, a transition which is weakly coupled to light. The scattering rate resulting from this interaction is proportional to the *photonic* component of the final state and also peaked at an energy which is one phonon below that of the reservoir. As noted in the previous section, even below threshold, a peak in PL at an emission angle (θ) corresponding to $E_{\mathrm{LP}}(\theta) \approx 2.94\,\mathrm{eV}$ is observed [30], independent of detuning, highlighting the effectiveness of this scattering mechanism. This behavior is in striking contrast to inorganic semiconductors, where relaxation occurs via multistep relaxation along the lower branch and where exciton–exciton interactions play a crucial role in overcoming the polariton bottleneck [48].

The emission lifetime was measured with a streak camera using the same detection geometry. The time dependence of the integrated intensity originating from $\mathrm{LP_b}$ is shown in Fig. 14.16a. At $0.018P_{\mathrm{th}}$, the emission follows single-exponential decay at long times, with a lifetime $\tau = 1.04 \pm 0.02\,\mathrm{ns}$. At higher pump fluences, the emission lifetime is further reduced at short times, along with an increase in the influence of bimolecular quenching. Above threshold, the emission decay time collapses to $< 30\,\mathrm{ps}$ and is limited by the resolution of our experimental apparatus. The short emission lifetime above threshold is a result of effective scattering from the exciton reservoir to the bottom of the lower polariton branch, and the short cavity photon lifetime compared to the nonradiative polariton decay rate.

From the photon fraction (α) obtained using the CHO model and the integrated PL intensity $I_{\mathrm{LP}}(\theta)$, the lower polariton population density $N_{\mathrm{LP}}(\theta)$ below and above threshold can be extracted using:

$$N_{\mathrm{LP}}(\theta) \propto \frac{I_{\mathrm{LP}}(\theta)}{\alpha(\theta)^2} \tag{14.8}$$

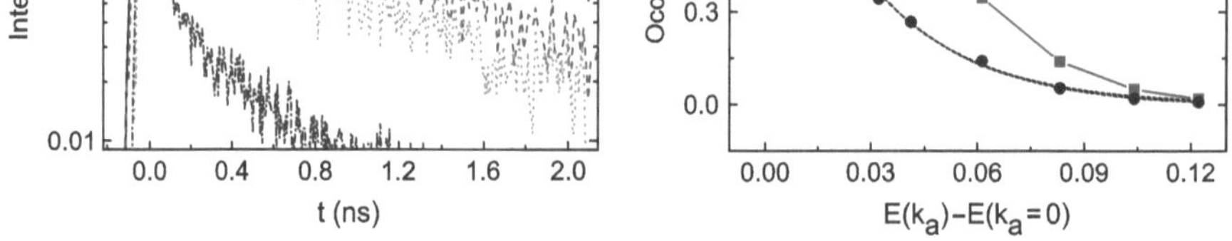

Fig. 14.16 (**a**) Photoluminescence transient intensity taken in the normal direction for increasing pump fluencies. The transients were normalized at $t = 0$ and show the evolution from weakly quenched, to strongly quenched, to lasing. The single-exponential decay time constant at low pump fluence (or long times) varies between 0.7 and 2.5 ns, depending on the probed spot due to variations in impurity and trap densities on the sample. The decay time, however, always collapses to $< 30\,\mathrm{ps}$ above threshold. (**b**) Polariton distribution functions below and above threshold. Above threshold, the distribution can be fit to a Boltzmann distribution: $n(E) \propto \exp(-\Delta E/k_{\mathrm{B}}T_{\mathrm{lattice}})$ (where $\Delta E \equiv E(k_a) - E(k_a = 0)$ and k_{B} is Boltzman's constant). The fit results in a lattice temperature $T_{\mathrm{lattice}} = 326\,\mathrm{K}$. Reprinted with permission from [45]

To convert to occupation number, the data is then divided by cos θ to correct for the fraction of detected states in momentum space.

The shape of the polariton population density function shown in Fig. 14.16b is determined by the various scattering rates and may or may not reach thermal equilibrium depending on the decay rate of polaritons compared to these rates. Below threshold, the system appears to be far from thermal equilibrium. Above threshold, the peak emission is found slightly off-normal at $\theta = 2.5°$. Indeed, this angle provides the resonance condition for direct pumping from the reservoir. However, for higher angles, the population distribution changes to a shape which is indistinguishable from a Boltzmann distribution with an activation energy corresponding to a lattice temperature $T_{\text{lattice}} = 326\,\text{K}$. This suggests that relaxation above threshold is greatly enhanced and that even under pulsed excitation, some thermalization occurs. The population near $\boldsymbol{k} = 0$, however, differs from that expected for a Bose–Einstein condensate at thermal equilibrium, and from Fig. 14.13b, it is apparent that in anthracene, nonradiative decay by bimolecular annihilation severely limits the possibility of achieving extremely large ground-state populations.

The observation of stimulated emission from crystalline anthracene has been reported at low temperature [49], but to our knowledge, this is the first demonstration of lasing from crystalline anthracene. We can, nevertheless, estimate the pump density required for conventional photon lasing in the same structure. Figure 14.17 shows the amplified spontaneous emission (ASE) spectrum emitted from the facet of a neat anthracene crystal. Using the variable stripe length measurement [50], an effective modal gain coefficient $g_{\text{eff}} = 19.6 \pm 0.2\,\text{cm}^{-1}$ at a pump density of $43\,\mu\text{J/cm}^2$ ($15.9\,\mu\text{J/cm}^2$ absorbed) was measured. In the microcavity, the mirror

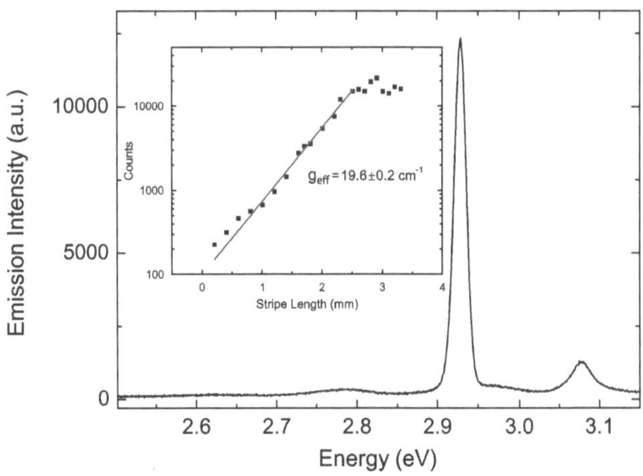

Fig. 14.17 Amplified spontaneous emission (ASE) spectrum collected from the edge of a 120-nm-thick anthracene single crystal [45]. The *inset* shows the peak ASE intensity measured as a function of pump stripe length. To extract the effective modal gain $g_{\text{eff}}(\lambda)$, the data is fit to $I(\lambda) \propto \exp(g_{\text{eff}}(\lambda)L) - 1$ where $I(\lambda)$ is the ASE intensity and L is the stripe length [50]

losses alone are calculated to be at least $540\,\mathrm{cm}^{-1}$. Assuming the best case where the gain scales linearly with the pump density (ignoring bimolecular quenching), a threshold $P_{\mathrm{th}} = 430\,\mu\mathrm{J/cm}^2$ can be calculated for conventional lasing. At threshold for polariton lasing, however, an absorbed pump density of only $P_{\mathrm{th}} = 320\,\mu\mathrm{J/cm}^2$ was measured.

Still, the distinction between polariton and photon lasing remains a subtle point in this organic semiconductor. Unlike the case of inorganic semiconductors, the Mott density is never exceeded in organic semiconductors (we know of no demonstration of a free electron–hole plasma in an organic material), and as a result, the polariton emission is never bleached. However, "transparency" is *always* exceeded for vibronic sublevels of the ground state since these are unoccupied in thermal equilibrium. Absorption occurs from the relaxed ground state to vibronic sublevels of the excited state, while spontaneous emission occurs from the relaxed excited state to vibronic sublevels of the ground state. In the microcavity, the latter transitions, with the exception of the 0–0, are not strongly coupled to light. As a result, pumping of the lower branch can occur radiatively via these. The mechanism for polariton lasing is thus analogous to that giving rise to amplified spontaneous emission or conventional lasing with the important distinction that here the polariton ground-state density $\langle n_{\mathrm{gs}} \rangle$ must exceed one, rather than the photonic mode density.

In inorganic microcavities, the polariton lasing threshold has been found to be as much as one order of magnitude lower than that for conventional photon lasing. However, in organic crystals with inversion symmetry (e.g., anthracene), only high-order multipoles contribute to the dynamic exciton–exciton interaction which is responsible for the efficient polariton–polariton scattering of inorganic semiconductors, making it relatively inefficient [51]. It is the large intramolecular phonon energy characteristic of organic materials, however, that enables direct pumping of the lower branch from the reservoir. Organic materials with efficient nonradiative scattering mechanisms such as exciton–phonon and exciton–exciton interactions may further reduce the polariton lasing threshold compared to that of conventional lasing. For example, the phase diagram of the experimental structure was calculated in the thermodynamic limit [52] (assuming an infinite particle lifetime), and we find a threshold density for condensation $N_{\mathrm{th}} = 3 \times 10^{13}/\mathrm{cm}^3$ as shown in Fig. 14.18. Direct electrical pumping of such a structure beyond threshold would be feasible with current organic light-emitting diode architectures. Moreover, electrically driven polariton emission has already been demonstrated in organic microcavities which use metallic mirrors as contacts [54, 55].

Finally, we should note that the role of incoherent states in molecular crystals is not without importance. In anthracene, for example, energetic and positional disorder is small due to the crystalline nature of the structure. At room temperature, however, the exciton–phonon interaction still serves to quickly localize Frenkel excitons. The notion of states with definite $\boldsymbol{k}$ beyond the DBR stop band is thus inadequate. One then still expects a large number of incoherent states to occur in organic crystals at room temperature based on the arguments presented at the end

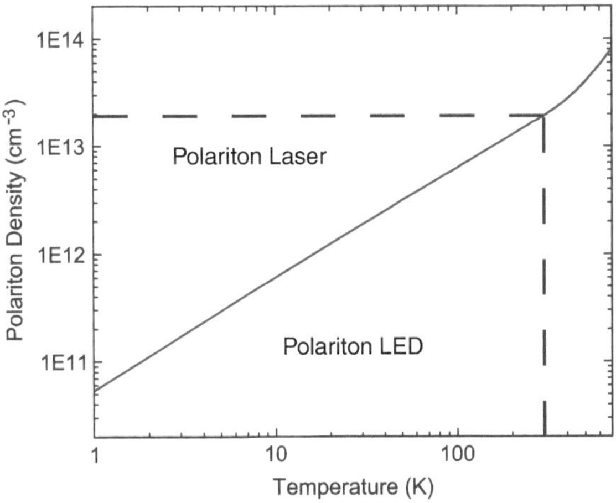

Fig. 14.18 Polariton phase diagram for quasicondensation [45] in the thermodynamic limit resulting from the finite system size [53]. This type of calculation assumes an infinite particle lifetime and ignores interactions between particles. Equivalently, similar results are obtained if the scattering rates are much faster than the particle decay rates. The critical density at room temperature, indicated by *dashed lines*, corresponds to approximately $3 \times 10^{13}/\text{cm}^3$

of Sect. 14.2. At low temperature, however, the phonon occupation number (and linewidth) of organic crystals is greatly reduced. One would expect the majority of the states to consist of wavelike excitons and polaritons. This delocalization may give rise to interesting physics due to the greater degree of interaction between such extended states.

14.5 Hybrid Organic–Inorganic Microcavities

The realization of an electrically pumped organic polariton laser would represent a substantial technological and scientific achievement. Unfortunately, this entails all of the difficulties of realizing a conventional electrically pumped organic laser [56, 57]. Electrical injection will result in additional decay pathways for polaritons and reservoir excitons such as singlet-polaron and singlet-triplet annihilation. Moreover, at high current densities, polaron absorption can be substantial and will reduce the cavity Q and thus the polariton lifetime. The only potential benefit of electrical injection is that exciton–polaron scattering may provide an additional relaxation pathway within the lower branch. For this reason, before moving to electrical injection, it is necessary to realize polariton lasers with thresholds much lower than that demonstrated in for anthracene.

An alternative route toward this goal is to use hybrid organic–inorganic microcavities where electrical injection is achieved via the inorganic semiconductor structure. Firstly, efficient nonradiative energy transfer from the inorganic polariton to a lower lying organic polariton may provide sufficient excitation densities to reach threshold. Secondly, the creation of a hybrid Frenkel–Wannier–Mott polariton has been predicted by Agranovich to result in strong nonlinearities [58] that, for a suitably designed structure, could be used to enhance relaxation toward $\boldsymbol{k} = 0$.

The realization of strongly coupled organic–inorganic microcavities has been demonstrated by both Holmes et al. [59] and Wenus et al. [60]. The first demonstration, which uses a conventional red vertical-cavity surface-emitting laser (VCSEL) architecture is perfectly suited for electrical injection. A schematic of the optically pumped structure which hybridizes 9 InGaP quantum wells (QWs) with a neat tetraphenylporphyrin (TPP) layer and the dispersion relation measured at 100 K are shown in Fig. 14.19. The middle branch corresponds to a hybrid Frenkel–Wannier–Mott polariton, while the lower branch is Frenkel-like. One drawback of this structure is that low temperatures are needed to reduce the InGaP QW damping to a level such that strong coupling can be observed. The use of a different material set such as GaN, for example, which possess a stable exciton at room temperature, could alleviate this restriction.

Angle-resolved PL from the hybrid microcavity under nonresonant excitation ($\lambda = 488$ nm) is shown in Fig. 14.20 for both 4 K and 100 K. In both cases, PL is observed from the LP branch and from a high-energy feature corresponding to the QW. The change in the relative intensity of both branches with temperature is concomitant with the increase in the QW PL efficiency at lower temperatures but also with increased trapping of excitation in the QW region [61]. Indeed, the absence of significant dispersion in the high-energy emission suggests that it is due to emission from localized excitons trapped by disorder in the ternary/quaternary well/barrier region, rather than true UP emission. This is consistent with a situation

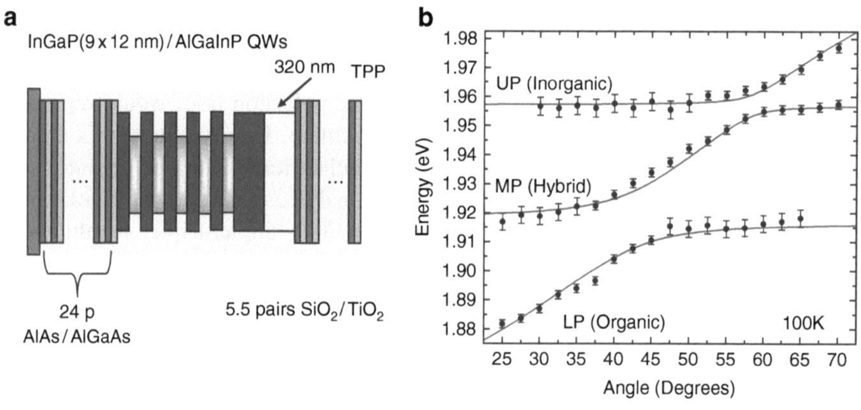

Fig. 14.19 (**a**) Schematic diagram of the hybrid inorganic–organic microcavity used in [59]. The structure is composed of 9 InGaP quantum wells (QWs) and 320-nm-thick TPP layer sandwiched between two DBRs. The QWs are situated at the antinodes of the cavity electric field. (**b**) Dispersion extracted from angle-resolved microcavity for the hybrid microcavity at 100 K. The LP branch has organic character, the UP inorganic, while the middle branch is hybridized in nature

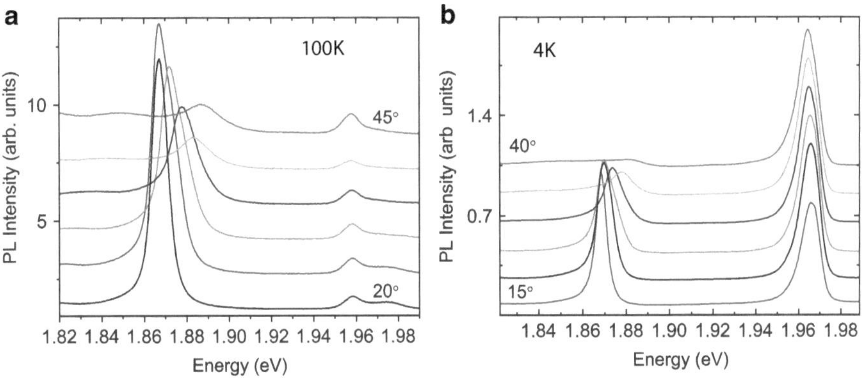

Fig. 14.20 Angle-resolved photoluminescence for the hybrid microcavity at (**a**) 100 K and (**b**) 4 K for a range of selected angles (shown in 5° increments). At 100 K organic polariton emission is dominant, while at 4 K inorganic emission dominates

where most UP emission is effectively trapped by the disorder, but where any hybrid branch population relaxes quickly to the LP. Here, both semiconductors are simultaneously excited by the pump light and further studies should allow for a determination of the relative transfer rates.

14.6 Conclusions

The study of organic cavity polaritons is still in its infancy as compared to the field of inorganic microcavities. A better understanding of the dynamics, however, is quickly emerging, providing new guidelines for the design of structures and the choice of materials [15, 47]. The development of polariton OLEDs [54, 55] and the demonstration of organic polariton lasing [45], for example, are significant steps forward, but the realization of ultralow thresholds and Bose–Einstein condensation will require materials possessing faster relaxation mechanisms. Cavities containing J-aggregates are known to exhibit rapid relaxation from the reservoir, but at the moment are plagued by high nonradiative losses. Organic materials exhibiting strong nonlinearities and hybrid organic–inorganic microcavities could potentially harness the strong exciton–exciton interactions that have been so critical in their inorganic counterparts. Indeed, the wealth of organic materials available suggests that the future is certain to unveil exciting new phenomena.

Acknowledgements The authors wish to acknowledge the Air Force Office of Scientific Research and Universal Display Corp. for partial financial support of this work.

References

1. S.R. Forrest, Chem. Rev. **97**(6), 1793–1896 (1997)
2. M. Pope, C.E. Swenberg, M. Pope, *Electronic Processes in Organic Crystals and Polymers*, 2nd edn. (Oxford University Press, New York, 1999)
3. F.F. So, S.R. Forrest, Phys. Rev. Lett. **66**(20), 2649–2652 (1991)
4. V.M. Agranovich, *Excitations in Organic Solids* (Oxford University Press, Oxford, 2009)
5. R.R. Lunt, K. Sun, M. Kröger, J.B. Benziger, S.R. Forrest, Phys. Rev. B: Condens. Matter Mater. Phys. **83**, 064114 (2011)
6. M.A. Baldo, V.G. Kozlov, P.E. Burrows, S.R. Forrest, V.S. Ban, B. Koene, M.E. Thompson, Appl. Phys. Lett. **71**(21), 3033–3035 (1997)
7. R. Butte, G. Christmann, E. Feltin, J.F. Carlin, M. Mosca, M. Ilegems, N. Grandjean, Phys. Rev. B: Condens. Matter Mater. Phys. **73**(3), 033315 (2006)
8. R. Shimada, J. Xie, V. Avrutin, U. Ozgur, H. Morkoc, Appl. Phys. Lett. **92**(1), 011127 (2008)
9. S. Christopoulos, G.B.H. von Hogersthal, A.J.D. Grundy, P.G. Lagoudakis, A.V. Kavokin, J.J. Baumberg, G. Christmann, R. Butte, E. Feltin, J.F. Carlin, N. Grandjean, Phys. Rev. Lett. **98**(12), 126405 (2007)
10. G. Christmann, R. Butte, E. Feltin, J.F. Carlin, N. Grandjean, Appl. Phys. Lett. **93**(5), 1102 (2008)
11. D.G. Lidzey, D.D.C. Bradley, M.S. Skolnick, T. Virgili, S. Walker, D.M. Whittaker, Nature **395**(6697), 53–55 (1998)
12. D.G. Lidzey, D.D.C. Bradley, T. Virgili, A. Armitage, M.S. Skolnick, S. Walker, Phys. Rev. Lett. **82**(16), 3316–3319 (1999)
13. T. Kobayashi, *J-Aggregates* (World Scientific, Singapore, 1996)
14. M. Litinskaya, P. Reineker, V.M. Agranovich, J. Luminesc. **110**(4), 364–372 (2004)
15. P. Michetti, G.C. La Rocca, Phys. Rev. B: Condens. Matter Mater. Phys. **77**(19), 195301 (2008)
16. T. Virgili, D. Coles, A.M. Adawi, C. Clark, P. Michetti, S.K. Rajendran, D. Brida, D. Polli, G. Cerullo, D.G. Lidzey, Phys. Rev. B: Condens. Matter Mater. Phys. **83**(24), 245309 (2011)
17. G.C. Schatz, M.A. Ratner, *Quantum Mechanics in Chemistry* (Dover, NY, USA, 2002)
18. R.J. Holmes, S.R. Forrest, Phys. Rev. Lett. **93**(18), 186404 (2004)
19. L. Fontanesi, L. Mazza, G.C. La Rocca, Phys. Rev. B: Condens. Matter Mater. Phys. **80**(23), 235313 (2009)
20. A. Kavokin, J.J. Baumberg, G. Malpuech, F.P. Laussy, *Microcavities* (Oxford University Press, Oxford, 2007)
21. G. Panzarini, L.C. Andreani, A. Armitage, D. Baxter, M.S. Skolnick, V.N. Astratov, J.S. Roberts, A.V. Kavokin, M.R. Vladimirova, M.A. Kaliteevski, Fiz. Tverd. Tela (Leningrad) [Sov. Phys. Solid State] **41**(8), 1223–1238 (1999)
22. V.M. Agranovich, M. Litinskaia, D.G. Lidzey, Phys. Rev. B: Condens. Matter Mater. Phys. **67**(8), 085311 (2003)
23. P. Michetti, G.C. La Rocca, Phys. Rev. B: Condens. Matter Mater. Phys. **71**(11), 115320 (2005)
24. S. Kena-Cohen, S.R. Forrest, Phys. Rev. B: Condens. Matter Mater. Phys. **77**(7), 073205 (2008)
25. A.S. Davydov, *Theory of Molecular Excitons* (Plenum Press, New York, 1971)
26. M. Litinskaya, P. Reineker, V.M. Agranovich, Phys. Status Solidi A: Appl. Res. **201**(4), 646–654 (2004)
27. H. Zoubi, G.C. La Rocca, Phys. Rev. B: Condens. Matter Mater. Phys. **71**(23), 235316 (2005)
28. S. Kena-Cohen, M. Davanco, S.R. Forrest, Phys. Rev. Lett. **101**(11), 116401 (2008)
29. C. Kim, P.E. Burrows, S.R. Forrest, Science **288**(5467), 831–833 (2000)
30. S. Kena-Cohen, M. Davanco, S.R. Forrest, Phys. Rev. B: Condens. Matter Mater. Phys. **78**(15), 153102 (2008)
31. R.R. Lunt, S. Kena-Cohen, J.B. Benziger, S.R. Forrest, Phys. Rev. Lett. **102**(6), 065504 (2009)
32. H. Kondo, A. Takeda, T. Tomikawa, H. Kurisu, S. Yamamoto, M. Matsuura, J. Luminesc. **119**, 137–141 (2006)
33. S. Hashimoto, N. Ohno, M. Itoh, Phys. Status Solidi B **165**(1), 277–286 (1991)

34. G.C. Morris, M.G. Sceats, Chem. Phys. **3**(2), 164–179 (1974)
35. D.W. Berreman, J. Opt. Soc. Am. **62**(4), 502–510 (1972)
36. T. Freixanet, B. Sermage, J. Bloch, J.Y. Marzin, R. Planel, Phys. Rev. B: Condens. Matter Mater. Phys. **60**(12), R8509–R8512 (1999)
37. R. Houdre, C. Weisbuch, R.P. Stanley, U. Oesterle, M. Ilegems, Phys. Rev. B: Condens. Matter Mater. Phys. **61**(20), R13333–R13336 (2000)
38. J. Shah, *Ultrafast Spectroscopy of Semiconductors and Semiconductor Nanostructures*, 2nd edn. (Springer, Berlin, 1999)
39. W. Langbein, J.M. Hvam, Phys. Rev. Lett. **88**(4), 047401 (2002)
40. M. Schubert, Phys. Rev. B: Condens. Matter Mater. Phys. **53**(8), 4265–4274 (1996)
41. A. Matsui, Y. Ishii, J. Phys. Soc. Jpn. **23**(3), 581 (1967)
42. A. Imamoglu, R.J. Ram, S. Pau, Y. Yamamoto, Phys. Rev. A **53**(6), 4250–4253 (1996)
43. J. Kasprzak, M. Richard, S. Kundermann, A. Baas, P. Jeambrun, J.M.J. Keeling, F.M. Marchetti, M.H. Szymanska, R. Andre, J.L. Staehli, V. Savona, P.B. Littlewood, B. Deveaud, L.S. Dang, Nature **443**(7110), 409–414 (2006)
44. H. Deng, G. Weihs, D. Snoke, J. Bloch, Y. Yamamoto, Proc. Natl. Acad. Sci. USA **100**(26), 15318–15323 (2003)
45. S. Kena-Cohen, S.R. Forrest, Nat. Photon. **4**(6), 371–375 (2010)
46. P. Debernardi, G.P. Bava, C. Degen, I. Fischer, W. Elsasser, IEEE J. Quantum Electron. **38**(1), 73–84 (2002)
47. L. Mazza, L. Fontanesi, G.C. La Rocca, Phys. Rev. B: Condens. Matter Mater. Phys. **80**(23), 235314 (2009)
48. F. Tassone, Y. Yamamoto, Phys. Rev. B: Condens. Matter Mater. Phys. **59**(16), 10830–10842 (1999)
49. O.S. Avanesyan, V.A. Bendersky, V.K. Brikenshtein, V.L. Broude, A.G. Lavrushko, I.I. Tartakovsky, P.G. Filippov, Quantum Electron. **7**(4), 403 (1977)
50. L. Dal Negro, P. Bettotti, M. Cazzanelli, D. Pacifici, L. Pavesi, Opt. Commun. **229**(1–6), 337–348 (2004)
51. M. Litinskaya, Phys. Rev. B: Condens. Matter Mater. Phys. **77**(15), 155325 (2008)
52. G. Malpuech, Y.G. Rubo, F.P. Laussy, P. Bigenwald, A.V. Kavokin, Semicond. Sci. Technol. **18**(10), S395–S404 (2003)
53. G. Malpuech, A. Kavokin, F.P. Laussy, Phys. Status Solidi A **195**(3), 568–578 (2003)
54. G.H. Lodden, R.J. Holmes, Phys. Rev. B: Condens. Matter Mater. Phys. **82**(12), 125317 (2010)
55. J.R. Tischler, M.S. Bradley, V. Bulovic, J.H. Song, A. Nurmikko, Phys. Rev. Lett. **95**(3), 036401 (2005)
56. M.A. Baldo, R.J. Holmes, S.R. Forrest, Phys. Rev. B: Condens. Matter Mater. Phys. **66**(3), 035321 (2002)
57. N.C. Giebink, S.R. Forrest, Phys. Rev. B: Condens. Matter Mater. Phys. **79**(7), 073302–073306 (2009)
58. V.M. Agranovich, G.C. La Rocca, F. Bassani, H. Benisty, C. Weisbuch, Opt. Mater. **9**(1–4), 430–436 (1998)
59. R.J. Holmes, S. Kena-Cohen, V.M. Menon, S.R. Forrest, Phys. Rev. B: Condens. Matter Mater. Phys. **74**(23), 235211 (2006)
60. J. Wenus, R. Parashkov, S. Ceccarelli, A. Brehier, J.S. Lauret, M.S. Skolnick, E. Deleporte, D.G. Lidzey, Phys. Rev. B: Condens. Matter Mater. Phys. **74**(23), 235212 (2006)
61. P.G. Lagoudakis, M.D. Martin, J.J. Baumberg, G. Malpuech, A. Kavokin, J. Appl. Phys. **95**(5), 2487–2489 (2004)

Chapter 15
Electrically Driven Polariton Light Emitting Devices

Simeon I. Tsintzos, Nikolaos T. Pelekanos, and Pavlos G. Savvidis

Abstract The aim of this chapter is to highlight the recent progress in the rapidly developing field of electrically driven polariton devices. The unprecedented potential of polariton-based devices owes its origin mainly to the bosonic property of polaritons to condense in the same final state, thus requiring no population inversion to achieve lasing. Consequently, it is widely believed that the threshold of a polariton laser is at least two orders of magnitude lower than that of a conventional semiconductor photon laser operating in the weak-coupling regime [C. Weisbuch, M. Nishioka, A. Ishikawa, Y. Arakawa, Phys. Rev. Lett. **69**, 3314–3317 (1992)]. This makes polariton lasers extremely promising as ultralow threshold lasers or as low-power sources of coherent and nonclassical light. Electrical injection of these entangled light-matter states is a key step toward the realisation of practical, compact devices. Recent demonstrations of electrically pumped polariton LEDs in organic as well as inorganic material systems operating up to room temperature have highlighted the potential of such devices for real-world polariton-based applications in optoelectronics.

15.1 Introduction

The ability to control and tailor light-matter interactions at nanometer scale has proved pivotal in improving the performance of semiconductor lasers in the last two decades. Microcavity concept has been successfully employed to reduce the number of allowed optical cavity modes, thus suppressing the spontaneous emission rate off the cavity resonance, while enhancing it at the cavity resonance. The degree of electromagnetic field confinement, set by the cavity finesse, can vary among

S.I. Tsintzos · N.T. Pelekanos · P.G. Savvidis (✉)
Department of Materials Science and Technology, University of Crete, and FORTH,
P.O. Box 1385, 71110 Heraklion, Greece
e-mail: psav@materials.uoc.gr

D. Sanvitto and V. Timofeev (eds.), *Exciton Polaritons in Microcavities*,
Springer Series in Solid-State Sciences 172, DOI 10.1007/978-3-642-24186-4_15,

different devices. Thus, two distinct regimes of light-matter interaction can be achieved, namely, the strong- and the weak-coupling regimes.

Devices operating in the weak-coupling regime normally employ wavelength size optical cavities to enhance light emission and increase spectral purity and directionality of the emission. RCLEDs with efficiency over 27% for emission into air from a single facet have been reported [2]. Devices with enhanced mirror reflection operate as vertical-cavity surface-emitting lasers (VCSELs). Since VCSELs concentrate a number of advantages compared to edge-emitting laser diodes, such as surface-emitting geometry, superior beam quality, high-coupling efficiency to optical fibers, and the possibility for high-density laser arrays, they have attracted significant interest in spite of their increased complexity.

Perhaps the ultimate optimization is realized in semiconductor microcavities, in which strong coupling between quantum well (QW) excitons and cavity photons gives rise to new hybrid half-light half-matter polariton quasiparticles [1]. Considerable work has been directed toward development of *optically pumped* polariton lasers which rely on efficient relaxation of polaritons into the lowest energy state via stimulated scattering, first identified by Savvidis et al [3]. Polariton lasing under optical pumping has been observed by the CNRS laboratories in planar and micro-sized micropillars [4] using GaAs-based microcavities and at room temperature in GaN-based microcavities by Cambridge in 2007 [5]. In a GaAs *polariton laser*, lasing is typically achieved at an injected carrier density of $3 \times 10^9\,\mathrm{cm}^{-2}$ per QW, which is well below the Mott transition density ($3 \times 10^{10}\mathrm{cm}^{-2}/\pi {a_\mathrm{B}}^2$, where a_B is the exciton Bohr radius) and the transparency condition of $10^{11}\,\mathrm{cm}^{-2}$ electron-hole pairs per QW.

This feature of polariton lasers is very promising for future applications. However, until now, polariton lasing and nonlinearities have only been demonstrated in optical experiments, which have shown the potential of polariton lasers for reduced lasing thresholds. In the context of practical light sources, polariton devices based on electrical injection, rather than optical pumping, would be by far more suited for real-world applications.

15.2 Electrically Injected Polariton LEDs

The first demonstration of an electrically pumped polariton emission has been reported by JR. Tischler in 2005 in organic semiconductor OLED [6] where the coupling strength is very large with a Rabi splitting of several hundreds of meV. Both room temperature operation and availability of well-established electrical injection technology of OLEDs contributed to their realization at this early stage, despite the fact that vast volume of the polariton physics-related research was performed on GaAs-based systems. In these organic structures, strong-coupling regime is achieved by embedding large oscillator strength J-aggregate PDAC/TDBC films inside low Q metallic mirror cavity formed by two Ag electrodes. For electrical excitation of the J-aggregates, wider band gap electron and hole-transporting BCP and poly-TPD layers were respectively inserted on either side of the device akin to a p-i-n LED design. Angle-dependent electroluminescence measurements showed anticrossing behavior and confirmed that emission arises from polariton states. However, one potential drawback in organic microcavities is that the electroluminescence spectra are superposed with emission from the hole-injection TPD layer and localized excitonic states that do not strongly couple to cavity photons, restricting the purity and efficiency of polaritonic emission.

Around that time, strong-coupling regime based on intersubband (ISB) transitions had been predicted and observed in a rather different types of system, namely, quantum cascade structures embedded in a planar microcavity [9]. In these systems, strong coupling between ISB transitions of the QW to the cavity mode of a ridge waveguide gives rise to the new eigenstates called ISB polaritons. Furthermore, in 2007, Sirtori et al. demonstrated the first electrically injected mid-IR polariton QC LED device operating at a temperature of 78 K [7, 8]. Electroluminescence measurements showed that electrical injection into polariton states persists up to room temperature. Under electrical injection, the spectral features of the emitted light change drastically, as electrons are resonantly injected in a reduced part of the polariton branches. The strong-coupling regime is particularly interesting for light emitting devices (LEDs) based on ISB transitions. This is because nonradiative processes control the lifetime of electrons in excited subbands affecting adversely their radiative efficiency. However, in the strong-coupling regime, the excitation dynamics is instead dominated by the Rabi oscillation frequency rather than the nonradiative relaxation rates. Stimulated emission of ISB polaritons could also lead to inversionless mid- and far-infrared lasers, with lower thresholds with respect to quantum cascade lasers.

However, until now to our knowledge, no nonlinearities have been observed using either of the above injection schemes. On the other hand, the GaAs system, when, referring to interband transitions, is very attractive since its strong nonlinearities could be exploited in an electrically pumped device. However, electrical injection of organic microcavities and unipolar ISB polariton quantum cascade devices is very different to electrical injection of GaAs planar microcavities and faces several technological challenges owing mainly to high resistivity of hole-transporting DBR mirror. Furthermore, electric field penetration inside highly doped DBR mirrors, due to free carrier absorption can reduce significantly microcavity finesse which is crucial for retaining strong-coupling regime. Therefore, it is clear that before any electrically pumped GaAs microcavity polariton LED device was to be realized, these issues had to be resolved.

15.3 Polariton LED Design Considerations

15.3.1 Resistivity of DBRs

It is well known that the strength of light-matter coupling depends critically on reducing the cavity mode volume and therefore on the ability to fabricate high reflectivity DBR mirrors. This is normally achieved by resorting to alternating $\lambda/4$

layers with high refractive index contrast, which in GaAs microcavities are typically some high and low refractive index AlGaAs layers, with different Al content. From an optical standpoint, the larger the refractive index constrast Δn between the two layers is, the fewer periods are required to obtain high reflectivity. However, large Δn contrasts are usually accompanied by significant energy band offsets at the DBR hetero-interfaces. In terms of carrier transport and electrical injection into the active region of the device, these potential barriers are usually responsible for large series resistance in the structure. Owing to comparatively higher mass of holes with respect to electrons, their passing through the interfaces of the heterojunctions is substantially restricted. As a result, the series resistance of the p-doped DBR is about ten times higher than the corresponding n-DBR resistance. This problem becomes even more intense as temperature decreases, since the thermal component of current flowing over the potential barriers is drastically reduced. Hence, at low temperatures, the tunneling current accounts for the majority of the current flowing across the DBR. This limits the electrical characteristics of the device and causes excess power consumption.

The resistance can be reduced by several methods. One method is to inject current into the active layer from the side conductive layer, bypassing the DBR mirror. Other methods that have been developed aim at lowering the DBR series resistance by minimizing the energy barriers through the use of interface composition grading techniques. The simplest approach is the insertion of a single barrier reduction layer between the high and low index DBR layers [10, 11], so that instead of one larger barrier, there are two smaller ones. More sophisticated approaches use continuous composition graded segments such as piece-wise linear [12], linear [13], parabolic [14], sinusoidal [15], and super lattice grating [11]. All these interface grading schemes have demonstrated at least two orders of magnitude improvement in the DBR series resistance compared to simple abrupt interface designs. As we will see later in this chapter, the ability to electrically pump microcavities at low temperatures is very crucial for the demonstration of GaAs-based electrically pumped polariton laser devices, since until now, polariton lasing under nonresonant optical excitation has been demonstrated up to 50 K in these systems.

15.4 Doping-Related Losses

Another important issue toward realization of electrically pumped microcavity device operating in the strong-coupling regime concerns cavity losses associated with the doping of the DBRs. Increasing the number of the DBR periods results in better optical confinement and higher photon lifetime, which in turn increases polariton lifetime and leads to breakdown of relaxation bottleneck. On the other hand, as already mentioned, DBRs should also serve as low-resistance contacts for current injection and allow effective heat dissipation from the active region. This imposes requirements on reflectance, series resistance, and thermal conductivity that are partially in conflict. For instance, while a high doping concentration is

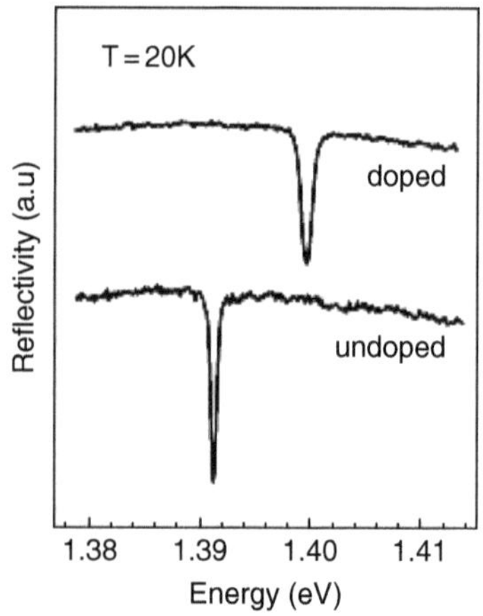

Fig. 15.1 Cavity mode comparison between a doped and an undoped microcavity sample

effective in reducing the series resistance, it is also connected to increased optical losses due to free carrier or interband absorption. Such losses are known to be more pronounced for p-type material, while on the other hand, it is generally assumed that the situation is less critical for n-type mirrors with relaxed requirements on the doping level.

The influence of the doping on the microcavity finesse and strong-coupling regime in general was performed by our group comparing the quality factor through reflectivity measurements between doped and undoped microcavities, which where otherwise structurally identical.

Figure 15.1 shows the reflectivity spectra from such samples, obtained in conditions of strong negative detuning, at an angle of incidence of 5° and at 20K. As seen from the figure, the linewidth of the cavity mode in the doped sample is broadened to 1.1 meV, compared to 0.7 meV in the undoped sample. This corresponds respectively to Q factors of 1,300 and 2,000. The reflectivity spectra close to zero-detuning condition, however, show that the strong-coupling regime remains intact in both samples, despite the moderate reduction of the cavity Q-factor, caused by the doping of the DBR mirrors in the doped sample. This represented an important step forward toward realization of the electrically injected polariton device [16].

15.5 Room Temperature GaAs Polariton LED

The basic concept of electrical injection of polaritons in strongly coupled microcavities has many similarities to the electrical injection in vertical-cavity surface-emitting lasers (VCSELs). To realize electrical injection of polaritons, we fabricate p-i-n diode MC in which holes injected from the top DBR meet electrons arriving through the bottom DBR to form excitons inside QWs located in the i-region of the device.

The sample used in the experiments presented in this chapter is a MC-LED structure grown by molecular beam epitaxy on a GaAs (001) n^+ substrate [17]. The bottom DBR mirror consists of 21 periods of GaAs/AlAs n-type doped to $2\times10^{18}\,cm^{-3}$ Si dopants, whereas the top DBR consists of 17 periods of GaAs/AlAs p-typed doped to $4\times10^{18}\,cm^{-3}$ Be dopants. The $5\lambda/2$ GaAs cavity which is embedded between the two doped DBRs contains three pairs of $In_{0.1}Ga_{0.9}As/GaAs$ QWs. To enhance the light-matter interaction, the QWs are placed at the antinodes of the electric field, as shown in the Fig. 15.2b. In this approach, polariton LED mesas with 400 μm diameter were dry etched using BCl_3/Cl_2 reactive-ion etching (RIE). The RIE was deep enough to remove the whole top p-DBR from the exposed areas, ensuring electrical isolation between adjacent polariton devices. N-type (Au/Ge) ohmic contact was evaporated on the backside of the substrate, while a ring-shaped p-type (Ti/Pt) contact was deposited on the top of the mesa. The fabricated polariton LED structure is shown schematically in Fig. 15.2a.

To avoid injection problems associated with the high resistivity of p-DBR at low temperatures, the sample was designed to exhibit strong coupling at 220 K. The first half periods of the DBRs where the penetration of the electric field into the mirror is considerable, were kept undoped to reduce free carrier absorption losses and carrier spillage into the active region of the device. Characteristic current–voltage (I–V) curve measured on the polariton diode device is shown in figure 15.3. Current injection and polariton LED operation is achieved under forward bias conditions with nearly complete screening of the diode field at the QW sites, whereas application of the reverse bias tunes the QW exciton energy via quantum confines Stark effect due to the applied electric field in the structure (Fig. 15.3).

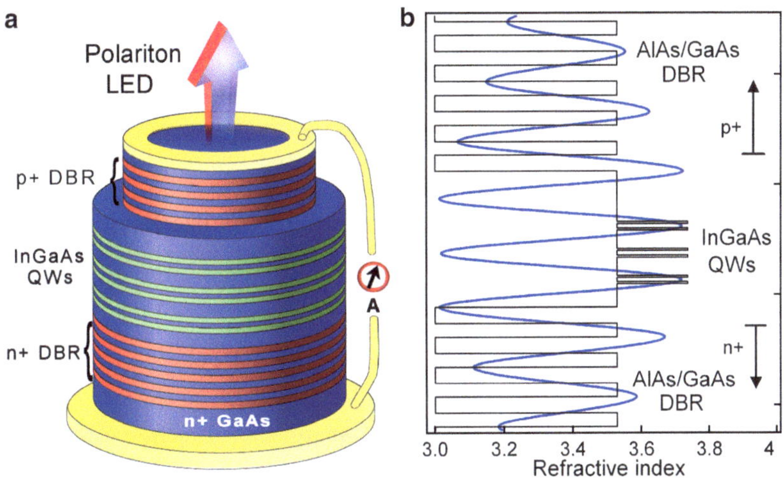

Fig. 15.2 Schematic sketch of the polariton microcavity LED. (**a**) To enable light emission from the front side, a ring shaped p-ohmic contact was deposited on the top of the mesa while the n-contact deposited on the backside of the substrate. (**b**) Electric field and refractive index distribution along the structure.

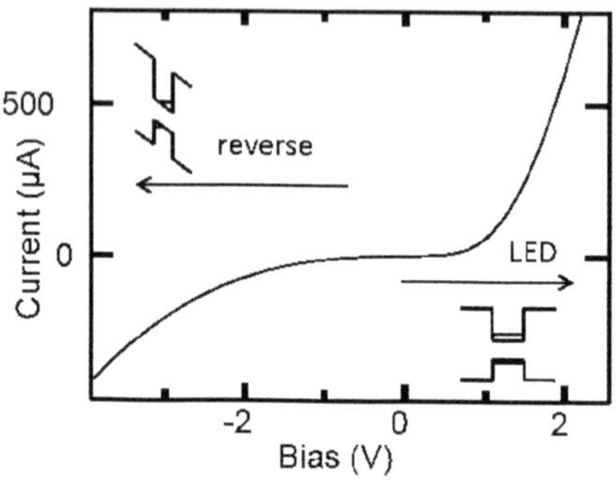

Fig. 15.3 Room temperature I-V characteristics of a polariton diode device

Reflectivity spectra on the sample at 220 K show well-separated (Rabi splitting of 4.4 meV) polariton states confirming that cavity mode broadening due to doping of the mirrors is not a restricting factor for operation in the strong-coupling regime [16].

The demonstration of strong coupling regime in a variety of semiconductor systems via reflectivity, transmission, and photo- or electroluminescence measurements involves the ability to tune either excitonic or cavity mode resonances with respect to the other, showing the characteristic anticrossing behavior between the two modes. To prove that emission from microcavity polariton LED device under electrical injection arises truly from polariton states both temperature and angular resolved electroluminescence (EL) measurements were performed. QW exciton energy can be tuned by $\sim -0.38\,\mathrm{meV\,K^{-1}}$ when varying sample temperature due to changes in the GaAs bandgap energy, whereas the corresponding cavity mode energy shift caused by variation of the refractive index is much weaker ($\sim -0.098\,\mathrm{meV\,K^{-1}}$ at 200 K). This allows tuning of the exciton mode through the cavity resonance.

Figure 15.4a shows electroluminescence spectra from the polariton LED at different temperatures between 180 K and 260 K collected normal to the sample. The emission at 180 K consists of a highenergy QW exciton emission line X at 1.357 eV and a lowenergy cavity mode line C at 1.347 eV. The increase in sample temperature by 80 K tunes the exciton energy by 29 meV down to 1.328 eV, and the cavity mode by 7.4 meV down to 1.34 eV, respectively. As seen in the figure, the exciton and cavity modes are well resolved over the entire temperature range and exhibit the characteristic anticrossing behavior of the strong coupling regime. This shows that EL emission in this intermediate temperature range arises unambiguously from exciton polariton states in the strong-coupling regime under electrical injection. Until recently, it was widely accepted that the relatively small exciton binding energies in InGaAs QWs were responsible for fast dissociation of excitons into free electron-hole pairs and that the observation of strong-coupling regime at room temperature in this material system would be impossible prompting many groups in search of other systems such as GaN, ZnO with larger exciton binding energies. Indeed, although excitonic transitions in bare InGaAs QWs are resolved in absorption measurements up to room temperature, the emission from these structures is largely dominated at elevated temperatures by free electron-hole

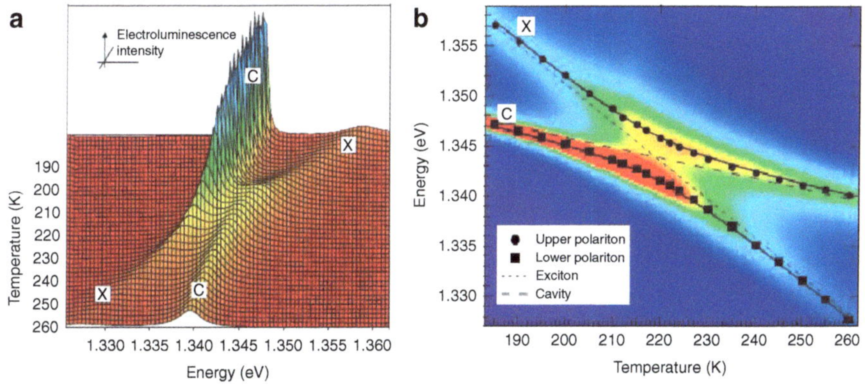

Fig. 15.4 (**a**) EL spectra of the processed sample were recorded as a function of energy for different values of temperature at a fixed angle of 12° and an injected current of 0.2 mA. Temperature changes induce a strong shift on the exciton energy due to bandgap temperature dependence. On the contrary, cavity mode temperature dependence is solely due to refractive index temperature change, which is much weaker effect. As a result, changing the temperature induces the characteristic anticrossing behavior, which revealed clearly in the figure, between exciton and cavity mode. (**b**) Contour plot and extracted peaks of the polariton emission spectra shown in (**a**) resolving the two polariton branches as a function of temperature. The *solid lines* are theoretical fits, while the *dashed and dotted lines* present the bare exciton and cavity modes, respectively. Rabi splitting of 4.8 meV was observed at 203K

pair recombination. Remarkably these results show that by dressing exciton with a photon in a semiconductor microcavity the excitonic contribution to the emission becomes dominant even at high temperatures.

In Fig. 15.4b the peak positions from the EL data are plotted. Excellent fit of the upper (UP) and lower (LP) polariton branches is obtained by applying a coupled harmonic oscillator model. In this model, the temperature dependence of the exciton and cavity modes has been taken into account as described above. The fits reveal that the condition of zero exciton–photon detuning at zero angle is reached at a temperature of 219 K with a normal mode Rabi splitting of 4.4 meV, in agreement with the reflectivity spectra taken on the same mesa at the same temperature.

Following this first demonstration of GaAs polariton LED at 220 K, we explored the possibility of room temperature operation. For this purpose, a new sample with larger Rabi splitting incorporating eight $In_{0.1}Ga_{0.9}As/GaAs$ QWs to increase the total exciton oscillator strength was grown [18]. In order to reduce the p-DBR series resistance in the electrical injection, the p-contact were deposited following an ohmic recess step when processing the polariton diodes. In this way, the top 11 periods of the p-DBR are electrically bypassed. 100 μm wide ring-shaped Ti/Pt p-type contacts were fabricated at the edge of the mesa shown in Fig. 15.5.

Figure 15.6a shows EL spectra, collected at zero angle, for different temperatures in the range of 260–320 K from a polariton LED containing four pairs of QWs, exhibiting the characteristic anticrossing behavior. The energy dispersion curves

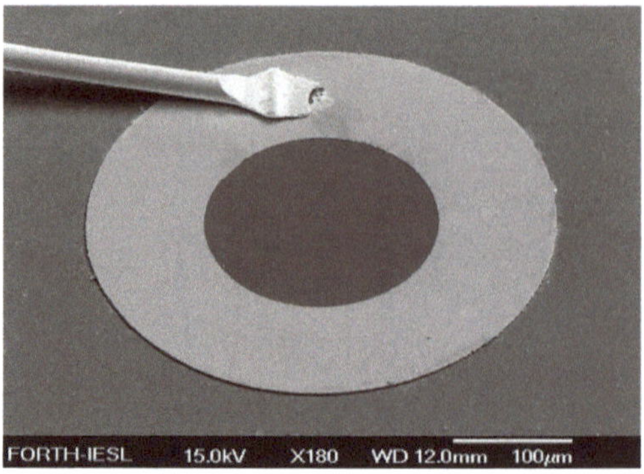

Fig. 15.5 SEM photo of a room temperature polariton LED

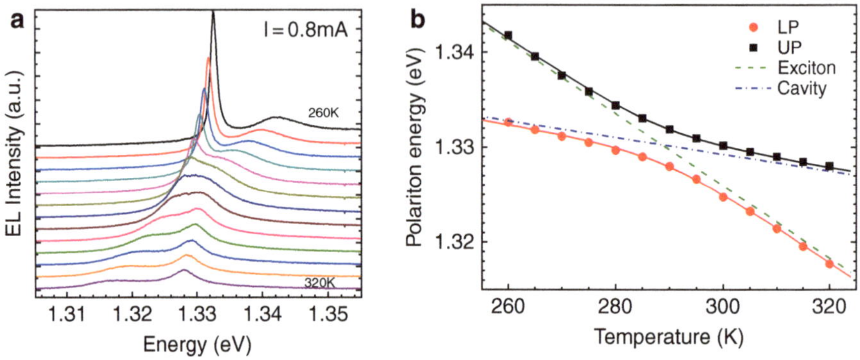

Fig. 15.6 Electroluminescence spectra recorded from a polariton LED at zero angle for different values of temperature between 260 K and 320 K, with a step of 5 K. (**a**) Polariton energy dispersion deduced from adjacent spectra. The *circular and square points* are extracted electroluminescence peaks, and the *solid lines* are fits for the upper and lower polariton branches

of the upper and lower polaritons are shown in Fig. 15.6b. The data points are extracted from the EL peaks in Fig. 15.6a, while the solid lines are theoretical fits which are generated by applying a coupled harmonic oscillator model to our data [19]. An excellent fitting of the upper and lower polariton branches is obtained, demonstrating that EL at these high temperatures is in the strong-coupling regime. The normal mode Rabi splitting is found to be 4 meV at 288 K.

The endurance of polariton's LED emission at even higher temperatures was tested by performing angle-resolved EL measurements on a polariton LED structure, with four pairs of QWs, which was designed to exhibit strong-coupling at temperatures higher than 300 K. Figure 15.7a presents selected angle-resolved EL spectra obtained at 315 K under electrical injection of 0.52 mA ($\sim$0.54 A/cm^2), from a diode which operates in a negative detuning of 11.5 meV at zero angle at this temperature. The corresponding energy dispersion curves show clear anticrossing behavior around 30° with a Rabi splitting of 4 meV at 315 K and are in good agreement with the model.

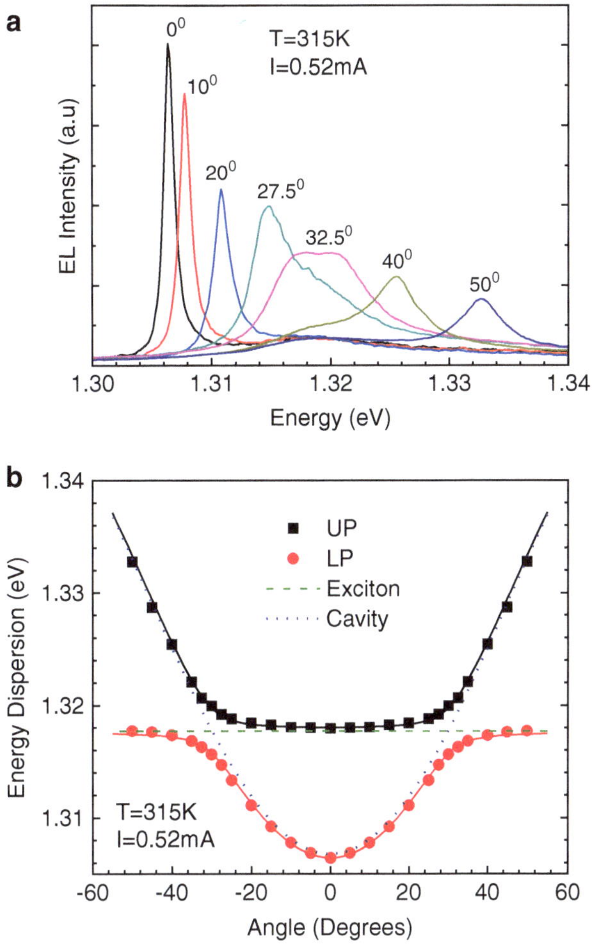

Fig. 15.7 (**a**) Selected EL spectra at various angles for 0.52 mA at 315 K. (**b**) Polariton energy dispersion curves vs. collection angle. The *circular and square points* are extracted EL peaks, while the *solid lines* are theoretical fittings

For in-depth analysis of the mechanism which allows operation of GaAs polariton LEDs in the strong-coupling regime at such high temperatures, we fabricated series of MC diodes, with operating temperatures ranging from 20 K up to room temperature. In Fig. 15.8, all zero-detuning Rabi splittings measured on various polariton LEDs as a function of temperature are plotted. Each data point is extracted from a set of temperature-dependent EL experiments under normal collection. The data points are grouped into a set of circles for MCs with six QWs ($N = 6$) and to a set of squares for MCs with eight QWs ($N = 8$).

As expected, sharp decrease of the Rabi splitting is observed when approaching room temperature due to phonon-induced broadening of the exciton. To account for

these results, a simplified model was employed, calculating the zero-detuning Rabi splitting in the strong-coupling regime using the equation [19]

$$\Omega\left(T\right) = \sqrt{4V^2 - \left(\gamma_{ex}\left(T\right) - \gamma_c\right)^2} \tag{15.1}$$

The parameter V represents the coupling strength between the cavity mode and the exciton, while γ_{ex} and γ_c are the exciton and photon linewidths, respectively. The expression in the square root has to remain positive, which is a precondition for the strong-coupling regime. In this model, the temperature dependence of $\Omega(T)$ comes solely through the exciton linewidth

$$\gamma_{ex}\left(T\right) = \gamma_{inh} + \gamma_{ac}T + \gamma_{LO}[\exp(\hbar\omega_{LO}/kT) - 1]^{-1}, \tag{15.2}$$

where γ_{inh} is the low-temperature inhomogeneous linewidth, γ_{ac} and γ_{LO} are constants representing the strength of exciton coupling with acoustic and LO phonons, respectively, and $\hbar\omega_{LO}$ is the LO phonon energy [20]. The determination of γ_{ex} was done experimentally, by performing transmittance measurements on a separate sample containing two InGaAs/GaAs QWs identical to those used in our MC structures. In the inset of Fig. 15.8, the measured exciton linewidth as a function of temperature is plotted. The solid line running through the experimental points was obtained by fitting the equation for $\gamma_{ex}(T)$ to the experimental results, using γ_{ac} = 4.4 μeV/K and γ_{LO} = 15.2 meV. Using these values for the exciton linewidth and taking measured value gor γ_c = 1 meV, both sets of data are fitted by applying the equation for Ω(T) using the $V(N)$ as the only adjustment parameter. The solid

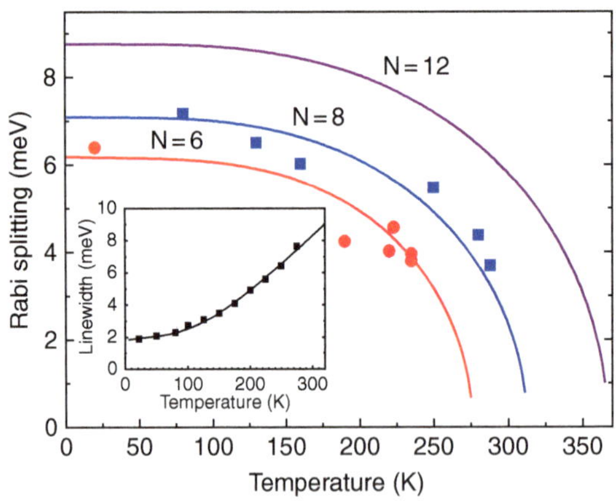

Fig. 15.8 Comparison of theoretical and experimental zero-detuning Rabi splitting as a function of temperature. The *circular points* refer to MCs with six QWs, while the squares to MCs with eight QWs. *Solid lines* are theoretical curves for 6, 8, and 12 QWs. The *inset* shows the temperature dependence of the exciton FWHM linewidth extracted from transmission measurements on a sample containing two 10 nm $In_{0.1}Ga_{0.9}As/GaAs$ QWs

lines running through the data points are best fits obtained for $V(6) = 3.13\,\text{meV}$ and $V(8) = 3.63\,\text{meV}$. As expected, the ratio between these two values follows $V(N) = \sqrt{N} \times V(1)$ relation, where $V(1)$ is the coupling constant per QW. As seen in Fig. 15.8, the model agrees well with the observed Rabi splitting and confirms the possibility of room temperature operation. It is worth noting that similar values for the constant V were obtained from transmittance experiments where the oscillator strength is measured directly from the optical density. The values obtained from these experiments are $V(6) = 3.25\,\text{meV}$ and $V(8) = 3.63\,\text{meV}$ which are very close to those obtained from the fit.

Figure 15.8 shows that by increasing the number of QWs in a microcavity to 12, it is possible to raise further operation temperature due to increase in exciton oscillator strength and coupling constant V. Other approaches to increase exciton oscillator strength include raising the QW barriers to increase carrier confinement. However, in the case of InGaAs/AlGaAs QW system, this comes at an expense of deteriorating exciton linewidths, with measured values, for instance, for 10 nm $In_{0.1}Ga_{0.9}As/Al_{0.3}Ga_{0.7}As$ QW close to 4.5 meV at 22 K. Instead, we resort to GaAs QWs the use of which allows both increasing the exciton oscillator strength and reducing the exciton linewidth.

Figure 15.9a shows the reflectivity spectrum of a 20-nm $GaAs/Al_{0.3}Ga_{0.7}As$ QW. Of the three reflectivity dips, the lower refers to the heavy hole exciton transition, the middle to the light hole exciton, and the third corresponds to the first excited heavy hole state.

Figure 15.9b records the measured heavy hole exciton linewidths from low-power PL experiments on this sample at various temperatures. Comparing the exciton linewidths with the corresponding InGaAs/GaAs QW, the reduced GaAs QW exciton linewidth is evident for the whole temperature range, confirming the overall better sample quality. From oscillator strengths measurements the coupling constant for the GaAs QWs was measured 1.74 meV per QW which is nearly 40% higher than in InGaAs QWs.

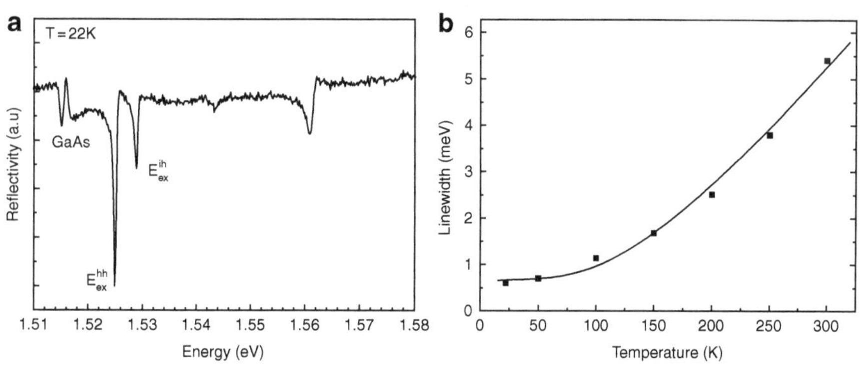

Fig. 15.9 (**a**) Reflectivity spectra and (**b**) linewidth as a function of temperature for a 20 nm $GaAs/Al_{0.3}Ga_{0.7}As$ QW

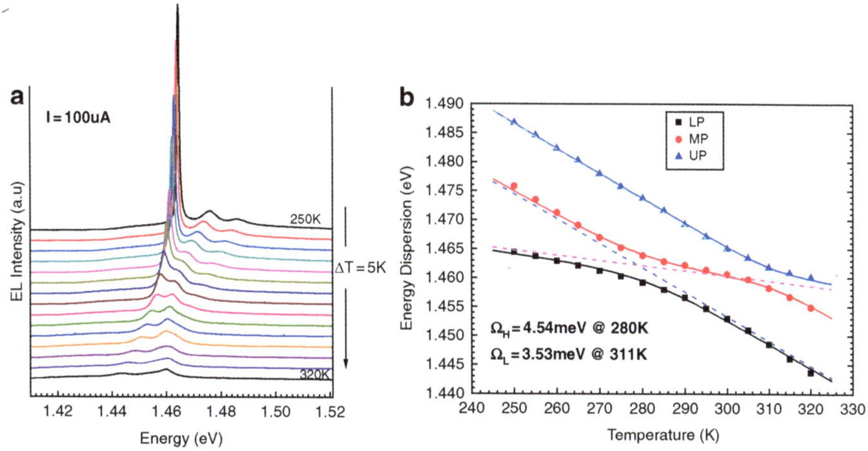

Fig. 15.10 (**a**) Electroluminescence spectra recorded from a polariton LED at zero angle for different values of temperature between 250 K and 320 K, with a step of 5 K. (**b**) Polariton energy dispersion curves extracted from the adjacent spectra

As expected, polariton LEDs based on GaAs QWs, exhibit enhanced stability at room temperature with Rabi splittings of 7.47 meV at 85 K and 6 meV at 300 K, shown in Fig. 15.10 where electroluminescence spectra and corresponding polariton branches under 100 μA electrical pumping for the temperature range of 250–320 K are plotted.

The importance of these results is that for the first time stable polariton LED device that operates at room temperature is demonstrated. Furthermore, it opens the way for new generation of polariton devices based on the well-established GaAs technology.

15.6 Toward Polariton Laser Diodes

The significance of demonstration of polariton LED device lies in the fact that it paves the way for the long awaited ultralow-threshold polariton laser which constitutes the holy grail of polariton research. Consequently, the exploration of polariton lasing in present GaAs polariton LEDs, by pushing electrical injection to the high-density regime in the current structures becomes more imminent than ever. However, several important considerations need to be taken into account: (1) the efficient injection of polariton which is hard to achieve at low temperatures due to high series resistance of the p-type DBR and (2) the difficulty of achieving polariton lasing in GaAs based MCs at high temperatures. The later is particularly challenging given that electrical injection creates exciton population with large in-plane $k_{//}$momentum. These "dark" long-living excitons severely limit the efficiency of the device, as they contribute negligibly to light emission.

We initially explore high-density regime of a polariton LED at high temperatures 235 K because of severe limitations of electrical injection at low temperatures due to high series resistance of the p-type DBR. In Fig. 15.11a, normalized EL spectra are presented with increasing injection current. At low injection currents, two clearly resolved polariton peaks are evident. With increasing injection current, the two polariton peaks progressively merge into a single emission line, underscoring the transition to the weak-coupling regime (Fig. 15.11b), i.e., that of a RCLED operation. The integrated EL emission from both polariton peaks shows a superlinear (quadratic) increase, as shown in Fig. 15.11c. The observed superlinear power dependence could be attributed to the enhanced relaxation of polaritons on the LP branch with increasing injection.

By measuring the optical power for which the PL emission intensity is identical with the EL intensity at $I = 1$ mA (Fig. 15.11c), we estimate that at the maximum current injection of 22 mA in our experiments, an estimated density of 10^{10} polaritons/cm^2 is injected per QW. This density compares well with the exciton saturation density of 3×10^{10} polaritons/cm^2 per QW [21], considering the much higher temperatures in our experiments. The above observations indicate that the present injection scheme and high operating temperatures are unlikely to yield a polariton laser, due to inefficient population of polariton states. Potential strategies could include utilization of fast LO phonon enhanced relaxation mechanisms to directly populate low-energy polariton states with simultaneous improvements of electrical characteristics of the diodes at low temperatures.

15.7 Bias Controlled Polariton Parametric Amplification

Since the discovery in 2000 of resonant polariton parametric amplification process [3] relying on efficient stimulated scattering of polaritons into final state, many new fundamental aspects of polaritons have been demonstrated using such parametric injection scheme. Parametric injection has been used to excite propagating polariton condensates and led to the demonstration of polariton superfluidity. Ciuti et al. showed that such strong nonlinearities are mediated by dipole–dipole interactions of polariton's excitonic constituent. In the recent years, the availability of electrically driven polariton LED device described in the previous chapters offered new opportunities to gain control over these nonlinear interactions through application of electrical bias. Thus, the first phase of all-optical experiments has now moved on to a second phase exploring electrical control of polariton-based nonlinearities. The pioneering work by G. Christmann et al. [22] demonstrates a new paradigm for manipulating polariton interactions, by incorporating electronic tunneling into p-i-n microcavities.

Polariton amplification is produced by exciting lower polariton branch with spectrally filtered pump pulses at the "magic" angle, while broadband probe pulse incident normal to the sample is used to initiate the parametric scattering process.

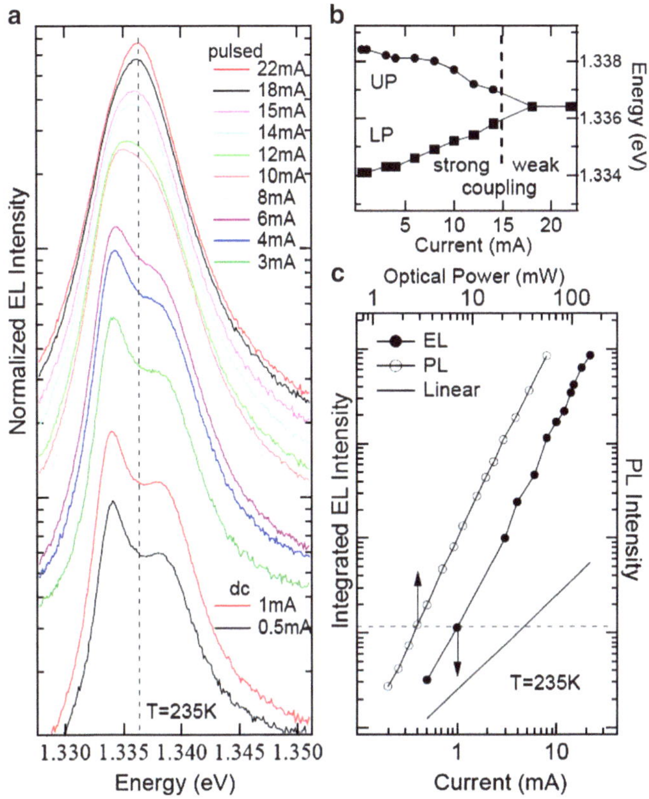

Fig. 15.11 (**a**) Normalized EL spectra with injection current, obtained in normal direction at 235 K for zero detuning. At higher currents, pulsed current source with a duty cycle of 1:200 is used to eliminate heating effects. (**b**) Extracted LP and UP peak positions showing the onset of the weak-coupling regime at high currents. (**c**) Integrated EL intensity of both the LP and UP peaks (*solid symbols*), showing a quadratic superlinear dependence. For comparison, the *solid line* represents a linear dependence. In open symbols, the integrated PL peak intensities obtained for various optical powers of excitation, showing the same superlinear dependence, are shown

The amplification scheme requires two pump polaritons at k_{p} to mutually scatter and yield signal polaritons at $k_{\mathrm{S}} = 0$, and idler polaritons at $k_{\mathrm{I}} = 2k_{\mathrm{p}}$, as shown in Fig. 15.12 When an electric bias is applied to the structure the exciton energy redshifts due to quantum-confined Stark effect allowing tuning of the polariton modes. The normalized probe reflectivity spectra (without pump) clearly show anticrossing behavior characteristic of strong-coupling regime. The reflected pump spectrum resonant with LP branch at 16 degrees is shown in Fig. 15.13. In the same graph, normalized probe reflectivity spectra when the pump is on are also presented when varying applied bias at zero pump-probe delay showing nearly tenfold probe gain due to parametric amplification.

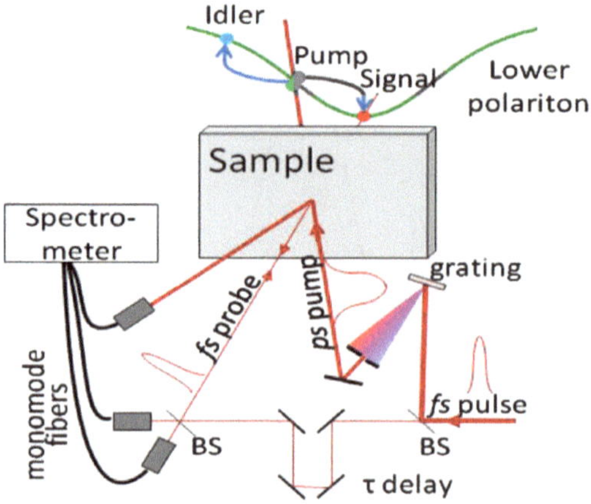

Fig. 15.12 Geometrical sketch of a pump-probe experiment

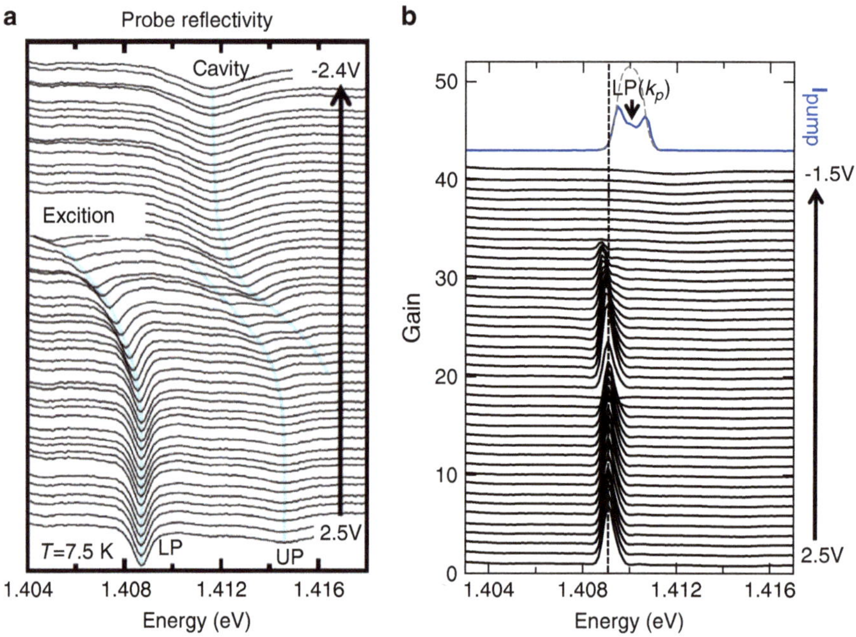

Fig. 15.13 (**a**) Probe reflectivity spectra for applied biases from −2.4 to 2.5 V shifted for clarity. *Blue lines* are guide to eyes. (**b**) Gain spectra for applied bias from 2.5 to −2.4 V (shifted for clarity), *vertical dashed line* is a guide to eyes. Reflected pump spectrum also shown [*blue (dark gray), top*], and *gray dashed line* is the full pump spectrum

The spectral positions of polariton modes and gain are extracted in Fig. 15.14, revealing an almost dispersionless gain peak (red circles). Three weak intensity peaks are also observed overlapping the bare polariton modes shown in Fig. 15.13a. This dispersionless gain which does not follow the LP branch field tuning in the

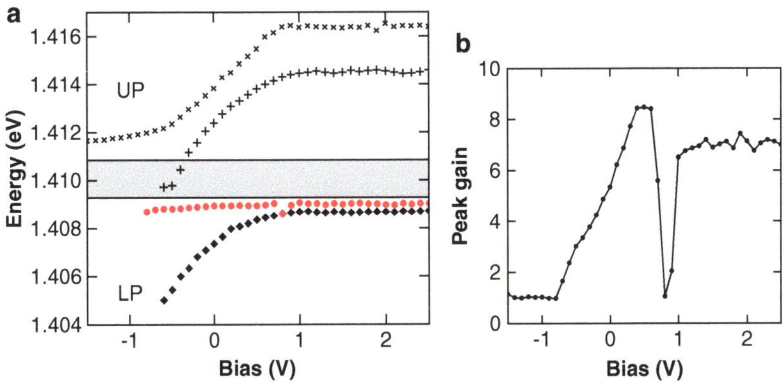

Fig. 15.14 (**a**) Peak positions for gain (*red circles*) and polariton branches extracted from reflectivity spectra (plus symbol, times symbol, filled diamond). (**b**) Peak probe gain vs. bias

absence of the pump indicates that the pump photo-injected carriers screen the electric field across the i-region by tunneling out of the QWs. Such drop in the applied field brings the LP back into resonance, thus maintaining the gain at the original LP energy.

Electrical control of parametric scattering of polaritons can be seen in Fig. 15.14b where bias-dependent peak gain is extracted. Notably at negative bias, the parametric gain is progressively lost, as the pump overlap with the LP dispersion becomes increasingly difficult to maintain. At positive bias, the device operates as a polariton LED, and interestingly, the gain is not reduced despite the simultaneous electrical injection of carriers into the DQWs. However, in addition to this behavior, a dramatic dip in gain centered at $V_t = 0.73\,\text{mV}$ is observed with gain reduction of more than 90%. At the same bias, a photocurrent anomaly is observed which corresponds to a local photocurrent peak superimposed on the monotonously increasing background originating from electron tunneling out of the DQWs through the triangular barriers. Field-tuning calculations of electronic levels of the DQWs show that at the specific bias, the electron $n = 1$ level of the left QW is exactly one LO phonon energy (36 meV) above the $n = 1$ level of the right QW. These observations indicate that the origin of the bias-controlled gain switching is LO phonon-assisted resonant tunneling of electrons into the RQW from polaritons in the LQW [23, 24]. The remaining bound electrons and holes continue to participate in the polariton dynamics; however, the resonant tunneling process destroys the gain process. The parametric scattering process depends on a coherent phase condition between signal ($k_S = 0$) and idler ($k_I = 2k_P$) polariton populations, the latter being rather sensitive to free carrier scattering [25, 26]. Because of their lighter effective mass, electrons are 10 times more efficient scatterers than holes, thus efficiently decreasing the gain in the RQW [27]. Because the polaritons cycle through the cavity photon every 700fs, the signal and idler lower polaritons are delocalized over both QWs, thus experiencing the full carrier scattering.

Interestingly, the speed of the quenching process suggests that this modulation, which can be both optically or electrically controlled, can approach 1 THz and can potentially be used in a variety of ultra high speed modulation schemes. Hence a range of novel devices and physics can emerge from the entangling of electrons and holes traveling through the structure both by quantum tunneling and entanglement with polaritons.

15.8 Conclusions

The results presented in this chapter record the latest progress in the development of a new generation of semiconductor microcavity polariton devices, utilizing the well-established and mature GaAs technology. The advantages of electrically driven polariton devices operating in the strong-coupling regime are multiple. Firstly, they can offer significant enhancement of spontaneous emission rate when compared to conventional RCLEDs. Furthermore, polariton lasers are expected to have at least one order of magnitude lower lasing thresholds. In addition, an original way to control parametric gain in semiconductor microcavities incorporating double quantum wells is demonstrated. Stark tuning and resonant tunneling between neighboring quantum wells allows dramatic changes in the strong-coupling-induced optical gain for minimal applied bias changes. The essential element for all the above devices is the ability to electrically pump polaritons without destroying strong-coupling regime at elevated temperatures, as demonstrated here.

References

1. C. Weisbuch, M. Nishioka, A. Ishikawa, Y. Arakawa, Observation of the coupled exciton-photon mode splitting in a semiconductor quantum microcavity. Phys. Rev. Lett. **69**, 3314–3317 (1992)
2. E. Burstein E and C. Weisbuch, *Confined Electrons and Photons: New Physics and Applications (*New York: Plenum*) 1995.*
3. P.G. Savvidis, J.J. Baumberg, R.M. Stevenson, M. Skolnick, D.M. Whittaker and J.S. Roberts, Angle-Resonant Stimulated Polariton Amplifier, Phys. Rev. Lett **84**, 1547 (2000)
4. D. Bajoni, P. Senellart, E Wertz, I Sagnes, A. Miard, A Lemaitre, J Bloch, Polariton laser using single micropillar GaAs-GaAlAs semiconductor cavities. Phys. Rev. Lett. **100**, 047401 (2008)
5. S. Christopoulos et al., Room-temperature polariton lasing in semiconductor microcavities. Phys. Rev. Lett. **98**, 126405 (2007)
6. J.R. Tischler, M.S. Bradley, V. Bulovic, J.H. Song, A. Nurmikko, A strong coupling in microcavity LED. Phys. Rev. Lett. **95**, 036401, (2005)
7. C. Ciuti, P. Schwendimann, B. Deveaud, A. Quattropani, Theory of the angle-resonant polariton amplifier. Phys. Rev. B. **62**, R4825–R4828 (2000)
8. L. Sapienza, A. Vasanelli, R. Colombelli, C. Ciuti, Y. Chassagneux, C. Manquest, U. Gennser, C. Sirtori, Electrically injected cavity polaritonsl. Phys. Rev. Lett. **100**, 136806 (2008)
9. L. Sapienza et al., Photovoltaic probe of cavity polaritons in a quantum cascade structure. Appl. Phys. Lett. **90**, 201101 (2007)

10. R.P. Schneider, J.A. Lott, Cavity design for improved electrical injection in InAlGaP/AlGaAs visible (639–661 nm) vertical cavity surface emitting laser diodes. Appl. Phys. Lett. **63**, 917 (1993)
11. K. Tai, L. Yang, Y.H. Wang, J.D. Wynn, A.Y. Cho, Drastic reduction of series resistance in doped semiconductor distributed Bragg reflectors for surface emitting lasers. Appl. Phys. Lett. **56**, 2496 (1990)
12. S.A. Chalmers, K.L. Lear, K.P. Killeen, Low resistance wavelength reproducible ***p***-type (Al,Ga)As distributed Bragg reflectors grown by molecular beam epitaxy. Appl. Phys. Lett. **62**, 1585 (1993)
13. M. Hong, J. Mannaerts, J. Hong, R. Fischer, K. Tai, J. Kwo, J. Vandenberg, Y. Wang, J. Gamelin, A simple way to reduce series resistance in p-doped semiconductor distributed Bragg reflectors. J. Cryst. Growth **111**, 1071 (1991)
14. M.G. Peters, B.J. Thibeault, D.B. Young, J.W. Scott, F.H. Peters, A.C. Gossard, L.A. Coldren, Band-gap engineered digital alloy interfaces for lower resistance vertical cavity surface emitting lasers. Appl. Phys. Lett. **63**, 3411 (1993)
15. M. Hong, D. Vakhshoori, J.P. Mannaerts, Y.F. Hsieh, Low resistivity vertical cavity surface emitting lasers grown by molecular beam epitaxy using sinusoidal composition grading in mirrors and *in situ* nonalloyed ohmic contacts. J. Vac. Sci. Technol. B **13**, 758 (1995)
16. S. Tsintzos, P.G. Savvidis, G. Konstantinidis, Z. Hatzopoulos, N.T. Pelekanos, Towards electrically-pumped microcavity polariton lasers. Phys. Stat. Sol. (c) **5**, 3594 (2008)
17. S. Tsintzos, N.T. Pelekanos, G. Konstantinidis, Z. Hatzopoulos, P.G. Savvidis, A GaAs polariton light-emitting diode operating near room temperature. Nature **453**, 372 (2008)
18. S.I. Tsintzos, P.G. Savvidis, G. Deligeorgis, Z. Hatzopoulos, N.T. Pelekanos, Room temperature GaAs exciton-polariton light emitting diode. Appl. Phys. Lett. **94**, 071109 (2009)
19. A.V. Kavokin, J.J. Baumberg, G. Malpuech, F.P. Laussy, *Microcavities* (Oxford University Press, Oxford, 2007)
20. D. Gammon, S. Rudin, T.L. Reinecke, D.S. Katzer, C.S. Kyono, Phonon broadening of excitons in $GaAs/Al_xGa_{1-x}As$ quantum wells. Phys. Rev. B **51**, 16785 (1995)
21. I. Carussotto, C. Ciuti, Probing microcavity polariton superfluidity through resonant Rayleigh scattering. Phys. Rev. Lett. **93**, 166401 (2004)
22. G. Christmann, C. Coulson, J.J. Baumberg, N.T. Pelekanos, Z. Hatzopoulos, S.I. Tsintzos, P.G. Savvidis, Control of polariton scattering in resonant-tunneling double-quantum-well semiconductor microcavities. Phys. Rev. B **82**, 113308 (2010)
23. D.Y. Oberli, J. Shah, T.C. Damen, J.M. Kuo, J.E. Henry, J. Lary, S.M. Goodnick, Optical phonon-assisted tunneling in double quantum well structures. Appl. Phys. Lett. **56**, 1239 (1990)
24. J.M. Feng, J.H. Park, S. Ozaki, H. Kubo, N. Mori, C. Hamaguchi, Resonant optical-phonon assisted tunnelling in an asymmetric double-quantum-well structure. Semicond. Sci. Technol. **12**, 1116 (1997)
25. P.G. Lagoudakis, M.D. Martin, J.J. Baumberg, A. Qarry, E. Cohen, L.N. Pfeiffer, Electron-polariton scattering in semiconductor microcavities. Phys. Rev. Lett. **90**, 206401 (2003)
26. M. Perrin, P. Senellart, A. Lemaître, J. Bloch, Polariton relaxation in semiconductor microcavities: efficiency of electron polariton scattering. Phys. Rev. B **72**, 075340 (2005)
27. G. Malpuech, A. Kavokin, A. Di Carlo, J.J. Baumberg, Polariton lasing by exciton-electron scattering in semiconductor microcavities. Phys. Rev. B **65**, 153310 (2002)

Index

D. Sanvitto and V. Timofeev (eds.), *Exciton Polaritons in Microcavities*, Springer Series in Solid-State Sciences 172, DOI 10.1007/978-3-642-24186-4,

MIX
Papier aus verantwortungsvollen Quellen
Paper from responsible sources
FSC® C105338

If you have any concerns about our products,
you can contact us on
ProductSafety@springernature.com

In case Publisher is established outside the EU,
the EU authorized representative is:
Springer Nature Customer Service Center GmbH
Europaplatz 3, 69115 Heidelberg, Germany

Printed by Libri Plureos GmbH
in Hamburg, Germany